Solutions Manual for *Real Analysis*, Part I, Royden and Fitzpatrick

Maurice Stadler

Twelfth printing, September 28, 2023

ISBN 979-8691588273

This book provides solutions and errata through Chapter 13 of textbook *Real Analysis*, fourth edition, 2018 reissue, by Halsey Royden and Patrick Fitzpatrick.

Email corrections, improvements, or better solutions to robtstad@sbcglobal.net.

Contents

Chapter 1

The Real Numbers: Sets, Sequences, and Functions

1.1 The Field, Positivity, and Completeness Axioms

On page 9, to avoid confusion in the definition of $[a, b]$, the condition that $a \leq b$ should be explicit to override the previously imposed condition that $a < b$ and make it clear that singleton sets are (closed) intervals.

1. We are asked to show that the multiplicative inverse of ab is $a^{-1}b^{-1}$. Using the associativity, commutativity, and identity axioms of multiplication,

$$(ab)a^{-1}b^{-1} = a\big(ba^{-1}b^{-1}\big) = a\big(bb^{-1}a^{-1}\big) = a\big(1a^{-1}\big) = aa^{-1} = 1.$$

   Thus, $(ab)^{-1} = a^{-1}b^{-1}$.

2. (i) Because $a \neq 0$, we need consider only two of P2's alternatives: a is positive, or else $-a$ is positive. If a is positive, then a^2 is positive by P1. If $-a$ is positive, then $(-a)^2$ is positive by P1 again; moreover, $(-a)^2 = (-1a)^2 = (-1)^2a^2 = 1a^2 = a^2$. Hence, the two alternatives reduce to consideration of a^2. Now $a^2 - 0$ is positive, so $a^2 > 0$.

   (ii) Suppose, to get a contradiction, that a^{-1} were not positive. Then by P2, $-a^{-1}$ would be positive, or else $a^{-1} = 0$. In the former case, $a\big(-a^{-1}\big)$ would be positive by P1, but $a\big(-a^{-1}\big) = a\big(-1a^{-1}\big) = -1aa^{-1} = (-1)1 = -1$, a contradiction. In the latter case, $a\big(-a^{-1}\big) = a0 = 0$, also a contradiction. Thus, a^{-1} is positive.

   (iii) By hypothesis, $a - b$ is positive. If $c > 0$, then c is also positive, and $(a - b)c$, or $ac - bc$, is positive by P1. Thus, $ac > bc$. If $c < 0$, then $0 - c$, or $-c$, is positive, and $(a - b)(-c)$, or $bc - ac$, is positive. Thus, $ac < bc$.

3. Assume first that E consists of a single point x. Then x itself is an upper bound for E because $x \leq x$. There is no smaller upper bound for E because there is no number b less than x such that $x \leq b$. Hence, $\sup E = x$. Analogously, $\inf E = x$. Thus, $\inf E = \sup E$.

   Conversely, assume that $\inf E = \sup E$. Then $\sup E \leq x \leq \sup E$ for all x in E. It follows that $x = \sup E$, and that E consists of a single point.

4. (i) If neither a nor b were zero, then by the positivity axioms, either ab or $-(ab)$ would be positive; either case contradicts that $ab = 0$. Thus, $a = 0$, or $b = 0$.

   (ii) $(a - b)(a + b) = (a - b)a + (a - b)b = a^2 - ba + ab - b^2 = a^2 - b^2$.
   Rewrite equality $a^2 = b^2$ as $(a - b)(a + b) = 0$. By part (i), $a - b = 0$, or $a + b = 0$. Thus, $a = b$, or $a = -b$.

   (iii) E is nonempty because E contains 0.
   If $c \leq 1$, then 1 is an upper bound for E because compound inequality $x^2 < c \leq 1$ implies that $x < 1$. If on the other hand $c > 1$, then c is an upper bound for E because inequality $x^2 < c$ implies that $x < c$.
   We are given that x_0 is the least number such that $x \leq x_0$ for all x in E. It follows that $x^2 < x_0^2$ for all x in E. Comparing to the definition of E, we conjecture that $x_0^2 = c$, which is true if there are no numbers y in $\mathbf{R} \sim E$ such that $y^2 < x_0^2$. Now numbers y in $\mathbf{R} \sim E$ are those such that $y^2 \geq c$ and include numbers greater than all those in E, that is, $y > \sup E$, a contradiction. Thus, $x_0^2 = c$.
   There is a unique positive x for which $x^2 = c$ by the conclusion of part (ii) with a replaced by x and b^2 by c.

5. (i) We have that

$$
\begin{aligned}
x^2 + bx/a + c/a &= 0 && \text{multiplicative inverse,}\\
x^2 + bx/a &= -c/a && \text{additive inverse,}\\
x^2 + bx/a + (b/(2a))^2 &= -c/a + (b/(2a))^2 && \text{completing the square,}\\
(x + b/(2a))^2 &= \left(b^2 - 4ac\right)/(2a)^2 && \text{commutativity of addition, distribtn.,}\\
x + b/(2a) &= \pm\sqrt{\left(b^2 - 4ac\right)/(2a)^2} && \text{Problems 4(ii),(iii),}\\
x &= \frac{-b \pm \sqrt{b^2 - 4ac}}{2a} && \text{additive inverse, distribution, commutativity of addition.}
\end{aligned}
$$

(ii) We include two more parts to the problem. If $b^2 - 4ac = 0$, then by the fourth equation above, $(x + b/(2a))^2 = 0$. By Problem 4(ii), $x + b/(2a) = 0$. By the additive inverse property, $x = -b/(2a)$. Thus, the quadratic equation has exactly one solution.

(iii) Assume $b^2 - 4ac < 0$. Then $\left(b^2 - 4ac\right)/(2a)^2 < 0$ by Problems 2(i) and (iii) so that $(x + b/(2a))^2 < 0$. Thus, the quadratic equation has no real solutions by Problem 2(i).

6. Because E is bounded below, there is a real number b such that $x \geq b$ for all x in E. Then $-x \leq -b$ for all x in E. Name $\{-x \,|\, x \in E\}$ set F so that we can rewrite the previous statement as $y \leq -b$ for all y in F, which shows that F is bounded above. By the completeness axiom, F has a supremum. Rewrite $y \leq \sup F$ for all y in F as $-y \geq -\sup F$ for all y in F. Equivalently, $x \geq -\sup F$ for all x in E. Now $-\sup F$ is the *greatest* lower bound of E because if there were a greater, then there would be a lesser lower bound than $\sup F$ for F. Thus, E has an infimum, and it is equal to $-\sup F$.

7. (i) If $a, b \geq 0$, then the claim is true. If $a, b < 0$, then $ab > 0$ so that $|ab| = ab$, and $|a||b| = (-a)(-b) = ab$. If $a \geq 0$ and $b < 0$, then $ab \leq 0$ so that $|ab| = -ab$, and $|a||b| = a(-b) = -ab$. Finally, if $a < 0$ and $b \geq 0$, then $ab \leq 0$ so that $|ab| = -ab$, and $|a||b| = (-a)b = -ab$. In all cases, $|ab| = |a||b|$.

(ii) If $a, b \geq 0$, then the claim is true (equality holds). If $a, b < 0$, then $a + b < 0$ so that $|a + b| = -(a + b)$, and $|a| + |b| = (-a) + (-b) = -(a + b)$. If $a \geq 0$, $b < 0$, and $a + b \geq 0$, then $|a + b| = a + b < a - b = |a| + |b|$. If $a \geq 0$, $b < 0$, and $a + b < 0$, then $|a + b| = -(a + b) < a - b = |a| + |b|$. If $a < 0$ and $b \geq 0$, the steps are analogous to those in the previous two sentences. In all cases, $|a + b| \leq |a| + |b|$.

Here is another answer, more of a proof than a verification. Add the two compound inequalities

$$
\begin{aligned}
-|a| \leq a \leq |a|,\\
-|b| \leq b \leq |b|
\end{aligned}
$$

to get compound inequality $-(|a| + |b|) \leq a + b \leq |a| + |b|$, from which it follows that $|a + b| \leq \big||a| + |b|\big| = |a| + |b|$.

(iii) In the following, symbol $\Leftrightarrow$ denotes *if and only if:*

$$
|x - a| < \epsilon \Leftrightarrow \left\{ \begin{array}{ll} x - a < \epsilon \Leftrightarrow x < a + \epsilon & \text{if } x - a \geq 0 \\ -(x - a) < \epsilon \Leftrightarrow a - \epsilon < x & \text{if } x - a < 0 \end{array} \right\} \Leftrightarrow a - \epsilon < x < a + \epsilon.
$$

Here is another, proof-like answer: $|x - a| < \epsilon$ if and only if $-\epsilon < x - a < \epsilon$, which upon adding a to all three parts is compound inequality $a - \epsilon < x < a + \epsilon$.

1.2 The Natural and Rational Numbers

In footnote 2, replace both occurrences of *Archimedeas* with *Archimedes*.

8. By the third paragraph of the section and the proof of Theorem 1, the least natural number is 1, so we start by showing that interval $(1,2)$ fails to contain a natural number. If $(1,2)$ contained a natural number ℓ, then by the first part of Problem 9, $\ell - 1$ would be a natural number, but $\ell < 2$, so that $\ell - 1 < 1$, which contradicts that 1 is the least natural number. Hence, the assertion is true when $n = 1$. Assume that k is a natural number for which $(k, k+1)$ contains no natural numbers. Consider $(k+1, (k+1)+1)$. If $(k+1, k+2)$ contained a natural number ℓ, then again by Problem 9, $\ell - 1$ would be a natural number. But $k+1 < \ell < k+2$ so that $k < \ell - 1 < k+1$, which contradicts our assumption, so $(k+1, k+2)$ fails to contain a natural number. Hence, the assertion is also true when $n = k+1$. Thus, by the principle of mathematical induction, $(n, n+1)$ fails to contain a natural number for each n.

9. The assertion is true vacuously when $n = 1$. Assume that k is a natural number for which $k - 1$ is a natural number. Because k belongs to inductive set $\mathbf{N}$, number $k+1$ belongs to $\mathbf{N}$. Then $(k+1) - 1$, or k, belongs to $\mathbf{N}$ by assumption. The assertion is therefore also true when $n = k+1$. Thus, by the principle of mathematical induction, $n - 1$ is a natural number if $n = 2, 3, \ldots$.

   For the second assertion, use induction on m. The assertion is true when $m = 1$ by the first part of this problem. Assume that if k and n are natural numbers such that $n > k$, then $n - k$ is a natural number. By the first part of this problem, $n - (k+1)$, or $(n-k) - 1$, also is a natural number. The assertion is therefore also true when $m = k+1$. Thus, by the induction principle, $n - m$ is a natural number for every m such that $n > m$.

10. First assume that $r > 0$. By the Archimedean property of $\mathbf{R}$, the set S of natural numbers defined by $S = \{n \in \mathbf{N} \,|\, n \geq r\}$ is nonempty. According to Theorem 1, S has a smallest member m. Because $m \in S$, we have that $m \geq r$. Moreover, $m < r+1$. Otherwise, $m \geq r+1$, or $m - 1 \geq r$, which shows that natural number $m - 1$ (see Problem 9) is in S, contradicting that m is the smallest natural number in S. Hence, integer m is in $[r, r+1)$.

    Next assume that r is an integer. Then integer r is in $[r, r+1)$.

    Lastly, assume that $r < 0$. By the Archimedean property of $\mathbf{R}$, there is a natural number n for which $n > -r$. We infer from the first case considered that there is an integer m in $[-r, -r+1)$ which is equivalent to $1-m$'s being in $(r, r+1]$. We show in the next paragraph that $1 - m$ is an integer. Moreover, we previously handled the case for which r is an integer. Hence, we can claim that there is an integer in $[r, r+1)$.

    To show that $1 - m$ is an integer, we first note that $-m$ is an integer because it follows from definition of an integer that the negative of an integer is an integer. Now we show that if n is an integer, then $n+1$ is also an integer. Indeed, that is true if n is a natural number because the set $\mathbf{N}$ is inductive. If n is 0 or -1, then $n+1 = 1$ or 0, both integers. Lastly, if n is the negative of a natural number greater than 1, express $n+1$ as $-(-n-1)$; $-n-1$ is a natural number by the first part of Problem 9, so $-(-n-1)$ is the negative of a natural number and therefore an integer.

    To conclude our solution to the problem, we show that integer p in $[r, r+1)$ is unique. Otherwise, there would be an integer q different from p in $[r, r+1)$. If $q > p$, then we have that $r \leq p < q < r+1$ and therefore $q - p < 1$. Therefore, integer q belongs to $(p, p+1)$. We show in the next paragraph that for every integer p, we have that $(p, p+1) \cap \mathbf{Z} = \emptyset$. If we suppose that $q < p$, then we reach analogously the same contradiction and so confirm that there is exactly one integer in $[r, r+1)$.

We now show that for each integer p, interval $(p, p+1)$ contains no integers. If p is a natural number, then the claim is true by Problem 8 and the fact that neither 0 nor the negatives of natural numbers can be in $(p, p+1)$. If p is the negative of a natural number, then express compound inequality $p < q < p+1$ as $-(p+1) < -q < -p$, which is equivalent to $-p < -q+1 < -p+1$. Again, the claim is true using the same reasoning and the fact that adding 1 to an integer yields another integer. Lastly, if $p = 0$, then the interval is $(0, 1)$, which contains no integers because it does not contain 0, nor a natural number since 1 is the least natural number, nor the negative of a natural number because 0 is the lower bound of the interval.

11. Mimic the proof of Theorem 1. Let E be a nonempty set of integers that is bounded above. By the completeness axiom, E has a supremum; denote $\sup E$ by c. Because $c-1$ is not an upper bound for E, there is an m in E for which $m > c-1$. We claim that m is the largest member of E. Otherwise, there is an n in E for which $n > m$. Because $n \in E$, we have that $n \leq c$. Thus, $c-1 < m < n \leq c$, and therefore $n - m < 1$. Therefore, integer n belongs to $(m, m+1)$. That contradiction (see the last paragraph of our solution to Problem 10) confirms that m is the largest member of E.

12. Let a and b be real numbers such that $a < b$. By Theorem 2, there is a rational number r that lies between $a - \sqrt{2}$ and $b - \sqrt{2}$ so that $r + \sqrt{2}$ lies between a and b. Number $r + \sqrt{2}$ is irrational. Otherwise, it would be equal to some rational number s, in which case $\sqrt{2}$ would equal rational number $s - r$, contradicting the third paragraph on page 12. Thus, the irrational numbers are dense in $\mathbf{R}$.

13. Take a real number c. We claim that $\{x \in \mathbf{Q} \,|\, x \leq c\}$ is such a set. By the Archimedean property of $\mathbf{R}$, the set is nonempty, which we see by expressing the set as $\{x \in \mathbf{Q} \,|\, -x \geq -c\}$. Not only is c is an upper bound for the set, c is *the least* upper bound. Otherwise, there would be a smaller upper bound c'. By Theorem 2, there is a rational number r between c' and c so that $c' < r < c$, which shows that r is in the set, and that therefore c' is not an upper bound, a contradiction. Thus, each real number is the supremum of a set of rational numbers.

    The proof for irrational numbers is essentially the same. The corresponding set of irrational numbers is nonempty because it contains $x - \sqrt{2}$ where x is an integer; $x - \sqrt{2}$ is irrational by our solution to Problem 12. Secondly, Problem 12 takes the place of Theorem 2.

14. Our proof holds if $r \geq -1$. The assertion is true when $n = 1$ (equality holds). Assume the assertion is true when $n = k$. Multiply both sides of inequality $(1+r)^k \geq 1 + kr$ by $1 + r$:

$$\begin{aligned} (1+r)^{k+1} &\geq 1 + kr + r + kr^2 && 1 + r \geq 0 \\ &\geq 1 + (k+1)r && kr^2 \geq 0. \end{aligned}$$

    Hence, the assertion is also true when $n = k+1$. Thus, by the principle of mathematical induction, if $r \geq -1$, then $(1+r)^n \geq 1 + nr$ for each n. This is Bernoulli's inequality.

15. (i) The assertion is true when $n = 1$. Assume that the assertion is true when $n = k$. If $n = k+1$, then

$$\begin{aligned} \sum_{j=1}^{k+1} j^2 &= \sum_{j=1}^{k} j^2 + (k+1)^2 \\ &= \frac{k(k+1)(2k+1)}{6} + k^2 + 2k + 1 && \text{assumption} \\ &= \frac{(k+1)((k+1)+1)(2(k+1)+1)}{6}. \end{aligned}$$

Hence, the assertion is also true when $n = k+1$. Thus, by the principle of mathematical induction, the assertion is true for every n.

(ii) We first prove by induction the lemma that $1 + 2 + \cdots + n = n(n+1)/2$. The lemma is true when $n = 1$. Assume that the lemma is true when $n = k$. If $n = k+1$, then

$$\begin{aligned} 1 + 2 + \cdots + k + (k+1) &= k(k+1)/2 + k + 1 \\ &= (k+1)((k+1)+1)/2, \end{aligned}$$

so the lemma is also true when $n = k+1$. Hence, by the principle of mathematical induction, the lemma is true for every n.

Now the problem's assertion is true when $n = 1$. Assume that the assertion is true when $n = k$. If $n = k+1$, then

$$\begin{aligned} 1^3 + 2^3 + \cdots + k^3 + (k+1)^3 &= (1 + 2 + \cdots + k)^2 + (k+1)^3 && \text{assumption} \\ &= (k(k+1)/2)^2 + k^3 + 3k^2 + 3k + 1 && \text{lemma} \\ &= ((k+1)((k+1)+1)/2)^2 \\ &= (1 + 2 + \cdots + (k+1))^2 && \text{lemma.} \end{aligned}$$

Hence, the assertion is also true when $n = k+1$. Thus, by the principle of mathematical induction, $1^3 + 2^3 + \cdots + n^3 = (1 + 2 + \cdots + n)^2$ for every n.

(iii) The assertion is true when $n = 1$ because $1+r = (1+r)(1-r)/(1-r) = (1-r^2)/(1-r)$. Assume that the assertion is true when $n = k$. If $n = k+1$, then

$$\begin{aligned} 1 + r + \cdots + r^k + r^{k+1} &= \frac{1 - r^{k+1}}{1 - r} + r^{k+1} \\ &= \frac{1 - r^{k+1} + r^{k+1} - r^{k+2}}{1 - r} \\ &= \frac{1 - r^{(k+1)+1}}{1 - r}. \end{aligned}$$

Hence, the assertion is also true when $n = k+1$. Thus, by the principle of mathematical induction, if $r \neq 1$, then $1 + r + \cdots + r^n = (1 - r^{n+1})/(1 - r)$ for every n.

1.3 Countable and Uncountable Sets

In the proof of Corollary 4(i), n has two connotations.

16. By definition (page 12), the set of integers is the union of the set of natural numbers, the set of their negatives, and $\{0\}$. The set of natural numbers is countable by Theorem 3. The set of negative natural numbers is equipotent to the set of natural numbers and so is also countable. The set $\{0\}$ is countable because it is finite (equipotent to $\{1\}$). Thus, by Corollary 6, the set of integers is countable.

17. If A is empty, then the claim is true vacuously. Consider the case for which A is nonempty. Assume first that A is countable. Then A is either finite or countably infinite. If A is finite, then there is a natural number n for which A is equipotent to $\{1, \ldots, n\}$, that is, there is a one-to-one mapping of A onto $\{1, \ldots, n\}$. Because $\{1, \ldots, n\}$ is a subset of $\mathbf{N}$, we can also say that the mapping is from A to $\mathbf{N}$. Lastly, if A is countably infinite, then A is equipotent to $\mathbf{N}$; hence, there is again a one-to-one mapping of A to $\mathbf{N}$.

    Conversely, assume that there is a one-to-one mapping f of A to $\mathbf{N}$. Then A is equipotent to a set of natural numbers which, by Theorem 3, is countable. Thus, by the first sentence of the third paragraph of the section, A is countable.

18. When $n = 1$, the assertion is true because $\mathbf{N}$ is equipotent to $\mathbf{N}$. Assume the Cartesian product of k sets $\mathbf{N}$ is countably infinite. By composing with a one-to-one correspondence between $\overbrace{\mathbf{N} \times \cdots \times \mathbf{N}}^{k \text{ times}}$ and $\mathbf{N}$, we can assume that $\overbrace{\mathbf{N} \times \cdots \times \mathbf{N}}^{k \text{ times}} = \mathbf{N}$. Then if $n = k+1$, we have that $\overbrace{\mathbf{N} \times \cdots \times \mathbf{N}}^{k+1 \text{ times}} = \overbrace{\mathbf{N} \times \cdots \times \mathbf{N}}^{k \text{ times}} \times \mathbf{N} = \mathbf{N} \times \mathbf{N}$, which is countably infinite by the proof of Corollary 4(i) when $n = 2$. Hence, whenever the assertion is true for k, it is true for $k+1$. Thus, by the principle of mathematical induction, the Cartesian product of n sets $\mathbf{N}$ is countably infinite for each n.

    We take the opportunity here to complete also the proof of part (ii) of Corollary 4. The first exercise is to verify that g is one-to-one. When g is restricted to the set of positive rational numbers, g is one-to-one by the same reasoning as in the proof of part (i). Likewise, when g is restricted to the set of negative rational numbers, g is one-to-one. Furthermore, g restricted to the set of positive rational numbers takes on even values only whereas g restricted to the set of negative rational numbers takes on odd values only, and neither takes on 1, which is the value of $g(0)$. Thus, g is one-to-one.

    The second exercise is to use the pigeonhole principle to show that $\mathbf{N} \times \mathbf{N}$ is not finite. Cartesian product $\mathbf{N} \times \mathbf{N}$ is not empty. So if $\mathbf{N} \times \mathbf{N}$ were finite, then there would be an n for which $\mathbf{N} \times \mathbf{N}$ were equipotent to $\{1, \ldots, n\}$, which is equipotent to $\{1\} \times \{1, \ldots, n\}$. But $\mathbf{N} \times \mathbf{N}$ contains all ordered pairs in set $\{1\} \times \{1, \ldots, n+1\}$, which is not equipotent to $\{1\} \times \{1, \ldots, n\}$ and therefore not equipotent to $\{1, \ldots, n\}$ because equipotence is an equivalence relation. Thus, $\mathbf{N} \times \mathbf{N}$ is not finite.

    A similar proof shows that neither is $\mathbf{Q}$ finite: Set $\mathbf{Q}$ is not empty. So if $\mathbf{Q}$ were finite, then there would be an n for which $\mathbf{Q}$ were equipotent to $\{1, \ldots, n\}$. But $\mathbf{Q}$ contains all natural numbers in $\{1, \ldots, n+1\}$, which is not equipotent to $\{1, \ldots, n\}$. Thus, $\mathbf{Q}$ is not finite.

19. Let Λ be finite, and for each λ in Λ, let E_λ be a countable set. We show that union $\bigcup_{\lambda \in \Lambda} E_\lambda$, which we denote by E, is countable. If E is empty, then it is countable. So assume that $E \neq \emptyset$. Because Λ is finite, there is a natural number n for which Λ is equipotent to $\{1, \ldots, n\}$. Let $\{\lambda_m \mid m \in \{1, \ldots, n\}\}$ be a "finite enumeration" of Λ where $\lambda_i \neq \lambda_j$ if $i \neq j$. Fix an m in $\{1, \ldots, n\}$. If E_{λ_m} is finite and nonempty, choose a natural number $N(m)$ and a one-to-one mapping f_m of $\{1, \ldots, N(m)\}$ onto E_{λ_m}; if E_{λ_m} is countably infinite, choose a one-to-one mapping f_m of $\mathbf{N}$ onto E_{λ_m}. Define set E' by

    $$E' = \{(m, k) \in \{1, \ldots, n\} \times \mathbf{N} \mid E_{\lambda_m} \neq \emptyset, \text{ and } 1 \leq k \leq N(m) \text{ if } E_{\lambda_m} \text{ is also finite}\}.$$

    Define mapping f of E' to E by $f(m, k) = f_m(k)$. Then f is a mapping of E' onto E. However, E' is a subset of countable set $\mathbf{N} \times \mathbf{N}$ and hence, by Theorem 3, is countable. Theorem 5 tells that E is also countable.

    We can devise a shorter proof if we are allowed to use Corollary 6 as proved for countably infinite Λ by noting that the union of a finite collection of countable sets is a subset of the union of a countably infinite collection of countable sets and then invoking Theorem 3.

20. Composition $g \circ f$ should be $g \circ f : A \to C$.

    Refer to the first two paragraphs of subsection "Mappings between sets" on page 4 for notation and justifications used in this solution. Assume that $g(f(a_1)) = g(f(a_2))$ for two members a_1, a_2 of A. Because g is one-to-one, $f(a_1) = f(a_2)$. Because f is one-to-one, $a_1 = a_2$. Thus, $g \circ f$ is one-to-one. Next, $g(f(A)) = g(B) = C$. Thus, $g \circ f$ is onto.

    Let $f(a)$ equal b. Then $f\big(f^{-1}(b)\big) = f(a) = b$, and $f^{-1}(f(a)) = f^{-1}(b) = a$; in other words, f is a mapping for which $f \circ f^{-1} = id_B$, and $f^{-1} \circ f = id_A$. Hence, f^{-1} is invertible and thus one-to-one and onto.

21. We want to prove that for each pair of natural numbers n and m, set $\{1,\dots,n+m\}$ is not equipotent to $\{1,\dots,n\}$. When $n=1$, no mapping f from the first set to the second is one-to-one because $f(1), f(2) = 1$. Hence, the sets are not equipotent. Assume that the pigeonhole principle is true when $n=k$, that is, $\{1,\dots,k+m\}$ is not equipotent to $\{1,\dots,k\}$. If $n=k+1$, then the two sets are $\{1,\dots,k+m+1\}$ and $\{1,\dots,k+1\}$. Give the k members of set $\{1,\dots,k+1\} \sim \{f(k+m+1)\}$ the names $1,\dots,k$ where f is a mapping from the first set to the second. Now consider how f maps members $1,\dots,k+m$ of the first set to the renamed members $1,\dots,k$ of the second. By our supposition, those two subsets are not equipotent, so f cannot be one-to-one. Hence, $\{1,\dots,k+m+1\}$ is not equipotent to $\{1,\dots,k+1\}$, i.e., the pigeonhole principle is true when $n=k+1$. Thus, by the principle of mathematical induction, the pigeonhole principle is true for every n.

22. Suppose, to get a contradiction, that $2^{\mathbf{N}}$ were countable. Let $A_1, A_2, \dots$ be the sets in $2^{\mathbf{N}}$. Now each A_i can be expressed as a sequence $\{a_n^i\}$ of 0s and 1s where the nth term in $\{a_n^i\}$ is 0 if $n \notin A_i$ or 1 if $n \in A_i$. For example, the sequence corresponding to $\{1,2,3,5,8,13\}$ is $1,1,1,0,1,0,0,1,0,0,0,0,1,0,\dots$ where the ellipses indicate 0s. Furthermore, each $\{a_n^i\}$ corresponds to a set in $2^{\mathbf{N}}$. Hence, the set S of all such sequences and $2^{\mathbf{N}}$ are equipotent and therefore indistinguishable from the set-theoretic point of view so that in that sense, $2^{\mathbf{N}} = S$, or $\{A_1, A_2, \dots\} = \{\{a_n^1\}, \{a_n^2\}, \dots\}$. Next, construct a sequence $\{b_n\}$ of 0s and 1s such that b_m is different from the mth term of $\{a_n^m\}$. We have a contradiction because $\{b_n\}$ is in S but is different from every $\{a_n^i\}$. Thus, $2^{\mathbf{N}}$ is uncountable.

Here is a solution that uses Theorem 7 and Problem 44. Define function $f : 2^{\mathbf{N}} \to [0,1]$ as follows. For a set A in $2^{\mathbf{N}}$, construct a real number denoted as a binary numeral $0.\dots$ such that the nth bit to the right of the binary point is 0 if $n \notin A$ or 1 if $n \in A$. For example, $f(\{1,2,3,5,8,13\}) = 0.11101001000010\dots$. The function is onto: Denote a real number in $[0,1]$ by a binary numeral, and make a set of natural numbers that includes n if and only if the nth digit to the right of the binary point is 1; that set is in $2^{\mathbf{N}}$. Thus, $2^{\mathbf{N}}$ is uncountable by Theorem 7 because f is a function from $2^{\mathbf{N}}$ onto a nondegenerate interval of real numbers. That conclusion is not invalidated by the fact that f is not one-to-one (for example, $f(\{1\}) = 0.1 = 1/2 = 0.0111\dots = f(\{2,3,4,\dots\})$).

23. Each set E_i in such a collection $\{E_1,\dots,E_n\}$ is either empty, equipotent to $\{0,\dots,m\}$ where $m \in \mathbf{N}$, or equipotent to $\mathbf{N}$. Then $E_1 \times \cdots \times E_n$ is equipotent to a subset of $\overbrace{\mathbf{N} \times \cdots \times \mathbf{N}}^{n \text{ times}}$, which is countable by Corollary 4(i). Thus, $E_1 \times \cdots \times E_n$ is countable by Theorem 3.

Next, the range of every such mapping must be a set of natural numbers, and because we are considering *all* mappings of $\mathbf{N}$ into $\mathbf{N}$, there exists a mapping onto each set of natural numbers. Thus, by Problem 22, the collection of those mappings is uncountable.

24. This is the second sentence in the proof of Theorem 7. A nondegenerate interval I is not empty. So if I were finite, then there would be a natural number n for which I would be equipotent to $\{1,\dots,n\}$. Let $[a,b]$ be a nondegenerate subinterval of I. Let δ equal $(b-a)/n$. Subinterval $[a,b]$ contains the set A of numbers $\{a, a+\delta, a+2\delta, \dots, a+(n-1)\delta, b\}$. Set A contains exactly $n+1$ numbers and is therefore equipotent to $\{1,\dots,n,n+1\}$. But I contains A and so cannot be equipotent to $\{1,\dots,n\}$. Thus, I is not finite.

25. Take two nondegenerate open intervals (a_1,b_1) and (a_2,b_2) of real numbers. Define function $f : (a_1,b_1) \to \mathbf{R}$ by $f(x) = a_2 + (b_2-a_2)(x-a_1)/(b_1-a_1)$. As shown in the following graph, f is an invertible function from (a_1,b_1) onto (a_2,b_2). Hence, (a_1,b_1) and (a_2,b_2) are equipotent. Likewise, $[a_1,b_1)$ and $[a_2,b_2)$ are equipotent, $(a_1,b_1]$ and $(a_2,b_2]$ are equipotent, and $[a_1,b_1]$ and $[a_2,b_2]$ are equipotent.

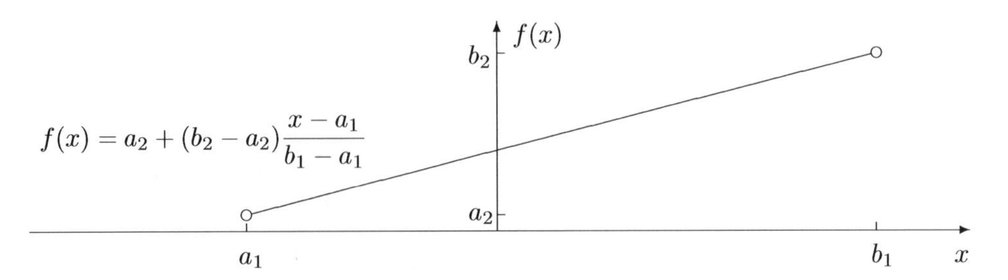

We show next that an open interval (a, b) is equipotent to $[a, b)$, $(a, b]$, and $[a, b]$. Our result above allows us to simplify our work by replacing a, b with $0, 1$. Consider the function $g : (0, 1) \to [0, 1)$ defined by

$$g(x) = \begin{cases} (p-1)/(q-1) & x \text{ can be expressed as } p/q \text{ where } p, q \in \mathbf{N}, \text{ and } p = q - 1, \\ x & \text{otherwise.} \end{cases}$$

For example, $g(0.5) = 0$, $g(0.66\ldots) = 0.5$, $g(0.75) = 0.66\ldots$, but $g(0.1) = 0.1$. Function g is onto because if a number y in $[0, 1)$ can be expressed as p/q where $p, q \in \mathbf{N} \cup \{0\}$, and $p = q - 1$, then $y = g((p + 1)/(q + 1))$ where $(p + 1)/(q + 1) \in (0, 1)$; otherwise, $y = g(y)$. Function g is also one-to-one because for each y in $g((0, 1))$, there is exactly one number x in $(0, 1)$ for which $y = g(x)$, in other words, if $g(x_1) = g(x_2)$, then $x_1 = x_2$. Hence, (a, b) is equipotent to $[a, b)$. We can devise an analogous function from $(0, 1)$ onto $(0, 1]$, and if necessary, compose that function with another analogous function from $(0, 1]$ onto $[0, 1]$.

We next consider the five types of unbounded intervals. Interval $(-\infty, \infty)$ is equipotent to $(0, 1)$ by the function from $(-\infty, \infty)$ to $(0, 1)$ defined by $x \mapsto 1/(1 + e^{-x})$, which is invertible as shown in the following graph.

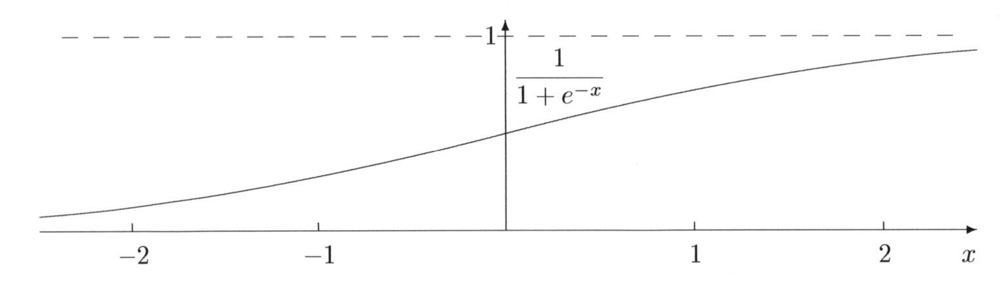

Intervals of type—

- (a, ∞) are equipotent to $(0, 1)$ by the function from $(0, 1)$ to (a, ∞) defined by

$$x \mapsto 1/x - 1 + a,$$

- $[a, \infty)$ are equipotent to $(0, 1]$ by the same function but from $(0, 1]$ to $[a, \infty)$,
- $(-\infty, a)$ are equipotent to $(0, 1)$ by the function from $(0, 1)$ to $(-\infty, a)$ defined by

$$x \mapsto 1 - 1/x + a,$$

- $(-\infty, a]$ are equipotent to $(0, 1]$ by the same function but from $(0, 1]$ to $(-\infty, a]$.

Those functions are invertible.

Thus, because equipotence is an equivalence relation, each two intervals of types (a, b), $[a, b)$, $(a, b]$, $[a, b]$, $(-\infty, \infty)$, (a, ∞), $[a, \infty)$, $(-\infty, a)$, $(-\infty, a]$ are equipotent.

26. Yes. We present two similar justifications. By Problem 25, $\mathbf{R}$, or $(-\infty, \infty)$, is equipotent to $(0,1]$. It therefore suffices to show that $(0,1] \times (0,1]$ is equipotent to $(0,1]$. Denote numbers in $(0,1]$ by their (unique) infinite decimal expansion $0.x_1x_2\ldots$ where x_i is a block of digits consisting of exactly one nonzero digit and the consecutive 0s that come before it. For example, $1/2 = 0.4\ 9\ 9\ldots$, $13/22 = 0.59\ 09\ 09\ldots$, and $e/3 = 0.9\ 06\ 09\ 3\ 9\ 4\ldots$. Consider function $f : (0,1] \times (0,1] \to (0,1]$ given by $f((0.x_1x_2\ldots, 0.y_1y_2\ldots)) = 0.x_1y_1x_2y_2\ldots$. That function is one-to-one because for each number z in $f((0,1] \times (0,1])$, there is exactly one pair (x,y) in $(0,1] \times (0,1]$ for which $z = f((x,y))$. The function is also onto because for a number in $(0,1]$, we can reverse the block-interleaving process of f to get a pair in $(0,1] \times (0,1]$. We ensure that f establishes a one-to-one correspondence by not allowing numbers to be denoted two different ways nor with terminating 0s. Thus, $(0,1] \times (0,1]$ is equipotent to $(0,1]$.

For our second (similar) justification, due to Georg Cantor, we first show that $(0,1)$ is equipotent to the set A of irrational numbers in $(0,1)$ to get that $\mathbf{R}$ is equipotent to A. Let $\{q_n \mid n \in \mathbf{N}\}$ be an enumeration of the rational numbers in $(0,1)$. Define a function $g : (0,1) \to A$ by

$$g(x) = \begin{cases} \sqrt{2}/2^{2n} & x = q_n \text{ for some } n \in \mathbf{N}, \\ \sqrt{2}/2^{2n-1} & x = \sqrt{2}/2^n \text{ for some } n \in \mathbf{N}, \\ x & \text{otherwise.} \end{cases}$$

Function g is onto because if a y in A is of the form $\sqrt{2}/2^n$, then $y = g(q_{n/2})$ if n is even or $g(\sqrt{2}/2^{(n+1)/2})$ if n is odd; otherwise, $y = g(y)$. The function is also one-to-one because for each y in $g((0,1))$, there is exactly one x in $(0,1)$ for which $y = g(x)$. Hence, $(0,1)$ is equipotent to A. It now suffices to show that $A \times A$ is equipotent to A because $\mathbf{R}$ is equipotent to $(0,1)$ by Problem 25.

Note that every irrational number u in $(0,1)$ is uniquely expressible as an infinite continued fraction:

$$u = \cfrac{1}{u_1 + \cfrac{1}{u_2 + \cfrac{1}{\ddots}}} = [0; u_1, u_2, \ldots].$$

Moreover, every such infinite continued fraction is irrational and in $(0,1)$. Hence, function $h : A \times A \to A$ defined by $h(([0; u_1, u_2, \ldots], [0; v_1, v_2, \ldots])) = [0; u_1, v_1, u_2, v_2, \ldots]$ is invertible. Thus, $A \times A$ is equipotent to A. You can find proofs of the two statements regarding irrational numbers and infinite continued fractions in Sections 7.3 and 7.4 of *An Introduction to the Theory of Numbers*, John Wiley and Sons, 1991 by Niven, Zuckerman, and Montgomery.

1.4 Open Sets, Closed Sets, and Borel Sets of Real Numbers

In the proof of the Heine-Borel theorem, the third sentence of the second paragraph should begin "Proposition 11 tells us ..."

In the nested set theorem, the first sentence should end "... F_1 is bounded."

27. No. Every open interval contains a rational number by Theorem 2 and an irrational number by Problem 12, so neither $\mathbf{Q}$ nor its complement, the set of irrational numbers, is open. By Proposition 11, $\mathbf{Q}$ is not closed. A more direct explanation of why $\mathbf{Q}$ is not closed is that it does not contain all of its points of closure. For example, $\sqrt{2}$ is a point of closure because every open interval that contains $\sqrt{2}$ contains a rational number, but $\sqrt{2}$ is not in $\mathbf{Q}$.

28. By Propositions 8 and 12, the empty-set $\emptyset$ and $\mathbf{R}$ are both open and closed. Let E be an open set different from $\emptyset$ and $\mathbf{R}$. Then E has an upper or lower bound. Assume first that E has an upper bound and therefore a supremum by the completeness axiom. By Proposition 9, E is the union of a countable, disjoint collection of open intervals. The interval with the greatest upper bound has the form $(a, \sup E)$. The supremum of E is a point of closure of E because every open interval that contains $\sup E$ also contains a point in E. The supremum of E, however, is not in E; therefore, E is not closed. We reach the same conclusion if E has a lower bound. Thus, $\emptyset$ and $\mathbf{R}$ are the only sets of real numbers that are both open and closed.

29. Put A to $(0, 1)$, and B to $(1, 2)$. The definitions on page 17 imply that $\overline{A} = [0, 1]$, and $\overline{B} = [1, 2]$. Then $(0, 1) \cap (1, 2) = \emptyset$, and $[0, 1] \cap [1, 2] = \{1\} \neq \emptyset$.

30. (i) If $E' = \emptyset$, then we are done by Proposition 12. Otherwise, let y be a point of closure of E'. If y is in E', then we are done. Otherwise, there is a different point x in E' such that $y - a_1 < x < y + b_1$ for all positive a_1, b_1. Because x is in E', x is a point of closure of $E \sim \{x\}$, so there is a different point z in E such that $x - a_2 < z < x + b_2$ for all positive a_2, b_2. Choose z to be also different from y. Such a z exists because we can reduce a_2 or b_2 to exclude y from $(x - a_2, x + b_2)$. Combining the two compound inequalities, $y - (a_1 + a_2) < z < y + (b_1 + b_2)$, which shows that every open interval that contains y contains a point z in $E \sim \{y\}$. Hence, y is a point of closure of $E \sim \{y\}$ and therefore also in E'. Thus, E' is closed because it contains all of its points of closure.

    (ii) $E \subseteq \overline{E}$ by the sentence after the definition on page 17, and $E' \subseteq \overline{E}$ by the definition given in this problem. Hence, $E \cup E' \subseteq \overline{E}$. Next, we get that inclusion in the other direction. If all points in $\overline{E}$ are in E, then $\overline{E} \subseteq E \cup E'$. Otherwise, let x be a point in $\overline{E}$ but not in E. Then x is a point of closure of $E \sim \{x\}$ and therefore in E', and again $\overline{E} \subseteq E \cup E'$. Thus, $\overline{E} = E \cup E'$.

31. Mimic the last part of the proof of Proposition 9. By definition of an isolated point, interval $(x - r, x + r)$ contains exactly one isolated point. By the density of the rationals, Theorem 2, each of those intervals contains a rational number. That establishes a one-to-one correspondence between the points of E and a subset of the rational numbers. We infer from Theorem 3 and Corollary 4(ii) that every set of rational numbers is countable. Therefore, E is countable.

32. The second sentence of the problem statement should be as in Problem 20 of Section 9.2: "The set of interior points of E is called the **interior** of E and denoted by $\operatorname{int} E$."

    (i) Subset E is open if and only if for each point y in E, there is a positive r for which $(y - r, y + r)$ is contained in E if and only if $y \in \operatorname{int} E$ if and only if $E \subseteq \operatorname{int} E$. It remains to replace that last inclusion with equality $E = \operatorname{int} E$, which we can do if $\operatorname{int} E \subseteq E$. But that is true for every set E because for each x in $\operatorname{int} E$, there is an open ball entered at x (and hence containing x) that is contained in E.

    (ii) Assume first that $\operatorname{int}(\mathbf{R} \sim E) = \emptyset$. Suppose, to get a contradiction, that E were not dense. Then there would be a y outside of E and a positive r such that $(y - r, y + r)$ would not be contained in E. But then y would be an isolated point in $\mathbf{R} \sim E$, a contradiction. Hence, E is dense.

    Conversely, assume that E is dense, i.e., between each pair of two real numbers there lies a point of E. Suppose, to get a contradiction, that $\mathbf{R} \sim E$ contained an interior point y. Then there would be a positive r such that $(y - r, y + r)$ would be contained in $\mathbf{R} \sim E$, which contradicts that there must be point of E in that interval. Hence, $\operatorname{int}(\mathbf{R} \sim E) = \emptyset$. A special case of this problem is Problem 27.

33. For example, for each n, put F_n to $[n, \infty)$. The intersection of those sets is empty because a real number x is not in $F_{\lfloor x \rfloor + 1}$ where $\lfloor \cdot \rfloor$ denotes the floor function.

34. The authors use the completeness axiom to prove the Heine-Borel theorem, so we can claim that the completeness axiom implies the Heine-Borel theorem. In turn, they use the Heine-Borel theorem to prove the nested set theorem, so we can also claim that the completeness axiom implies the nested set theorem.

We next show that the Heine-Borel theorem implies the completeness axiom. Let E be a nonempty set of real numbers that is bounded above. If E is not bounded below, replace E with a nonempty subset of E formed by removing numbers less than a sufficiently-low lower bound. Set $\overline{E}$ is closed by Proposition 10. Set $\overline{E}$ is also bounded (otherwise, we could choose a number in $\overline{E}$ greater than an upper bound for E and that is in an open interval that contains no point of E; likewise for the lower bound). Suppose, to get a contradiction, that $\overline{E}$ had no maximum number. Then $\{(-\infty, x)\}_{x \in \overline{E}}$ would be an open cover of $\overline{E}$ that had no finite subcover, contradicting the Heine-Borel theorem. Hence, $\overline{E}$ has a maximum number c. Now, c is an upper bound for E because $E \subseteq \overline{E}$. Moreover, c is the least upper bound for E. For if there were a smaller upper bound, then there would be a positive ϵ such that $c - \epsilon$ is an upper bound for E, and the open interval $(c - \epsilon/2, c + \epsilon)$ would contain a point, namely c, in $\overline{E}$ but no points in E, which contradicts the definition of the closure of E. Thus, E has a least upper bound.

It is *not* true that the nested set theorem implies the completeness axiom. See James Propp, "Real Analysis in Reverse," *The American Mathematical Monthly* 120, no. 5 (2013): 392–408 (arxiv.org/pdf/1204.4483.pdf). For the sake of argument, however, we assume the Archimedean property even though it is a consequence of the completeness axiom. Let E be as previously. Construct a descending countable collection $\{F_n\}_{n=1}^{\infty}$ of nonempty closed and bounded sets as follows. Let F_1 be $[a_1, b_1]$ where a_1 is a point in E, and b_1 an upper bound for E. For n equal to $2, 3, \ldots$, define F_n by

$$F_n = [a_n, b_n] = \begin{cases} [a_{n-1}, (a_{n-1} + b_{n-1})/2] & (a_{n-1} + b_{n-1})/2 \text{ is an upper bound for } E, \\ [(a_{n-1} + b_{n-1})/2, b_{n-1}] & \text{otherwise.} \end{cases}$$

The intersection of those sets is closed by Proposition 12 and not empty by the nested set theorem. The length of the intervals approaches zero, so by the Archimedean property, the intersection contains a single point, say c. Now c is an upper bound for E. Otherwise, there would be a y in E such that $y > c$. Because b_n is an upper bound for E, we also have that $c < y \leq b_n$, or $0 < y - c \leq b_n - c$. Because c is in $[a_n, b_n]$, difference $b_n - c$ is no more than the length $(b_1 - a_1)/2^{n-1}$ of the interval, which is no more than $(b_1 - a_1)/n$; in symbols,

$$0 < y - c \leq b_n - c \leq \frac{b_1 - a_1}{2^{n-1}} \leq \frac{b_1 - a_1}{n}.$$

By the Archimedean property, there is an n for which $(b_1 - a_1)/n$ is less than fixed number $y - c$, which contradicts the compound inequality. Hence, c is an upper bound for E. Moreover, c is the least upper bound for E. Otherwise, there would be a positive ϵ such that $c - \epsilon$ would be the least upper bound. In particular, the point a_n in E would be no greater than $c - \epsilon$ so that $a_n \leq c - \epsilon < c$, or $0 < \epsilon \leq c - a_n$. Using the same reasoning as previously,

$$0 < \epsilon \leq c - a_n \leq \frac{b_1 - a_1}{2^{n-1}} \leq \frac{b_1 - a_1}{n}.$$

By the Archimedean property, there is an n such that $(b_1 - a_1)/n < \epsilon$, which contradicts the compound inequality. Thus, E has a least upper bound.

35. The collection $\mathcal{B}$ of Borel sets includes also the closed sets by part (ii) of the definition on page 19 and Proposition 11. Furthermore, $\mathcal{B}$ is the smallest σ-algebra that contains the closed sets because every σ-algebra $\mathcal{A}$ that contains the closed sets contains the open sets (same reasoning as above), and $\mathcal{B}$ is the smallest of all such $\mathcal{A}$ by the definition on page 20.

36. By Problem 35 and the definition on page 20, the collection $\mathcal{B}$ of Borel sets contains all closed and open sets, in particular $\{a\}$ and (a, b). Then $[a, b)$ is in $\mathcal{B}$ by part (iii) of the definition of a σ-algebra because $[a, b) = \{a\} \cup (a, b)$. It remains to show that $\mathcal{B}$ is the smallest σ-algebra that contains such intervals. Every σ-algebra that contains intervals of the form $[a, b)$ also includes (a, b) because (see Problem 37)

$$(a, b) = \bigcup_{n=2}^{\infty} \left[a + \frac{b-a}{n}, b\right)$$

as well as $(-\infty, b)$, (a, ∞), $(-\infty, \infty)$ because

$$(-\infty, b) = \bigcup_{n=1}^{\infty} (a - n, b), \qquad (a, \infty) = \bigcup_{n=1}^{\infty} (a, b + n), \qquad (-\infty, \infty) = (-\infty, b) + (a, \infty).$$

Hence, every σ-algebra that contains intervals of the form $[a, b)$ contains all open intervals and therefore all open sets by Proposition 9. Collection $\mathcal{B}$ is the smallest of all such σ-algebras.

37. Open sets $\emptyset$ and $\mathbf{R}$ are closed sets by Proposition 12. Then by Proposition 9 and Corollary 6, it suffices to show that each open interval is countable union of closed sets. We can express bounded open intervals (a, b) as

$$\bigcup_{n=2}^{\infty} \left[a + \frac{b-a}{n}, b - \frac{b-a}{n}\right]$$

because by the Archimedean property, there is an n for which a number in (a, b) is in $[a + (b-a)/n, b - (b-a)/n]$. Open intervals $(-\infty, b)$ unbounded on the left can be expressed as $(-\infty, b-1] \cup (b-1, b)$, and those (a, ∞) on the right as $(a, a+1) \cup [a+1, \infty)$. Thus, each open set is an F_σ set.

1.5 Sequences of Real Numbers

In the proof of Theorem 15, the seventh sentence should begin "Since a is an upper bound ..."

Proof of Proposition 14 Suppose that $\{a_n\}$ converges to real number b also. Then for every positive ϵ, there is an index N for which if $n \geq N$, then $|b - a_n| < \epsilon/2$ and $|a - a_n| < \epsilon/2$. By the triangle inequality, $|b - a| \leq |b - a_n| + |a_n - a| < \epsilon$. Because that holds for ϵ arbitrarily small, $b = a$. Thus, the limit is unique.

We now show that $\{a_n\}$ is bounded. Because $\{a_n\} \to a$, there is an index N for which if $n \geq N$, then $|a - a_n| < 1$, which implies that $|a_n| < |a| + 1$. Put M to $\max\{|a_1|, \ldots, |a_{N-1}|, |a| + 1\}$. Then $|a_n| \leq M$ for all n. Thus, $\{a_n\}$ is bounded.

Lastly, take the limit as $n \to \infty$ of both sides of inequality $a_n \leq c$ to get that $a \leq c$, which concludes the proof. Note that we can strengthen this part of the proposition by changing the hypothesis to "if $a_n \leq c$ for all but finitely many n." $\square$

Theorem 18 for extended real numbers To simplify some of our solutions, we state and prove a more general version of Theorem 18 (see the discussion of extended real numbers on pages 9–10): Let $\{a_n\}$ and $\{b_n\}$ be sequences of real numbers that converge to extended real numbers. Then for each pair of real numbers α and β, if $\alpha \lim_{n\to\infty} a_n + \beta \lim_{n\to\infty} b_n$ is not of the form $\infty - \infty$, and if $\alpha \lim_{n\to\infty} a_n$ or $\beta \lim_{n\to\infty} b_n$ are not of the form $0 \cdot \infty$, then sequence $\{\alpha a_n + \beta b_n\}$ converges to an extended real number, and

$$\lim_{n\to\infty}(\alpha a_n + \beta b_n) = \alpha \lim_{n\to\infty} a_n + \beta \lim_{n\to\infty} b_n.$$

Moreover,

$$\text{if } a_n \leq b_n \text{ for all } n, \text{ then } \lim_{n\to\infty} a_n \leq \lim_{n\to\infty} b_n.$$

Proof Define a and b by

$$a = \lim_{n\to\infty} a_n, \text{ and } b = \lim_{n\to\infty} b_n.$$

- If a and b are real numbers, the proof is the same as the authors' proof of their Theorem 18.
- Next, assume that $a = \infty$, $b \in \mathbf{R}$, and $\alpha > 0$. If $\beta = 0$, ignore terms involving β in this part of the proof. For each real number c, there is an index N such that if $n \geq N$, then

$$a_n \geq \frac{c - (\beta b - 1)}{\alpha}, \text{ and } |b_n - b| < \frac{1}{\beta}, \text{ in particular, } b_n > b - \frac{1}{\beta}.$$

  Then if $n \geq N$,

$$\begin{aligned} \alpha a_n + \beta b_n &> \alpha a_n + \beta(b - 1/\beta) \\ &= \alpha a_n + (\beta b - 1) \\ &\geq c. \end{aligned}$$

  Hence, $\lim_{n\to\infty}(\alpha a_n + \beta b_n) = \infty$. If $a = -\infty$ or $\alpha < 0$, we have analogous results and proofs.

- Next, assume that $a, b = \infty$ and $\alpha, \beta > 0$. There is an N such that if $n \geq N$, then

$$a_n \geq \frac{c}{2\alpha}, \text{ and } b_n \geq \frac{c}{2\beta}.$$

  Then $\alpha a_n + \beta b_n \geq c$ if $n \geq N$. Hence, $\lim_{n\to\infty}(\alpha a_n + \beta b_n) = \infty$. Permissible cases where a or b are $-\infty$, and α or β are negative have analogous results and proofs.

We now prove the last part of the theorem.

- If $a = -\infty$ or $b = \infty$, then the claim follows directly from definition of $\pm\infty$.
- If $a = \infty$, then there is an N such that if $n \geq N$, then $a_n \geq c$. So if $n \geq N$, it follows that $b_n \geq c$ because $b_n \geq a_n$. Hence, $b = \infty$. If $b = -\infty$, the result and proof are analogous. □

38. Denote $\liminf\{a_n\}$ by ℓ. If $\ell = -\infty$, then by Proposition 19(ii)(iii), $\{a_n\}$ is not bounded below, so $\{a_n\}$ itself is a subsequence that converges to $-\infty$; hence, ℓ is a cluster point and the smallest of $\{a_n\}$. If $\ell = \infty$, then there is only one cluster point, ∞, so ℓ is the smallest.

    Now assume that $\{a_n\}$ is bounded below. Take a positive ϵ. By definition of *limit inferior*, we can choose an index N such that if $n \geq N$, then $a_n > \ell - \epsilon$. By Proposition 19(i)(iii), there are infinitely many indices n for which $a_n < \ell + \epsilon$, so we can choose natural numbers $n_1, n_2, \ldots$ such that $N \leq n_1 < n_2 < \cdots$, and $a_{n_k} < \ell + \epsilon$. Therefore, if $n_k \geq N$, then $|\ell - a_{n_k}| < \epsilon$. Hence, by the definition on page 21, $\{a_{n_k}\}$ converges to ℓ making it a cluster point.

Next suppose, to get a contradiction, that ℓ' were a smaller cluster point. Put ℓ' equal to $\ell - (\ell - \ell')$, or $\ell - \epsilon$. By Proposition 19(i)(iii), there are only finitely many indices n for which $a_n < \ell - \epsilon/2 = \ell' + \epsilon/2$, so no subsequence can converge to ℓ', contradicting that it is a cluster point. Thus, ℓ is the smallest cluster point of $\{a_n\}$. The proof is analogous for the limit superior.

39. (i) Assume first that $\limsup\{a_n\} = \ell$. By the definitions of convergence and limit on page 21, there is an index N for which if $n \geq N$, then $\ell - \epsilon < \sup\{a_k \,|\, k \geq n\} < \ell + \epsilon$. The first inequality shows that there are infinitely many n for which $\sup\{a_k \,|\, k \geq n\} > \ell - \epsilon$. The numbers in $\{a_k \,|\, k \geq n\}$ are taken from the terms in $\{a_n\}$, and $\ell - \epsilon$ is not an upper bound for $\{a_k \,|\, k \geq n\}$, so we can also claim that there are infinitely many n for which $a_n > \ell - \epsilon$. The second inequality shows that there are no $n \geq N$ for which $\sup\{a_k \,|\, k \geq n\} > \ell + \epsilon$, so there are only finitely many n for which that inequality holds. The numbers in $\{a_k \,|\, k \geq n\}$ are no greater than $\sup\{a_k \,|\, k \geq n\}$ and are taken from all but a finite number of terms in $\{a_n\}$, so we can also claim that there are only finitely many a_n for which $a_n > \ell + \epsilon$. Thus, there are only finitely many n for which $a_n > \ell + \epsilon$.

Conversely, assume that for each positive ϵ, there are infinitely many n for which $a_n > \ell - \epsilon/2$. Then it is certainly true that there is a number in $\{a_k \,|\, k \geq n\}$ greater than $\ell - \epsilon/2$ so that $\sup\{a_k \,|\, k \geq n\} > \ell - \epsilon$ for all n. We assume also that there are only finitely many n for which $a_n > \ell + \epsilon/2$. Then there is an N for which if $n \geq N$, then $a_n \leq \ell + \epsilon/2$ so that $\sup\{a_k \,|\, k \geq n\} < \ell + \epsilon$. Combining both results, we have that there is an N for which if $n \geq N$, then $|\ell - \sup\{a_k \,|\, k \geq n\}| < \epsilon$. Thus, $\limsup\{a_n\} = \ell$.

(ii) Assume first that $\{\sup\{a_k \,|\, k \geq n\}\} \to \infty$. By the definition of convergence to infinity on page 23, for each c, there is an index N such that if $n \geq N$, then $\sup\{a_k \,|\, k \geq n\} \geq c$. The terms in $\{\sup\{a_k \,|\, k \geq n\}\}$ are taken from $\{a_n\}$, so it is not true that $a_n \leq c$ for all n. Hence, $\{a_n\}$ is not bounded above.

Conversely, assume that $\{a_n\}$ is not bounded above. Then $\sup\{a_k \,|\, k \geq n\} = \infty$ for each n. Otherwise, $\{a_k \,|\, k \geq n\}$ would be bounded above, and the remaining finite number of a_n, i.e., $\{a_k \,|\, 1 \leq k < n\}$, would also be bounded (as in the proof of Theorem 17) making $\{a_n \,|\, n \in \mathbf{N}\}$ altogether bounded above, a contradiction. Thus, $\{\sup\{a_k \,|\, k \geq n\}\}$ is not bounded above, so $\{\sup\{a_k \,|\, k \geq n\}\} \to \infty$.

(iii) By Problem 6, $\sup\{a_k \,|\, k \geq n\} = -\inf\{-a_k \,|\, k \geq n\}$, so their limits as sequences as functions of n must also be the same. Thus, $\limsup\{a_n\} = -\liminf\{-a_n\}$ where we also made use our Theorem 18 (linearity) for extended real numbers.

(iv) $\{a_n\}$ converges to ∞ if and only if $\limsup\{a_n\} = \infty$ by part (ii) and the definition of convergence to infinity on page 23. In this case, $\{-a_n\}$ converges to $-\infty$ by Theorem 18 (linearity); by part (iii), $\liminf\{a_n\} = -\limsup\{-a_n\} = -(-\infty) = \infty$.

Analogously, $\{a_n\}$ converges to $-\infty$ if and only if $\liminf\{a_n\} = \limsup\{a_n\} = -\infty$.

The remaining case is that $a \in \mathbf{R}$. Assume first that $\limsup\{a_n\} = a = \liminf\{a_n\}$, which also equals $-\limsup\{-a_n\}$. Then by part (i), there are only finitely many indices n for which $a_n > a + \epsilon/2$, and $-a_n > -a + \epsilon/2$. It follows that there is an index N for which if $n \geq N$, then $a_n \leq a + \epsilon/2$, and $-a_n \leq -a + \epsilon/2$ so that $|a - a_n| \leq \epsilon/2 < \epsilon$. Thus, $\{a_n\}$ converges to a.

Conversely, assume that $\{a_n\}$ converges to a. Then by the definition of convergence on page 21, there are infinitely many n for which $a_n > a - \epsilon$, and only finitely many n for which $a_n > a + \epsilon$. By part (i), $\limsup\{a_n\} = a$. Because $\{-a_n\}$ converges to $-a$, we have analogously that $\limsup\{-a_n\} = -a$, or $\liminf\{a_n\} = a$.

(v) Because $b_n \geq a_n$ for all n, the least upper bound for $\{b_k \,|\, k \geq n\}$ is an upper bound for $\{a_k \,|\, k \geq n\}$. It follows that $\sup\{b_k \,|\, k \geq n\} \geq \sup\{a_k \,|\, k \geq n\}$ for all n. Thus, by Theorem 18 (monotonicity), $\limsup\{b_k \,|\, k \geq n\} \geq \limsup\{a_k \,|\, k \geq n\}$.

40. Sequence $\{a_n\}$ converges to an extended real number if and only if $\liminf\{a_n\} = \limsup\{a_n\}$ (Proposition 19(iv)) if and only if the smallest cluster point is also the largest cluster point (Problem 38) if and only if there is exactly one cluster point.

41. The problem statement should read "Show that $\liminf\{a_n\} \leq \limsup\{a_n\}$." Referring to Problem 38, the smallest cluster point of $\{a_n\}$ cannot be greater than the largest cluster point.

42. The displayed equation should have braces: $\limsup\{a_n \cdot b_n\} \leq (\limsup\{a_n\}) \cdot (\limsup\{b_n\})$.

    Denote $\{\sup\{a_k \,|\, k \geq n\}\}$ by $\{c_n\}$, and $\{\sup\{b_k \,|\, k \geq n\}\}$ by $\{d_n\}$. We first prove a lemma.

    Lemma If $\{c_n\} \to c$, and $\{d_n\} \to d$ where c and d are extended real numbers, then $\{c_n d_n\} \to cd$ provided cd is not of the form $0 \cdot \infty$.

    Proof Break the proof into cases.

    - First, assume that $c, d \in \mathbf{R}$. For every positive ϵ, there is an index N for which if $n \geq N$, then $|c - c_n| < \sqrt{\epsilon}$, and $|d - d_n| < \sqrt{\epsilon}$. Multiplying, $|(c - c_n)(d - d_n)| < \epsilon$, which shows that $\{(c - c_n)(d - d_n)\} \to 0$. Then

    $$\begin{aligned}
    \lim_{n\to\infty} c_n d_n - cd &= \lim_{n\to\infty} c_n d_n - cd + d \lim_{n\to\infty} (c - c_n) + c \lim_{n\to\infty} (d - d_n) \\
    &= \lim_{n\to\infty} (c_n d_n - cd + d(c - c_n) + c(d - d_n)) && \text{Theorem 18} \\
    &= \lim_{n\to\infty} (c - c_n)(d - d_n) \\
    &= 0.
    \end{aligned}$$

      Hence, $\{c_n d_n\} \to cd$.

    - Next, assume that $c = \infty$ and $0 < d < \infty$. For each real number M, there is an N such that if $n \geq N$, then $c_n \geq 2M/d$ and $|d - d_n| < d/2$, in particular, $d_n > d/2$. Then $c_n d_n > c_n d/2 \geq M$ if $n \geq N$. Hence, $\{c_n d_n\} \to \infty$. For the remaining cases where one of c or d is $\pm\infty$ and the other nonzero, we have analogous results and proofs.

    - Lastly, assume that $c, d = \infty$. There is an N such that if $n \geq N$, then $c_n \geq \sqrt{M}$ and $d_n \geq \sqrt{M}$. Then $c_n d_n \geq M$ if $n \geq N$. Hence $\{c_n d_n\} \to \infty$. For the other cases where c and d are $\pm\infty$, we have analogous results and proofs. □

    Now for indices j such that $j \geq n$, we have that $a_j b_j \leq \sup\{a_k \,|\, k \geq n\} \cdot \sup\{b_k \,|\, k \geq n\}$. It follows that $\sup\{a_k b_k \,|\, k \geq n\} \leq \sup\{a_k \,|\, k \geq n\} \cdot \sup\{b_k \,|\, k \geq n\}$ for all n. Thus, $\limsup\{a_n b_n\} \leq \limsup\{a_n\} \cdot \limsup\{b_n\}$ provided the product on the right is not of the form $0 \cdot \infty$ by our Theorem 18 (monotonicity) for extended real numbers and our lemma.

43. Take a real sequence $\{a_n\}$. Call n a *peak* index of $\{a_n\}$ if $a_n > a_m$ whenever $m > n$, i.e., if a_n is greater than every subsequent term in the sequence. If $\{a_n\}$ has infinitely many peaks $n_1, n_2, \ldots$ where $n_1 < n_2 < \cdots$, then the subsequence $\{a_{n_k}\}$ corresponding to those peaks is decreasing. Otherwise, let N be the last peak, and put n_1 to $N + 1$. Then n_1 is not a peak because $n_1 > N$, which implies the existence of an n_2 greater than n_1 such that $a_{n_2} \geq a_{n_1}$. Again, n_2 is not a peak; hence there is an n_3 greater than n_2 such that $a_{n_3} \geq a_{n_2}$. Repeating that process leads to an increasing subsequence of $\{a_n\}$.

    Next, let $\{a_n\}$ be bounded. By the above, $\{a_n\}$ has a necessarily bounded monotone subsequence. By Theorem 15, that subsequence converges, proving the Bolzano-Weierstrass theorem.

44. Let a_1 be the largest integer less than p such that $a_1/p \leq x$. Having chosen $a_1, a_2, \ldots, a_{n-1}$, let a_n be the largest integer less than p such that

$$\frac{a_1}{p} + \frac{a_2}{p^2} + \cdots + \frac{a_n}{p^n} \leq x.$$

As shown below in our answer to the converse part of this problem, $\sum_{n=1}^{\infty} a_n/p^n$ converges to some nonnegative real number y where $y \leq x$ by construction. The limit y, however, cannot be less than x; otherwise, y would equal $x - \epsilon$ for some positive ϵ, but the Archimedean property tells us that there is some p^n for which $a_n/p^n < \epsilon$, a contradiction. Thus, there is a $\{a_n\}$ such that $x = \sum_{n=1}^{\infty} a_n/p^n$. As for uniqueness,

$$\begin{aligned}
\frac{a_1}{p} &\leq \frac{a_1}{p} + \sum_{n=2}^{\infty} \frac{a_n}{p^n} && \text{right side is } x \\
&\leq \frac{a_1}{p} + \sum_{n=2}^{\infty} \frac{p-1}{p^n} && \text{equality if and only if } a_n = p-1 \text{ for all } n \geq 2 \\
&= \frac{a_1}{p} + \frac{1}{p} && \text{sum of a geometric series.}
\end{aligned}$$

That shows that if $a_n \neq p-1$ whenever $n \geq 2$, then $a_1/p \leq x < (a_1+1)/p$. Intervals $[k/p, (k+1)/p)$ where $k = 0, \ldots, p-1$ are disjoint, and their union is $[x, 1)$, so a_1 is uniquely determined by where x falls in one of those intervals . Repeating the argument for $x - a_1/p$, $x - a_1/p - a_2/p^2$, ..., we find that $\{a_n\}$ is unique as long as the terms do not all equal $p-1$ from some point onward. If they do, say starting when $n = m+1$, then

$$\sum_{n=m+1}^{\infty} (p-1)/p^n = 1/p^m = 1/p^m + 0/p^{m+1} + 0/p^{m+2} + \cdots,$$

which tells us that we get a second sequence $\{b_n\}$ of integers such that $0 \leq b_n < p$, and $x = \sum_{n=1}^{\infty} b_n/p^n$ where

$$b_n = \begin{cases} a_n & n < m, \\ a_n + 1 & n = m, \\ 0 & n > m. \end{cases}$$

Because the terms of $\{b_n\}$ are 0 if $n \geq m+1$, sum x is a finite sum of rational numbers b_n/p^n where $n = 1, \ldots, m$ and so is of the form q/p^m where q is a natural number in interval $(0, p^m)$. By reversing the argument, we find that if x is of the form q/p^m, then we have exactly two sequences that satisfy the given requirements.

For the converse, each term a_n/p^n is nonnegative, and the sequence of partial sums is bounded because

$$0 \leq \sum_{n=1}^{m} \frac{a_n}{p^n} \leq \sum_{n=1}^{m} \frac{p-1}{p^n} < \sum_{n=1}^{\infty} \frac{p-1}{p^n} = 1$$

where the equality comes from the formula for the sum of a geometric series. By Proposition 20(iii), the series converges to a number, which we showed is between 0 and 1.

45. (i) Referring to the second paragraph before the proposition and Theorem 17, $\sum_{k=1}^{\infty} a_k$ is summable if and only if $\{s_n\} \to s$ if and only if for each positive ϵ, there is an index $N-1$ for which

$$\text{if } n-1,\ n+m \geq N-1, \text{ then } \left|\sum_{k=n}^{n+m} a_k\right| = \left|\sum_{k=1}^{n+m} a_k - \sum_{k=1}^{n-1} a_k\right| = |s_{n+m} - s_{n-1}| < \epsilon.$$

(ii) In the following, we use the triangle inequality in the first step and part (i) in the last.

$$\left|\sum_{k=n}^{n+m} a_k\right| \leq \sum_{k=n}^{n+m} |a_k| = \left|\sum_{k=n}^{n+m} |a_k|\right| < \epsilon \text{ if } n \geq N \text{ and for each natural number } m.$$

Thus, $\sum_{k=1}^{\infty} a_k$ is summable.

(iii) Because each a_k is nonnegative, sequence $\{s_n\}$ of partial sums is increasing. By Theorem 15, $\{s_n\}$ converges if and only if it is bounded.

46. The authors use the nested set theorem and the Archimedean property of **R**, both of which depend on the completeness axiom, to prove the Bolzano-Weierstrass theorem, so we can claim that the completeness axiom implies the Bolzano-Weierstrass theorem. For the converse, it suffices by the second part of this problem to prove that the Bolzano-Weierstrass theorem implies the monotone convergence theorem. Let $\{a_n\}$ be an increasing sequence. If that sequence converges, say to a, then every term of $\{a_n\}$ is at most a. Otherwise, if some term a_N were greater than a, say $a_N = a + 2\epsilon$, then we have an ϵ such that $a_n \nless a + \epsilon$ if $n \geq N$, contradicting that $\{a_n\} \to a$. Hence, $\{a_n\}$ is bounded above (by a) (this is a special case of Proposition 14). Now assume that $\{a_n\}$ is bounded. By the Bolzano-Weierstrass theorem, there is a subsequence $\{a_{n_k}\}$, necessarily increasing, that converges to some number a. We claim that $\{a_n\} \to a$. Indeed, take a positive ϵ. Because $\{a_{n_k}\} \to a$, there is an index K for which if $k \geq K$, then $a - \epsilon < a_{n_k} \leq a$. Because $\{a_n\}, \{a_{n_k}\}$ are increasing, there is a k such that $k \geq K$ and $a - \epsilon < a_{n_k} \leq a_n \leq a_{n_{k+1}} \leq a$ if $n \geq n_K$. Therefore $\{a_n\} \to a$. The proof for the case when the sequence is decreasing is the same. Thus, the Bolzano-Weierstrass theorem implies the monotone convergence theorem.

The proof of Theorem 15 shows that the completeness axiom implies the monotone convergence theorem. For the converse, we first prove the Archimedean property assuming the monotone convergence theorem (we cannot use the authors' proof because it depends on the completeness axiom). Suppose, to get a contradiction, that **R** did not satisfy the Archimedean property. Then there is a real number greater than every term of increasing sequence $\{n\}$ so that $\{n\}$ is bounded. By the monotone convergence theorem, $\{n\}$ converges, say to a. That implies that $\{n - 1\}$ converges to a also. By Theorem 18, $\{n - (n-1)\} \to a - a = 0$, but $\{n - (n-1)\} = \{1\} \to 1$, a contradiction. Hence, **R** satisfies the Archimedean property.

Now let E be a nonempty set of real numbers that is bounded above. Let B be the set of upper bounds for E and A the complement of B. Then A, B are nonempty disjoint subsets whose union is **R** such that every number in A is less than every number in B. Define sequences $\{a_n\}, \{b_n\}$ such that a_n is the largest among $0, 1/2^n, 2/2^n, \ldots$ in A, and b_n is the smallest among those numbers in B. Sequences $\{a_n\}, \{b_n\}$ are bounded and monotone, so by the monotone convergence theorem, they converge. Moreover, they converge to the same limit c because $\{b_n - a_n\}$ converges to 0 by the Archimedean property. Now $a_n \leq c \leq b_n$ for all n, so $c - a_n \leq 1/2^n$, and $b_n - c \leq 1/2^n$. It follows from the Archimedean property that for every positive ϵ, there is a 2^n such that $1/2^n < \epsilon$ so that $a_n > c - \epsilon$, and $b_n < c + \epsilon$. Hence, every number less than or equal to $c - \epsilon$ is in A, and every number greater than or equal to $c + \epsilon$ is in B; therefore, every number less than c is in A, and every number greater than c is in B. Now c is an upper bound for E. Otherwise, some x in E exceeds c. Then $(x + c)/2 > c$, so $(x + c)/2$ is in B and is therefore an upper bound for E, which contradicts that $x > (x + c)/2$. Lastly, c is the least upper bound of E. If not, then some b less than c is an upper bound for E. But b (being less than c) is in A and so cannot be an upper bound for E. Thus, the monotone convergence theorem implies the completeness axiom.

1.6 Continuous Real-Valued Functions of a Real Variable

In the proof of the extreme value theorem, the tenth sentence should end "... choice of M."

Proof of Proposition 21 This is a straightforward combination of definitions of convergence and continuity. Assume first that f is continuous at x_* and that $\{x_n\}$ converges to x_*. Because f is continuous at x_* in E, for each positive ϵ, there is a δ for which if $x_n \in E$ and $|x_n - x_*| < \delta$, then $|f(x_n) - f(x_*)| < \epsilon$. Because $\{x_n\} \to x_*$, there is an index N for which if $n \geq N$, then $|x_n - x_*| < \delta$. Those two statements together say that for each ϵ, if $n \geq N$, then $|f(x_n) - f(x_*)| < \epsilon$. Thus, $\{f(x_n)\} \to f(x_*)$.

Conversely, assume that whenever $\{x_n\} \to x_*$, image sequence $\{f(x_n)\}$ converges to $f(x_*)$. In other words, whenever there is for every δ an N_1 for which if $n \geq N_1$, then $|x_n - x_*| < \delta$, there is for every ϵ an N_2 for which if $n \geq N_2$, then $|f(x_n) - f(x_*)| < \epsilon$. Now if $N_1 < N_2$, replace N_1 with the greater N_2, which does not change the truth of the previous sentence. Then we can say that for every ϵ, there is a δ for which if $|x_n - x_*| < \delta$, then $|f(x_n) - f(x_*)| < \epsilon$. Thus, f is continuous at x_*. □

47. If $E = \emptyset$, put g to 0. If $E = \mathbf{R}$, put g to f. Otherwise, $\mathbf{R} \sim E$ is nonempty and, by Proposition 11, open. By Proposition 9, $\mathbf{R} \sim E$ is composed of a countable, disjoint collection of open intervals $(a_1, b_1), (a_2, b_2), \ldots$. Define g by

$$g(x) = \begin{cases} f(x) & x \in E, \\ f(b_k) & x \in (a_k, b_k), \text{ and } a_k = -\infty, \\ f(a_k) + (f(b_k) - f(a_k))\dfrac{x - a_k}{b_k - a_k} & x \in (a_k, b_k),\ a_k \neq -\infty, \text{ and } b_k \neq \infty, \\ f(a_k) & x \in (a_k, b_k), \text{ and } b_k = \infty. \end{cases}$$

(The function used in the third case is the same as the function we used in Problem 25 where we also provide a graph.) Function g is continuous on E by hypothesis. Function g is also continuous on each of the open intervals (a_k, b_k). Indeed, take an x in (a_k, b_k) and a positive ϵ. Put δ equal to $\left|\dfrac{b_k - a_k}{f(b_k) - f(a_k)}\right|\epsilon$. If $x' \in (a_k, b_k)$, and $|x' - x| < \delta$, then

$$\begin{aligned} |g(x') - g(x)| &= \left| f(a_k) + (f(b_k) - f(a_k))\frac{x' - a_k}{b_k - a_k} - f(a_k) - (f(b_k) - f(a_k))\frac{x - a_k}{b_k - a_k} \right| \\ &= \left|\frac{f(b_k) - f(a_k)}{b_k - a_k}\right| |x' - x| \\ &< \left|\frac{f(b_k) - f(a_k)}{b_k - a_k}\right| \delta \\ &= \epsilon. \end{aligned}$$

In intervals where $f(b_k) = f(a_k)$, or $a_k = -\infty$, or $b_k = \infty$, function g is constant so that $|g(x') - g(x)| = 0 < \epsilon$. In all cases, g is continuous on each (a_k, b_k). It remains to consider the situation in which x, x' are on different sides of the a_k or b_k. For the sake of concreteness, assume that $x < a_k < x'$ where $x \in E$, $x' \in (a_k, b_k)$, and $a_k \neq -\infty$. For each ϵ, there are δ_1, δ_2 for which if $|a_k - x| < \delta_1$ and $|x' - a_k| < \delta_2$, then $|g(a_k) - g(x)| < \epsilon/2$ and $|g(x') - g(a_k)| < \epsilon/2$. It follows that if $|x' - x| \leq |x' - a_k| + |a_k - x| < \delta_1 + \delta_2 = \delta$, then $|g(x') - g(x)| \leq |g(x') - g(a_k)| + |g(a_k) - g(x)| < \epsilon/2 + \epsilon/2 = \epsilon$. Thus, g is continuous on $\mathbf{R}$.

48. Function f is continuous at zero: Take a positive ϵ. Put δ to ϵ. It follows that if x' is a real number and $|x' - 0| < \delta$, then

$$|f(x') - f(0)| = |f(x')| = \left\{ \begin{array}{ll} |x'| & x' \text{ irrational} \\ |p/q - p/(6q^3) + \cdots| < |p/q| = |x'| & x' = p/q \end{array} \right\} < \delta = \epsilon$$

where we replaced $p\sin(1/q)$ with its Maclaurin series.

Function f is not continuous at nonzero rational numbers x: Let ϵ equal $|x - f(x)|$. If x is positive, then for each positive δ there is by Problem 12 an irrational number x' in interval $(x, x+\delta)$ so that $|f(x') - f(x)| = x' - f(x) > x - f(x) = \epsilon$. If x is negative, take x' from $(x-\delta, x)$ so that $|f(x') - f(x)| = f(x) - x' > f(x) - x = \epsilon$.

Function f is continuous at irrational numbers x: First we claim that for each natural number N, there is a positive δ such that $q \geq N$ for a p/q in $(x-\delta, x+\delta)$. Otherwise, there would be an N such that for all n, there would be a rational p_n/q_n in $(x-1/n, x+1/n)$, and $0 < q_n < N$. Then $|p_n| \leq q_n \max\{|x-1|, |x+1|\} < N \max\{|x-1|, |x+1|\}$ for all n. Hence, there would be only finitely many choices of p_n, q_n for each n. That implies that there is would be a rational p/q in $(x-1/n, x+1/n)$ for infinitely many n, which contradicts the last paragraph on page 16.

Now, choose N such that $N^2 > \max\{|x-1|, |x+1|\}/(3\epsilon)$ and δ such that $0 < \delta < \min\{1, \epsilon/2\}$ and $q \geq N$ for a p/q in $(x-\delta, x+\delta)$. If an x' in $(x-\delta, x+\delta)$ is irrational, then

$$|f(x) - f(x')| = |x - x'| < \delta < \epsilon/2 < \epsilon.$$

Else if x' is a rational p/q, then

$$\begin{aligned}
|f(x) - f(x')| &\leq |f(x) - x'| + |x' - f(x')| && \text{triangle inequality}\\
&= |x - x'| + \left|p/q - \left(p/q - p/(6q^3) + \cdots\right)\right| && \text{Maclaurin series again}\\
&< \delta + \left|p/\left(6q^3\right)\right|\\
&< \epsilon/2 + \max\{|x-1|, |x+1|\}/\left(6q^2\right)\\
&\leq \epsilon/2 + \max\{|x-1|, |x+1|\}/\left(6N^2\right) && N \leq q \text{ by choice of } \delta\\
&< \epsilon/2 + \epsilon/2 = \epsilon && \text{choice of } N.
\end{aligned}$$

49. (i) Because f and g are continuous, for each x in E and positive ϵ, there are positive δ_1, δ_2 for which if $x' \in E$, and $|x' - x| < \delta_1$ and $|x' - x| < \delta_2$, then $|f(x') - f(x)| < \epsilon/2$ and $|g(x') - g(x)| < \epsilon/2$. Put δ to $\min\{\delta_1, \delta_2\}$. If $|x' - x| < \delta$, then

$$|(f+g)(x') - (f+g)(x)| \leq |f(x') - f(x)| + |g(x') - g(x)| < \epsilon/2 + \epsilon/2 = \epsilon.$$

Thus, $f+g$ is continuous.

For fg, we use the following three inequalities. As in the previous proof, there is a δ for which all three hold simultaneously if $|x' - x| < \delta$:

$$\begin{aligned}
&|f(x') - f(x)| < \epsilon \text{ so that } |f(x')| \leq |f(x') - f(x)| + |f(x)| < \epsilon + |f(x)|,\\
&|g(x') - g(x)| < \epsilon/(2(|f(x)| + \epsilon)),\\
&|f(x') - f(x)| < \epsilon/(2|g(x)|).
\end{aligned}$$

If $g(x) = 0$, ignore the second term on the right in the following:

$$\begin{aligned}
|(fg)(x') - (fg)(x)| &\leq |f(x')g(x') - f(x')g(x)| + |f(x')g(x) - f(x)g(x)|\\
&= |f(x')||g(x') - g(x)| + |f(x') - f(x)||g(x)|\\
&< (\epsilon + |f(x)|)\frac{\epsilon}{2(|f(x)| + \epsilon)} + \frac{\epsilon}{2|g(x)|}|g(x)|\\
&= \epsilon/2 + \epsilon/2\\
&= \epsilon \text{ or if } g(x) = 0, \text{ then } \epsilon/2.
\end{aligned}$$

Thus fg is continuous.

Another solution takes advantage of Proposition 21. Whenever a sequence $\{x_n\}$ converges to x_*, then $\lim_{n\to\infty} f(x_n) = f(x_*)$, and $\lim_{n\to\infty} g(x_n) = g(x_*)$. By Theorem 18,

$$\lim_{n\to\infty}(f+g)(x_n) = \lim_{n\to\infty} f(x_n) + \lim_{n\to\infty} g(x_n) = (f+g)(x_*).$$

Thus, by Proposition 21 in the other direction, $f+g$ is continuous.

That proof also works for fg by invoking our lemma to Problem 42 instead of Theorem 18. That lemma holds for real sequences in general and gives us here that $\{f(x_n)g(x_n)\}$ converges to $f(x_*)g(x_*)$.

(ii) Let A be the domain of h. Because f is continuous on E, for each y in A, there is a positive η for which if $y' \in A$ and $|h(y') - h(y)| < \eta$, then $|f(h(y')) - f(h(y))| < \epsilon$. Because h is continuous on A, for that η, there is a δ for which if $|y' - y| < \delta$, then $|h(y') - h(y)| < \eta$. Those two statements together say that if $|y' - y| < \delta$, then $|(f \circ h)(y') - (f \circ h)(y)| < \epsilon$. Thus, $f \circ h$ is continuous on A.

(iii) Because $\max\{f, g\} = (f + g + |f - g|)/2$, it is continuous by parts (i) and (iv).

(iv) We first prove the reverse triangle inequality.

Reverse triangle inequality For all pairs of real numbers a and b,

$$||a| - |b|| \leq |a - b|.$$

Proof Apply the triangle inequality twice:

$$|a| \leq |a - b| + |b|, \text{ or } |a| - |b| \leq |a - b|, \text{ and}$$
$$|b| \leq |b - a| + |a|, \text{ or } |b| - |a| \leq |a - b|.$$

Thus, $||a| - |b|| \leq |a - b|$. The normed-linear-space version is $|\|a\| - \|b\|| \leq \|a - b\|$, which we use to solve problems in subsequent chapters. □

As before, there is a positive δ for which if $|x' - x| < \delta$, then $|f(x') - f(x)| < \epsilon$. But by the reverse triangle inequality, $||f(x')| - |f(x)|| \leq |f(x') - f(x)|$, so $||f(x')| - |f(x)|| < \epsilon$. Thus, $|f|$ is continuous.

50. The authors' remarks after the definition of *Lipschitz* hold also for uniform continuity. Indeed, in the notation of the definitions of *Lipschitz* and *uniform continuity*, for all x, x' in E and each ϵ, a δ of ϵ/c responds to the ϵ challenge regarding the criterion for the uniform continuity of f at x. Explicitly, $|f(x) - f(x')| \leq c|x - x'| < c\delta = c(\epsilon/c) = \epsilon$. If $c = 0$, then f is a constant function and so is again uniformly continuous.

Next, function $x \mapsto \sqrt{x}$ is uniformly continuous if $x \geq 0$. Indeed, put δ to ϵ^2. Then for all x, x' in $[0, \infty)$,

$$\begin{aligned}
|\sqrt{x} - \sqrt{x'}| &= \sqrt{|\sqrt{x} - \sqrt{x'}||\sqrt{x} - \sqrt{x'}|} \\
&\leq \sqrt{|\sqrt{x} - \sqrt{x'}||\sqrt{x} + \sqrt{x'}|} \\
&= \sqrt{|x - x'|} \\
&< \sqrt{\delta} \\
&= \epsilon.
\end{aligned}$$

On the other hand, function $x \mapsto \sqrt{x}$ is not Lipschitz because there is no constant c such that

$$c \geq (\sqrt{x'} - \sqrt{0})/(x' - 0) = 1/\sqrt{x'}$$

for all x' in $[0, \infty)$ since we can make $1/\sqrt{x'}$ larger than c by choosing an x' less than $1/c^2$.

51. Put $\varphi(x_i)$ to $f(x_i)$ with $x_1, \ldots, x_{n-1}$ yet to be determined. Because $[a, b]$ is a closed, bounded set of real numbers, f, φ are uniformly continuous on $[a, b]$ by Theorem 23. Then there is a positive δ such that for all x, x_i in $[a, b]$, if $|x - x_i| < \delta$, then $|f(x) - f(x_i)| < \epsilon/2$, $|\varphi(x_i) - \varphi(x)| < \epsilon/2$ so that

$$|f(x) - \varphi(x)| \leq |f(x) - f(x_i)| + |\varphi(x_i) - \varphi(x)| < \epsilon/2 + \epsilon/2 = \epsilon.$$

That shows that if the difference between consecutive x_is is less than δ, then our φ satisfies the desired constraint $|f(x) - \varphi(x)| < \epsilon$ for all x in $[a, b]$. For example, $x_i = x_{i-1} + \delta/2$ where $i = 1, \ldots, n-1$ will do.

52. If E is closed and bounded, then every continuous real-valued function f on E takes a maximum value by the extreme value theorem. Conversely, assume that every such f takes a maximum value. Then E must be bounded; otherwise, $f(x) = |x|$ would define an f that takes no maximum value on E, a contradiction. Set E must also be closed; otherwise, E would have a point of closure x' that is not in E so that $f(x) = 1/|x - x'|$ would define an f that takes no maximum value on E. Here is another way to prove that E is closed using that same x': Function f defined by $f(x) = -|x - x'|$ is continuous, bounded above by 0, and takes a maximum on E by assumption. Because every open interval that contains x' also contains a point in E, we can find an x in E to make $f(x)$ as close to 0 as we want. It follows that the maximum value that f takes is 0; that happens when $x = x'$, which contradicts that $x' \notin E$.

53. If E is closed and bounded, then every open cover of E has a finite subcover by the Heine-Borel theorem.

Conversely, assume that every open cover of E has a finite subcover. Suppose, to get a contradiction, that E were not bounded. Consider open cover $\{(x-1, x+1)\}_{x \in E}$ of E. Take a finite subcover $\{(x_n - 1, x_n + 1)\}_{n=1}^{N}$. Because E is not bounded, there is an x in E such that $|x| > \max\{|x_n \pm 1|\}_{n=1}^{N}$ so that x is not in the finite subcover. Hence, there is no finite subcover of E, which contradicts our assumption. Thus, E is bounded.

Now suppose, to get another contradiction, that E were not closed. Then E would have a point of closure x' not in E. Consider open cover $\{(-\infty, x' - 1/n) \cup (x' + 1/n, \infty)\}_{n=1}^{\infty}$ of E. The contradiction is that the open cover has no finite subcover because for all n, interval $(x' - 1/n, x' + 1/n)$ contains a point in E. Thus, E is closed.

54. Assume first that E is an interval. Suppose, to get a contradiction, that f is a continuous real-valued function on E, but $f(E)$ were not an interval. Then there would be real numbers a, b, c such that $[a, b] \subseteq E$ and $a \leq b$ for which c is between $f(a)$ and $f(b)$ but no x_0 in $[a, b]$ at which $c = f(x_0)$. That contradicts the intermediate value theorem, which holds also if $f(b) < c < f(a)$. Thus, $f(E)$ is an interval.

Conversely, assume that $f(E)$ is an interval for every continuous real-valued function f. It suffices to consider just one such f. Identity function id_E, for example, is continuous, and $id_E(E) = E$. Thus, if $f(E)$ is an interval, then E is too.

55. Assume first that a monotone function on an open interval is continuous. Then the function's image is an interval by the first part of our solution to Problem 54. That proof does not require the monotone and open stipulations, but the proof in the last paragraph of the section does use them.

Conversely, assume that the image of a monotone function on open interval (a, b) is an interval. The last paragraph of the section starts this part of the proof. It remains to show that $f(x_0^-) = f(x_0) = f(x_0^+)$. If not, then say for concreteness that $f(x_0^-) \neq f(x_0)$. Because f is monotone, say increasing, $f(x_0^-) < f(x_0) \leq f(x_0^+)$. Then there are points in interval

$[f(x_0^-), f(x_0)]$ that are not in $(f(a), f(b))$ which contradicts that $f((a,b))$ is an interval. Thus, f is continuous. The proof is similar if $f(x_0) \neq f(x_0^+)$ or if f is decreasing.

56. For each natural number n, define set G_n by

$$G_n = \{x \,|\, \text{there is a positive } \delta \text{ such that if } x', x'' \in (x-\delta, x+\delta), \text{ then } |f(x')-f(x'')| < 1/n\}.$$

G_n is open because for each x in G_n, there is an r given by $r = |x' - x''|/2 > 0$ where $x'' \neq x'$ for which interval $(x - r, x + r)$ is contained in G_n.

We now show that f is continuous at x if and only if x is in G_n for all n, that is, $x \in \bigcap_{n=1}^{\infty} G_n$. Assume first that $x \in \bigcap_{n=1}^{\infty} G_n$. Take a positive ϵ. We can choose n such that $1/n < \epsilon$ by the Archimedean property. Replace x'' in the definition of G_n with x. Then the statement defining which points are in G_n satisfies the definition of the continuity of f at x.

Conversely, let f be continuous at x. Then for each ϵ, there are a δ and an n for which if

$$x' \in G_n, \text{ and } |x' - x| < \delta, \text{ then } |f(x') - f(x)| < 1/(2n) < \epsilon, \text{ and if}$$
$$x'' \in G_n, \text{ and } |x'' - x| < \delta, \text{ then } |f(x'') - f(x)| < 1/(2n) < \epsilon.$$

By the triangle inequality,

$$|f(x') - f(x'')| \leq |f(x') - f(x)| + |f(x) - f(x'')| < 1/(2n) + 1/(2n) = 1/n.$$

Hence, x is in G_n for all n. Thus, the set of points at which f is continuous is a countable intersection of opens sets, a G_δ set.

Here is a second solution. Let G be the set of points at which f is continuous. By the second De Morgan identity on page 4 and Proposition 11, it suffices to show that $\mathbf{R} \sim G$ is an F_σ set. By Propositions 21 and 19(iv), f is continuous at x if and only if whenever $\{x_n\} \to x$, then $\liminf\{f(x_n)\} = \limsup\{f(x_n)\} = f(x)$. In the following, W stands for the statement "whenever $\{x_n\} \to x$."

$$\begin{aligned}
\mathbf{R} \sim G &= \{x \,|\, W,\ \liminf\{f(x_n)\} < \limsup\{f(x_n)\}\} \qquad \text{Problem 41} \\
&= \{x \,|\, W,\ \text{there are } p, q \text{ in } \mathbf{Q} \text{ such that } \liminf\{f(x_n)\} \leq p < q \leq \limsup\{f(x_n)\}\} \\
&= \bigcup_{\substack{p,q \in \mathbf{Q} \\ p<q}} \{x \,|\, W,\ \liminf\{f(x_n)\} \leq p\} \cap \{x \,|\, W,\ q \leq \limsup\{f(x_n)\}\},
\end{aligned}$$

which is a countable union of sets by Corollaries 4(ii) and 6. The union is an F_σ set if each set in the intersection is closed (Proposition 12). To show that $\{x \,|\, W,\ \liminf\{f(x_n)\} \leq p\}$ is closed, it suffices to show that its complement $\{x \,|\, W,\ \liminf\{f(x_n)\} > p\}$ is open. Express it equivalently as

$$\begin{aligned}
&\{x \,|\, \text{whenever for each } \delta, \text{ there is an index } N \text{ for which if } n \geq N, \text{ then } |x - x_n| < \delta, \\
&\quad \text{it follows that } \liminf\{f(x_n)\} > p\} = E, \text{ say,}
\end{aligned}$$

where "it follows that" is the "outer" implication. Let x be a point in E. For each x' in $(x - \delta, x + \delta)$, there is a δ' such that $(x' - \delta', x' + \delta') \subseteq (x - \delta, x + \delta)$ by the definition of an open set (interval) on page 16. Consider the image sequence $\{f(x_n')\}$ of some $\{x_n'\}$ that takes terms x_n' from $(x' - \delta', x' + \delta')$. We can consider $\{f(x_n')\}$ to be a subsequence of some $\{f(x_n)\}$ by the set inclusion above; it follows that $\liminf\{f(x_n')\} \geq \liminf\{f(x_n)\}$. Hence, $\liminf\{f(x_n')\} > p$, which shows that x' is in E, which in turn shows that the open interval $(x - \delta, x + \delta)$ is contained in E. Set E is therefore open by definition. The proof that $\{x \,|\, W,\ q \leq \limsup\{f(x_n)\}\}$ is closed is analogous. Thus, G is a G_δ set.

57. Starting with Proposition 21, Theorem 17, and the Archimedean property, we can express the set of points at which $\{f(x_n)\}$ converges to a real number as

$$\begin{aligned}
&\{x \mid \text{for all } k \text{ in } \mathbf{N}, \text{ there is an } N \text{ for which if } n, m \geq N, \text{ then } |f(x_m) - f(x_n)| \leq 1/k\} \\
&\quad = \bigcap_{k \in \mathbf{N}} \{x \mid \text{there is an index } N \text{ for which if } n, m \geq N, \text{ then } |f(x_m) - f(x_n)| \leq 1/k\} \\
&\quad = \bigcap_{k \in \mathbf{N}} \bigcup_{N \in \mathbf{N}} \{x \mid \text{if } m \geq N, \text{ then } |f(x_m) - f(x_N)| \leq 1/k\} \\
&\quad = \bigcap_{k \in \mathbf{N}} \bigcup_{N \in \mathbf{N}} \bigcap_{m \geq N} \{x \mid |f(x_m) - f(x_N)| \leq 1/k\} \\
&\quad = \bigcap_{k \in \mathbf{N}} \bigcup_{N \in \mathbf{N}} \bigcap_{m \geq N} \{x \mid f(x_m) \in [f(x_N) - 1/k, f(x_N) + 1/k]\}.
\end{aligned}$$

Recall that intersection expresses the condition "for all" as in "for all k in $\mathbf{N}$" and "for all m greater than or equal to N" whereas union expresses the condition "for one" as in "there is an index N." The innermost set is closed by the second part of Exercise 58. The inner intersection is closed by Proposition 12. The union, therefore, is a countable union of closed sets, an F_σ set. Thus, the set of points at which $\{f(x_n)\}$ converges is the intersection of a countable union of F_σ sets.

58. The first claim is true by Proposition 22 with E replaced by $\mathbf{R}$ so that $\mathbf{R} \cap \mathcal{U}$ is open set $\mathcal{U}$.

Next, let F be a closed set. Then $\mathbf{R} \sim F$ is open by Proposition 11 so that $f^{-1}(\mathbf{R} \sim F)$ is open by the previous. Now $f^{-1}(\mathbf{R} \sim F) = \mathbf{R} \sim f^{-1}(F)$; otherwise, a domain point x in $f^{-1}(F)$ would be mapped to points in F and $\mathbf{R} \sim F$, which violates the definition of a function. Thus, $f^{-1}(F)$ is closed because $\mathbf{R} \sim f^{-1}(\mathbf{R} \sim F)$ is closed.

We use the hint of Problem 42 of Section 2.7. Let $\mathcal{B}$ denote the collection Borel sets. Define $\mathcal{A}$ to be the collection of sets E for which $f^{-1}(E)$ is Borel—in symbols, $\mathcal{A} = \{E \mid f^{-1}(E) \in \mathcal{B}\}$. By the first part of this problem and definition of a Borel set, $\mathcal{A}$ contains all of the open sets. Next, we show that $\mathcal{A}$ is a σ-algebra. Referring to the definition on page 19, (i) the empty-set, $\emptyset$, belongs to $\mathcal{A}$ because $\emptyset$ is open by Proposition 8; (ii) the complement of a set E in $\mathcal{A}$ belongs to $\mathcal{A}$ because $f^{-1}(\mathbf{R} \sim E) = \mathbf{R} \sim f^{-1}(E)$, which is Borel since it is the complement of a Borel set; (iii) the union of a countable collection of sets E_n in $\mathcal{A}$ belongs to $\mathcal{A}$ because

$$f^{-1}\left(\bigcup_{n=1}^{\infty} E_n\right) = \bigcup_{n=1}^{\infty} f^{-1}(E_n),$$

which is Borel since it is the union of a countable collection of Borel sets. Hence, $\mathcal{A}$ is a σ-algebra that contains the open sets. By definition, $\mathcal{B}$ is the smallest σ-algebra that contains the open sets, so $\mathcal{B} \subseteq \mathcal{A}$. Thus, if a set B is Borel, then it is in $\mathcal{A}$ so that $f^{-1}(B)$ is Borel by definition of $\mathcal{A}$. For a proof of the displayed identity above, see our solution to Problem 23 of Section 11.4.

59. There is an N such that $|f_N(x) - f(x)| < \epsilon/3$, and $|f(x') - f_N(x')| < \epsilon/3$ for all x, x' in E. Also, because f_N is continuous, there is a positive δ for which if $|x' - x| < \delta$, then $|f_N(x') - f_N(x)| < \epsilon/3$. Altogether, if $|x' - x| < \delta$, then

$$\begin{aligned}
|f(x') - f(x)| &\leq |f(x') - f_N(x')| + |f_N(x') - f_N(x)| + |f_N(x) - f(x)| \quad \text{triangle inequality} \\
&< \epsilon/3 + \epsilon/3 + \epsilon/3 \\
&= \epsilon.
\end{aligned}$$

Thus, f is continuous on E.

Chapter 2

Lebesgue Measure

2.1 Introduction

1. $m(B) = m((B \sim A) \cup A) = m(B \sim A) + m(A) \geq m(A)$ by the first two sentences on page 31.

2. By definition of disjoint sets, A is disjoint from $\emptyset$. Then $m(A) = m(A \cup \emptyset) = m(A) + m(\emptyset)$. Thus, $m(\emptyset) = 0$.

3. Let A_1 equal E_1, and A_n equal $E_n \sim \bigcup_{k=1}^{n-1} E_k$ where $n = 2, 3, \ldots$. Then $\bigcup_{k=1}^{\infty} A_k = \bigcup_{k=1}^{\infty} E_k$, $\{A_k\}_{k=1}^{\infty}$ is a countable disjoint collection of sets in $\mathcal{A}$, and $A_k \subseteq E_k$ for each k. Thus, $m(\bigcup_{k=1}^{\infty} E_k) = m(\bigcup_{k=1}^{\infty} A_k) = \sum_{k=1}^{\infty} m(A_k) \leq \sum_{k=1}^{\infty} m(E_k)$ by Problem 1.

4. Set function c is translation invariant because adding a number to each element in a set does not change the number of elements in the set.

   Next, let $\{E_k\}_{k=1}^{\infty}$ be a (countable) disjoint collection of sets in $\mathbf{R}$. If some E_k has infinitely many members, then so does $\bigcup_{k=1}^{\infty} E_k$. Hence, $c(\bigcup_{k=1}^{\infty} E_k) = \infty = \sum_{k=1}^{\infty} c(E_k)$ because c is nonnegative. If each E_k is finite, and only a finite number of E_k are nonempty, then $\bigcup_{k=1}^{\infty} E_k$ is finite so that $c(\bigcup_{k=1}^{\infty} E_k) = \sum_{k=1}^{\infty} c(E_k)$ because $c(\emptyset) = 0$. If an infinite number of E_k are nonempty, then $\bigcup_{k=1}^{\infty} E_k$ is countably infinite by the proof of Corollary 6 of Section 1.3 so that $c(\bigcup_{k=1}^{\infty} E_k) = \infty$. Also, $\sum_{k=1}^{\infty} c(E_k) = \infty$ because there are a countably infinite number of E_k with at least one element. Thus, c is countably additive over countable disjoint unions of sets.

2.2 Lebesgue Outer Measure

5. If $[0,1]$ were countable, its outer measure would be zero by the example. But that contradicts Proposition 1 that $m^*([0,1]) = 1$. Thus, $[0,1]$ is not countable.

6. Because A is a subset of $[0,1]$, we have that $m^*(A) \leq m^*([0,1]) = 1$ by monotonicity. Next, we show that $m^*(A) \geq 1$. Let B be the set of rational numbers in $[0,1]$; because B is countable by Corollary 4(ii) of Section 1.3, $m^*(B) = 0$. Then

   $$m^*(A) = m^*(A) + m^*(B) \geq m^*(A \cup B) = m^*([0,1]) = 1$$

   where the inequality comes from Proposition 3. Thus, $m^*(A) = 1$. This problem is a special case of Problem 9 (interchange A and B).

7. As in the proof of Proposition 3, for each natural number n, there is a countable collection $\{I_{n,k}\}_{k=1}^{\infty}$ of open, bounded intervals for which

   $$E \subseteq \bigcup_{k=1}^{\infty} I_{n,k} \text{ and } \sum_{k=1}^{\infty} \ell(I_{n,k}) < m^*(E) + 1/n.$$

   Put G to $\bigcap_{n=1}^{\infty} \bigcup_{k=1}^{\infty} I_{n,k}$. Set G is a G_δ set (because the union is open by Proposition 8 of Section 1.4) for which $E \subseteq G$. Furthermore, $m^*(G) = m^*(E)$ because the following holds for all n:

   $$m^*(E) \leq m^*(G) \leq m^*\left(\bigcup_{k=1}^{\infty} I_{n,k}\right) \leq \sum_{k=1}^{\infty} \ell(I_{n,k}) < m^*(E) + 1/n.$$

8. We establish four preliminary facts. (i) The closure $\overline{B}$ of B is $[0,1]$ because every open interval that contains a number in $[0,1]$ also contains a rational number in B by Theorem 2 of Section 1.2; on the other hand, every point x outside of $[0,1]$ is not in $\overline{B}$ because we can construct an open interval that contains x and is small enough to be disjoint from B. (ii) By the last sentence in the proof of Proposition 10 of Section 1.4 or Problem 23 of Section 9.2, $\overline{B} \subseteq \overline{\bigcup_{k=1}^{n} I_k}$. (iii) It follows from definition of a point of closure that $\bigcup_{k=1}^{n} \overline{I_k} = \overline{\bigcup_{k=1}^{n} I_k}$ by checking set inclusions. (iv) The outer measure of closed interval $\overline{I_k}$ is the same as that of I_k because they have the same length (Proposition 1). Thus,

$$1 = m^*([0,1]) = m^*(\overline{B}) \le m^*\left(\overline{\bigcup_{k=1}^{n} I_k}\right) = m^*\left(\bigcup_{k=1}^{n} \overline{I_k}\right) \le \sum_{k=1}^{n} m^*(\overline{I_k}) = \sum_{k=1}^{n} m^*(I_k).$$

9. By Proposition 3, $m^*(A \cup B) \le m^*(A) + m^*(B) = m^*(B)$. On the other hand, we have that $m^*(B) \le m^*(A \cup B)$ because $B \subseteq A \cup B$. Thus, $m^*(A \cup B) = m^*(B)$.

10. We have three solutions, similar in that each is based on the idea that we can make the intervals that cover A disjoint from those that cover B because all points in B are separated from all points in A by a distance of at least α. (i) Referring to the definition of m^*, let the $I_{A,k}$ be intervals of length less than $\alpha/2$ that cover A, and the $I_{B,k}$ those that cover B. The first step below is the definition of $m^*(A \cup B)$. The second step is valid because we need not consider intervals whose lengths are $\alpha/2$ or greater since we are taking the infimum, and because no $I_{A,k}$ contains a point b and vice versa since $|b-a| \ge \alpha$. The latter reason also justifies the third step:

$$\begin{aligned} m^*(A \cup B) &= \inf\left\{ \sum_{k=1}^{\infty} \ell(I_k) \,\middle|\, A \cup B \subseteq \bigcup_{k=1}^{\infty} I_k \right\} \\ &= \inf\left\{ \sum_{k=1}^{\infty} \ell(I_{A,k}) + \sum_{k=1}^{\infty} \ell(I_{B,k}) \,\middle|\, A \subseteq \bigcup_{k=1}^{\infty} I_{A,k},\ B \subseteq \bigcup_{k=1}^{\infty} I_{B,k} \right\} \\ &= \inf\left\{ \sum_{k=1}^{\infty} \ell(I_{A,k}) \,\middle|\, A \subseteq \bigcup_{k=1}^{\infty} I_{A,k} \right\} + \inf\left\{ \sum_{k=1}^{\infty} \ell(I_{B,k}) \,\middle|\, B \subseteq \bigcup_{k=1}^{\infty} I_{B,k} \right\} \\ &= m^*(A) + m^*(B). \end{aligned}$$

(ii) Let A' be the set $\bigcup_{a \in A}(a - \alpha/2, a + \alpha/2)$. Then A' contains all of A's points, but none of B's points because the distance from B's points to A's points is more than $\alpha/2$. In symbols, $A \subseteq A'$ so that $A \cap (\mathbf{R} \sim A') = \emptyset$, and $A' \cap B = \emptyset$ so that $B \subseteq (\mathbf{R} \sim A')$. In the following, we start with an equality that separates the parts of the intervals in the definition of m^* that cover $A \cup B$ and lie in A' from the parts that lie in $\mathbf{R} \sim A'$; that equality holds because lengths of intervals are countably additive over countable disjoint unions of sets:

$$\begin{aligned} m^*(A \cup B) &= m^*((A \cup B) \cap A')) + m^*((A \cup B) \cap (\mathbf{R} \sim A')) \\ &= m^*((A \cap A') \cup (B \cap A')) + m^*((A \cap (\mathbf{R} \sim A')) \cup (B \cap (\mathbf{R} \sim A'))) \\ &= m^*(A \cup \emptyset) + m^*(\emptyset \cup B) \\ &= m^*(A) + m^*(B). \end{aligned}$$

(iii) By Proposition 3, $m^*(A \cup B) \le m^*(A) + m^*(B)$. We now prove the other direction of that inequality. Take a positive ϵ. There is a countable collection $\{I_k\}_{k=1}^{\infty}$ of open, bounded intervals for which $A \cup B \subseteq \bigcup_{k=1}^{\infty} I_k$, and $\sum_{k=1}^{\infty} \ell(I_k) < m^*(A \cup B) + \epsilon/2$. For each k, there is a finite collection $\{J_{k,i}\}_{i=1}^{N(k)}$ of open subintervals for which $\ell(J_{k,i}) < \alpha/2$, $I_k \subseteq \bigcup_{i=1}^{N(k)} J_{k,i}$, and $\sum_{i=1}^{N(k)} \ell(J_{k,i}) < \ell(I_k) + \epsilon/2^{k+1}$. Summing the latter inequality over k gives that

$$\sum_{k=1}^{\infty}\sum_{i=1}^{N(k)} \ell(J_{k,i}) < \sum_{k=1}^{\infty} \ell(I_k) + \epsilon/2.$$

Let $\mathcal{J}_A$ equal $\{J_{k,i} \mid J_{k,i} \cap A \neq \emptyset\}$, and $\mathcal{J}_B$ equal $\{J_{k,i} \mid J_{k,i} \cap B \neq \emptyset\}$. Then $\mathcal{J}_A, \mathcal{J}_B$ are disjoint collections of open, bounded intervals and hence open covers of A, B. Then altogether,

$$m^*(A)+m^*(B) \leq \sum_{J\in\mathcal{J}_A} \ell(J) + \sum_{J\in\mathcal{J}_B} \ell(J) \leq \sum_{k=1}^{\infty}\sum_{i=1}^{N(k)} \ell(J_{k,i}) < \sum_{k=1}^{\infty} \ell(I_k) + \frac{\epsilon}{2} < m^*(A\cup B)+\epsilon.$$

Because that holds for ϵ arbitrarily small, $m^*(A) + m^*(B) = m^*(A \cup B)$.

2.3 The σ-Algebra of Lebesgue Measurable Sets

In footnote 6, the last sentence should read "..., we let $A \sim B$ denote $\{x \in A \mid x \notin B\}$ ..."

11. By definition (page 38), the σ-algebra contains $\mathbf{R}$, hence $(-\infty, \infty)$. Moreover,

$$\begin{aligned}
(a,\infty)^C &= (-\infty, a], \\
\bigcup_{n=1}^{\infty}\left(-\infty, a - \frac{1}{n}\right] &= (-\infty, a), \\
(-\infty, a)^C &= [a, \infty), \\
((-\infty, a] \cup [b,\infty))^C &= (a, b) \qquad\qquad b \geq a, \\
((-\infty, a) \cup [b,\infty))^C &= [a, b), \\
((-\infty, a] \cup (b,\infty))^C &= (a, b], \\
((-\infty, a) \cup (b,\infty))^C &= [a, b],
\end{aligned}$$

so the σ-algebra contains the intervals on the right sides of the above equalities. It remains to show that all intervals I are of the above type, proof of which the authors leave as an exercise (page 10). If I is bounded above, then by the completeness axiom, I has a least upper bound $\sup I$. By the definitions at the top of page 9, $\sup I = b$, and that if $\sup I$ is in I, then the right side of the interval has the form $b]$, and if not, $b)$. If I is not bounded above, then by the "The extended real numbers" subsection on pages 9–10, the right side of the interval has the form $\infty)$. The left side of the interval has three analogous forms. The nature of the left side of an interval is independent of that of the right, so there are 3×3, or 9, types of intervals. Thus, the eight listed above along with the given (a, ∞) encompass all types of intervals.

12. Intervals of the form (a, ∞) are open and therefore Borel sets by definition (page 39). By Problem 11, a σ-algebra of subsets of $\mathbf{R}$ that contains such intervals contains all intervals. The collection of Borel sets is a σ-algebra. Thus, every interval is a Borel set.

13. (i) We want to show that if set F equals $\bigcup_{k=1}^{\infty} F_k$ (abbreviated $\cup F_k$ in the following) where each F_k is closed, then $F + y$ is also F_σ for a real number y. We first show that $(\cup F_k)+y = \cup(F_k+y)$. Let w be a point in $(\cup F_k)+y$. Then $w-y$ is in $\cup F_k$ so that $w-y$ is in some F_k, or equivalently, w is in some $F_k + y$. Hence, $((\cup F_k) + y) \subseteq \cup(F_k + y)$. Next, let x be a point in $\cup(F_k+y)$. Then x is in some F_k+y so that $x-y$ is in some F_k. Then $x-y$ is in $\cup F_k$, or equivalently, x is in $(\cup F_k)+y$. Hence, $\cup(F_k+y) \subseteq ((\cup F_k)+y)$. By those two set inclusions, $(\cup F_k) + y = \cup(F_k + y)$.

It remains to show that $F_k + y$ is closed for each k. Because F_k is closed, F_k contains all of its points of closure. In other words, every open interval that contains a point of closure z of F_k also contains a point in F_k. That implies that every open interval that contains $z + y$ contains a point in $F_k + y$ so that $z + y$ is a point of closure of $F_k + y$, and also that $F_k + y$ contains all of its points of closure. Hence, $F_k + y$ is closed so that $\cup(F_k + y) = F + y$ is F_σ. Thus, the translate of an F_σ set is F_σ.

(ii) We can prove this similarly. Instead, we use a De Morgan identity and part (i). Let G equal $\bigcap_{k=1}^{\infty} G_k$ where each G_k is open. Then the translate of G^C is F_σ because $G^C = (\cap G_k)^C = \cup G_k^C$ and each G_k^C is closed by Proposition 11 of Section 1.4. In other words, $G^C + y$ is F_σ. It remains to show that $(G + y)^C = G^C + y$ because then $G + y$ would be G_δ. Indeed, $w \in G^C + y$ if and only if $w - y \in G^C$ if and only if $w - y \notin G$ if and only if $w \notin G + y$ if and only if $w \in (G + y)^C$. Thus, the translate of a G_δ set is G_δ.

(iii) Let E be a set of measure zero. Then along with Proposition 2, $m^*(E+y) = m^*(E) = 0$. Thus, the translate of a set of measure zero also has measure zero.

14. For each integer k, define interval I_k by $I_k = [k, k+1)$. Intersection $E \cap I_k$ is a bounded subset of E. Moreover,

$$0 < m^*(E) = m^*\left(\bigcup_{k=-\infty}^{\infty} (E \cap I_k)\right) \leq \sum_{k=-\infty}^{\infty} m^*(E \cap I_k),$$

which shows that at least one of the $E \cap I_k$ has positive outer measure.

15. Let E_k equal $E \cap [k\epsilon, (k+1)\epsilon)$ for each integer k. Each E_k is measurable, which is apparent after we replace A in the definition on page 35 with $[k\epsilon, (k+1)\epsilon)$ to get the result that $m^*([k\epsilon, (k+1)\epsilon)) = m^*(E_k) + m^*(E_k^C)$ and note that E_k is measurable if and only if E_k^C is measurable by the symmetry of the definition and that every interval is measurable by Proposition 8. Moreover, each E_k has measure at most ϵ because E_k is a subset of an interval of length ϵ.

Now because $m^*(E)$ is finite, series $\sum_{k=1}^{\infty}(m^*(E_k) + m^*(E_{-k}))$ is summable to some positive real number. By Proposition 20(i) of Section 1.5, there is an index N for which $\sum_{k=n}^{n+m}(m^*(E_k) + m^*(E_{-k})) < \epsilon$ if $n \geq N$ and for every natural number m. That implies that the measure of the set defined by $E_* = \bigcup_{k=N}^{\infty}(E_k \cup E_{-k})$ is less than ϵ. Then E is the disjoint union of the finite collection of measurable sets $E_*, E_{-N+1}, E_{-N+2}, \ldots, E_{N-2}, E_{N-1}$, each of which has measure at most ϵ.

2.4 Outer and Inner Approximation of Lebesgue Measurable Sets

16. The second sentence of the proof of Theorem 11 explains a way to use parts (i) and (ii) to prove the remaining two quickly. F is a closed set contained in E if and only if F^C is an open set containing E^C. Moreover, because $E \sim F = F^C \sim E^C$, we have that $m^*(E \sim F) < \epsilon$ if and only if $m^*(F^C \sim E^C) < \epsilon$. Statement (iii) of Theorem 11 is therefore equivalent to measurability of E^C, and thus E, by Theorem 11(i).

Similarly for (iv), F is an F_σ set contained in E if and only if F^C is a G_δ set containing E^C. Moreover, $m^*(E \sim F) = 0$ if and only if $m^*(F^C \sim E^C) = 0$. Statement (iv) is therefore equivalent to measurability of E by Theorem 11(ii).

17. Assume first that E is measurable. Then by Theorem 11, there is a closed set F and open set $\mathcal{O}$ for which $F \subseteq E \subseteq \mathcal{O}$, $m^*(E \sim F) < \epsilon/2$, and $m^*(\mathcal{O} \sim E) < \epsilon/2$ so that

$$m^*(\mathcal{O} \sim F) = m^*((\mathcal{O} \sim E) \cup (E \sim F))$$

$$\begin{aligned}
&\leq m^*(\mathcal{O} \sim E) + m^*(E \sim F) && \text{Proposition 3}\\
&< \epsilon/2 + \epsilon/2\\
&= \epsilon.
\end{aligned}$$

Conversely, assume that there is a closed set F and open set $\mathcal{O}$ for which $F \subseteq E \subseteq \mathcal{O}$, and $m^*(\mathcal{O} \sim F) < \epsilon$. Because $\mathcal{O} \sim E \subseteq \mathcal{O} \sim F$, we have that $m^*(\mathcal{O} \sim E) \leq m^*(\mathcal{O} \sim F) < \epsilon$ by monotonicity. Thus, E is measurable by Theorem 11(i).

18. For each natural number n, there is a countable collection $\{I_{n,k}\}_{k=1}^\infty$ of open, bounded intervals for which $E \subseteq \bigcup_{k=1}^\infty I_{n,k}$ and $\sum_{k=1}^\infty \ell(I_{n,k}) < m^*(E) + 1/n$. Put G_δ set G to $\bigcap_{n=1}^\infty \bigcup_{k=1}^\infty I_{n,k}$, which is measurable by Theorem 9. Then for each n, we have inequality $m(G) \leq m(\bigcup_{k=1}^\infty I_{n,k}) = \sum_{k=1}^\infty \ell(I_{n,k}) < m^*(E) + 1/n$. Hence, $m(G) \leq m^*(E)$. On the other hand, $m^*(E) \leq m(G)$ because $E \subseteq G$. Thus, $m(G) = m^*(E)$.

    Next, by Theorem 11(iv), E is measurable if and only if there is an F_σ set F contained in E for which $m^*(E \sim F) = m^*(E) - m(F) = 0$. We can use the excision property here because F is measurable by Theorem 9 and has finite outer measure since $F \subseteq E$ and $m^*(E) < \infty$.

19. By the contrapositive of Theorem 11(i), there is a positive ϵ for which $m^*(\mathcal{O} \sim E) \geq \epsilon$ for all open sets $\mathcal{O}$ that contain E. Choose such an $\mathcal{O}$ that is a countable collection $\{I_k\}_{k=1}^\infty$ of open, bounded intervals for which $m^*(\mathcal{O}) = \sum_{k=1}^\infty \ell(I_k) < m^*(E) + \epsilon$. Consequently, $m^*(\mathcal{O} \sim E) \geq \epsilon > m^*(\mathcal{O}) - m^*(E)$.

20. Assume first that E is measurable. In the definition on page 35, replace A with (a, b):

$$\begin{aligned}
m^*((a,b)) &= m^*((a,b) \cap E) + m^*\big((a,b) \cap E^C\big), \text{ or}\\
b - a &= m^*((a,b) \cap E) + m^*((a,b) \sim E).
\end{aligned}$$

    Conversely, assume that $\ell((a,b)) = m^*((a,b) \cap E) + m^*((a,b) \sim E)$ for each (a,b). Because E has finite outer measure, there is a countable collection of open, bounded intervals $\{I_k\}_{k=1}^\infty$ that covers E and for which $\sum_{k=1}^\infty \ell(I_k) - m^*(E) < \epsilon$ for each positive ϵ. Now,

$$\begin{aligned}
\ell(I_k) &= m^*(I_k \cap E) + m^*(I_k \sim E) && I_k \text{ is an } (a,b),\\
\sum_{k=1}^\infty \ell(I_k) &= \sum_{k=1}^\infty m^*(I_k \cap E) + \sum_{k=1}^\infty m^*(I_k \sim E)\\
&\geq m^*\left(\bigcup_{k=1}^\infty (I_k \cap E)\right) + m^*\left(\bigcup_{k=1}^\infty (I_k \sim E)\right) && \text{Proposition 3}\\
&\geq m^*(E) + m^*\left(\bigcup_{k=1}^\infty I_k \sim E\right) && \bigcup_{k=1}^\infty I_k \sim E \subseteq \bigcup_{k=1}^\infty (I_k \sim E)\\
&= m^*(E) + m^*(G \sim E) && \text{denote } \bigcup_{k=1}^\infty I_k \text{ by } G.
\end{aligned}$$

    Hence, G is an open set containing E such that $m^*(G \sim E) < \epsilon$, which by Theorem 11(i) is equivalent to the measurability of E.

21. Take two measurable sets E_1 and E_2. Then for k equal to 1 and 2, there is a G_δ set G_k containing E_k for which $m^*(G_k \sim E_k) = 0$. We first show that $G_1 \cup G_2$ is a G_δ set. Let G_2 be the countable intersection $\bigcap_{i=1}^\infty G_{2i}$ of open sets. Then

$$G_1 \cup G_2 = G_1 \bigcup \left(\bigcap_{i=1}^\infty G_{2i}\right) = \bigcap_{i=1}^\infty (G_1 \cup G_{2i}),$$

a countable intersection of open sets $G_1 \cup G_{2i}$. Hence, $G_1 \cup G_2$ is a G_δ set. Now

$$\begin{aligned}(G_1 \cup G_2) \sim (E_1 \cup E_2) &\subseteq (G_1 \sim E_1) \cup (G_2 \sim E_2),\\ m^*((G_1 \cup G_2) \sim (E_1 \cup E_2)) &\leq m^*((G_1 \sim E_1) \cup (G_2 \sim E_2)) && \text{monotonicity}\\ &\leq m^*(G_1 \sim E_1) + m^*(G_2 \sim E_2) = 0 + 0 && \text{Proposition 3.}\end{aligned}$$

Hence, there is a G_δ set $G_1 \cup G_2$ containing $E_1 \cup E_2$ for which $m^*((G_1 \cup G_2) \sim (E_1 \cup E_2)) = 0$. Thus, $E_1 \cup E_2$ is measurable.

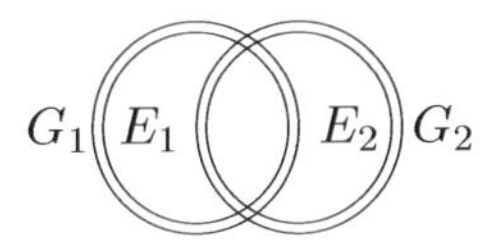

Analogously for part (iv), there is an F_σ set F_k contained in E_k for which $m^*(E_k \sim F_k) = 0$. $F_1 \cup F_2$ is an F_σ set because $F_1 \cup F_2$ is the union of two countable unions of closed sets. Then $E_1 \cup E_2$ is measurable because

$$\begin{aligned}(E_1 \cup E_2) \sim (F_1 \cup F_2) &\subseteq (E_1 \sim F_1) \cup (E_2 \sim F_2),\\ m^*((E_1 \cup E_2) \sim (F_1 \cup F_2)) &\leq m^*((E_1 \sim F_1) \cup (E_2 \sim F_2))\\ &\leq m^*(E_1 \sim F_1) + m^*(E_2 \sim F_2) = 0 + 0.\end{aligned}$$

22. Set $\mathcal{O}$ covers A and, by Proposition 9 of Section 1.4, is the union $\bigcup_{k=1}^{\infty} I_k$ of a countable collection of open intervals. By Propositions 3 and 1, $m^*(\mathcal{O}) \leq \sum_{k=1}^{\infty} \ell(I_k)$. Then the definitions of m^{**} and m^* coincide. Thus, $m^{**} = m^*$.

23. If A is measurable, $m^{***}(A) \geq m^*(F) > m^*(A) - \epsilon$ where the first inequality comes from definition of m^{***}, and the second from Theorem 11(iii), which guarantees the existence of such a set F. Because that inequality holds for ϵ arbitrarily small, $m^{***}(A) \geq m^*(A)$. By monotonicity of outer measure and Problem 22, $m^{***}(A) \leq m^{**}(A) = m^*(A)$ because $F \subseteq A \subseteq \mathcal{O}$. Thus, $m^{***} = m^*$.

    If A is not measurable, then by the contrapositive of Problem 17, there is a positive δ such that $m^*(\mathcal{O} \sim F) \geq \delta$ for all $F, \mathcal{O}$, that is, $m^*(F) + \delta \leq m^*(\mathcal{O})$. For positive ϵ_1, ϵ_2, there are particular $F, \mathcal{O}$ for which $F \subseteq A \subseteq \mathcal{O}$ and $m^{***}(A) + \epsilon_1 \leq m^*(F)$, $m^*(\mathcal{O}) < m^*(A) + \epsilon_2$ by definitions of m^{***}, m^* (see the remark on page 43). Compounding those last three inequalities, $m^{***}(A) + \epsilon_1 < m^*(F) + \delta \leq m^*(\mathcal{O}) < m^*(A) + \epsilon_2$. Because that holds for ϵ_1, ϵ_2 arbitrarily small, $m^{***}(A) < m^*(A)$ if $m^{***}(A) < \infty$, and $m^{***}(A) = m^*(A)$ if $m^{***}(A) = \infty$.

2.5 Countable Additivity, Continuity, and the Borel-Cantelli Lemma

24. The equation holds if E_1 or E_2 has infinite measure. Otherwise, express $E_1 \cup E_2$ as disjoint union $(E_1 \cap E_2) \cup (E_1 \sim (E_1 \cap E_2)) \cup (E_2 \sim (E_1 \cap E_2))$. By Proposition 13 and the excision property,

$$\begin{aligned}m(E_1 \cup E_2) &= m(E_1 \cap E_2) + m(E_1) - m(E_1 \cap E_2) + m(E_2) - m(E_1 \cap E_2)\\ &= m(E_1) + m(E_2) - m(E_1 \cap E_2).\end{aligned}$$

25. If $m(B_1) = \infty$, then it may be that $m(B_k) = \infty$ for all k. Then the difference $m(B_1) - m(B_k)$ in the proof is not defined. For example, put B_k to (k, ∞). Then $m(\bigcap_{k=1}^{\infty} B_k) = m(\emptyset) = 0$, but $\lim_{k \to \infty} m(B_k) = \infty$.

26. By the distributive property of intersection over union and Proposition 3,

$$m^*\left(A \cap \bigcup_{k=1}^{\infty} E_k\right) = m^*\left(\bigcup_{k=1}^{\infty}(A \cap E_k)\right) \leq \sum_{k=1}^{\infty} m^*(A \cap E_k).$$

On the other hand, by Proposition 6 and monotonicity,

$$\sum_{k=1}^{n} m^*(A \cap E_k) = m^*\left(A \cap \bigcup_{k=1}^{n} E_k\right) \leq m^*\left(A \cap \bigcup_{k=1}^{\infty} E_k\right).$$

The right side is independent of n. Therefore,

$$\sum_{k=1}^{\infty} m^*(A \cap E_k) \leq m^*\left(A \cap \bigcup_{k=1}^{\infty} E_k\right).$$

Thus, in view of our first inequality, we see that the two sides are equal.

27. (i) Because m' is countably additive, it is finitely additive by definition of *countable* (page 13). See also the last sentence of Section 2.2.
    To prove excision, consider sets A and B in $\mathcal{M}'$ such that $m'(A) < \infty$ and $A \subseteq B$. Because m' is finitely additive, $m'(B) = m'((B \sim A) \cup A) = m'(B \sim A) + m'(A)$.
    Proof of monotonicity is the same as that of Problem 1.
    To prove countable monotonicity, consider a countable collection $\{E_k\}_{k=1}^{\infty}$ of sets in $\mathcal{M}'$ that covers another set E in $\mathcal{M}'$. Define A_1 by $A_1 = E_1$, and define A_k where $k = 2, 3, \ldots$ by $A_k = E_k \sim \bigcup_{i=1}^{k-1} E_i$. Because $\mathcal{M}'$ is a σ-algebra, $\{A_k\}_{k=1}^{\infty}$ is a disjoint collection of sets in $\mathcal{M}'$ whose union is the same as that of $\{E_k\}_{k=1}^{\infty}$ such that $A_k \subseteq E_k$. By monotonicity and countable additivity,

$$m'(E) \leq m'\left(\bigcup_{k=1}^{\infty} E_k\right) = m'\left(\bigcup_{k=1}^{\infty} A_k\right) = \sum_{k=1}^{\infty} m'(A_k) \leq \sum_{k=1}^{\infty} m'(E_k).$$

    (ii) Having established the properties of part (i) along with the hypothesis that $m'(\emptyset) = 0$, the proof is the same as that of Theorem 15 with Lebesgue measure replaced by m'.

28. Let $\{E_k\}_{k=1}^{\infty}$ be a countable disjoint collection of measurable sets. For each natural number n, denote $\bigcup_{k=1}^{n} E_k$ by A_n. Then $\{A_n\}_{n=1}^{\infty}$ is ascending, and $\bigcup_{n=1}^{\infty} A_n = \bigcup_{k=1}^{\infty} E_k$. Applying property (11) and Proposition 6,

$$\begin{aligned} m\left(\bigcup_{k=1}^{\infty} E_k\right) = m\left(\bigcup_{n=1}^{\infty} A_n\right) &= \lim_{n\to\infty} m(A_n) = \lim_{n\to\infty} m\left(\bigcup_{k=1}^{n} E_k\right) \\ &= \lim_{n\to\infty} \sum_{k=1}^{n} m(E_k) = \sum_{k=1}^{\infty} m(E_k). \end{aligned}$$

2.6 Nonmeasurable Sets

29. (i) If the set is empty, the claim is true vacuously. Now consider a nonempty set with members x, y, z, not necessarily distinct. In the following, we use the basic fact that the difference, sum, or product of two rational numbers is again rational. Rational equivalence is reflexive because $x - x = 0$; symmetric because if $x - y$ belongs to $\mathbf{Q}$, then so does $-(x - y)$, hence, $y - x$; and transitive because if $x - y$ and $y - z$ belong to $\mathbf{Q}$, then so does $(x - y) + (y - z)$, hence, $x - z$. Thus, rational equivalence defines an equivalence relation.

(ii) Properties (i) and (ii) on page 48 imply that $\mathcal{C}_{\mathbf{Q}}$ contains exactly one rational number. Thus, $\{0\}$ will do for $\mathcal{C}_{\mathbf{Q}}$.

(iii) Irrational equivalence is not transitive on $\mathbf{R}$ and thus not an equivalence relation on $\mathbf{R}$. For example, $1-\sqrt{2}$ and $\sqrt{2}-0$ are irrational, but $1-0$ is not irrational or zero.

On the other hand, irrational equivalence is an equivalence relation on $\mathbf{Q}$. Because the difference of two rational numbers is never irrational, we need consider only differences that are zero. Irrational equivalence is reflexive because $x-x=0$; symmetric because if $x-y$ is zero, then so is $y-x$; and transitive because if $x-y$ and $y-z$ are each zero, then so is $x-z$ because $x-z=(x-y)+(y-z)$.

30. Such a choice set cannot be countable because by Proposition 4, the set would be measurable, contradicting the proof of Theorem 17. Thus, such a choice set must be uncountable.

31. The authors justify the assertion by "the countable subadditivity of outer measure." That is how we solved Problem 14, which is a more explicit justification. Every set E of real numbers with positive outer measure contains a bounded subset E' that also has positive outer measure. Thus, if E' contains a subset that fails to be measurable, then so does E.

32. If Λ is allowed to be finite, Lemma 16 becomes false. For example, if $m(E)=1$ and $\Lambda=\{0\}$, then the collection of translates of E is a single set and therefore disjoint, but $m(E)\neq 0$.

If Λ is allowed to be uncountable, Lemma 16 remains true because there exists a countably infinite subset of Λ that satisfies the conditions of the lemma.

If Λ is allowed to be unbounded, Lemma 16 becomes false. For example, if $E=(0,1)$ and $\Lambda=\mathbf{N}$, then the collection of translates $\{n+E\}_{n\in\mathbf{N}}$ is disjoint, but $m(E)=1\neq 0$.

33. For each natural number k, there is an open set G_k that contains E and for which

$$m^*(G_k)<m^*(E)+1/k$$

(see the beginning of the proof of Theorem 11(i)). Define set G by $G=\bigcap_{k=1}^{\infty}G_k$. Then G is a G_δ set that contains E. Moreover, because for each k, set G_k contains G, by the monotonicity of outer measure, $m^*(G)\leq m^*(E)+1/k$. Thus, $m^*(G)=m^*(E)$. By the contrapositive of Theorem 11(ii), $m^*(G\sim E)>0$.

2.7 The Cantor Set and the Cantor-Lebesgue Function

34. Consider ψ^{-1} of Proposition 21 but from $[0,1]$ onto $[0,1/2]$. Function ψ^{-1} exists and is continuous by Problem 41. To show that ψ^{-1} is strictly increasing, consider y,y' in $[0,1]$ such that $y<y'$. Then $\psi^{-1}(y)<\psi^{-1}(y')$; otherwise, $\psi^{-1}(y)\geq\psi^{-1}(y')$, which implies that $\psi\psi^{-1}(y)\geq\psi\psi^{-1}(y')$ because ψ is increasing, which implies that $y\geq y'$, a contradiction. Next, our domain and range are well defined because $\psi(0)=0$ and $\psi(1/2)=1$. Finally, it follows from Proposition 19 and Proposition 21(i) considering just the first halves of the two sets described in its proof (which are nonempty and satisfy the requirements of this problem by the symmetry of the Cantor set) that ψ^{-1} maps a set of positive measure onto a set of measure zero.

35. Assume first that f is continuous at x_0. Because I is open, there is a positive r for which interval (x_0-r,x_0+r) is contained in I. Choose a natural number M such that $1/M<r$. Then the sequences defined by $a_n=x_0-1/(M+n)$, and $b_n=x_0+1/(M+n)$ are in I such that $a_n<x_0<b_n$ for each n. Moreover, $\{a_n\}$ and $\{b_n\}$ converge to x_0 so that by Proposition 21 of Section 1.6,

$$\lim_{n\to\infty}f(a_n)=f(x_0)=\lim_{n\to\infty}f(b_n).$$

Thus, by Theorem 18 (linearity) of Section 1.5,

$$\lim_{n\to\infty} (f(b_n) - f(a_n)) = f(x_0) - f(x_0) = 0.$$

Conversely, assume that there are such sequences $\{a_n\}$ and $\{b_n\}$. Then by definition (page 21), for every positive ϵ, there is an index N for which if $n \geq N$, then $f(b_n) - f(a_n) < \epsilon$. Put δ to $\min\{x_0 - a_N, b_N - x_0\}$, and assume that $|x - x_0| < \delta$. Then $a_N < x < b_N$. In the following, we repeatedly use the hypothesis that f is increasing:

$$\begin{aligned} f(a_N) &\leq f(x) \leq f(b_N), \\ f(a_N) - f(x_0) &\leq f(x) - f(x_0) \leq f(b_N) - f(x_0), \\ f(a_N) - f(b_N) \leq f(a_N) - f(x_0) &\leq f(x) - f(x_0) \leq f(b_N) - f(x_0) \leq f(b_N) - f(a_N). \end{aligned}$$

Thus, $|f(x) - f(x_0)| < \epsilon$, proving that f is continuous at x_0.

36. No. Put f^{-1} to the ψ of Proposition 21. By Problem 41, f is continuous. Let A be the measurable set of Proposition 21(ii) so that $f^{-1}(A)$ is nonmeasurable.

37. Take a set E of measure zero in $[a, b]$. Take a positive ϵ. By the definition of outer measure, there is a countable collection of open intervals $\{I_k\}_{k=1}^{\infty}$ that covers E and for which $\sum_{k=1}^{\infty} \ell(I_k) < \epsilon/c$. Let y_k equal $\inf\{f(u) \mid u \in I_k\}$, and z_k equal $\sup\{f(u) \mid u \in I_k\}$. Points y_k and z_k exist by the extreme value theorem (Section 1.6) because f is continuous on the nonempty, closed, bounded set $[a, b]$. Then $f(I_k) \subseteq [y_k, z_k]$ so that by monotonicity and the Lipschitz property, we deduce that

$$\begin{aligned} m(f(I_k)) &\leq m([y_k, z_k]) \\ &= z_k - y_k \\ &\leq c\ell(I_k). \end{aligned}$$

That result with subadditivity and monotonicity applied to inclusion $f(E) \subseteq \bigcup_{k=1}^{\infty} f(I_k)$ gives us that

$$\begin{aligned} m(f(E)) &\leq m\left(\bigcup_{k=1}^{\infty} f(I_k)\right) \\ &\leq \sum_{k=1}^{\infty} m(f(I_k)) \\ &\leq \sum_{k=1}^{\infty} c\ell(I_k) < \epsilon. \end{aligned}$$

Because that holds for ϵ arbitrarily small, $m(f(E)) = 0$.

Next, let F be F_σ set $\bigcup_{k=1}^{\infty} F_k$ where each F_k is closed and also bounded because $F_k \in [a, b]$. We claim that $f(F_k)$ is closed. Let $\mathcal{U}$ be an open cover of $f(F_k)$. By Proposition 22 of Section 1.6, $f^{-1}(\mathcal{U})$ is an open cover of F_k. By the Heine-Borel theorem (Section 1.4), $f^{-1}(\mathcal{U})$ has a finite subcover $\{f^{-1}(U_1), \ldots, f^{-1}(U_n)\}$ where $U_i \in \mathcal{U}$. Then

$$f(F_k) \subseteq f\left(\bigcup_{i=1}^{n} f^{-1}(U_i)\right) \subseteq \bigcup_{i=1}^{n} U_i.$$

By Problem 53 of Section 1.6, F_k is closed. Thus,

$$f(F) = f\left(\bigcup_{k=1}^{\infty} F_k\right) = \bigcup_{k=1}^{\infty} f(F_k),$$

an F_σ set.

Assume now that E is measurable. By Theorem 11(iv) and Proposition 4, there is an F_σ set F contained in E for which $m(E \sim F) = 0$. Express E as the disjoint union $F \cup (E \sim F)$ so that $f(E) = f(F) \cup f(E \sim F)$. By Proposition 7 and the first two parts of this problem, $f(E)$ is measurable.

38. The intersection of every collection of closed sets is closed; therefore, F is closed.

    Next, let a and b be in $[0,1]$ such that $a < b$. At the nth stage, 2^n disjoint, equal-length intervals remain, so each must have length less than $1/2^n$. Choose a natural number N such that $1/2^N < b - a$. Then at the Nth stage, no remaining interval contains (a,b). Because those remaining intervals are closed and disjoint, (a,b) contains a point not in a remaining interval, i.e., (a,b) contains a point in $[0,1] \sim F$. Thus, $[0,1] \sim F$ is dense in $[0,1]$.

    Finally, note that at the nth stage, we remove 2^{n-1} disjoint intervals each of length $\alpha/3^n$. Then

$$\begin{aligned} m(F) &= m([0,1]) - \sum_{n=1}^{\infty} 2^{n-1}\alpha/3^n \\ &= 1 - (\alpha/2)\sum_{n=1}^{\infty}(2/3)^n \\ &= 1 - (\alpha/2)(2/3)/(1-2/3) \qquad \text{sum of a geometric series} \\ &= 1 - \alpha. \end{aligned}$$

39. *Boundary* is defined in Section 10.2. Take the hint. Complement F^C is open by Proposition 11 of Section 1.4. We claim that F is in the boundary of F^C. Let x be a point in F. Then for each positive r, interval $(x-r, x+r)$ contains not only point x in F but also a point in F^C because F^C contains $[0,1] \sim F$, which is dense in $[0,1]$. Thus, F^C is an open set of real numbers whose boundary is of measure at least $1-\alpha$, which is positive.

40. By Proposition 9 of Section 1.4, it suffices to assume that $\mathcal{O}$ is an open interval. By construction,

$$\mathbf{C} = \bigcap_{k=1}^{\infty} C_k$$

    where C_k is the disjoint union of closed intervals each of length $1/3^k$. Choose k large enough so that $1/3^k < \ell(\mathcal{O})$. Then $\mathcal{O}$ has a nonempty open subset $\mathcal{O} \cap C_k^C$ that is disjoint from $\mathbf{C}$. Thus, $\mathbf{C}$ is nowhere dense.

    An even simpler solution is possible using the definitions in Section 10.2. We need to show that $\operatorname{int}\overline{\mathbf{C}} = \emptyset$. Because $\mathbf{C}$ is closed (Proposition 19), $\overline{\mathbf{C}} = \mathbf{C}$, so it suffices to show that $\operatorname{int}\mathbf{C} = \emptyset$. Take a point x in $[0,1]$. Open interval $(x-r, x+r)$ cannot be contained in $\mathbf{C}$; otherwise, by monotonicity,

$$0 = m(\mathbf{C}) \geq m((x-r, x+r)) = 2r > 0,$$

    a contradiction. Thus, $\operatorname{int}\mathbf{C} = \emptyset$, so $\mathbf{C}$ is nowhere dense.

    To prove that the two definitions of *nowhere dense* are equivalent, it suffices to show that characterization of nowhere dense in Problem 16 of Section 10.2 is equivalent to the definition in this problem. In Problem 16, let E be the A of this problem, and let X be $\mathbf{R}$. Assume first that every open set $\mathcal{O}$ has a nonempty open subset that is disjoint from A. Then there is an open interval (a,b) in $\mathcal{O}$ such that no point of A is between a and b. Thus, $A \cap \mathcal{O}$ is not dense in $\mathcal{O}$. Conversely, assume that for each open subset $\mathcal{O}$ of $\mathbf{R}$, intersection $A \cap \mathcal{O}$ is not dense in $\mathcal{O}$. Then there are two points a and b in $\mathcal{O}$ between which there are no points of $A \cap \mathcal{O}$. Thus, $\mathcal{O}$ has a nonempty open subset (a, b) that is disjoint from A.

41. Let f be the function and I the interval. Define f to be from I to $f(I)$ so that f is onto, a requirement for the existence of f^{-1}. The other requirement is that f be one-to-one, which it is because for each number y of $f(I)$, there is exactly one number x of I for which $y = f(x)$; otherwise, there would be distinct x and x' such that $f(x')$ and $f(x)$ would be equal, which contradicts that f is strictly increasing.

    We now prove that f^{-1} is continuous. Let the endpoints of I be a and b where $a < b$. Take a positive ϵ. Let ϵ' equal $\min\{\epsilon, (b-x)/2, (x-a)/2\}$. Define δ_1, δ_2, and δ by $f(x+\epsilon') = y+\delta_1$, $f(x-\epsilon') = y - \delta_2$, and $\delta = \min\{\delta_1, \delta_2\}$. Because f is strictly increasing, f^{-1} is also strictly increasing using the same reasoning as in our solution to Problem 34. Then $f^{-1}(y') < x + \epsilon'$ whenever $y' < y + \delta_1$, and $f^{-1}(y') > x - \epsilon'$ whenever $y' > y - \delta_2$. We can now justify the condition for f^{-1} to be continuous: For each $\epsilon > 0$, there is a $\delta > 0$ for which if $y, y' \in f(I)$ and $|y' - y| < \delta$, then

$$\begin{aligned}|f^{-1}(y') - f^{-1}(y)| &= |f^{-1}(y') - x| \\ &< \epsilon' \\ &\leq \epsilon.\end{aligned}$$

    Another way to prove that f^{-1} is continuous is—by Propositions 22 and 9 of Chapter 1—to show that for each open interval (c, d) in I, we have that $f((c, d)) = f(I) \cap U$ where U is an open set. We claim that $U = (f(c), f(d))$. Indeed, for every x in (c, d), we have that $f(c) < f(x) < f(d)$; hence,

$$f((c, d)) \subseteq f(I) \cap (f(c), f(d)).$$

    Conversely, for every y in $f(I) \cap (f(c), f(d))$, there is an x in I such that $f(x) = y$. It follows from compound inequality $f(c) < f(x) < f(d)$ that $x \in (c, d)$ so that $y \in f((c, d))$; hence

$$f(I) \cap (f(c), f(d)) \subseteq f((c, d)).$$

    Thus, $f((c, d)) = f(I) \cap U$.

42. This is the third part of Problem 58 of Section 1.6.

43. Let g be such a function. Then g^{-1} is continuous by Problem 41. So $(g^{-1})^{-1}(B)$ is a Borel set by Problem 42. Thus, $g(B)$ is a Borel set.

Chapter 3

Lebesgue Measurable Functions

3.1 Sums, Products, and Compositions

1. $f - g$ is continuous at each point x in $[a, b]$ by Problem 49(i) of Section 1.6. So for each positive ϵ, there is a δ for which if $x' \in (x-\delta, x+\delta) \cap [a, b]$, then $|(f-g)(x)-(f-g)(x')| < \epsilon$. Let E_0 be the subset of $[a, b]$ where $f - g \neq 0$. Restrict x' to be in $(x - \delta, x + \delta) \cap [a, b] \sim E_0$. Such points x' exists because $m((x - \delta, x + \delta) \cap [a, b]) \geq \min\{\delta, b - a\} > 0$ and $m(E_0) = 0$ by definition of a.e. With that restriction, $(f - g)(x') = 0$ so that $|(f - g)(x)| < \epsilon$. Because that holds for ϵ arbitrarily small, $f - g = 0$ on $[a, b]$.

   Here is another solution. Use Propositions 22 and 9 of Chapter 1 to express the set of points in $[a, b]$ on which $f - g \neq 0$, that is, $(f - g)^{-1}((-\infty, 0) \cup (0, \infty))$, as $[a, b] \cap \mathcal{U}$ where $\mathcal{U}$ is a collection of open intervals. The measure of $[a, b] \cap \mathcal{U}$ is zero because $[a, b] \cap \mathcal{U} \subseteq [a, b]$ and $f - g = 0$ a.e. on $[a, b]$. Now $[a, b] \cap \mathcal{U}$ must be empty; otherwise, it would consist of intervals, which contradicts that $m([a, b] \cap \mathcal{U}) = 0$. Thus, $f - g = 0$ on $[a, b]$.

   No, such an assertion is not true. For example, let E be singleton set $\{0\}$ and define f by $f(0) = 1$, and g by $g(0) = 2$. Then f and g are continuous on $\{0\}$, and $f = g$ a.e. on $\{0\}$ because $m(\{0\}) = 0$, but $f \neq g$ on $\{0\}$.

2. No. For example, let $f : [0, 2) \to \{0, 1\}$ have a jump discontinuity (page 27):

   $$f(x) = \begin{cases} 0 & x \in [0, 1) = D, \\ 1 & x \in [1, 2) = E. \end{cases}$$

   f's restrictions to measurable intervals D, E are continuous, but f is not continuous on $[0, 2)$, which is $D \cup E$. Another counterexample is in our solution to Problem 37 of Section 11.4.

   The converse, however, holds: If f is continuous on $D \cup E$, then $f|_D$ and $f|_E$ are continuous.

3. Yes. Let the domain be E and the points of discontinuity $x_1, x_2, \ldots, x_n$ in order from least to greatest. Then f is continuous on sets defined by—

   $$\begin{aligned} A_0 &= E \cap (\infty, x_1), \\ A_k &= E \cap (x_k, x_{k+1}) \text{ where } k = 1, 2, \ldots, n - 1, \text{ and} \\ A_n &= E \cap (x_n, \infty). \end{aligned}$$

   For each real number c, sets A'_k defined by $A'_k = \{x \in A_k \,|\, f(x) > c\} = A_k \cap f^{-1}((c, \infty))$ where $k = 0, 1, 2, \ldots, n$ are measurable by the proof of Proposition 3. Observe that set $\{x \in E \,|\, f(x) > c\}$ is equal to $(\bigcup_{k=0}^{n} A'_k) \cup X$ where X is some subset of $\{x_1, x_2, \ldots, x_n\}$. Because X and each A'_k are measurable, $\{x \in E \,|\, f(x) > c\}$ is measurable. Thus, f is measurable by definition using Proposition 1(i).

4. No. By Vitali's theorem, there is a nonmeasurable subset W of $(0, 1)$, say. Define f by

   $$f(x) = \begin{cases} e^x & x \in W, \\ -e^x & x \notin W. \end{cases}$$

   Because f is one-to-one, $f^{-1}(c)$ is either the empty-set or a singleton set and therefore measurable. By construction, however, $f^{-1}((0, \infty)) = \{x \in \mathbf{R} \,|\, f(x) > 0\} = W$. Therefore, f is not measurable.

5. Yes. For each real number a and natural number k, there is a rational number c_k for which $a < c_k < a + 1/k$ because $\mathbf{Q}$ is dense in $\mathbf{R}$. Then

$$\{x \in E \,|\, f(x) > a\} = \bigcup_{k=1}^{\infty} \{x \in E \,|\, f(x) \geq a + 1/k\} = \bigcup_{k=1}^{\infty} \{x \in E \,|\, f(x) > c_k\},$$

which shows that $\{x \in E \,|\, f(x) > a\}$ is measurable because it is the union of a countable collection of measurable sets. Thus, f is measurable.

6. Assume first that g is measurable. Then f is measurable because for each real number c, $\{x \in D \,|\, f(x) > c\} = \{x \in \mathbf{R} \,|\, g(x) > c\} \cap D$, which is measurable since it is the intersection of two measurable sets. Conversely, assume that f is measurable. Then g is measurable by Proposition 5(ii) because $g|_D$ is measurable by assumption, and $g|_{\mathbf{R}\sim D}$ is measurable since $\{x \in \mathbf{R} \sim D \,|\, g(x) > c\}$ is either $\emptyset$ or $\mathbf{R} \sim D$, each a measurable set.

7. Assume first that for each Borel set B, set $f^{-1}(B)$ is measurable. By the definition on page 20, every open set is Borel. Thus, f is measurable by Proposition 2.

   Conversely, assume that f is measurable. We first prove the hint. Let $\mathcal{A}$ be the collection of sets A such that $f^{-1}(A)$ is measurable. We need to verify that the conditions of a σ-algebra defined on page 38 hold for $\mathcal{A}$. First, $\mathcal{A}$ contains $\mathbf{R}$ because $f^{-1}(\mathbf{R}) = E$, which is measurable by hypothesis. Next, let $\{A_i\}_{k=1}^{\infty}$ be a countable collection of sets in $\mathcal{A}$. As in the proof of Proposition 2 or by our solution to Problem 23 of Section 11.4, $f^{-1}(\bigcup_{k=1}^{\infty} A_i) = \bigcup_{k=1}^{\infty} f^{-1}(A_i)$, which is measurable. Likewise, $f^{-1}(\mathbf{R} \sim A) = \mathbf{R} \sim f^{-1}(A)$, which is measurable. Hence, $\mathcal{A}$ is closed with respect to the formation of countable unions and complements and therefore is a σ-algebra. Now because f is measurable, for each open set $\mathcal{O}$, set $f^{-1}(\mathcal{O})$ is measurable. Each $\mathcal{O}$ is therefore in $\mathcal{A}$, so $\mathcal{A}$ is a σ-algebra containing all of the open sets, therefore all Borel sets. Thus, for each Borel set B, set $f^{-1}(B)$ is measurable.

8. Proposition 1 remains valid for Borel sets because (a) by Theorem 9 of Section 2.3, the collection $\mathcal{M}$ of Lebesgue measurable sets contains the collection $\mathcal{B}$ of Borel sets and (b) $\mathcal{B}$ is a σ-algebra, and every deduction in the proof of Proposition 1 follows from the fact that $\mathcal{M}$ is a σ-algebra. The same applies to Theorem 6, but there are other deductions in its proof that we need to verify. They are related to Lebesgue measurability, which we are not to replace; properties of rational numbers; and basic algebra. None of those becomes invalid if we replace *(Lebesgue) measurable set* with *Borel set*. Thus, Theorem 6 also remains valid for Borel sets.

   (i) Let f be Borel measurable with domain E. Then sets E and $\{x \in E \,|\, f(x) > c\}$ for each c are Borel sets. Because Borel sets are Lebesgue measurable sets, f is a Lebesgue measurable function by definition using Proposition 1(i).

   (ii) Let $\mathcal{A}$ be the collection of sets A such that $f^{-1}(A)$ is a Borel set. By the same argument we used to prove the hint of Problem 7, $\mathcal{A}$ is a σ-algebra. Now apply the converse part of the proof of Proposition 2 to Borel measurable functions and Borel sets. Each $f^{-1}(B_k)$, that is, $\{x \in E \,|\, f(x) < b_k\}$, and $f^{-1}(A_k)$, that is, $\{x \in E \,|\, f(x) > a_k\}$, in that proof are Borel sets by definition of a Borel measurable function and Proposition 1 applied to Borel sets. Hence, we have the result analogous to Proposition 2, which is that for each open set $\mathcal{O}$, set $f^{-1}(\mathcal{O})$ is a Borel set. Each $\mathcal{O}$ is therefore in $\mathcal{A}$, so $\mathcal{A}$ is a σ-algebra containing all of the open sets, therefore all Borel sets. Thus, for each Borel set B, set $f^{-1}(B)$ is a Borel set.

   (iii) Let E be the domain of g and hence of $f \circ g$. Then

$$\{x \in E \,|\, (f \circ g)(x) > c\} = (f \circ g)^{-1}((c, \infty)) = g^{-1}\big(f^{-1}((c, \infty))\big),$$

which is a Borel set by two applications of part (ii). Thus, $f \circ g$ is Borel measurable.

(iv) Similarly, $\{x \in E \mid (f \circ g)(x) > c\} = g^{-1}(f^{-1}((c, \infty)))$, which is Lebesgue measurable by application of part (ii) to the inner inverse and Problem 7 to the outer inverse. Thus, $f \circ g$ is Lebesgue measurable.

9. Yes. For each x in E_0 and natural number k, there is an index N for which if $m, n \geq N$, then $|f_m(x) - f_n(x)| < 1/k$ by Theorem 17 of Section 1.5. We can therefore express E_0 as

$$\bigcap_{k=1}^{\infty} \bigcup_{N=1}^{\infty} \bigcap_{m,n \geq N} \{x \in E \mid |(f_m - f_n)(x)| < 1/k\}.$$

Thus, E_0 is measurable because $\{x \in E \mid |(f_m - f_n)(x)| < 1/k\}$ is measurable by definition of a measurable function and Theorem 6 (linearity).

10. No, $f \circ g$ is not necessarily measurable by the example on pages 57–58.

11. By Proposition 2, $f^{-1}(\mathcal{O})$ is a measurable set for each open set $\mathcal{O}$. Because g is one-to-one and onto, g^{-1} exists. By the hint, g^{-1} maps a measurable set in $\mathbf{R}$ to a measurable set. Then $(f \circ g)^{-1}(\mathcal{O}) = g^{-1}(f^{-1}(\mathcal{O}))$, which is a measurable set. Thus, $f \circ g$ is measurable.

3.2 Sequential Pointwise Limits and Simple Approximation

12. Mimic the proof of the simple approximation lemma. Let (c, d) contain $f(E)$, and let $c = y_0 < y_1 < \cdots < y_{m-1} < y_m = d$ be a partition of $[c, d]$ such that $y_k - y_{k-1} < 1/2$ where $1 \leq k \leq m$. Define E_k by $E_k = f^{-1}([y_{k-1}, y_k))$. Put φ_1 to $\sum_{k=1}^{m} y_{k-1}\chi_{E_k}$, and ψ_1 to $\sum_{k=1}^{m} y_k\chi_{E_k}$. For n equal to $2, 3, \ldots$, make a new partition by cutting in half each interval of the $(n-1)$st's partition so that the length of each interval is less than $1/2^n$, and define φ_n and ψ_n analogously to the definitions of φ_1 and ψ_1. By our construction of the partitions, $\{\varphi_n\}$ is increasing, and $\{\psi_n\}$ decreasing. Now for every positive ϵ, there is an index N such that $1/2^N < \epsilon$. Then the simple approximation lemma and its proof imply that $f - \varphi_n < 1/2^n < \epsilon$, and $\psi_n - f < 1/2^n < \epsilon$ on E if $n \geq N$. Thus, $\{\varphi_n\}$ and $\{\psi_n\}$ converge to f uniformly on E.

13. Once again, mimic the proof of the simple approximation lemma. Let

$$\cdots < y_{n,-2} < y_{n,-1} < y_{n,0} < y_{n,1} < y_{n,2} < \cdots$$

be partitions of $\mathbf{R}$ such that $y_{n,k} - y_{n,k-1} < 1/n$ for all integers k. Define sets $E_{n,k}$ by $E_{n,k} = f^{-1}([y_{n,k-1}, y_{n,k}))$. Put f_n to $\sum_{k=-\infty}^{\infty} y_{n,k-1}\chi_{E_{n,k}}$. Let x belong to E. For each n, there is a unique k for which $y_{n,k-1} \leq f(x) < y_{n,k}$ and therefore

$$f_n(x) = y_{n,k-1} \leq f(x) < y_{n,k} < y_{n,k-1} + 1/n = f_n(x) + 1/n.$$

Now for every positive ϵ, there is an index N such that $1/N < \epsilon$ so that $f - f_n < 1/n < \epsilon$ on E if $n \geq N$. Thus, $\{f_n\}$ converges to f uniformly on E.

14. Let E_0 be the subset of E on which f is not finite; then $m(E_0) = 0$. For each natural number n, define F_n by $F_n = (\{x \in E \mid -n < f(x)\} \cap \{x \in E \mid f(x) < n\}) \sim E_0$. Then $\{F_n\}$ is an ascending sequence such that f is bounded on F_n, and such that both F_n and $E \sim F_n$ are measurable because they are intersections and complements of measurable sets. Therefore, $m(\bigcap_{n=1}^{\infty}(E \sim F_n)) = m(\emptyset) = 0 = \lim_{n \to \infty} m(E \sim F_n)$. By definition of convergence, there is an index N such that $m(E \sim F_N) < \epsilon$. Put F to F_N. Then F satisfies all requested properties.

15. Put F to be the F of Problem 14. By Problem 12, such a sequence $\{\varphi_n\}$ exists on F. Extend the φ_n to all of E by putting $\varphi_n(x)$ to 0 if x is in $E \sim F$. Then F and $\{\varphi_n\}$ satisfy all requested properties.

16. Because E is a subset of a bounded interval, $m(E)$ is finite. Then by the hint, there is a finite collection $\mathcal{O}$ of open intervals such that $m(E \sim \mathcal{O}) + m(\mathcal{O} \sim E) < \epsilon$. Put F equal to $(E \cap \mathcal{O}) \cup (I \sim (E \cup \mathcal{O}))$. Sets F and $I \sim F$ are measurable because $E, \mathcal{O}, I$ are measurable and the collection of measurable sets is a σ-algebra. Then $m(I \sim F) = m(\mathcal{O} \sim E) + m(E \sim \mathcal{O}) < \epsilon$. Next, put h to $\chi_{\mathcal{O}}$. Function h is a step function (see the definition on page 69) because $\mathcal{O}$ is a finite collection of open intervals. Then on F, we have that $h = \chi_E$.

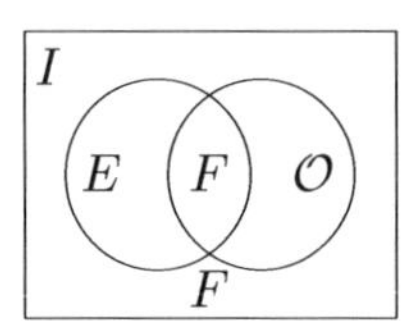

17. Per the hint, represent ψ by $\sum_{k=1}^{n} c_k \chi_{E_k}$ in the notation after the definition on page 61. By Proposition 1, each E_k is measurable. By Theorem 12 of Section 2.4, there is a finite collection $\mathcal{O}_k$ of open intervals such that $m(E_k \sim \mathcal{O}_k) + m(\mathcal{O}_k \sim E_k) < \epsilon/n$. Put F equal to union $\bigcup_{k=1}^{n}(E_k \cap \mathcal{O}_k)$. Then because I is a disjoint union of the E_k, we can write that $m(I \sim F) = \sum_{k=1}^{n} m(E_k \sim \mathcal{O}_k) \leq \sum_{k=1}^{n}(m(E_k \sim \mathcal{O}_k) + m(\mathcal{O}_k \sim E_k)) < \epsilon$. Next, put h to $\psi\chi_F$. Function h is a step function because F consists of open intervals (taken from the $\mathcal{O}_k$). Then $h = \psi$ on F.

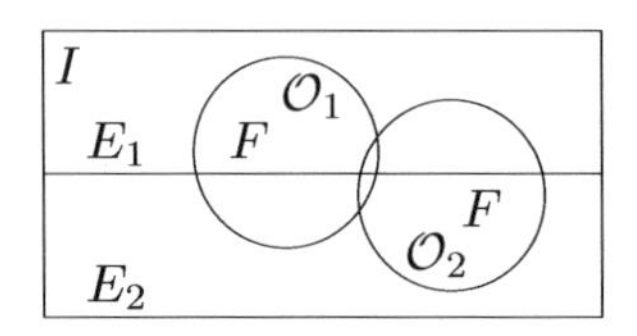

18. Put F to be to the same F of Problem 15 with E replaced by I. Then $m(I \sim F) < \epsilon$. By Problem 15 with n large enough, there is a simple function φ_N such that $|\varphi_N - f| < \epsilon$ on F. By Problem 17, there is a step function h equal to φ_N on F so that $|h - f| < \epsilon$ on F.

19. The claims are true by the paragraph after the definition on page 61. We give a more explicit answer. In the notation of that paragraph, consider two simple functions φ and ψ defined on equal sets $\bigcup_{j=1}^{m} A_j$ and $\bigcup_{k=1}^{n} B_k$ by $\varphi = \sum_{j=1}^{m} a_j \chi_{A_j}$ and $\psi = \sum_{k=1}^{n} b_k \chi_{B_k}$. Then

$$\begin{aligned}
\varphi + \psi &= \sum_{j=1}^{m} a_j \chi_{A_j} + \sum_{k=1}^{n} b_k \chi_{B_k} = \sum_{j=1}^{m}\sum_{k=1}^{n}(a_j + b_k)\chi_{A_j \cap B_k}, \\
\varphi\psi &= \sum_{j=1}^{m} a_j \chi_{A_j} \cdot \sum_{k=1}^{n} b_k \chi_{B_k} = \sum_{j=1}^{m}\sum_{k=1}^{n} a_j b_k \chi_{A_j \cap B_k}, \\
\max\{\varphi, \psi\} &= \max\left\{\sum_{j=1}^{m} a_j \chi_{A_j}, \sum_{k=1}^{n} b_k \chi_{B_k}\right\} = \sum_{j=1}^{m}\sum_{k=1}^{n} \max\{a_j, b_k\}\chi_{A_j \cap B_k}, \\
\min\{\varphi, \psi\} &= -\max\{-\varphi, -\psi\}.
\end{aligned}$$

Each right side takes only a finite number (mn at most) of values. Thus, the sum, product, maximum, and minimum of two simple functions are simple.

20. Note that

$$\chi_A\chi_B = \begin{cases} 1\cdot 1 = 1 & x\in A \text{ and } x\in B, \text{ that is, } x\in A\cap B,\\ 1\cdot 0 = 0 & x\in A \text{ and } x\notin B,\\ 0\cdot 1 = 0 & x\notin A \text{ and } x\in B,\\ 0\cdot 0 = 0 & x\notin A \text{ and } x\notin B.\end{cases}$$

Express the last three cases succinctly with the statement that $\chi_A\chi_B = 0$ if $x\notin A\cap B$. Thus, $\chi_A\chi_B = \chi_{A\cap B}$. Next,

$$\chi_A+\chi_B-\chi_A\chi_B = \begin{cases} 1+1-1\cdot 1 = 1 & x\in A \text{ and } x\in B,\\ 1+0-1\cdot 0 = 1 & x\in A \text{ and } x\notin B,\\ 0+1-0\cdot 1 = 1 & x\notin A \text{ and } x\in B,\\ 0+0-0\cdot 0 = 0 & x\notin A \text{ and } x\notin B, \text{ that is, } x\notin A\cup B.\end{cases}$$

Express the first three cases succinctly with the statement that $\chi_A+\chi_B-\chi_A\chi_B = 1$ if $x\in A\cup B$. Thus, $\chi_A+\chi_B-\chi_A\chi_B = \chi_{A\cup B}$. Last,

$$1-\chi_A = \begin{cases} 1-1 = 0 & x\in A, \text{ that is, } x\notin A^C,\\ 1-0 = 1 & x\notin A, \text{ that is, } x\in A^C.\end{cases}$$

Thus, $1-\chi_A = \chi_{A^C}$.

21. For each real number c,

$$\{x\in E \mid \inf\{f_n(x)\}\geq c\} = \bigcap_{k=1}^{\infty}\{x\in E \mid f_k(x)\geq c\},$$
$$\{x\in E \mid \sup\{f_n(x)\}\geq c\} = \bigcup_{k=1}^{\infty}\{x\in E \mid f_k(x)\geq c\}.$$

Because the f_n are measurable, each $\{x\in E \mid f_k(x)\geq c\}$ is a measurable set by definition of a measurable function using Proposition 1(ii). Because the collection of measurable sets is a σ-algebra, the intersection and union are measurable sets. Thus, again by definition of a measurable function using Proposition 1(ii), $\inf\{f_n\}$ and $\sup\{f_n\}$ are measurable. Next,

$$\liminf\{f_n(x)\} = \lim_{n\to\infty}\inf\{f_k(x) \mid k\geq n\} = \sup\{\inf\{f_k(x) \mid k\geq n\}\},$$
$$\limsup\{f_n(x)\} = \lim_{n\to\infty}\sup\{f_k(x) \mid k\geq n\} = \inf\{\sup\{f_k(x) \mid k\geq n\}\}.$$

By our solution to the first half of this exercise, $\inf\{f_k \mid k\geq n\}$ and $\sup\{f_k \mid k\geq n\}$ are each a sequence of measurable functions. Then appealing again to the first half of this exercise, the supremum of the first and infimum of the second are measurable functions. Thus, $\liminf\{f_n\}$ and $\limsup\{f_n\}$ are measurable.

22. Each E_n is open because $E_n = \{x\in[a,b] \mid (f-f_n)(x)\in(-1,\epsilon)\} = (f-f_n)^{-1}((-1,\epsilon))$, which is open by Proposition 22 of Section 1.6 along with the fact that $f-f_n$ is continuous by that section's Problem 49(i). To see that $\{E_n\}$ covers $[a,b]$, take a point x in $[a,b]$. Because $\{f_n\}$ is increasing and converges to f pointwise, there is an index N for which $f(x)-f_N(x)<\epsilon$. Hence, x is in set E_N of $\{E_n\}$. Next, by the Heine-Borel theorem, $[a,b]$ has a finite subcover $\{E_n\}_{n=1}^M$. Again because $\{f_n\}$ is increasing, $f(x)-f_{n+1}(x)\leq f(x)-f_n(x)$ so that E_{n+1} contains E_n. It follows that $\{E_n\}_{n=1}^M = E_M$ so that $[a,b]\subseteq E_M$; in other words, $f-f_n<\epsilon$ on $[a,b]$ if $n\geq M$. Thus, convergence is uniform on $[a,b]$.

23. By the last paragraph of Section 3.1 and its notation, $f = f^+ - f^-$. Because f^+, f^- are each measurable and nonnegative, there are corresponding sequences $\{\varphi_n^+\}, \{\varphi_n^-\}$ from the text's simple approximation theorem proof. Define sequence $\{\varphi_n\}$ by $\varphi_n = \varphi_n^+ - \varphi_n^-$, a simple function by Problem 19. Then because for each x in E, at least one of the pairs $f^+, \{\varphi_n^+\}$ or $f^-, \{\varphi_n^-\}$ is zero, we conclude that $\{\varphi_n^+\} - \{\varphi_n^-\}$ converges to $f^+ - f^-$ pointwise and has the property that $|\varphi_n^+ - \varphi_n^-| \leq |f^+ - f^-|$, that is, $\{\varphi_n\}$ converges to f pointwise and has the property that $|\varphi_n| \leq |f|$, which is the general simple approximation theorem.

24. Take a real number c. Consider set $\{x \in I \,|\, f(x) + x/n > c\}$. Because $f(x) + x/n$ is strictly increasing, $f(x) + x/n > f(y) + y/n$ when $x > y$. If there is a y such that $f(y) + y/n = c$, then $\{x \in I \,|\, f(x) + x/n > c\} = I \cap (y, \infty)$, a measurable set. If no such y exists, then c is outside the bounds of $f(I)$ or at a jump discontinuity as described in the last paragraph of Section 1.6. In either case, there is an x_0 for which $\{x \in I \,|\, f(x) + x/n > c\} = I \cap (x_0, \infty)$ or possibly $I \cap [x_0, \infty)$, both measurable sets. Then $f(x) + x/n$ is a measurable function by definition using Proposition 1(i), and therefore $\{f(x) + x/n\}$ is a sequence of measurable functions. Next, take an x in I and a positive ϵ. Choose an index N greater than $|x|/\epsilon$. If $n \geq N$, then $|f(x) - f(x) - x/n| = |x|/n < \epsilon$. Hence, $\{f(x) + x/n\}$ converges to $f(x)$. By Proposition 9, f is measurable.

3.3 Littlewood's Three Principles, Egoroff's Theorem, and Lusin's Theorem

25. This is Problem 47 of Section 1.6.

26. Quoting from the theorem's proof, the restriction of f to F is a continuous function because "the uniform limit of continuous functions is continuous," which is Problem 59 of Section 1.6.

    No, there need not be any points at which f as a function on E is continuous. For example, put f to $\chi_{\mathbf{Q}}$, and put E to $\mathbf{R}$. Function $\chi_{\mathbf{Q}}$ is measurable by the sentence after the definition of χ (page 61) and the fact that $\mathbf{Q}$ is measurable by Proposition 4 of Section 2.3 and Corollary 4(ii) of Section 1.3. Function $\chi_{\mathbf{Q}}$, however, is not continuous. Indeed, if x is rational, then for every positive δ, there is by Problem 12 of Section 1.2 an irrational number x' in $(x - \delta, x + \delta)$ so that $|\chi_{\mathbf{Q}}(x') - \chi_{\mathbf{Q}}(x)| = |0 - 1| = 1 \not< 1/2$. Similarly, if x is irrational, then there is by Theorem 2 of Section 1.2 a rational number x' in $(x - \delta, x + \delta)$ so that $|\chi_{\mathbf{Q}}(x') - \chi_{\mathbf{Q}}(x)| = |1 - 0| = 1 \not< 1/2$. Hence, there is no point at which $\chi_{\mathbf{Q}}$ is continuous.

27. A counterexample is sequence $\{\chi_{[n,\infty)}\}$ of measurable functions on domain $(-\infty, \infty)$. Sequence $\{\chi_{[n,\infty)}\}$ converges to 0 pointwise. If the theorem were true in this case, then there would be a set F for which $\{\chi_{[n,\infty)}\} \to 0$ uniformly on F, and $m(F^C) < 1$. By definition of uniform convergence, there would be an index N for which $\chi_{[N,\infty)} - 0 = \chi_{[N,\infty)} < 1$ on F. But $\chi_{[N,\infty)} < 1$ only on $(-\infty, N)$, so F must be a subset of $(-\infty, N)$, which leads to the contradiction that $1 > m(F^C) \geq m([N, \infty)) = \infty$.

28. By Proposition 9, f is measurable. Let E' be the set of points at which convergence is pointwise. By Problem 9, E' is measurable. By Proposition 5(ii), f and each f_n are measurable on E'. By Problem 14, there is a measurable set E'' contained in E' such that f is bounded on E'' and $m(E' \sim E'') < \epsilon/2$. By Proposition 5(ii), each f_n is measurable on E''. By Egoroff's theorem, there is a closed set F contained in E'' for which $\{f_n\} \to f$ uniformly on F and $m(E'' \sim F) < \epsilon/2$. Finally,

$$m(E \sim F) = m(E \sim E') + m(E' \sim E'') + m(E'' \sim F) < 0 + \epsilon/2 + \epsilon/2 = \epsilon.$$

    Thus, F has all the required properties.

29. Break E into a collection of sets $E_k = E \cap [k, k+1)$, where k is an integer. Each E_k is measurable, being the intersection of two measurable sets. By Proposition 5(ii), f is measurable on each E_k. By Lusin's theorem, for each $\epsilon > 0$, there is a continuous function g_k and a closed set F_k contained in E_k for which

$$f = g_k \text{ on } F_k \text{ and } m(E_k \sim F_k) < \epsilon/2^{|k|+2} \text{ for each } k.$$

Put F to $\bigcup_k F_k$, and g to $\sum_k g_k \chi_{F_k}$. Set F is contained in E, function g equals f on F, and g is continuous on F because g is the sum of continuous functions $g_k \chi_{F_k}$ on F. To show that F is closed, consider a point x of closure of F. Then every open interval I that contains x also contains a point in F. In particular, those Is with length less than 1 contain a point in some union $F_k \cup F_{k+1}$. By Proposition 12 of Section 1.4, $F_k \cup F_{k+1}$ is closed, so it contains x. Hence, $\bigcup_k F_k$, which is F, contains its points of closure and so is closed. By Problem 25, g has a continuous extension to all of $\mathbf{R}$. Finally,

$$m(E \sim F) = m\left(\bigcup_{k=-\infty}^{\infty} (E_k \sim F_k)\right) = \sum_{k=-\infty}^{\infty} m(E_k \sim F_k) < \sum_{k=-\infty}^{\infty} \frac{\epsilon}{2^{|k|+2}} = 3\epsilon/4 < \epsilon.$$

Thus, g and F have all the required properties.

30. By Problem 14, there is a measurable set E' contained in E such that f is bounded on E' and $m(E \sim E') < \epsilon/2$. By Proposition 5(ii), f is measurable on E'. By Lusin's theorem, there is a continuous function g on $\mathbf{R}$ and a closed set F contained in E' for which $f = g$ on F and $m(E' \sim F) < \epsilon/2$. Lastly,

$$m(E \sim F) = m(E \sim E') + m(E' \sim F) < \epsilon/2 + \epsilon/2 = \epsilon.$$

Thus, g and F have all the required properties.

31. For k equal to $2, 3, \ldots$, define A_k by $A_k = E \cap (-k, k)$. Each A_k is measurable with measure no more than $2k$. By Proposition 5(ii), each f_n is measurable on each A_k. Egoroff's theorem allows us to put E_k to be a measurable (closed) set contained in A_k for which $\{f_n\} \to f$ uniformly on E_k and $m(A_k \sim E_k) < \epsilon$ for each positive ϵ. Put E_1 to $E \sim \bigcup_{k=2}^{\infty} E_k$ so that E_1 is measurable, and $E = \bigcup_{k=1}^{\infty} E_k$. Finally,

$$\begin{aligned}
m(E_1) &= m\left(E \sim \bigcup_{k=2}^{\infty} E_k\right) \\
&= m\left(\bigcap_{k=2}^{\infty} (E \sim E_k)\right) && \text{De Morgan identity} \\
&= m\left(\bigcap_{k=2}^{\infty} ((E \sim A_k) \cup (A_k \sim E_k))\right) && E_k \subseteq A_k \subseteq E \\
&= m\left(\left(\bigcap_{k=2}^{\infty} (E \sim A_k)\right) \cup \left(\bigcap_{k=2}^{\infty} (A_k \sim E_k)\right)\right) && \text{distributive property} \\
&= m\left(\bigcap_{k=2}^{\infty} (A_k \sim E_k)\right) && \bigcap_{k=2}^{\infty} (E \sim A_k) = \emptyset \text{ by definition of } A_k \\
&\le m(A_k \sim E_k) \text{ for every } k && \text{monotonicity} \\
&< \epsilon.
\end{aligned}$$

Because that holds for ϵ arbitrarily small, $m(E_1) = 0$.

Chapter 4

Lebesgue Integration

4.1 The Riemann Integral

In the example, the seventh sentence should end "..., and $f_n(x) = 0$ otherwise."

1. The rational numbers are countably infinite in $[0, 1]$, so for each n, there is a q_i where $i > n$ so that $(f - f_n)(q_i) = 1 - 0 = 1$. That shows that there is no index N for which $f - f_n < 1/2$ on $[0, 1]$ if $n \geq N$. Thus, $\{f_n\}$ fails to converge to f uniformly on $[0, 1]$.

2. If $P' = P$, the claim is trivially true. Otherwise, there is a point x' in P' between two consecutive points x_{i-1}, x_i of P. Now

$$\begin{aligned} m_1' &= \inf\{f(x) \mid x_{i-1} < x < x'\} \geq \inf\{f(x) \mid x_{i-1} < x < x_i\} = m_i, \text{ and} \\ m_2' &= \inf\{f(x) \mid x' < x < x_i\} \quad \geq \inf\{f(x) \mid x_{i-1} < x < x_i\} = m_i \end{aligned}$$

so that if P' has just one more partition point than P, then

$$\begin{aligned} L(f, P') - L(f, P) &= m_1'(x' - x_{i-1}) + m_2'(x_i - x') - m_i(x_i - x_{i-1}) \\ &= (m_1' - m_i)(x' - x_{i-1}) + (m_2' - m_i)(x_i - x') \\ &\geq 0. \end{aligned}$$

Repeating that reasoning for the other points that are in P' but not in P proves that $L(f, P') \geq L(f, P)$. The proof that upper Darboux sums decrease under refinement is analogous.

3. Let P_1, P_2 be partitions of the interval. Partition $P_1 \cup P_2$, say P', is a refinement of each of P_1, P_2 so that by definition of Darboux sums and Problem 2,

$$L(f, P_1) \leq L(f, P') \leq U(f, P') \leq U(f, P_2),$$

and so $L(f, P_1) \leq U(f, P_2)$. Thus, each lower Darboux sum is no greater than each upper Darboux sum. Now fix P_2 in that inequality and take the supremum over partitions P_1 to get that $\sup\{L(f, P) \mid P\} \leq U(f, P_2)$ where P is a partition of the interval (note that the subscript 1 is no longer needed). The left side of that inequality is a fixed real number, so taking the infimum over the partitions in that inequality does not change the left side and so yields that $\sup\{L(f, P) \mid P\} \leq \inf\{U(f, P) \mid P\}$. Thus, the lower Riemann integral is no greater than the upper Riemann integral.

4. By hypothesis, $\sup\{L(f, P) \mid P\} = \inf\{U(f, P) \mid P\}$ where P is a partition of $[a, b]$. Then for each n, there is a P_n such that $U(f, P_n) - L(f, P_n) < 1/n$. Taking the limit yields that $\lim_{n \to \infty}(U(f, P_n) - L(f, P_n)) = 0$.

5. By the definition of *convergence*, for every positive ϵ, there is an index N for which $U(f, P_N) - L(f, P_N) < \epsilon$. Then Problem 3 and the definitions of supremum and infimum imply that $\inf\{U(f, P) \mid P\} - \sup\{L(f, P) \mid P\} < \epsilon$ where P is a partition of $[a, b]$. That inequality holds for every ϵ. Therefore, $\inf\{U(f, P) \mid P\} = \sup\{L(f, P) \mid P\}$, and thus f is Riemann integrable over $[a, b]$.

6. Because f is uniformly continuous (Theorem 23 of Section 1.6), for each natural number n, there is a δ_n such that for all x, x' in $[a, b]$, if $|x - x'| < \delta_n$, then $|f(x) - f(x')| < 1/(n(b - a))$.

Let P_n be a partition of $[a, b]$ with subintervals of length less than δ_n. Then in the notation of the section, $M_i - m_i < 1/(n(b-a))$ for each i so that

$$U(f, P_n) - L(f, P_n) = \sum_i (M_i - m_i)(x_i - x_{i-1}) < \sum_i \frac{1}{n(b-a)}(x_i - x_{i-1}) = 1/n.$$

Then $\lim_{n\to\infty} (U(f, P_n) - L(f, P_n)) = 0$. Thus, by Problem 5, f is Riemann integrable over $[a, b]$.

7. In the second step below, M_i equals the value of f at the upper bound x_i of subinterval (x_{i-1}, x_i) because f is increasing; analogously, $m_i = f(x_{i-1})$.

$$\begin{aligned} U(f, P_n) - L(f, P_n) &= \sum_{i=1}^{n} (M_i - m_i)(x_i - x_{i-1}) && \text{notation of the section} \\ &= \frac{1}{n}\sum_{i=1}^{n} (f(x_i) - f(x_{i-1})) && x_i - x_{i-1} = 1/n \\ &= \frac{f(1) - f(0)}{n} && \text{sum is telescoping.} \end{aligned}$$

Taking the limit yields $\lim_{n\to\infty}(U(f, P_n) - L(f, P_n)) = 0$. Thus, by Problem 5, f is Riemann integrable over $[0, 1]$.

8. Each integral symbol in the problem statement should have a prefix (R). Because $\{f_n\} \to f$ uniformly, for each positive ϵ, there is an index N for which $|f - f_n| < \epsilon/(b-a)$ on $[a, b]$ if $n \geq N$. Hence, $f_n - \epsilon/(b-a) < f < f_n + \epsilon/(b-a)$, and along with Problem 3 and the fact that f_N is Riemann integrable, if $n \geq N$, then

$$(R)\int_a^b \left(f_n - \frac{\epsilon}{b-a}\right) < \sup\{L(f, P) \,|\, P\} \leq \inf\{U(f, P) \,|\, P\} < (R)\int_a^b \left(f_n + \frac{\epsilon}{b-a}\right), \quad (4.1)$$

where P is a partition of $[a, b]$. Consequently,

$$0 \leq \inf\{U(f, P) \,|\, P\} - \sup\{L(f, P) \,|\, P\} < 2\epsilon.$$

That inequality holds for every ϵ. Thus, f is Riemann integrable over $[a, b]$. Now that we know f is Riemann integrable, we can rewrite inequality (4.1) as

$$(R)\int_a^b \left(f_n - \frac{\epsilon}{b-a}\right) < (R)\int_a^b f < (R)\int_a^b \left(f_n + \frac{\epsilon}{b-a}\right) \text{ if } n \geq N,$$

or

$$\left|(R)\int_a^b f - (R)\int_a^b f_n\right| < \epsilon \text{ if } n \geq N.$$

Thus, by the definition on page 21,

$$\lim_{n\to\infty} (R)\int_a^b f_n = (R)\int_a^b f.$$

4.2 The Lebesgue Integral of a Bounded Measurable Function over a Set of Finite Measure

In the last displayed equation on page 77, the second f should have a subscript n.

In the bounded convergence theorem, the second sentence should begin "Suppose $\{f_n\}$ is uniformly bounded on E, ..."

9. For each real number c, set $\{x \in E \mid f(x) > c\}$ has measure zero by monotonicity and is therefore measurable. Thus, f is measurable by the definition on page 55 using Proposition 1(i). By Theorem 4, f is integrable over E.

   Next, because f is bounded, there is a number M for which $|f| \leq M$ on E. Then by Corollary 7, the second part of Theorem 5, and definition of the integral of a simple function, $0 \leq \left|\int_E f\right| \leq \int_E |f| \leq \int_E M = Mm(E) = M0 = 0$. Thus, $\int_E f = 0$.

   We can also prove that $\int_E f = 0$ from the definition of the integral of a bounded function:

$$\begin{aligned}
\int_E f &= \inf\left\{\int_E \psi \,\middle|\, \psi \text{ simple and } f \leq \psi \text{ on } E\right\} \\
&= \inf\left\{\sum_{i=1}^n a_i m(E_i) \,\middle|\, f \leq \sum_{i=1}^n a_i \chi_{E_i} \text{ on } E\right\} && \text{definition of } \int_E \psi \text{ and equation (1)} \\
&= 0 && \text{each } m(E_i) \text{ is 0 because } E_i \subseteq E.
\end{aligned}$$

10. $f\chi_A$ is measurable by Theorem 6 of Section 3.1 and the sentence after the definition of χ_A on page 61. Both f and $f\chi_A$ are integrable by Theorem 4.

    Let φ be a simple function such that $\varphi \leq f$ on E and hence on A. Then $f\chi_A \geq \varphi\chi_A$, and

$$\begin{aligned}
\int_E f\chi_A &\geq \int_E \varphi\chi_A && \text{second part of Theorem 5} \\
&= \int_E \left(\sum_{i=1}^n a_i\chi_{E_i}\right)\chi_A && \text{notation from page 71} \\
&= \int_E \sum_{i=1}^n a_i\chi_{E_i \cap A} && \text{Problem 20 of Section 3.2} \\
&= \sum_{i=1}^n a_i m(E_i \cap A) && \text{Lemma 1} \\
&= \sum_{E_i \cap A \neq \emptyset} a_i m(E_i \cap A) && m(\emptyset) = 0\text{; allows us to ignore } E \sim A \\
&= \int_A \sum_{E_i \cap A \neq \emptyset} a_i\chi_{E_i \cap A} && \text{definition of integral of simple function} \\
&= \int_A \varphi && \text{canonical representation of } \varphi \text{ on } A.
\end{aligned}$$

    Taking the supremum over φ,

$$\begin{aligned}
\int_E f\chi_A &\geq \sup\left\{\int_A \varphi \,\middle|\, \varphi \text{ simple and } \varphi \leq f \text{ on } A\right\} \\
&= \int_A f && \text{definition of } \int_A f.
\end{aligned}$$

    We get the opposite inequality $\int_E f\chi_A \leq \int_A f$ with analogous reasoning starting with a simple function ψ such that $\psi \geq f$ on A and taking the infimum. Thus, $\int_A f = \int_E f\chi_A$.

11. No, it does not by the example on page 70.

12. Let $E \sim E_0$ where $E_0 \subseteq E$ be the set on which $f = g$. By hypothesis, $m(E_0) = 0$. We assume that g is bounded on E although the problem could be read as g's being bounded a.e. on E

(in which case a solution would require material subsequent to this section). By Problem 9, g is measurable on E_0, and $\int_{E_0} f$, $\int_{E_0} g$ each equal 0. Then f, g are each integrable on each of $E, E_0, E \sim E_0$ by Theorem 4 and the fact that the collection of measurable sets is a σ-algebra. Then by Corollary 6,

$$\int_E f = \int_{E \sim E_0} f + \int_{E_0} f = \int_{E \sim E_0} f + 0 = \int_{E \sim E_0} g + \int_{E_0} g = \int_E g.$$

13. No, it does not by the example before the theorem.

14. We need to show that the premises of Proposition 8 that each f_n is bounded, and that $\{f_n\} \to f$ uniformly imply the premises of the bounded convergence theorem that $\{f_n\}$ is uniformly bounded and that $\{f_n\} \to f$ pointwise. Because f_n is bounded, there is a number M_n for which $|f_n| \leq M_n$. By the first sentence of the proof of Proposition 8, there is a number M for which $|f| \leq M$. Because $\{f_n\} \to f$ uniformly, for each positive ϵ, there is an index N for which $|f_n - f| < \epsilon$ if $n \geq N$. Then,

$$\begin{aligned} |f_n| &< |f| + \epsilon,\ n \geq N && \text{restatement of uniform convergence} \\ &\leq M + \epsilon,\ n \geq N \\ &\leq \max\{M, M_1, \ldots, M_{N-1}\} + \epsilon,\ n \geq 1, \end{aligned}$$

which shows that $\max\{M, M_1, \ldots, M_{N-1}\} + \epsilon$ is a number M' for which $|f_n| < M'$ for all n. Hence, $\{f_n\}$ is uniformly bounded.

Next, because $\{f_n\} \to f$ uniformly, it follows directly from the definitions on pages 60 and 21 that $\{f_n\} \to f$ pointwise. Thus, Proposition 8 is a special case of the bounded convergence theorem.

15. To verify the assertions, read the proofs of the referenced theorems and lemma and note where finite and countable additivity are used.

16. Consider E as a union of disjoint sets A and B defined by

$$\begin{aligned} A &= \{x \in E \mid f(x) = 0\}, \\ B &= \{x \in E \mid f(x) > 0\}. \end{aligned}$$

Then

$$\begin{aligned} 0 &= \int_E f && \text{assumption} \\ &= \int_A f + \int_B f && \text{Corollary 6} \\ &= \int_A 0 + \inf\left\{\int_B \psi \,\middle|\, \psi \text{ simple and } f \leq \psi \text{ on } B\right\} && f = 0 \text{ on } A;\ \text{definition of } \int_B f \\ &= \inf\left\{\sum_{i=1}^n a_i m(B_i) \,\middle|\, f \leq \sum_{i=1}^n a_i \chi_{B_i} \text{ on } B\right\} && \text{definition of } \int_B \psi \text{ and (1).} \end{aligned}$$

By definition of infimum, for each positive ϵ, there is a ψ such that $\int_B \psi < \epsilon$. Because that holds for every ϵ, it also holds for ϵ equal to 0. Furthermore, because $\sum_{i=1}^n a_i \chi_{B_i} \geq f > 0$ on each B_i, each a_i must be greater than zero. Hence, for $\int_B \psi$, that is, $\sum_{i=1}^n a_i m(B_i)$, to be zero, each $m(B_i)$ must be zero. By the countable additivity of measure, we have that $m(B) = \sum_{i=1}^n m(B_i) = \sum_{i=1}^n 0 = 0$. Thus, $f = 0$ a.e. on E.

4.3 The Lebesgue Integral of a Measurable Nonnegative Function

In the last sentence of first paragraph on page 80, replace "of a subset of E" with "on a subset of E."

"Chebychev" is misspelled here and in Section 18.2. A correct spelling, "Chebyshev," is in the index. Although there are many correct transliterations of the Russian name, *Chebychev* is not one of them.

In the proof of Proposition 9, the third sentence should begin "By the continuity of Lebesgue measure, ..."

On pages 82–85, 89, 90, and 94, arguments of lim inf and lim sup should be enclosed in braces. For example in Fatou's lemma, $\displaystyle\int_E f \leq \liminf\left\{\int_E f_n\right\}$. Similarly in Problem 33, correct notation is $\{f_n\} \to f$, and $\left\{\int_E |f - f_n|\right\} \to 0$.

17. In definition (8) of $\int_E f$, Problem 9 tells us that $\int_E h = 0$. Thus, $\int_E f = 0$ because the supremum of $\{0\}$ is 0.

    Here is another solution: $\{n\}$ is an increasing sequence of nonnegative measurable (simple) functions on E such that $\{n\} \to f$ pointwise on E. Then by the monotone convergence theorem and definition of the integral of a simple function,

$$\int_E f = \lim_{n\to\infty} \int_E n = \lim_{n\to\infty} nm(E) = \lim_{n\to\infty} n0 = \lim_{n\to\infty} 0 = 0.$$

18. By the first paragraph of the section, we need to show that $\int_{E_0} f$ is independent of the choice of set of finite measure E_0 outside of which f vanishes. Let E_1 be another set of finite measure outside of which f vanishes. Let set A be the (finite) support of f. By the definitions of E_0, E_1, A, we have that $A \subseteq E_0$, $A \subseteq E_1$, and $f \equiv 0$ on both $E_0 \sim A$ and $E_1 \sim A$. Then using Corollary 6,

$$\begin{aligned}\int_{E_1} f - \int_{E_0} f &= \left(\int_A f + \int_{E_1\sim A} f\right) - \left(\int_A f + \int_{E_0\sim A} f\right)\\ &= \int_A f + 0 - \int_A f - 0\\ &= 0.\end{aligned}$$

    Thus, $\int_{E_1} f = \int_{E_0} f$, which shows that the integral is properly defined.

19. Consider the increasing sequence $\left\{x^\alpha \chi_{[1/n,1]}\right\}$ of nonnegative bounded measurable (Proposition 3 of Section 3.1) functions on $[0,1]$. That sequence converges to f pointwise on $[0,1]$. Each function in the sequence is Riemann integrable and therefore Lebesgue integrable by Theorem 3, and the two integrals are equal. Then

$$\begin{aligned}\int_0^1 f &= \lim_{n\to\infty} \int_0^1 x^\alpha \chi_{[1/n,1]} && \text{monotone convergence theorem}\\ &= \lim_{n\to\infty} (R)\int_{1/n}^1 x^\alpha\\ &= \lim_{n\to\infty} \begin{cases}(1-n^{-\alpha-1})/(\alpha+1) & \alpha \neq -1\\ \log n & \alpha = -1\end{cases}\\ &= \begin{cases}1/(\alpha+1) & \alpha > -1\\ \infty & \alpha \leq -1.\end{cases}\end{aligned}$$

20. By Fatou's lemma,

$$\int_E f \leq \liminf\left\{\int_E f_n\right\} \leq M$$

which also shows that Fatou's lemma implies "this property." (Starting with Beppo Levi's lemma leads to the same conclusions.) For the converse, assume the premises of Fatou's lemma and that "this property" holds. By the first sentence in the proof of Fatou's lemma, we can assume that $\{f_n\} \to f$ pointwise on E. Let $\ell = \liminf\{\int_E f_n\}$. By Proposition 19(i)(iii) of Section 1.5, for each positive ϵ, there are infinitely many indices n for which $\int_E f_n < \ell + \epsilon$. Considering now just those infinitely many $\int_E f_n$ as a sequence, there is by Proposition 16 of Section 1.5 a subsequence $\{\int_E f_{n_k}\}$ that converges. Because $\{f_n\}$ converges to f pointwise, so does $\{f_{n_k}\}$. Then by "this property," $\int_E f \leq \ell + \epsilon$. Because that holds for every ϵ, it holds also if $\epsilon = 0$. Thus, $\int_E f \leq \ell = \liminf\{\int_E f_n\}$, proving Fatou's lemma.

21. By definition (8), definition of a supremum, and the fact that $\int_E f < \infty$ because f is integrable, there is a bounded, measurable function g of finite support such that $0 \leq g \leq f$ on E and $\int_E g > \int_E f - \epsilon/2$. By the supremum definition of the integral of a bounded function defined on a set of finite measure (page 73), there is an η such that $\eta \leq g$ and $\int_E \eta > \int_E g - \epsilon/2$. Because f, g are nonnegative, we can choose a nonnegative η. Combining the inequalities gives that $0 \leq \eta \leq f$ on E and $\int_E (f - \eta) < \epsilon$.

    Next, choose η such that $\int_E (f - \eta) < \epsilon/2$. Let $\sum_{i=1}^{n} a_i \chi_{E_i}$ be the canonical representation of η on E. By Theorem 12 of Section 2.4, for each i, there is a finite disjoint collection of open intervals $\{I_{ik}\}_{k=1}^{n_i}$ for which

$$\begin{aligned} m\left(E_i \Delta \bigcup_{k=1}^{n_i} I_{ik}\right) &= \int_{\mathbf{R}} \left|\chi_{E_i} - \sum_{k=1}^{n_i} \chi_{I_{ik}}\right| \\ &< \frac{\epsilon}{2a_i n} \end{aligned}$$

    where the integral comes from the definition of the integral of a simple function. Define h on E by

$$h = \sum_{i=1}^{n} \sum_{k=1}^{n_i} a_i \chi_{I_{ik}}.$$

    The partition for step function h consists of the endpoints of intervals E, I_{ik} excluding endpoints of the I_{ik} that are outside of interval E. Then

$$\begin{aligned} \int_E |\eta - h| &= \int_E \left|\sum_{i=1}^{n} a_i \chi_{E_i} - \sum_{i=1}^{n} \sum_{k=1}^{n_i} a_i \chi_{I_{ik}}\right| \\ &\leq \sum_{i=1}^{n} a_i \int_E \left|\chi_{E_i} - \sum_{k=1}^{n_i} \chi_{I_{ik}}\right| \\ &< \sum_{i=1}^{n} a_i \frac{\epsilon}{2a_i n} = \epsilon/2. \end{aligned}$$

    Thus, h is a step function on E that has finite support, and $\int_E |f - h| < \epsilon$.

22. According to Fatou's lemma,

$$\int_E f \leq \liminf\left\{\int_E f_n\right\}.$$

However,

$$\begin{aligned}
\int_E f &= \int_{\mathbf{R}} f - \int_{\mathbf{R}\sim E} f && \text{Theorem 11 (additivity)}\\
&\geq \int_{\mathbf{R}} f - \liminf\left\{\int_{\mathbf{R}\sim E} f_n\right\} && \text{Fatou's lemma}\\
&= -\liminf\left\{-\left(\int_{\mathbf{R}} f - \int_{\mathbf{R}\sim E} f_n\right)\right\}\\
&= -\liminf\left\{-\left(\int_{\mathbf{R}} f_n - \int_{\mathbf{R}\sim E} f_n\right)\right\} && \text{hypothesis}\\
&= -\liminf\left\{-\int_E f_n\right\} && \text{Theorem 11 (additivity)}\\
&= \limsup\left\{\int_E f_n\right\} && \text{Proposition 19(iii) of Section 1.5.}
\end{aligned}$$

We have reached the same point in the proof of the monotone convergence theorem at which the authors jump to the conclusion. To elaborate, the two inequalities combined give that

$$\limsup\left\{\int_E f_n\right\} \leq \int_E f \leq \liminf\left\{\int_E f_n\right\}.$$

By Section 1.5's Problem 41, Proposition 19(iv), and definition of *convergence*,

$$\int_E f = \lim_{n\to\infty} \int_E f_n.$$

23. Consider the sequence $\{a_n\chi_{[n,n+1)}\}$ of nonnegative measurable functions on E. We have that $f = \sum_{n=1}^{\infty} a_n\chi_{[n,n+1)}$ on E. Then by Corollary 12 and Lemma 1,

$$\int_E f = \sum_{n=1}^{\infty}\int_E a_n\chi_{[n,n+1)} = \sum_{n=1}^{\infty} a_n m([n,n+1)) = \sum_{n=1}^{\infty} a_n 1 = \sum_{n=1}^{\infty} a_n.$$

24. (i) By the simple approximation theorem (Section 3.2), there is a sequence $\{\psi_n\}$ of functions that has all the required properties except possibly finite support. Put φ_n equal to $\psi_n\chi_{[-n,n]}$ for each n. Then $\{\varphi_n\}$ still has the properties of the theorem along now with finite support for each φ_n because φ_n vanishes outside $[-n,n]$.

    (ii) If $\int_E f = \infty$, the claim is trivially true. Otherwise, using part (i),

$$\int_E f = \lim_{n\to\infty}\int_E \varphi_n = \sup\left\{\int_E \varphi \,\middle|\, \varphi \text{ simple, of finite support and } 0 \leq \varphi \leq f \text{ on } E\right\}$$

    where the monotone convergence theorem justifies the first equality and the proof of Theorem 15 of Section 1.5 justifies the second. Note that $\varphi_n \leq f$ for each n because $\{\varphi_n\}$ is increasing and converges to f.

25. By inequality (12), $\int_E f_n \leq \int_E f$ for each n so that $\limsup\{\int_E f_n\} \leq \int_E f$. By Fatou's lemma, $\int_E f \leq \liminf\{\int_E f_n\}$. Thus, by the same reasoning as in the end of our solution to Problem 22, $\lim_{n\to\infty}\int_E f_n = \int_E f$.

26. Consider the decreasing sequence $\{\chi_{[n,\infty)}\}$ of nonnegative measurable functions on $[1,\infty)$. That sequence converges to 0 pointwise on $[1,\infty)$. By definition of the integral of a simple function,

$$\lim_{n\to\infty}\int_{[1,\infty)}\chi_{[n,\infty)} = \lim_{n\to\infty} m([n,\infty)) = \infty,$$

but by Proposition 9 (or by the first paragraph of the section with $E_0 = \emptyset$),

$$\int_{[1,\infty)} 0 = 0.$$

Thus, in this case, the monotone convergence theorem does not hold.

27. The authors have not yet defined limit inferior for a sequence of functions; we assume that lim inf is defined pointwise, analogous to the definition of max and min before Proposition 8 of Section 3.1: $\liminf\{f_n\}(x) = \liminf\{f_n(x)\}$ for $x \in E$. We also need to prove that $\liminf\{f_n\}$ is measurable. Indeed, for each c,

$$\{x \in E \mid \liminf\{f_n\}(x) \le c\} = \bigcup_{n=1}^{\infty}\{x \in E \mid f_n(x) \le c\},$$

so that set is measurable because it is the countable union of measurable sets. We analogously define $\inf\{f_k \mid k \ge n\}$ and use similar reasoning to show that it is measurable. Then,

$$\begin{aligned}
\int_E \liminf\{f_n\} &= \int_E \lim_{n\to\infty}\inf\{f_k \mid k\ge n\} && \text{definition of lim inf}\\
&\le \liminf\left\{\int_E \inf\{f_k \mid k \ge n\}\right\} && \text{Fatou's lemma}\\
&\le \liminf\left\{\int_E f_n\right\} && \text{inequalities (7) of Section 1.5 and (12).}
\end{aligned}$$

4.4 The General Lebesgue Integral

28. Starting and ending with definition (8),

$$\begin{aligned}
\int_E f^+\chi_C &= \sup\left\{\int_E h \,\middle|\, h \text{ bounded, measurable, of finite support; } 0 \le h \le f^+\chi_C \text{ on } E\right\}\\
&= \sup\left\{\int_E h\chi_C \,\middle|\, h\chi_C \text{ bounded, measurable, of finite support; } 0 \le h\chi_C \le f^+\chi_C \text{ on } E\right\}\\
&= \sup\left\{\int_C h \,\middle|\, h \text{ bounded, measurable, of finite support; } 0 \le h \le f^+ \text{ on } C\right\}\\
&= \int_C f^+.
\end{aligned}$$

Replacing h with $h\chi_C$ in the second step changes nothing because $0 \le h \le f^+\chi_C$. Problem 10 justifies the third step. Likewise, $\int_E f^-\chi_C = \int_C f^-$. Thus,

$$\int_E f\chi_C = \int_E (f\chi_C)^+ - \int_E (f\chi_C)^- = \int_E f^+\chi_C - \int_E f^-\chi_C = \int_C f^+ - \int_C f^- = \int_C f.$$

29. Neither statement is true. We disprove the second statement which also disproves the first because absolute convergence implies convergence. Consider counterexample function f defined by $f = \chi_{[n,n+0.5)} - \chi_{[n+0.5,n+1)}$. Series $\sum_{n=1}^{\infty} a_n$ converges absolutely because

$$\begin{aligned}\sum_{n=1}^{\infty}|a_n| &= \sum_{n=1}^{\infty}\left|\int_n^{n+1} f\right| \\ &= \sum_{n=1}^{\infty}\left|\int_n^{n+1} (\chi_{[t,t+0.5)} - \chi_{[t+0.5,t+1)})\right| \\ &= \sum_{n=1}^{\infty}|m([n,n+0.5)) - m([n+0.5,n+1))| \\ &= \sum_{n=1}^{\infty}|0.5 - 0.5| \\ &= 0.\end{aligned}$$

But by the definitions on pages 86 and 84 and Proposition 14, f is not integrable over $[1,\infty)$ because $\int_{[1,\infty)} f^+ = \infty$ since $f^+ = \chi_{[n,n+0.5)}$ takes the constant value 1 on a subset of $[1,\infty)$ of infinite measure, namely, subset $\bigcup_{n=1}^{\infty}[n, n+0.5)$ (see the first paragraph on page 80).

30. Refer to our solution of Problem 27 for definitions and measurability of $\liminf\{f_n\}$ and $\inf\{f_k \,|\, k \geq n\}$ and apply them analogously to $\limsup\{f_n\}$ and $\sup\{f_k \,|\, k \geq n\}$. We infer from Proposition 15 that, by possibly excising from E a countable collection of sets of measure zero and using the countable additivity of Lebesgue measure, we can assume that $|f_n| \leq g$ on E. Then,

$$\begin{aligned}\int_E \liminf\{f_n\} &= \int_E \lim_{n\to\infty} \inf\{f_k \,|\, k \geq n\} && \text{definition of } \liminf \\ &= \lim_{n\to\infty} \int_E \inf\{f_k \,|\, k \geq n\} && \text{Lebesgue dominated convergence theorem} \\ &\leq \lim_{n\to\infty} \inf\left\{\int_E f_k \,\middle|\, k \geq n\right\} && \text{Theorem 17 (monotonicity)} \\ &= \liminf\left\{\int_E f_n\right\} && \text{definition of } \liminf.\end{aligned}$$

Inequality $\liminf\{\int_E f_n\} \leq \limsup\{\int_E f_n\}$ is true by Problem 41 of Section 1.5 applied to $f_n(x)$ for each x in E. Lastly, inequality $\limsup\{\int_E f_n\} \leq \int_E \limsup\{f_n\}$ is proved analogously to our proof above for the first inequality.

31. Note that h is measurable because it is the difference $f - g$ of two measurable functions where g is measurable because it is integrable.

Consider another choice of a finite integrable function g' and nonnegative function h' whose sum is f. If $\int_E h = \infty$, then by linearity from Theorem 10 and the proof of Theorem 17, $\int_E h' = \int_E (h + g - g') = \int_E h + \int_E g - \int_E g' = \infty$ because $\int_E g$ and $\int_E g'$ are finite, g, g' being integrable. Hence, $\int_E f = \infty$ using either choice.

On the other hand, if $0 \leq \int_E h < \infty$, then h is integrable and so is h' using the same reasoning as above. Then by Theorem 17, $\int_E g' + \int_E h' = \int_E (g' + h') = \int_E f = \int_E (g+h) = \int_E g + \int_E h$ which shows once again that the definition of $\int_E f$ is independent of the choice of the two functions.

32. Because $\lim_{n\to\infty} \int_E g_n = \int_E g$, there is an index N for which $\int_E g_n < \int_E g + 1 < \infty$ if $n \geq N$, which shows that g_n is integrable over E if $n \geq N$.

    Because $|f_n| \leq g_n$ on E, and $|f| \leq g$ a.e. on E (inequality (7) of Section 1.5 applied pointwise on E), and g, g_n are integrable over E if $n \geq N$, the integral comparison test tells us that f, f_n also are integrable over E if $n \geq N$. We infer from Proposition 15 that, by possibly excising from E a countable collection of sets of measure zero and using the countable additivity of Lebesgue measure, we can assume that f, f_n are finite on E if $n \geq N$. Functions $g - f$, $g_n - f_n$ are properly defined, nonnegative, and measurable if $n \geq N$. Moreover, $\{g_n - f_n\}$ converges pointwise a.e. on E to $g - f$. Fatou's lemma holds if $n \geq N$ and tells us that

$$\int_E (g - f) \leq \liminf \left\{ \int_E (g_n - f_n) \right\}.$$

    Hence, by linearity of integration for integrable functions and hypothesis,

$$\begin{aligned} \int_E g - \int_E f = \int_E (g - f) &\leq \liminf \left\{ \int_E (g_n - f_n) \right\} \\ &= \int_E g - \limsup \left\{ \int_E f_n \right\}, \end{aligned}$$

    that is,

$$\limsup \left\{ \int_E f_n \right\} \leq \int_E f.$$

    Similarly, considering $\{g_n + f_n\}$,

$$\int_E f \leq \liminf \left\{ \int_E f_n \right\}.$$

    The proof is complete.

33. Assume first that $\{\int_E |f - f_n|\} \to 0$. Then for every positive ϵ, there is an index N for which if $n \geq N$, then

$\left\| \int_E \|f\| - \int_E \|f_n\| \right\| = \left\| \int_E (\|f\| - \|f_n\|) \right\|$	Theorem 17 (linearity); definition on page 86
$\leq \int_E \big\| \|f\| - \|f_n\| \big\|$	Proposition 16
$\leq \int_E \|f - f_n\|$	Th. 17 (monotonicity); reverse triangle inequality
$< \epsilon$	assumption.

    Thus, $\lim_{n\to\infty} \int_E |f_n| \to \int_E |f|$. (Recall that we proved the reverse triangle inequality in our solution to Problem 49(iv) of Section 1.6.)

    Conversely, assume that $\lim_{n\to\infty} \int_E |f_n| \to \int_E |f|$. By hypothesis, $\{|f - f_n|\}$ is a sequence of measurable functions on E that converges to 0 pointwise a.e. on E. Now $\{|f| + |f_n|\}$ is a sequence of nonnegative measurable functions on E that converges pointwise a.e. on E to $2|f|$ and dominates $\{|f - f_n|\}$ on E in the sense that $|f - f_n| \leq |f| + |f_n|$ on E for all n by the triangle inequality. By equation (6) of Section 1.5 and our assumption, we have that $\lim_{n\to\infty} \int_E (|f| + |f_n|) = \int_E 2|f| < \infty$. Thus, by Theorem 19 and Proposition 9, $\lim_{n\to\infty} \int_E |f - f_n| = \int_E 0 = 0$.

34. Sequence $\{f\chi_{[-n,n]}\}$ is an increasing sequence of nonnegative measurable functions on $\mathbf{R}$ that converges pointwise to f on $\mathbf{R}$. Then using the monotone convergence theorem,

$$\lim_{n\to\infty}\int_{-n}^{n} f = \lim_{n\to\infty}\int_{\mathbf{R}} f\chi_{[-n,n]} = \int_{\mathbf{R}} f.$$

35. For each natural number n, $\{f(x,y_n)\}$ where $0 \le y_n \le 1$ and $\{y_n\} \to 0$, is a sequence of measurable functions of x that converges to $f(x)$. By supposition, $|f(x,y_n)| \le g(x)$ for all n. By Lebesgue's dominated convergence theorem,

$$\lim_{n\to\infty}\int_0^1 f(x,y_n)\,dx = \int_0^1 f(x)\,dx.$$

Because that is true for every sequence $\{y_n\}$ such that $0 \le y_n \le 1$ and $\{y_n\} \to 0$, we conclude that

$$\lim_{y\to 0}\int_0^1 f(x,y)\,dx = \int_0^1 f(x)\,dx.$$

To show that h is continuous, now let $\{y_n\} \to y$ where $0 \le y \le 1$. Then by hypothesis and Proposition 21 of Section 1.6, $\{f(x,y_n)\}$ converges to $f(x,y)$. Appealing again to Lebesgue's dominated convergence theorem,

$$\lim_{n\to\infty} h(y_n) = \lim_{n\to\infty}\int_0^1 f(x,y_n)\,dx = \int_0^1 f(x,y)\,dx = h(y).$$

Thus, by Proposition 21 in the other direction, h is continuous.

36. Sequence $\left\{\dfrac{f(x,y+1/(Mn)) - f(x,y)}{1/(Mn)}\right\}$, where M is a constant that is chosen to keep $(x,y+1/(Mn))$ in Q for all n, is a sequence of measurable functions of x that converges to $\partial f(x,y)/\partial y$ by definition of a partial derivative. We have three ways to show that the sequence is dominated by an integrable function of x.

(i) Because $\partial f(x,y)/\partial y$ exists, f is a continuous function of y for each fixed value of x. By the mean value theorem, there is a point y_n in $(0,1)$ such that

$$\left|\frac{f(x,y+1/(Mn)) - f(x,y)}{1/(Mn)}\right| = \left|\frac{\partial f}{\partial y}(x,y_n)\right| \le g(x) \text{ for all } (x,y) \text{ in } Q \text{ and for all } n.$$

(ii) Starting with the fundamental theorem of calculus, and assuming that $M > 0$,

$$\begin{aligned}
\left|\frac{f(x,y+1/(Mn)) - f(x,y)}{1/(Mn)}\right| &= \left|\int_y^{y+1/(Mn)} \frac{1}{Mn}\frac{\partial f(x,t)}{\partial t}\,dt\right| \\
&\le \int_y^{y+1/(Mn)} \left|\frac{1}{Mn}\frac{\partial f(x,t)}{\partial t}\right| dt && \text{Proposition 16} \\
&\le \int_y^{y+1/(Mn)} \left|\frac{1}{Mn}\right| g(x)\,dt && \text{Theorem 19} \\
&= \left|\frac{1}{Mn}\right| g(x)\left(y + \frac{1}{Mn} - y\right) \\
&= g(x).
\end{aligned}$$

If $M < 0$, we need to swap the limits of integration, but the end result remains the same.

(iii) Because the sequence converges to $\partial f(x,y)/\partial y$, there is an index N for which if $n \geq N$, then

$$\left|\frac{f(x,y+1/(Mn)) - f(x,y)}{1/(Mn)}\right| < \left|\frac{\partial f(x,y)}{\partial y}\right| + 1 \leq g(x) + 1.$$

In this case, we note that the justifications in the remainder of the proof hold when $n \geq N$.

We now conclude the proof. For all y in $[0,1]$,

$$\begin{aligned}
\frac{d}{dy}\int_0^1 f(x,y)\,dx &= \lim_{n\to\infty} \frac{\int_0^1 f(x,y+1/(Mn))\,dx - \int_0^1 f(x,y)\,dx}{1/(Mn)} && \text{definition of derivative} \\
&= \lim_{n\to\infty} \int_0^1 \frac{f(x,y+1/(Mn)) - f(x,y)\,dx}{1/(Mn)}\,dx && \text{Theorem 17 (linearity)} \\
&= \int_0^1 \frac{\partial f(x,y)}{\partial y}\,dx && \text{Lebesgue dominated co.}
\end{aligned}$$

4.5 Countable Additivity and Continuity of Integration

37. $\{E_n\}_{n=1}^\infty$ is a descending countable collection of subsets of E. That each E_n is measurable is easier to see using the equivalent definition $E_n = E \sim (-n,n)$ and the fact that the complement of two measurable sets is measurable because the collection of measurable sets is a σ-algebra. We also have that $\bigcap_{n=1}^\infty E_n = \emptyset$ so that $m(\bigcap_{n=1}^\infty E_n) = m(\emptyset) = 0$. Then by Theorem 21(ii),

$$\lim_{n\to\infty}\int_{E_n} f = \int_{\bigcap_{n=1}^\infty E_n} f = 0.$$

Thus, by inequality (3) of Section 1.5, there is an N for which if $n \geq N$, then $\left|\int_{E_n} f\right| < \epsilon$.

38. (i) Although we give a real-number value for $\lim_{n\to\infty}\int_1^n f$ in the following, it is sufficient to know that the series converges by, for example, the alternating series test.

$$\begin{aligned}
\lim_{n\to\infty}\int_1^n f = \sum_{n=1}^\infty \int_{[n,n+1)} \frac{(-1)^n}{n} &= \sum_{n=1}^\infty \frac{(-1)^n}{n} && \text{integral of simple function (page 71)} \\
&= -\log 2.
\end{aligned}$$

On the other hand, f fails to be integrable as we see upon testing it in the definition of an integrable function (pages 84 and 86):

$$\int_{[1,\infty)} |f| = \sum_{n=1}^\infty \frac{1}{n} = \infty$$

by the comparison or integral tests.

(ii) If we are allowed to use the well-known result that $\int_0^\infty (\sin x)/x\,dx = \pi/2$, then existence of $\lim_{n\to\infty}\int_1^n (\sin x)/x\,dx$ follows directly. Here is another proof. Using integration by parts,

$$\lim_{n\to\infty}\int_1^n \frac{\sin x}{x}\,dx = \lim_{n\to\infty}\left(-\frac{\cos x}{x}\Bigg|_1^n - \int_1^n \frac{\cos x}{x^2}\,dx\right) = \cos 1 - \int_1^\infty \frac{\cos x}{x^2}\,dx.$$

Moreover, by Proposition 16 and Theorem 17 (monotonicity),

$$\left|\int_1^\infty \frac{\cos x}{x^2}\,dx\right| \le \int_1^\infty \left|\frac{\cos x}{x^2}\right| dx \le \int_1^\infty \frac{1}{x^2}\,dx = 1 < \infty.$$

Thus, $\lim_{n\to\infty} \int_1^n (\sin x)/x\,dx$ exists. On the other hand, f fails to be integrable:

$$\begin{aligned}
\lim_{n\to\infty} \int_1^n \left|\frac{\sin x}{x}\right| dx &= \int_1^\pi \left|\frac{\sin x}{x}\right| dx + \sum_{k=1}^\infty \int_{k\pi}^{(k+1)\pi} \left|\frac{\sin x}{x}\right| dx \\
&\ge \sum_{k=1}^\infty \int_{k\pi}^{(k+1)\pi} \frac{|\sin x|}{(k+1)\pi}\,dx && \text{Theorem 17} \\
&= \sum_{k=1}^\infty \frac{2}{(k+1)\pi} = \infty.
\end{aligned}$$

Those results do not contradict the continuity of integration because continuity of integration assumes that f is integrable on, in our case, $[1,\infty)$, which is not true.

39. (i) Define $E_0 = \emptyset$. Because $\{E_n\}_{n=1}^\infty$ is ascending, $\{E_n \sim E_{n-1}\}_{n=1}^\infty$ is a disjoint countable collection of measurable subsets of $\bigcup_{n=1}^\infty E_n$ whose union is $\bigcup_{n=1}^\infty E_n$. Then

$$\begin{aligned}
\int_{\bigcup_{n=1}^\infty E_n} f &= \sum_{n=1}^\infty \int_{E_n \sim E_{n-1}} f && \text{Theorem 20} \\
&= \sum_{n=1}^\infty \left(\int_{E_n} f - \int_{E_{n-1}} f \right) && \text{Corollary 18} \\
&= \lim_{n\to\infty} \sum_{k=1}^n \left(\int_{E_k} f - \int_{E_{k-1}} f \right) && \text{definition of a series} \\
&= \lim_{n\to\infty} \left(\int_{E_n} f - \int_{E_0} f \right) && \text{sum is telescoping} \\
&= \lim_{n\to\infty} \int_{E_n} f && m(E_0) = m(\emptyset) = 0.
\end{aligned}$$

(ii) Because $\{E_n\}_{n=1}^\infty$ is descending, $\{E_1 \sim E_n\}_{n=1}^\infty$ is ascending. Then

$$\begin{aligned}
\int_{\bigcap_{n=1}^\infty E_n} f &= \int_{E_1} f - \int_{E_1 \sim \bigcap_{n=1}^\infty E_n} f && \text{Corollary 18} \\
&= \int_{E_1} f - \int_{\bigcup_{n=1}^\infty (E_1 \sim E_n)} f && \text{De Morgan's identity} \\
&= \int_{E_1} f - \lim_{n\to\infty} \int_{E_1 \sim E_n} f && \text{part (i)} \\
&= \int_{E_1} f - \lim_{n\to\infty} \left(\int_{E_1} f - \int_{E_n} f \right) && \text{Corollary 18} \\
&= \lim_{n\to\infty} \int_{E_n} f.
\end{aligned}$$

If we disregard the statement before Theorem 21, we can devise shorter proofs. For part (i), note that $\{f\chi_{E_n}\}$ is a sequence of measurable functions such that $\{f\chi_{E_n}\} \to f\chi_{\bigcup_{n=1}^\infty E_n}$

pointwise on E. Furthermore, $|f\chi_{E_n}| \leq |f|$. Then we can use Lebesgue's dominated convergence theorem to write

$$\lim_{n\to\infty}\int_{E_n} f = \lim_{n\to\infty}\int_E f\chi_{E_n} = \int_E f\chi_{\bigcup_{n=1}^\infty E_n} = \int_{\bigcup_{n=1}^\infty E_n} f.$$

The proof of part (ii) is the same but with the fact that $\{f\chi_{E_n}\}$ converges to $f\chi_{\bigcap_{n=1}^\infty E_n}$.

4.6 Uniform Integrability: The Vitali Convergence Theorem

40. To show that F is properly defined, we need to show that $F(x)$ exists and is uniquely defined for all real numbers x. First, $\int_{-\infty}^x |f| = \int_{\mathbf{R}} |f\chi_{(-\infty,x]}| \leq \int_{\mathbf{R}} |f| < \infty$, which shows that f is integrable over $(-\infty, x]$ for all x. Thus, F exists. Secondly, uniqueness of F follows from uniqueness of the Lebesgue integral.

    For continuity, consider an increasing sequence $\{x_n\}$ that converges to x. Then $\{f\chi_{(-\infty,x_n]}\}$ is a sequence of measurable functions that converges to $f\chi_{(-\infty,x]}$ and is dominated by $|f|$ on $(-\infty, x]$. Using Lebesgue's dominated convergence theorem,

$$\lim_{n\to\infty} F(x_n) = \lim_{n\to\infty}\int_{-\infty}^{x_n} f = \lim_{n\to\infty}\int_{-\infty}^{x} f\chi_{(-\infty,x_n]} = \int_{-\infty}^{x} f\chi_{(-\infty,x]} = \int_{-\infty}^{x} f = F(x)$$

    for all x. Thus, by Proposition 21 of Section 1.6, F is continuous.

    No, F is not necessarily Lipschitz. Consider counterexample function defined by

$$F(x) = \int_{-\infty}^{x} \frac{\chi_{[0,\infty]}}{2\sqrt{t}}\,dt = \sqrt{x}\chi_{[0,\infty]}.$$

    Function $\chi_{[0,\infty]}/(2\sqrt{t})$ is integrable over $\mathbf{R}$, but $\sqrt{x}\chi_{[0,\infty]}$ is not Lipschitz by Problem 50 of Section 1.6.

41. The sequence of functions $\{\chi_{[-n,n]}\}$ is uniformly integrable over $\mathbf{R}$. Indeed, in the definition on page 93, put δ to ϵ. Then for each $\chi_{[-n,n]}$, if $A \subseteq \mathbf{R}$ is measurable and $m(A) < \delta$, then $\int_A |\chi_{[-n,n]}| \leq m(A) < \epsilon$. However, $\{\chi_{[-n,n]}\} \to 1$ pointwise on $\mathbf{R}$, but 1 is not integrable on $\mathbf{R}$ because $\int_{\mathbf{R}} 1 = \infty$ by the first paragraph on page 80.

42. Our counterexample is similar to the example on page 83. In Theorem 26, put E to $[-1, 1]$ and h_n to $n(-\chi_{(-1/n,0)} + \chi_{(0,1/n)})$ so that $\{h_n\}$ is a sequence of integrable functions that converges to 0 pointwise on E. Moreover,

$$\lim_{n\to\infty}\int_E h_n = \lim_{n\to\infty}\int_{[-1,1]} n(-\chi_{(-1/n,0)} + \chi_{(0,1/n)}) = \lim_{n\to\infty}(-1+1) = 0.$$

    The sequence is not, however, uniformly integrable over E. Indeed, in the definition on page 93, put ϵ to $1/2$ and A to $(0, 1/N)$ where N is some index. Interval A is a subset of E and measurable, and $m(A) = 1/N$. Now for every positive δ, we can choose N large enough so that $m(A) < \delta$, but $\int_A |h_N| = \int_{(0,1/N)} |N(-\chi_{[-1/N,0)} + \chi_{(0,1/N]})| = 1 \not< \epsilon$. Thus, the "only if" direction of Theorem 26 is false.

43. Change h_n to f_n in the problem statement. Because $\{f_n\}$ is uniformly integrable over E, for each positive ϵ there is a positive δ_f such that if a subset A of E is measurable and $m(A) < \delta_f$, then $\int_A |f_n| < \epsilon/|2\alpha|$ for each n. We have the analogous statement for $\{g_n\}$. Put δ to $\min\{\delta_f, \delta_g\}$ so that $m(A) < \delta$. Then by the triangle inequality and Theorem 17,

$$\int_A |\alpha f_n + \beta g_n| \leq \int_A (|\alpha f_n| + |\beta g_n|) = |\alpha|\int_A |f_n| + |\beta|\int_A |g_n| < |\alpha|\frac{\epsilon}{|2\alpha|} + |\beta|\frac{\epsilon}{|2\beta|} = \epsilon$$

    for each n. Thus, $\{\alpha f_n + \beta g_n\}$ is uniformly integrable over E.

44. (i) If f is nonnegative, this is true by the first part of Problem 21. For the general case, express f as $f^+ - f^-$. Then there are simple functions η^+, η^- on $\mathbf{R}$ of finite support, and $\int_{\mathbf{R}} |f^+ - \eta^+| < \epsilon/2$ and $\int_{\mathbf{R}} |f^- - \eta^-| < \epsilon/2$. Put η to $\eta^+ - \eta^-$, which is a simple function of finite support. Then using the triangle inequality and Theorem 17,

$$\int_{\mathbf{R}} |f - \eta| = \int_{\mathbf{R}} |f^+ - f^- - \eta^+ + \eta^-| \leq \int_{\mathbf{R}} |f^+ - \eta^+| + \int_{\mathbf{R}} |f^- - \eta^-| < \frac{\epsilon}{2} + \frac{\epsilon}{2} = \epsilon.$$

(ii) By Problem 37, there is a natural number N_+ for which if $n \geq N_+$, then $\left|\int_{\mathbf{R}_n} f^+\right| < \epsilon/8$ where $\mathbf{R}_n = [-n, n]^C$. We have the analogous statement for f^-. Put I to $[-N, N]$ where $N = \max\{N_+, N_-\}$. Then,

$$\int_{I^C} |f| = \int_{I^C} |f^+ - f^-| \leq \left|\int_{I^C} f^+\right| + \left|\int_{I^C} f^-\right| < \frac{\epsilon}{8} + \frac{\epsilon}{8} = \epsilon/4.$$

By part (i) and Corollary 18, $\int_I |f - \eta| \leq \int_{\mathbf{R}} |f - \eta| < \epsilon/4$.
By Problem 18 of Section 3.2 and Proposition 15, there is an s that vanishes outside of I and a measurable subset F of I for which $|\eta - s| < \epsilon'$ on F and $m(I \sim F) < \epsilon'$ where $\epsilon' = \min\{\epsilon/(8N), \epsilon/(4M)\}$ and $M = \max\{|(\eta - s)(x)| \mid x \in I \sim F\}$. Note that M exists because $|\eta - s|$ is a simple function. Then altogether,

$$\begin{aligned}
\int_{\mathbf{R}} |f - s| &= \int_I |f - s| + \int_{I^C} |f - s| && \text{Corollary 18} \\
&= \int_I |f - \eta + \eta - s| + \int_{I^C} |f| && s \text{ vanishes outside of } I \\
&< \int_I |f - \eta| + \int_I |\eta - s| + \epsilon/4 && \text{triangle inequality, Theorem 17} \\
&< \epsilon/4 + \int_F |\eta - s| + \int_{I \sim F} |\eta - s| + \epsilon/4 && \text{Corollary 18} \\
&< \epsilon/4 + m(F)\frac{\epsilon}{8N} + m(I \sim F)M + \epsilon/4 && \\
&< \epsilon/4 + 2N\frac{\epsilon}{8N} + \frac{\epsilon}{4M}M + \epsilon/4 && m(F) \leq m(I) = 2N \\
&= \epsilon.
\end{aligned}$$

If we disregard the hint, we can shorten our proof. By the second part of Problem 21, there is a step function s^+ that vanishes outside of I and $\int_I |f^+ - s^+| < 3\epsilon/8$. We have the analogous statement for f^-. Put s to $s^+ - s^-$, which is a step function on $\mathbf{R}$ that vanishes outside I. Then similar to the previous,

$$\begin{aligned}
\int_{\mathbf{R}} |f - s| &= \int_I |f - s| + \int_{I^C} |f - s| \\
&= \int_I |f^+ - f^- - s^+ + s^-| + \int_{I^C} |f| \\
&< \int_I |f^+ - s^+| + \int_I |f^- - s^-| + \epsilon/4 \\
&< 3\epsilon/8 + 3\epsilon/8 + \epsilon/4 \\
&= \epsilon.
\end{aligned}$$

(iii) By part (ii), we can write that $\int_{\mathbf{R}} |f - s| < \epsilon/2$ where for the partition $\{x_0, \ldots, x_m\}$ of I, step function s can be expressed on $\mathbf{R}$ as $s(x) = c_i$ if $x_{i-1} < x < x_i$ where

$i = 0, \ldots, m+1$, $c_0 = 0 = c_{m+1}$, $x_{-1} = -\infty$, and $x_{m+1} = \infty$. We now construct g as a piecewise linear function (see Problem 51 of Section 1.6) that is equal to s except that we remove the jump discontinuities of s by inserting steep linear pieces around the partition points. Put δ equal to

$$\min\left\{\left\{\frac{\epsilon}{(m+1)\max\{|c_{i+1}-c_i|\}}\right\} \cup \left\{\frac{x_i - x_{i-1}}{2}\right\}\right\} \text{ where } i = 0, \ldots, m.$$

Define g by

$$g(x) = \begin{cases} c_i + (x - x_i)\dfrac{c_{i+1} - c_i}{\delta} & x_i \leq x \leq x_i + \delta,\ 0 \leq i \leq m-1, \\ c_m - (x - (x_m - \delta))\dfrac{c_m}{\delta} & x_m - \delta \leq x \leq x_m, \\ s(x) & \text{otherwise.} \end{cases}$$

Then the integral of $|s - g|$ is the sum of triangular areas:

$$\int_{\mathbf{R}} |s - g| = \sum_{i=0}^{m} \frac{\delta |c_{i+1} - c_i|}{2} \leq \sum_{i=0}^{m} \frac{\epsilon}{2(m+1)} = \frac{\epsilon}{2}.$$

Altogether, $\int_{\mathbf{R}} |f - g| = \int_{\mathbf{R}} |f - s + s - g| \leq \int_{\mathbf{R}} |f - s| + \int_{\mathbf{R}} |s - g| < \epsilon/2 + \epsilon/2 = \epsilon$.

45. Because f is integrable over E, set E is measurable, $\hat{f}$ restricted to E is a measurable function, and $\hat{f}$ is bounded on $\mathbf{R}$ (Proposition 15). Set $\mathbf{R} \sim E$ is also measurable because the collection of measurable sets is a σ-algebra. Function $\hat{f}$ restricted to $\mathbf{R} \sim E$ is also measurable by the definition on page 55 using Proposition 1(i) because for each real number c, set $\{x \in (\mathbf{R} \sim E) \mid \hat{f}(x) > c\}$ is either all of $\mathbf{R} \sim E$ (if $c < 0$) or the empty set (if $c \geq 0$). By Proposition 5(ii) of Section 3.1, $\hat{f}$ is measurable on $\mathbf{R}$. Thus,

$$\begin{aligned} \int_{\mathbf{R}} \hat{f} &= \int_{\mathbf{R}} \left(\hat{f}^+ - \hat{f}^-\right) \\ &= \int_{\mathbf{R}} \hat{f}^+ - \int_{\mathbf{R}} \hat{f}^- && \text{Theorem 10} \\ &= \int_E f^+ + \int_{\mathbf{R} \sim E} \hat{f}^+ - \int_E f^- - \int_{\mathbf{R} \sim E} \hat{f}^- && \text{Theorem 11} \\ &= \int_E \left(f^+ - f^-\right) + 0 - 0 && \text{Theorem 10; Proposition 9} \\ &= \int_E f \\ &< \infty && f \text{ is integrable over } E. \end{aligned}$$

Thus, $\hat{f}$ is integrable over $\mathbf{R}$, and $\int_{\mathbf{R}} \hat{f} = \int_E f$.

Next, by part (i), there is a simple function $\hat{\eta}$ on $\mathbf{R}$ that has finite support and $\int_{\mathbf{R}} |\hat{f} - \hat{\eta}| < \epsilon$. Because $\hat{f} \equiv 0$ outside of E, we can have that $\hat{\eta} \equiv 0$ outside of E. Put η to be $\hat{\eta}$ restricted to E. Then $|f - \eta|$ is integrable over E, and $|\hat{f} - \hat{\eta}|$ is the extension of $|f - \eta|$ to all of $\mathbf{R}$ with $|\hat{f} - \hat{\eta}| \equiv 0$ outside of E. Thus, by the previous, $\int_E |f - \eta| = \int_{\mathbf{R}} |\hat{f} - \hat{\eta}| < \epsilon$.

By part (iii), there is a continuous function $\hat{g}$ on $\mathbf{R}$ and $\int_{\mathbf{R}} |\hat{f} - \hat{g}| < \epsilon$. Put g to be $\hat{g}$ restricted to E. Then $\int_E |f - g| \leq \int_E |f - g| + \int_{\mathbf{R} \sim E} |\hat{f} - \hat{g}| = \int_{\mathbf{R}} |\hat{f} - \hat{g}| < \epsilon$.

46. By Problem 44(ii), there is a step function s on $\mathbf{R}$ that vanishes outside a closed, bounded interval and $\int_{-\infty}^{\infty}|f(x) - s(x)|\,dx = \int_{\mathbf{R}}|f - s| < \epsilon/2$. Express s as in our solution to Problem 44(iii). Then using Proposition 16 with inequality $|f(x)\cos nx| \le |f(x)|$, the triangle inequality, Problem 48 to justify that $f\cos$ and $s\cos$ are integrable, Theorem 17, and Corollary 18,

$$\begin{aligned}
\left|\int_{-\infty}^{\infty} f(x)\cos nx\,dx\right| &\le \int_{-\infty}^{\infty} |f(x)\cos nx|\,dx \\
&= \int_{-\infty}^{\infty} |f(x)\cos nx - s(x)\cos nx + s(x)\cos nx|\,dx \\
&\le \int_{-\infty}^{\infty} |(f(x) - s(x))\cos nx|\,dx + \int_{-\infty}^{\infty} |s(x)\cos nx|\,dx \\
&\le \int_{-\infty}^{\infty} |f(x) - s(x)|\,dx + \sum_{i=1}^{m}|c_i|\int_{x_{i-1}}^{x_i} |\cos nx|\,dx \\
&< \frac{\epsilon}{2} + \sum_{i=1}^{m}|c_i|\frac{|\sin nx_i - \sin nx_{i-1}|}{n} \\
&\le \epsilon/2 + 2m\max\{|c_i|\}/n \\
&\le \epsilon/2 + \epsilon/2 \text{ if } n \ge N \text{ for some index } N \text{ where } N > 4m\max\{|c_i|\}/\epsilon \\
&= \epsilon.
\end{aligned}$$

Thus, by definition of convergence (page 21), $\lim_{n\to\infty}\int_{-\infty}^{\infty} f(x)\cos nx\,dx = 0$.

47. (i) Express f as the difference $f^+ - f^-$ of two nonnegative measurable functions. By Problem 24(i), there is an increasing sequence $\{\varphi_n\}$ of nonnegative simple functions on $\mathbf{R}$, each of finite support, that converges pointwise on $\mathbf{R}$ to f^+. Then

$$\begin{aligned}
\int_{-\infty}^{\infty} f^+(x)\,dx &= \lim_{n\to\infty}\int_{-\infty}^{\infty} \varphi_n(x)\,dx && \text{monotone convergence theorem} \\
&= \lim_{n\to\infty}\sum_{i=1}^{r_n} a_{ni}m(E_{ni}) && \text{Lemma 1 with } \varphi_n = \sum_{i=1}^{r_n} a_{ni}\chi_{E_{ni}} \\
&= \lim_{n\to\infty}\sum_{i=1}^{r_n} a_{ni}m(E_{ni} - t) && \text{Proposition 2 of Section 2.2} \\
&= \lim_{n\to\infty}\int_{-\infty}^{\infty} \varphi_n(x+t)\,dx \\
&= \int_{-\infty}^{\infty} f^+(x+t)\,dx.
\end{aligned}$$

Similarly, $\int_{-\infty}^{\infty} f^-(x)\,dx = \int_{-\infty}^{\infty} f^-(x+t)\,dx$. Thus, $\int_{-\infty}^{\infty} f(x)\,dx = \int_{-\infty}^{\infty} f(x+t)\,dx$.

(ii) The integral in the problem statement wants the dx symbol. We note the following facts.

- Because g is bounded, $|g| \le M$ for some number M.
- Because f is integrable over $(-\infty, \infty)$, by our solution to Problem 44(iii), there is a piecewise linear function h on $\mathbf{R}$ that vanishes outside of an interval I and $\int_{-\infty}^{\infty}|f(x) - h(x)|\,dx = \int_{\mathbf{R}}|f - h| < \epsilon/(3M)$.
- Moreover, h is Lipschitz, which is evident if we let the c in the definition on page 25 be the maximum of the absolute values of the slopes of the linear pieces of h. By

Problem 50 of Section 1.6, h is uniformly continuous on $\mathbf{R}$. Hence, there is a positive δ such that if $|t| < \delta$, then $|h(x) - h(x+t)| < \epsilon/(3Mm(I))$.

Using those facts along with Problem 48; Theorem 17; Theorem 18 of Section 1.5 with the understanding that $\lim_{t\to 0}$ means that we replace t with t_n, and the sequence $\{t_n\}$ converges to 0; and the triangle inequality,

$$\begin{aligned}
&\lim_{t\to 0}\int_{-\infty}^{\infty} g(x)(f(x) - f(x+t))\,dx \\
&\quad\le M\lim_{t\to 0}\int_{-\infty}^{\infty}|f(x) - f(x+t) + h(x) - h(x) + h(x+t) - h(x+t)|\,dx \\
&\quad\le M\lim_{t\to 0}\bigg(\int_{-\infty}^{\infty}|f(x) - h(x)|\,dx + \int_{-\infty}^{\infty}|f(x+t) - h(x+t)|\,dx \\
&\qquad\qquad + \int_I|h(x) - h(x+t)|\,dx\bigg) \\
&\quad< M\left(\frac{\epsilon}{3M} + \frac{\epsilon}{3M} + \frac{\epsilon}{3Mm(I)}m(I)\right) \\
&\quad= \epsilon.
\end{aligned}$$

Because that holds for arbitrarily small ϵ, we are done.

48. Because f is integrable, it is measurable by definition and finite by Proposition 15. Then fg is measurable by Theorem 6 of Section 3.1, so $|fg|$ is measurable by the first paragraph on page 59. Then by Theorem 10 where $|g| \le M < \infty$, we have that $\int_E|fg| \le M\int_E|f| < \infty$. Thus, fg is integrable.

49. Assume (i) and the hypothesis of (ii). Product fg is integrable over $\mathbf{R}$ by Problem 48, and $fg = 0$ a.e. on $\mathbf{R}$. Let E_0 be the set of zero measure on which $f \ne 0$. Then by Propositions 15 and 9, $\int_{\mathbf{R}} fg = \int_{\mathbf{R}\sim E_0} fg = 0$, which proves (ii).

    Next, assume (ii). Put g to χ_A, which, by page 61, is a measurable and bounded function for every measurable set A. Then by hypothesis and Problem 28, $0 = \int_{\mathbf{R}} f\chi_A = \int_A f$, which proves (iii).

    Next, (iii) implies (iv) because every open set is measurable by Theorem 9 of Section 2.3.

    Finally, assume (iv). Define disjoint sets E_+, E_- by

$$\begin{aligned}
E_+ &= \{x \in \mathbf{R} \mid f(x) > 0\},\\
E_- &= \{x \in \mathbf{R} \mid f(x) < 0\},
\end{aligned}$$

    which are measurable by the definition on page 55 and the fact that f is a measurable function. By Theorem 11(i) of Section 2.4, for each positive η, there is an open set $\mathcal{O}$ containing E_+ for which $m(\mathcal{O} \sim E_+) < \eta$. Then for each positive ϵ,

$$\begin{aligned}
0 &= \int_{\mathcal{O}} f && \text{hypothesis}\\
&= \int_{E_+} f + \int_{\mathcal{O}\sim E_+} f && \text{Corollary 18}\\
&= \int_{E_+} f - \int_{\mathcal{O}\sim E_+} |f| && f = -|f| \text{ on } \mathcal{O}\sim E_+\text{; Theorem 17}\\
&> \int_{E_+} f - \epsilon && \text{Proposition 23.}
\end{aligned}$$

Because that holds for ϵ arbitrarily small, $\int_{E_+} f = 0$. Continuing with Chebyshev's inequality,

$$0 = \int_{E_+} f \geq \frac{1}{n} m\left\{x \in E_+ \,\middle|\, f(x) \geq \frac{1}{n}\right\} \geq 0,$$

which shows that $m\{x \in E_+ \,|\, f(x) \geq 1/n\} = 0$ so that

$$\begin{aligned} m(E_+) &= \lim_{n \to \infty} m\{x \in E_+ \,|\, f(x) \geq 1/n\} \\ &= 0. \end{aligned}$$

Analogously, $m(E_-) = 0$ so that $m(E_+ \cup E_-) = 0$. But $E_+ \cup E_-$ is precisely the set of points x for which $f(x) \neq 0$. Thus, $f = 0$ a.e. on $\mathbf{R}$, which proves (i).

50. If $\mathcal{F}$ is uniformly integrable over E, then the *only if* direction of the claim is true by statement (27) and Proposition 16. Conversely, assume the hypothesis of the *if* direction of the claim. Define subsets A_+, A_- of A by

$$\begin{aligned} A_+ &= \{x \in A \,|\, f(x) \geq 0\}, \\ A_- &= \{x \in A \,|\, f(x) < 0\}. \end{aligned}$$

Then A_+, A_- are measurable, disjoint sets whose union is A. By hypothesis, for each positive ϵ, there are positive δ_+, δ_- such that for each $f \in \mathcal{F}$,

$$\text{if } m(A_+) < \delta_+, \text{ then } \int_{A_+} |f| = \left|\int_{A_+} f\right| < \epsilon/2, \text{ and}$$
$$\text{if } m(A_-) < \delta_-, \text{ then } \int_{A_-} |f| = \left|\int_{A_-} f\right| < \epsilon/2.$$

Put δ equal to $\min\{\delta_+, \delta_-\}$ so that for each f in $\mathcal{F}$, if $m(A) < \delta$, then

$$\begin{aligned} \int_A |f| &= \int_{A_+} |f| + \int_{A_-} |f| \\ &< \epsilon/2 + \epsilon/2 \\ &= \epsilon. \end{aligned}$$

Thus, $\mathcal{F}$ is uniformly integrable over E.

51. If $\mathcal{F}$ is uniformly integrable over E, then the *only if* direction of the claim is true by statement (27) with set A replaced by $E \cap \mathcal{U}$ where $\mathcal{U}$ is open.

Conversely, assume the hypothesis of the *if* direction of the claim. If a subset A of E is measurable and $m(A) < \delta$, then by Theorem 11(i) of Section 2.4, there is an open set $\mathcal{U}$ containing A for which $m(\mathcal{U} \sim A) < \delta - m(A)$. Hence,

$$\begin{aligned} m(E \cap \mathcal{U}) &\leq m(\mathcal{U}) \\ &= m(A) + m(\mathcal{U} \sim A) \\ &< \delta. \end{aligned}$$

From Corollary 18, the fact that $A \subseteq (E \cap \mathcal{U})$, and hypothesis, it follows that

$$\int_A |f| \leq \int_{E \cap \mathcal{U}} |f| < \epsilon.$$

Thus, $\mathcal{F}$ is uniformly integrable over E.

Chapter 5

Lebesgue Integration: Further Topics

5.1 Uniform Integrability and Tightness: A General Vitali Convergence Theorem

1. $\{f_k\}_{k=1}^n$ is uniformly integrable over E by Proposition 24 of Section 4.6. We now show that $\{f_k\}_{k=1}^n$ is also tight. By the proof of Proposition 1, for each k and positive ϵ, there is a subset E_k of E of finite measure for which

$$\int_{E\sim E_k} |f_k| < \epsilon.$$

Define E_0 by $E_0 = \bigcup_{k=1}^n E_k$ so that $E \sim E_0 \subseteq E \sim E_k$. Then by additivity-over-domains properties of integration,

$$\int_{E\sim E_0} |f_k| \le \int_{E\sim E_k} |f_k| < \epsilon$$

for all k. Thus, $\{f_k\}_{k=1}^n$ is tight over E.

2. If $\{h_n\}$ is uniformly integrable and tight, then $\lim_{n\to\infty} \int_E h_n = 0$ by the Vitali convergence theorem. Conversely, assume that $\lim_{n\to\infty} \int_E h_n = 0$. Then $\{h_n\}$ is uniformly integrable over E by Theorem 26 of Section 4.6 noting that the converse part of the proof does not require that E be of finite measure. We now show that $\{h_n\}$ is also tight. For each positive ϵ, we can choose an index N for which if $n \ge N$, then $\int_E h_n < \epsilon$. Therefore, because $h_n \ge 0$ on E,

$$\text{if } E_0 \text{ is a subset of } E \text{ of finite measure and } n \ge N, \text{ then } \int_{E\sim E_0} h_n < \epsilon. \tag{5.1}$$

According to Problem 1, the finite collection $\{h_n\}_{n=1}^{N-1}$ is tight over E. Let E_0 respond to the ϵ challenge regarding the criterion for the tightness of $\{h_n\}_{n=1}^{N-1}$. We infer from (5.1) that E_0 also responds to the ϵ challenge regarding the criterion for the tightness of $\{h_n\}$.

3. Change h_n to f_n in the problem statement. We proved uniform integrability in our solution to Problem 43 of Section 4.6. We now prove tightness. Because $\{f_n\}$ is tight over E, for each positive ϵ, there is a subset E_f of finite measure such that $\int_{E\sim E_f} |f_n| < \epsilon/|2\alpha|$ for all n. We have the analogous statement for $\{g_n\}$. Put E_0 to $E_f \cup E_g$ so that $E \sim E_0$ is a subset of each of $E \sim E_f$ and $E \sim E_g$. Then by the triangle inequality, Theorem 17, and additivity-over-domains properties of integration,

$$\int_{E\sim E_0} |\alpha f_n + \beta g_n| \le |\alpha| \int_{E\sim E_f} |f_n| + |\beta| \int_{E\sim E_g} |g_n| < |\alpha| \frac{\epsilon}{|2\alpha|} + |\beta| \frac{\epsilon}{|2\beta|} = \epsilon$$

for all n. Thus, $\{\alpha f_n + \beta g_n\}$ is tight over E.

4. Assume that $\{f_n\}$ is tight and uniformly integrable. Then for each positive ϵ and all indices n, there is a measurable subset E_0 of E that has finite measure for which $\int_{E\sim E_0} |f_n| < \epsilon/2$, and there is a positive δ such that for each measurable subset A, and hence $A \cap E_0$, of E, if $m(A \cap E_0) < \delta$, then $\int_{A\cap E_0} |f_n| < \epsilon/2$. Because A is a subset of the union of disjoint sets $A \cap E_0$ and $E \sim E_0$, it follows from additivity-over-domains properties of integration that

$$\int_A |f_n| \le \int_{A\cap E_0} |f_n| + \int_{E\sim E_0} |f_n| < \frac{\epsilon}{2} + \frac{\epsilon}{2} = \epsilon.$$

Conversely, assume the hypothesis of the *if* direction of the claim. If $m(A) < \delta$, then $m(A \cap E_0) < \delta$ so that $\int_A |f_n| < \epsilon$. Thus, $\{f_n\}$ is uniformly integrable. Next, put A to $E \sim E_0$. Then because $m((E \sim E_0) \cap E_0) = m(\emptyset) = 0 < \delta$, we have that $\int_{E \sim E_0} |f_n| < \epsilon$. Thus, $\{f_n\}$ is also tight.

5. Assume first that $\{f_n\}$ is uniformly integrable and tight on $\mathbf{R}$. Then for each ϵ and all indices n, there is a δ such that if A is a measurable set and $m(A) < \delta$, then $\int_A |f_n| < \epsilon/3$, and there is a measurable set E_0 of finite measure for which $\int_{E_0^C} |f_n| < \epsilon/3$. Choose r large enough so that $m(E_0 \cap (-r,r)^C) < \delta$. Then we can use uniform integrability to write that $\int_{E_0 \cap (-r,r)^C} |f_n| < \epsilon/3$ and also $\int_{\mathcal{O} \cap (-r,r)} |f_n| < \epsilon/3$ if $m(\mathcal{O} \cap (-r,r)) < \delta$ for an open subset $\mathcal{O}$. We also have from our tightness statement above that $\int_{E_0^C \cap (-r,r)^C} |f_n| \le \int_{E_0^C} |f_n| < \epsilon/3$.

   We show next that the domains of integration of three of the above integrals are disjoint subsets of $\mathcal{O}$:

$$\begin{aligned}\mathcal{O} &= (\mathcal{O} \cap (-r,r)) \cup (\mathcal{O} \cap (-r,r)^C) \\ &\subseteq (\mathcal{O} \cap (-r,r)) \cup (-r,r)^C \\ &= (\mathcal{O} \cap (-r,r)) \cup (E_0 \cap (-r,r)^C) \cup (E_0^C \cap (-r,r)^C).\end{aligned}$$

   Thus, by additivity-over-domains properties of integration,

$$\int_{\mathcal{O}} |f_n| \le \int_{\mathcal{O} \cap (-r,r)} |f_n| + \int_{E_0 \cap (-r,r)^C} |f_n| + \int_{E_0^C \cap (-r,r)^C} |f_n| < \frac{\epsilon}{3} + \frac{\epsilon}{3} + \frac{\epsilon}{3} = \epsilon.$$

   Conversely, assume the hypothesis of the *if* direction of the claim. Consider a measurable set A for which $m(A) < \delta/2$. By Theorem 11(i) of Section 2.4, there is an $\mathcal{O}$ containing A for which $m(\mathcal{O} \sim A) < \delta/2$. Then

$$m(\mathcal{O} \cap (-r,r)) = m((\mathcal{O} \sim A) \cap (-r,r)) + m(A \cap (-r,r)) < \delta/2 + \delta/2 = \delta$$

   so that by additivity-over-domains properties of integration and hypothesis,

$$\int_A |f_n| \le \int_{\mathcal{O}} |f_n| < \epsilon.$$

   Thus, $\{f_n\}$ is uniformly integrable by the definition on page 93 with δ replaced by $\delta/2$. Next, put E_0 to $(-r,r)$. Then because E_0^C is an open set and $m(E_0^C \cap (-r,r)) = m(\emptyset) = 0 < \delta$, we have that $\int_{E_0^C} |f_n| < \epsilon$. Thus, $\{f_n\}$ is also tight.

5.2 Convergence in Measure

6. We first establish that if a sequence $\{a_n\}$ of real numbers converges to a, then every subsequence of $\{a_n\}$ converges to a. By hypothesis, for every positive ϵ, there is an index N for which if $n \ge N$, then $|a - a_n| < \epsilon$. For any subsequence $\{a_{n_k}\}$ of $\{a_n\}$, choose index K such that $n_k \ge N$ where $k \ge K$. It follows that if $k \ge K$, then $|a - a_{n_k}| < \epsilon$. Thus, $\{a_{n_k}\} \to a$. Applying that result to this problem, every subsequence of $\{f_n\}$ converges in measure to f, and that if $\{f_n\} \to f$ pointwise, every subsequence of $\{f_n\}$ converges to f pointwise.

   Assume first that $\{f_n\} \to g$ in measure. By Theorem 4, there is a subsequence $\{f_{n_k}\}$ that converges pointwise a.e. to g. Because $\{f_n\} \to f$ in measure, $\{f_{n_k}\} \to f$ in measure, so there is a further subsequence $\{f_{n_i}\}$ of $\{f_{n_k}\}$ that converges pointwise a.e. to f. By the previous paragraph, $\{f_{n_i}\}$ converges pointwise a.e. to g also. Thus, $f = g$ a.e. on E.

Conversely, assume that $f = g$ a.e. on E. Let E_0 be the subset of E on which $f \neq g$ so that $m(E_0) = 0$. Then for each $\eta > 0$,

$$\begin{aligned}\lim_{n\to\infty} m\{x \in E \,|\, |f_n(x) - g(x)| > \eta\} &= \lim_{n\to\infty} m\{x \in E \sim E_0 \,|\, |f_n(x) - g(x)| > \eta\} \\ &= \lim_{n\to\infty} m\{x \in E \sim E_0 \,|\, |f_n(x) - f(x)| > \eta\} \\ &= 0\end{aligned}$$

because $\{f_n\} \to f$ in measure on all of E. Thus, $\{f_n\} \to g$ in measure on E.

7. For each natural number k, define set B_k by $B_k = \{x \in E \,|\, |g(x)| \geq k\}$. Because $\{B_k\}_{k=1}^{\infty}$ is a descending collection of measurable sets, and $m(B_1) \leq m(E) < \infty$, we can use Theorem 15(ii) of Section 2.5 to write

$$\lim_{k\to\infty} m(B_k) = m\left(\bigcap_{k=1}^{\infty} B_k\right) = m\{x \in E \,|\, |g(x)| = \infty\} = 0$$

since g is finite a.e. on E. Hence, for every positive ϵ, we can assign an index K such that $m(B_K) < \epsilon/2$. Furthermore, because $\{f_n\} \to f$ in measure, for each positive η, there is an index N for which if $n \geq N$, then $m\{x \in E \,|\, |f_n(x) - f(x)| > \eta/K\} < \epsilon/2$. In the following, the first set in the third line is in case $|g(x)| < K$.

$$\begin{aligned}&m\{x \in E \,|\, |f_n(x)g(x) - f(x)g(x)| > \eta\} \\ &\quad = m\{x \in E \,|\, |f_n(x) - f(x)||g(x)| > \eta\} \\ &\quad \leq m(\{x \in E \,|\, |f_n(x) - f(x)| > \eta/K\} \cup \{x \in E \,|\, |g(x)| \geq K\}) \\ &\quad < \epsilon/2 + \epsilon/2 \qquad\qquad \text{if } n \geq N \\ &\quad = \epsilon,\end{aligned}$$

that is, $\lim_{n\to\infty} m\{x \in E \,|\, |f_n(x)g(x) - f(x)g(x)| > \eta\} = 0$. So $\{f_n g\} \to f_n g$ in measure on E.

Next, fix a nonnegative number M. Explanations for the following steps are given afterwards:

$$\begin{aligned}&\lim_{n\to\infty} m\{x \in E \,\big|\, \big|f_n^2(x) - f^2(x)\big| > \eta\} \\ &\quad = \lim_{n\to\infty} m\{x \in E \,|\, |f_n(x) - f(x)||f_n(x) + f(x)| > \eta\} \\ &\quad \leq \lim_{n\to\infty} m\{x \in E \,|\, |f_n(x) - f(x)||f_n(x)| + |f_n(x) - f(x)||f(x)| > \eta\} \\ &\quad \leq \lim_{n\to\infty} m\{x \in E \,|\, |f_n(x) - f(x)||f_n(x)| > \eta/2\} \\ &\quad + \lim_{n\to\infty} m\{x \in E \,|\, |f_n(x) - f(x)||f(x)| > \eta/2\} \\ &\quad \leq \lim_{n\to\infty} m\{x \in E \,|\, |f_n(x) - f(x)| > \eta/(2M)\} + \lim_{n\to\infty} m\{x \in E \,|\, |f_n(x)| \geq M\} + 0 \\ &\quad = 0 + \lim_{n\to\infty} m\{x \in E \,|\, |f_n(x)| \geq M \text{ and } |f_n(x) - f(x)| \leq \eta\} \\ &\qquad + \lim_{n\to\infty} m\{x \in E \,|\, |f_n(x)| \geq M \text{ and } |f_n(x) - f(x)| > \eta\} \\ &\quad \leq m\{x \in E \,|\, |f(x)| \geq M - \eta\} + 0.\end{aligned}$$

The first step is simply factoring. The second step uses the triangle inequality and monotonicity of Lebesgue measure and convergence. For the third step, note that if the sum of two nonnegative numbers is greater than η, then at least one of them is greater than $\eta/2$; we also use Proposition 3 of Section 2.2 and Theorem 18 of Section 1.5. In the fourth step, we break the first limit from the previous step into two limits using the same reasoning as in our solution to the first part of this problem; we also use that solution to put the second limit from the previous step to 0. In the fifth step, we put the first limit from the previous line to 0 because $\{f_n\} \to f$ in measure; we split the second set from the previous line into two disjoint sets. In the final step, we put the second limit from the previous step to 0 because $\{f_n\} \to f$ in measure; the remaining set contains the first set of the previous step (and does not depend on n). Because our reasoning holds for all nonnegative M, it holds also as $M \to \infty$, in which case the measure in the last line is zero because f is finite a.e. on E. Thus, $\{f_n^2\} \to f^2$ in measure.

For the last claim, define functions s_n, s, d_n, d on E by

$$s_n(x) = (f_n(x) + g_n(x))/4, \qquad s(x) = (f(x) + g(x))/4,$$
$$d_n(x) = (f_n(x) - g_n(x))/4, \qquad d(x) = (f(x) - g(x))/4,$$

all of which are finite a.e. on E and measurable by Theorem 6 (linearity) of Section 3.1. Moreover, $\{s_n\} \to s$ and $\{d_n\} \to d$ in measure by our solution to Problem 10. Then

$$\begin{aligned}
&\lim_{n\to\infty} m\{x \in E \,|\, |f_n(x)g_n(x) - f(x)g(x)| > \eta\} \\
&= \lim_{n\to\infty} m\{x \in E \,|\, |s_n^2(x) - d_n^2(x) - s^2(x) + d^2(x)| > \eta\} \\
&\le \lim_{n\to\infty} m\{x \in E \,|\, |s_n^2(x) - s^2(x)| > \eta/2\} + \lim_{n\to\infty} m\{x \in E \,|\, |d_n^2(x) - d^2(x)| > \eta/2\} \\
&= 0 + 0
\end{aligned}$$

by the second part of this problem. Thus, $\{f_n g_n\} \to fg$ in measure. Another way to prove that is to reason as in our second solution to Problem 10.

8. Let $\{f_n\}$ be a sequence of nonnegative measurable functions on E that converges in measure to f on E. By definition of lim inf (page 23), there is a subsequence $\{f_{n_k}\}$ such that $\lim_{k\to\infty} \int_E f_{n_k} = \liminf\{\int_E f_n\}$. Because $\{f_{n_k}\} \to f$ in measure, there is a further subsequence $\{f_{n_i}\}$ of $\{f_{n_k}\}$ that converges pointwise a.e. on E to f by Theorem 4. Then starting with the original version of Fatou's lemma,

$$\int_E f \le \liminf\left\{\int_E f_{n_i}\right\} = \lim_{k\to\infty} \int_E f_{n_k} = \liminf\left\{\int_E f_n\right\}.$$

For the three theorems, we first establish that if every subsequence $\{a_{n_k}\}$ of $\{a_n\}$ has a further subsequence $\{a_{n_i}\}$ that converges to a, then $\{a_n\}$ converges to a. Suppose, to get a contradiction, that $\{a_n\}$ did not converge to a. If a is a real number, then there is a positive ϵ such that for any N, there is an $n \ge N$ for which $|a_n - a| \ge \epsilon$. That allows us to extract a subsequence $\{a_{n_k}\}$ that has no further subsequence that converges to a, which contradicts our assumption. If $a = \pm\infty$, similar reasoning leads again to a contradiction. Thus, $\{a_n\} \to a$. In particular, if $\{\int_E f_n\}$ satisfies the hypothesis, then $\lim_{i\to\infty} \int_E f_{n_i} = \lim_{n\to\infty} \int_E f_n$.

Now let $\{f_n\}$ be a sequence of measurable functions on E that converges in measure to f. Consider any subsequence $\{f_{n_k}\}$ of $\{f_n\}$. Because $\{f_{n_k}\} \to f$ in measure, there is a further subsequence $\{f_{n_i}\}$ of $\{f_{n_k}\}$ that converges pointwise a.e. on E to f. If $\{f_n\}$ is nonnegative and increasing, then so is $\{f_{n_i}\}$ (monotone convergence theorem). If there is a function g that is integrable over E and dominates $\{f_n\}$ on E, then g also dominates $\{f_{n_i}\}$ on E (Lebesgue dominated convergence theorem). If $\{f_n\}$ is uniformly integrable and tight over E, then so is $\{f_{n_i}\}$ (Vitali convergence theorem). Then applying the original versions of the three theorems stated in parentheses and our result from the previous paragraph,

$$\int_E f = \lim_{i\to\infty} \int_E f_{n_i} = \lim_{n\to\infty} \int_E f_n.$$

The original versions of the Lebesgue dominated and Vitali convergence theorems also tell us that f is integrable. Thus, the three theorems remain valid if "pointwise a.e." is replaced by "in measure."

9. For example, $\{\chi_{[n,n+1]}\}$ is a sequence of measurable functions on $\mathbf{R}$ that converges pointwise on $\mathbf{R}$ to 0 (see the first paragraph of Section 5.1). But

$$\lim_{n\to\infty} m\{x \in \mathbf{R} \,|\, |\chi_{[n,n+1]}(x) - 0| > 1/2\} = \lim_{n\to\infty} 1 = 1 \neq 0.$$

Thus, $\{\chi_{[n,n+1]}\}$ does not converge in measure to 0.

10. Let $\{f_n\} \to f$ and $\{g_n\} \to g$ both in measure on a set E of finite measure. By Problem 7, $\{\alpha f_n\} \to \alpha f$ and $\{\beta g_n\} \to \beta g$ in measure on E for finite constants α, β. Then for each positive η,

$$\begin{aligned}
&\lim_{n\to\infty} m\{x \in E \,|\, |\alpha f_n(x) + \beta g_n(x) - \alpha f(x) + \beta g(x)| > \eta\} \\
&\leq \lim_{n\to\infty} m\{x \in E \,|\, |\alpha f_n(x) - \alpha f(x)| > \eta/2\} + \lim_{n\to\infty} m\{x \in E \,|\, |\beta g_n(x) - \beta g(x)| > \eta/2\} \\
&= 0 + 0
\end{aligned}$$

using reasoning similar to that of our solution to Problem 7. Thus, $\{\alpha f_n + \beta g_n\} \to \alpha f + \beta g$ in measure on E.

Here is a solution that uses Problem 11. Let $\{f_{n_k}\}, \{g_{n_k}\}$ be any subsequences of $\{f_n\}, \{g_n\}$ (they share the same indices n_k). By the previous and by the first paragraph of our solution to Problem 6, $\{f_{n_k}\} \to f$ in measure, so by Theorem 4, there is a further subsequence $\{f_{n_i}\}$ that converges pointwise a.e. to f. Applying Theorem 4 to $\{g_{n_i}\}$, there is a further subsequence $\{g_{n_j}\}$ of $\{g_{n_i}\}$ that converges pointwise a.e. to g. By the first paragraph of our solution to Problem 6, $\{f_{n_j}\} \to f$ pointwise a.e. By Theorem 18 (linearity) of Section 1.5, $\{\alpha f_{n_j} + \beta g_{n_j}\} \to \alpha f + \beta g$ pointwise a.e. By Problem 11, $\{\alpha f_n + \beta g_n\} \to \alpha f + \beta g$ in measure.

11. Assume first that $\{f_n\} \to f$ in measure. By the first paragraph of our solution to Problem 6, every subsequence of $\{f_n\}$ converges in measure to f. By Theorem 4, each of those subsequences has a further subsequence that converges pointwise a.e. on E to f.

Conversely, assume that every subsequence of $\{f_n\}$ has a further subsequence that converges pointwise a.e. on E to f. Suppose, to get a contradiction, that $\{f_n\}$ did not converge in measure to f. Then there are a positive ϵ and positive η such that for any N, there is an $n \geq N$ for which $m\{x \in E \,|\, |f_n(x) - f(x)| > \eta\} \geq \epsilon$. That allows us to extract a subsequence $\{f_{n_k}\}$ such that $m\{x \in E \,|\, |f_{n_k}(x) - f(x)| > \eta\} \geq \epsilon$ for all k. By assumption, $\{f_{n_k}\}$ has a further subsequence that converges pointwise a.e. on E to f. By Proposition 3, that further subsequence converges in measure to f, which contradicts our previous consequence that $m\{x \in E \,|\, |f_{n_k}(x) - f(x)| > \eta\} \geq \epsilon$ for all k. Thus, $\{f_n\} \to f$ in measure on E.

12. Refer to the definition on page 22 and subsequent Theorem 17. Take a positive number ϵ. Choose N large enough so that $1/2^{N-1} \leq \epsilon$. Then if $n > m \geq N$,

$$|a_m - a_n| = \left| \sum_{j=m}^{n-1} (a_j - a_{j+1}) \right| \leq \sum_{j=m}^{n-1} \frac{1}{2^j} = \frac{1}{2^{m-1}} - \frac{1}{2^{n-1}} < \frac{1}{2^{m-1}} \leq \frac{1}{2^{N-1}} \leq \epsilon$$

where we used the triangle inequality in the second step. If $n \leq m$, we conclude again that $|a_m - a_n| < \epsilon$. Thus, $\{a_j\}$ is Cauchy and therefore converges to a real number.

13. Take the hint. Then $\sum_{j=1}^{\infty} m(E_j) < \infty$. The Borel-Cantelli lemma (Section 2.5) tells us that for almost all x in E, there is an index $J(x)$ such that $x \notin E_j$ if $j \geq J(x)$, that is,

$$\left|f_{n_{j+1}}(x) - f_{n_j}(x)\right| \leq 1/2^j \text{ if } j \geq J(x).$$

Now if in our solution to Problem 12 we choose N so that it is also greater than or equal to $J(x)$, then the problem's claim remains valid if we replace "for all j" with "if $j \geq J(x)$." Then we can say that subsequence $\{f_{n_j}\}$ converges pointwise a.e. on E to some measurable function f, a result we use in the following where we replace $f(x)$ with $\lim_{i\to\infty} f_{n_i}(x)$. Take a positive number ϵ. Choose J' large enough so that $1/2^{J'-1} < \epsilon$. If $j \geq J'$, then

$$\begin{aligned} m\{x \in E \,|\, |f_{n_j}(x) - f(x)| > \epsilon\} &= m\left\{x \in E \,\middle|\, \left|f_{n_j}(x) - \lim_{i\to\infty} f_{n_i}(x)\right| > \epsilon\right\} \\ &\leq \sum_{i=j}^{\infty} m\left\{x \in E \,\middle|\, \left|f_{n_i}(x) - f_{n_{i+1}}(x)\right| > \frac{1}{2^i}\right\} \\ &< \sum_{i=j}^{\infty} \frac{1}{2^i} = \frac{1}{2^{j-1}} < \epsilon. \end{aligned}$$

Hence, $\lim_{j\to\infty} m\{x \in E \,|\, |f_{n_j}(x) - f(x)| > \epsilon\} = 0$. Finally, for each positive η,

$$\begin{aligned} &\lim_{n\to\infty} m\{x \in E \,|\, |f_n(x) - f(x)| > \eta\} \\ &\quad\leq \lim_{n\to\infty} m\{x \in E \,|\, |f_n(x) - f_{n_j}(x)| > \eta/2\} + m\{x \in E \,|\, |f_{n_j}(x) - f(x)| > \eta/2\} \\ &\quad= 0 + 0 \end{aligned}$$

using reasoning similar to that of our solution to Problem 7 along with the following: The first step holds for all f_{n_j}, so we can take $j \to \infty$ which allows us to set the first term to zero because $\{f_n\}$ is Cauchy in measure and the second to zero by the previous result. Thus, $\{f_n\} \to f$ in measure on E.

14. Assume first that $\lim_{n\to\infty} \rho(f_n, f) = 0$. Then for each positive η, by definition of the integral of a simple function, Theorems 10 and 11 of Section 4.3, and Problem 6 of Section 9.1,

$$\begin{aligned} \lim_{n\to\infty} m\{x \in E \,|\, |f_n(x) - f(x)| > \eta\} &= \lim_{n\to\infty} \int_E \chi_{\{x\in E \,|\, |f_n(x)-f(x)|>\eta\}} \\ &\leq \lim_{n\to\infty} \frac{\eta+1}{\eta} \int_E \frac{|f_n - f|}{1 + |f_n - f|} \\ &= \frac{\eta+1}{\eta} \lim_{n\to\infty} \rho(f_n, f) = 0. \end{aligned}$$

Thus, $\{f_n\} \to f$ in measure. Conversely, assume that $\{f_n\} \to f$ in measure. Then

$$\begin{aligned} &\lim_{n\to\infty} \int_E \frac{|f_n - f|}{1 + |f_n - f|} \\ &\quad= \lim_{n\to\infty} \int_{\{x\in E \,|\, |f_n(x)-f(x)|>\eta\}} \frac{|f_n - f|}{1 + |f_n - f|} + \lim_{n\to\infty} \int_{\{x\in E \,|\, |f_n(x)-f(x)|\leq\eta\}} \frac{|f_n - f|}{1 + |f_n - f|} \\ &\quad\leq \lim_{n\to\infty} \int_{\{x\in E \,|\, |f_n(x)-f(x)|>\eta\}} 1 + \lim_{n\to\infty} \int_{\{x\in E \,|\, |f_n(x)-f(x)|\leq\eta\}} \frac{\eta}{1+\eta} \\ &\quad\leq \lim_{n\to\infty} m\{x \in E \,|\, |f_n(x) - f(x)| > \eta\} + \int_E \frac{\eta}{1+\eta} = 0 + \frac{\eta}{1+\eta} m(E). \end{aligned}$$

Because that holds for each η, it holds also if $\eta = 0$. Thus, $\lim_{n\to\infty} \rho(f_n, f) = 0$.

Following is another proof of the converse. We first extend Theorem 6 of Section 3.1 by proving that if the f of that theorem is also positive, then $1/f$ is measurable on E. Observe that for a number c,

$$\{x \in E \,|\, 1/f(x) > c\} = \{x \in E \,|\, f(x) < 1/c\} \text{ if } c \geq 0$$

and

$$\{x \in E \,|\, 1/f(x) > c\} = \{x \in E \,|\, f(x) > 1/c\} \text{ if } c < 0.$$

Thus, measurability of f implies measurability of $1/f$. Applying that theorem and our extension of it to this problem, we have that

$$\left\{\frac{|f_n - f|}{1 + |f_n - f|}\right\}$$

is a sequence of measurable functions on E (a fact we used implicitly in our first solution).

Now since $\{f_n\} \to f$ in measure,

$$\left\{\frac{|f_n - f|}{1 + |f_n - f|}\right\} \to 0 \text{ in measure}$$

because by monotonicity of measure and convergence,

$$\begin{aligned}\lim_{n\to\infty} m\left\{x \in E \,\middle|\, \frac{|f_n(x) - f(x)|}{1 + |f_n(x) - f(x)|} > \eta\right\} &\leq \lim_{n\to\infty} m\{x \in E \,|\, |f_n(x) - f(x)| > \eta\} \\ &= 0.\end{aligned}$$

Moreover, the constant function 1 dominates

$$\frac{|f_n - f|}{1 + |f_n - f|}$$

and is integrable over E because $\int_E 1 = m(E) < \infty$. Thus, by the Lebesgue dominated convergence theorem and Problem 8,

$$\begin{aligned}\lim_{n\to\infty} \rho(f_n, f) &= \lim_{n\to\infty} \int_E \frac{|f_n - f|}{1 + |f_n - f|} \\ &= \int_E 0 \\ &= 0.\end{aligned}$$

5.3 Characterizations of Riemann and Lebesgue Integrability

15. By Theorem 8, the set E_{0f} of points in $[a, b]$ at which f fails to be continuous has measure zero. We have the analogous statement for g. Product fg is bounded on $[a, b]$ and continuous on $[a, b] \sim (E_{0f} \cup E_{0g})$ by Problem 49(i) of Section 1.6. The set of points in $[a, b]$ at which fg fails to be continuous is a subset of $E_{0f} \cup E_{0g}$. But $m(E_{0f} \cup E_{0g}) \leq m(E_{0f}) + m(E_{0g}) = 0 + 0$. Thus, fg is Riemann integrable over $[a, b]$.

16. By Theorem 8, f is Riemann integrable over $[a, b]$. By Theorem 3 of Section 4.2, f is Lebesgue integrable over $[a, b]$. Thus, f is measurable by definition of a Lebesgue integrable function.

Next, let E_0 be the set of discontinuities so that $m(E_0) = 0$. By Section 2.3's Proposition 4 and Theorem 9, sets $[a, b]$, E_0, and $[a, b] \sim E_0$ are measurable. Refer to pages 54–56 for the remainder of this solution. Function f restricted to E_0 is measurable because E_0 is measurable and for each real number c, set $\{x \in E_0 \mid f(x) > 0\}$ is measurable since it has measure zero, being a subset of E_0. Function f restricted to $[a, b] \sim E_0$ is measurable by Proposition 3. Thus, f is measurable on $[a, b]$ by Proposition 5(ii).

17. For example, define f by

$$f(x) = \left\lfloor \frac{1}{x} \right\rfloor \sin\left(2\pi \left\lfloor \frac{1}{x} \right\rfloor \left\lceil \frac{1}{x} \right\rceil x\right)$$

where $\lfloor\cdot\rfloor$ denotes the floor function and $\lceil\cdot\rceil$ the ceiling function. Because sine is a continuous function, to verify that f is continuous on $(0, 1]$, it suffices to check points where $\lfloor\cdot\rfloor$ and $\lceil\cdot\rceil$ are discontinuous, namely where $1/x$ is a natural number k. Whenever a sequence $\{x_n\}$ in $(0, 1]$ converges to $1/k$, its image sequence $\{f(x_n)\}$ converges to 0. But we have that $f(1/k) = k \sin(2\pi k) = 0$. Thus, by Proposition 21 of Section 1.6, f is continuous on $(0, 1]$.

Moreover, $\left\{\int_{[1/n,1]} f\right\}$ converges (to 0) because $\int_{[1/n,1]} f = 0$ for each n as shown here:

$$\begin{aligned}
\int_{1/n}^{1} \left\lfloor \frac{1}{x} \right\rfloor \sin\left(2\pi \left\lfloor \frac{1}{x} \right\rfloor \left\lceil \frac{1}{x} \right\rceil x\right) dx &= \sum_{k=2}^{n} \int_{1/k}^{1/(k-1)} (k-1) \sin(2\pi(k-1)kx)\, dx \\
&= \sum_{k=2}^{n} \int_{k-1}^{k} \frac{\sin(2\pi u)}{k}\, du \qquad u = (k-1)kx \\
&= \sum_{k=2}^{n} 0 \\
&= 0.
\end{aligned}$$

But f is not Lebesgue integrable over $[0, 1]$ because

$$\begin{aligned}
\int_0^1 |f(x)|\, dx &= \lim_{n\to\infty} \int_{1/n}^{1} |f(x)|\, dx \\
&= \lim_{n\to\infty} \sum_{k=2}^{n} \int_{k-1}^{k} \left| \frac{\sin(2\pi u)}{k} \right| du \\
&= \frac{2}{\pi} \sum_{k=2}^{\infty} \frac{1}{k} \\
&= \infty.
\end{aligned}$$

On the other hand, that cannot happen if f is nonnegative because in that case,

$$\begin{aligned}
\int_{[0,1]} |f| &= \int_{[0,1]} f \\
&= \lim_{n\to\infty} \int_{[1/n,1]} f \\
&< \infty
\end{aligned}$$

since $\left\{\int_{[1/n,1]} f\right\}$ converges.

Chapter 6

Differentiation and Integration

6.1 Continuity of Monotone Functions

In the ninth sentence on page 109, replace q_n with q_k. In the subsequent sentence, replace $1/2^n$ with $1/2^{n-1}$.

1. If C is finite, we can use the same function described for the countably-infinite case below, but here is a simpler solution: Form a partition $\{a = x_0, x_1, x_2, \ldots, x_{m-1}, x_m = b\}$ of $[a, b]$ using the points in C (see the bottom of page 69). The following step function ψ on $[a, b]$ will do:

$$\psi(x) = \begin{cases} 0 & x = a \text{ and } a \in C, \\ 1 & x = a \text{ and } a \notin C, \\ i & x \in (a, b);\ x_{i-1} \leq x < x_i, \\ m & x = b \text{ and } b \notin C, \\ m+1 & x = b \text{ and } b \in C. \end{cases}$$

   If C is countably infinite, use the same function f in the proof of Proposition 2 along with extensions $f(a) = 0$ and $f(b) = 1$. Function f satisfies the requirements on (a, b); it remains to check f at endpoints a and b. Function f is increasing on $[a, b]$ because by construction, $0 \leq f(x) \leq \sum_{n=1}^{\infty} 1/2^n = 1$ for x in (a, b). If $a \in C$, that is, a is some q_k, then $f(x) - f(a) = f(x) \geq 1/2^k$ if $x > a$. Therefore, f fails to be continuous at a. If $a \notin C$, there is an interval $[a, c)$ for which q_k does not belong to $[a, c)$ where $1 \leq k \leq n$ so that $f(x) - f(a) < 1/2^{n-1}$ for all x in $[a, c)$. Therefore, f is continuous at a. We have analogous results at b.

2. This is true by our solution to Problem 1 with C equal to $\mathbf{Q}$, which is countably infinite by Corollary 4(ii) of Section 1.3, and the fact that f happens to be strictly increasing in this case. Indeed, by Theorem 2 of Section 1.2, there is a rational number between each pair of numbers u, v so that by (1), if $u < v$, then $f(v) - f(u) > 0$.

   Here is another, more interesting, solution. Define function g on $[0, 1]$ by $g(0) = 0$ and

$$g(x) = \sum_{\{q \mid 0 \leq p/q \leq x\}} 1/q^3 \text{ where } 0 < x \leq 1, \text{ and } p/q \text{ is a rational number in lowest terms.}$$

   To see that g is properly defined, note that there are at most $q - 1$ rational numbers p/q in $(0, 1)$ so that

$$\begin{aligned} g(x) &\leq \frac{1}{1^3} + \frac{1}{1^3} + \sum_{q=2}^{\infty} \frac{q-1}{q^3} \\ &< 2 + \sum_{q=2}^{\infty} \frac{q}{q^3} \\ &= 2 + \sum_{q=2}^{\infty} \frac{1}{q^2} \\ &< 2 + \int_1^{\infty} \frac{1}{t^2}\, dt \\ &= 2 + 1, \end{aligned}$$

which shows that g is finite.

Function g is strictly increasing on $[0, 1]$ because

$$\text{if } 0 < u < v \leq 1, \text{ then } g(v) - g(u) = \sum_{\{q \,|\, u < p/q \leq v\}} 1/q^3 > 0$$

since there is a p/q between u, v.

Function g fails to be continuous at any rational number p/q in $(0, 1]$ because

$$g(p/q) - g(x) > 1/q^3 \text{ if } x < p/q.$$

At rational number 0, we have that $g(x) - g(0) = g(x) > 1$ if $x > 0$.

Now consider an irrational number x_0 in $[0, 1]$. Take a positive ϵ. Choose n large enough so that $\sum_{q=n}^{\infty} 1/q^2 \leq \epsilon$. There is an interval $(x_0 - \delta, x_0 + \delta)$ that does not contain any p/q for which $q < n$. Then if $|x - x_0| < \delta$, we infer that $|f(x) - f(x_0)| < \sum_{q=n}^{\infty} 1/q^2 \leq \epsilon$ using reasoning similar to that we used above to show that g is finite. Therefore, f is continuous at the irrational numbers.

3. If f is increasing, extend it to an increasing function on all of $\mathbf{R}$ by defining f on $\mathbf{R} \sim E$ by $f(x) = \sup\{f(x') \,|\, x' \in E \text{ and } x' < x\}$. If f is decreasing, use the infimum instead. By Theorem 1, f on $\mathbf{R}$, which is the open interval $(-\infty, \infty)$, is continuous except possibly at a countable number of points. Thus, because $E \subseteq \mathbf{R}$, function f on E is continuous except possibly at a countable number of points in E.

4. There is no such function if E has an isolated point x that is also in C because f is continuous at x. Indeed, by definition of an isolated point (page 20), there is a positive δ such that $(x - \delta, x + \delta) \cap E = \{x\}$. Now referring to the definition of continuity on page 25, if $x' \in E$ and $|x' - x| < \delta$, then $x' = x$ so that $|f(x') - f(x)| = |f(x) - f(x)| = 0 < \epsilon$.

6.2 Differentiability of Monotone Functions: Lebesgue's Theorem

The integrand of the last integral in the displayed inequality after (16) is f.

5. Put $\mathcal{F}$ to $\{[x, x] \,|\, x \in E\}$, which is a collection of closed, bounded, degenerate intervals that covers E in the sense of Vitali, the definition of which we expand here to collections that include degenerate intervals. Let F be the union of a finite (disjoint) subcollection of $\mathcal{F}$. By Propositions 1, 4, and 5 of Chapter 2, F is measurable. The Vitali covering lemma does not extend to this case because for each ϵ less than or equal to $m^*(E)$,

$$\begin{aligned} m^*(E \sim F) &= m^*(E \cap F^C) & \\ &= m^*(E) - m^*(E \cap F) & &\text{definition on page 35} \\ &= m^*(E) - m^*(F) & &F \subseteq E \\ &= m^*(E) & &m^*(F) = 0 \text{ by Proposition 3 of Section 2.2} \\ &\geq \epsilon. \end{aligned}$$

6. Give the name $\mathcal{C}$ to the covering collection of nondegenerate general intervals. Define a collection $\mathcal{F}$ of closed, bounded, nondegenerate intervals by

$$\mathcal{F} = \{[a, b] \,|\, (a, b) \subseteq I, \ I \in \mathcal{C}, \ a \geq \inf E, \ b \leq \sup E, \ a < b\}.$$

Collection $\mathcal{F}$ covers E in the sense of Vitali because for each point x in E and positive ϵ, there is an interval I in $\mathcal{C}$ that contains x and for which $\ell(I) < \epsilon$ so that there is an $[a,b]$ in $\mathcal{F}$ that contains x and for which $\ell([a,b]) = \ell((a,b)) \leq \ell(I) < \epsilon$.

By the Vitali covering lemma, for each positive ϵ, there is a finite disjoint subcollection $\{[a_k,b_k]\}_{k=1}^n$ of $\mathcal{F}$ for which

$$m^*\left(E \sim \bigcup_{k=1}^n [a_k,b_k]\right) < \epsilon.$$

Because $\{[a_k,b_k]\}_{k=1}^n$ is a finite collection of disjoint closed intervals, there is some minimum distance δ between the endpoints of each pair of them. Now each $[a_k,b_k]$ is derived from some I in $\mathcal{C}$ such that $(a_k,b_k) \subseteq I \subseteq (a_k - \delta/2, b_k + \delta/2)$ where we write the second inclusion in case a point covered by $[a_k,b_k]$ is an endpoint of $[a_k,b_k]$. Give index k to the interval I in $\mathcal{C}$ from which $[a_k,b_k]$ is derived. Then $\{I_k\}_{k=1}^n$ is a finite disjoint subcollection of $\mathcal{C}$ for which

$$m^*\left(E \sim \bigcup_{k=1}^n I_k\right) \leq m^*\left(E \sim \bigcup_{k=1}^n [a_k,b_k]\right) < \epsilon$$

because each I_k is at least as long as the corresponding $[a_k,b_k]$.

7. A counterexample is Weierstrass's function (see the remark on page 113), a function that is continuous on all open intervals and that fails to be differentiable at any point. By the contrapositive of Lebesgue's theorem, there is no open interval on which f is monotone.

8. Take a point x in J. As in the proof of the Vitali covering lemma, because x belongs to J and $I \cap J \neq \emptyset$, the distance from x to the midpoint of I is at most $\ell(J) + \ell(I)/2$ and hence, since $\ell(J) < \ell(I)/\gamma \leq 2\ell(I)$, the distance from x to the midpoint of I is less than $5\ell(I)/2$. That means that x belongs to $5 * I$. Thus, $J \subseteq 5 * I$.

   The same is not true if $0 < \gamma < 1/2$. For example, let $J = [0,5]$, $I = [5,7]$, and $\gamma = 1/3$. Then $\ell(I) = 2 > 5/3 = \gamma\ell(J)$, and $J \cap I = \{5\} \neq \emptyset$. However, point 0 is in J but not in $5 * I$, which is $[1,11]$. Thus, $J \not\subseteq 5 * I$.

9. Assume first that such a countable collection exists. By the Borel-Cantelli lemma (page 46), almost all x in $\mathbf{R}$ belong to at most finitely many of the I_ks; in other words, there is a set E_0 for which $m(E_0) = 0$ and all x in $\mathbf{R} \sim E_0$ belong to at most finitely many of the I_ks. Because each point in E belongs to infinitely many of the I_ks, we have that $E \subseteq E_0$. Thus, $m(E) = 0$.

   Conversely, assume that $m(E) = 0$. Then by Theorem 11(i) of Section 2.4, Proposition 9 of Section 1.4, and the excision property, for each natural number n, there is a set G_n consisting of a countable number of open intervals such that $E \subseteq G_n$ and $m(G_n) \leq 1/2^n$. By Corollary 6 of Section 1.3, we can arrange all of those open intervals into a single countable collection $\{I_k\}_{k=1}^\infty$. Furthermore,

$$\sum_{k=1}^\infty \ell(I_k) = \sum_{n=1}^\infty m(G_n) \leq \sum_{n=1}^\infty 1/2^n = 1 < \infty.$$

   Now every point x in E is in some I_k, say I_{k_1}. Choose n_1 large enough so that $1/2^{n_1} < \ell(I_{k_1})$. In G_{n_1} there is some interval I_{k_2} such that x is in I_{k_2}. Moreover, $I_{k_2} \neq I_{k_1}$ because $\ell(I_{k_2}) \leq 1/2^{n_1} < \ell(I_{k_1})$. Continuing in that fashion, we obtain infinitely many intervals $I_{k_1}, I_{k_2}, I_{k_3}, \ldots$ to which x belongs.

10. Function f is increasing because if $a < u < v < b$, then

$$\begin{aligned} f(v) - f(u) &= \sum_{k=1}^{\infty} \ell((c_k, d_k) \cap (-\infty, v)) - \sum_{k=1}^{\infty} \ell((c_k, d_k) \cap (-\infty, u)) \\ &= \sum_{k=1}^{\infty} \ell((c_k, d_k) \cap [u, v)) \\ &\geq 0. \end{aligned}$$

Now for all natural number K and $x \in E$, point x is in $(c_1, d_1) \cap (c_2, d_2) \cap \cdots \cap (c_K, d_K)$ by hypothesis. Because that intersection is open (Proposition 8 of Section 1.4), there is a positive r such that interval $[x, x + r)$ is in that intersection (definition of an open set on page 16) and hence in every (c_k, d_k) where $k = 1, \ldots, K$. Consequently,

$$\begin{aligned} [f(x + r) - f(x) &= \sum_{k=1}^{\infty} \ell((c_k, d_k) \cap [x, x + r)) \\ &\geq \sum_{k=1}^{K} \ell((c_k, d_k) \cap [x, x + r)) \\ &= Kr. \end{aligned}$$

Then,

$$\begin{aligned} \overline{D}f(x) = \lim_{h \to 0} \left(\sup_{0 < |t| \leq h} \frac{f(x + t) - f(x)}{t} \right) &\geq \lim_{h \to 0} \left(\sup_{0 < |t| \leq h} \frac{f(x + r) - f(x)}{t} \right) \\ &\geq \lim_{h \to 0} \left(\sup_{0 < |t| \leq h} \frac{Kr}{t} \right) \\ &\geq K. \end{aligned}$$

Because that holds for each K, we have that $\overline{D}f(x) = \infty$ so that f fails to be differentiable at each point in E.

11. For a simple function φ defined on $[\alpha + \gamma, \beta + \gamma]$,

$$\begin{aligned} \int_{\alpha+\gamma}^{\beta+\gamma} \varphi(t)\, dt &= \int_{\alpha+\gamma}^{\beta+\gamma} \left(\sum_{i=1}^{n} a_i \chi_{E_i}(t) \right) dt && \text{notation of Lemma 1 of Section 4.2} \\ &= \sum_{i=1}^{n} a_i m(E_i) && \text{Lemma 1 of Section 4.2} \\ &= \sum_{i=1}^{n} a_i m(E_i - \gamma) && \text{Proposition 2 of Section 2.2} \\ &= \int_{\alpha}^{\beta} \left(\sum_{i=1}^{n} a_i \chi_{E_i - \gamma}(t) \right) dt \\ &= \int_{\alpha}^{\beta} \left(\sum_{i=1}^{n} a_i \chi_{E_i}(t + \gamma) \right) dt \\ &= \int_{\alpha}^{\beta} \varphi(t + \gamma)\, dt. \end{aligned}$$

Next, consider nonnegative integrable function g^+ (consideration of bounded measurable functions is implicit in our use below of the monotone convergence theorem). Function g^+ is measurable by definition of an integrable function. By the simple approximation theorem (Section 3.2), there is an increasing sequence $\{\varphi_n\}$ of simple functions on E that converges pointwise on E to g^+. We can choose the φ_n to be nonnegative. By two applications of the monotone convergence theorem (Section 4.3) and our result above for simple functions,

$$\begin{aligned}\int_{\alpha+\gamma}^{\beta+\gamma} g^+(t)\,dt &= \lim_{n\to\infty}\int_{\alpha+\gamma}^{\beta+\gamma} \varphi_n(t)\,dt\\ &= \lim_{n\to\infty}\int_{\alpha}^{\beta} \varphi_n(t+\gamma)\,dt\\ &= \int_{\alpha}^{\beta} g^+(t+\gamma)\,dt.\end{aligned}$$

We have the analogous result for g^-. It follows from definition of a general integrable function g given by $g^+ - g^-$ and Theorem 17 (linearity), both of Section 4.4, that

$$\int_{\alpha}^{\beta} g(t+\gamma)\,dt = \int_{\alpha+\gamma}^{\beta+\gamma} g(t)\,dt.$$

In the following, we start with the definition of Diff, then use linearity of integration and the result above, then break each integral into two, and lastly cancel and apply the definition of Av:

$$\begin{aligned}\int_u^v \operatorname{Diff}_h f &= \int_u^v \frac{f(t+h)-f(t)}{h}\,dt\\ &= \frac{1}{h}\left(\int_{u+h}^{v+h} f(t)\,dt - \int_u^v f(t)\,dt\right)\\ &= \frac{1}{h}\left(\int_{u+h}^{v} f(t)\,dt + \int_{v}^{v+h} f(t)\,dt - \int_{u}^{u+h} f(t)\,dt - \int_{u+h}^{v} f(t)\,dt\right)\\ &= \operatorname{Av}_h f(v) - \operatorname{Av}_h f(u).\end{aligned}$$

12. For each positive h and at every rational point x, there is a t between $-h$ and 0 and another between 0 and h such that $x+t$ is irrational by Problem 12 of Section 1.2. Then

$$\overline{D}\chi_{\mathbf{Q}}(x) = \lim_{h\to 0}\left(\sup_{0<|t|\le h}\frac{\chi_{\mathbf{Q}}(x+t)-\chi_{\mathbf{Q}}(x)}{t}\right) = \lim_{\substack{h\to 0\\ -h\le t<0}}\frac{0-1}{t} = \infty,$$

$$\underline{D}\chi_{\mathbf{Q}}(x) = \lim_{h\to 0}\left(\inf_{0<|t|\le h}\frac{\chi_{\mathbf{Q}}(x+t)-\chi_{\mathbf{Q}}(x)}{t}\right) = \lim_{\substack{h\to 0\\ 0<t\le h}}\frac{0-1}{t} = -\infty.$$

Analogously, at every irrational point y, there is a t between 0 and h and another between $-h$ and 0 such that $y+t$ is rational (Theorem 2 of Section 1.2). Then

$$\overline{D}\chi_{\mathbf{Q}}(y) = \lim_{h\to 0}\left(\sup_{0<|t|\le h}\frac{\chi_{\mathbf{Q}}(y+t)-\chi_{\mathbf{Q}}(y)}{t}\right) = \lim_{\substack{h\to 0\\ 0<t\le h}}\frac{1-0}{t} = \infty,$$

$$\underline{D}\chi_{\mathbf{Q}}(y) = \lim_{h\to 0}\left(\inf_{0<|t|\le h}\frac{\chi_{\mathbf{Q}}(y+t)-\chi_{\mathbf{Q}}(y)}{t}\right) = \lim_{\substack{h\to 0\\ -h\le t<0}}\frac{1-0}{t} = -\infty.$$

In either case, $\overline{D}\chi_{\mathbf{Q}} = \infty$, and $\underline{D}\chi_{\mathbf{Q}} = -\infty$.

13. The disjoint countable subcollection $\{I_k\}_{k=1}^{\infty}$ that the authors construct in their proof of the Vitali covering lemma will do for this problem. Indeed, we deduce from (5) and the last paragraph of the proof that there is an index N for which if $n \geq N$, then

$$m^*\left(E \sim \bigcup_{k=1}^{n} I_k\right) \leq \sum_{k=n+1}^{\infty} 5\ell(I_k) < \epsilon.$$

Thus, by definition of *convergence* (page 21),

$$m^*\left(E \sim \bigcup_{k=1}^{\infty} I_k\right) = \lim_{n\to\infty} m^*\left(E \sim \bigcup_{k=1}^{n} I_k\right) \leq \lim_{n\to\infty} \sum_{k=n+1}^{\infty} 5\ell(I_k) = 0.$$

I am including the following alternative solution although I find it less than satisfying. The proof is similar to that of Theorem 11(ii) of Section 2.4. By the Vitali covering lemma and its notation, for each natural number p, there is a finite disjoint subcollection $\{I_{pi}\}_{i=1}^{n_p}$ of $\mathcal{F}$ for which

$$m^*\left(E \sim \bigcup_{i=1}^{n_p} I_{pi}\right) < \frac{1}{p}.$$

Combine all those I_{pi} into a single countable collection $\{I'_j\}_{j=1}^{\infty}$. That collection is countable because it is the union of a countable (index p) collection of countable sets (each with n_p intervals) (Corollary 6 of Section 1.3). Moreover, because for each p, we have that $E \sim \bigcup_{j=1}^{\infty} I'_j \subseteq E \sim \bigcup_{i=1}^{n_p} I_{pi}$, by monotonicity of outer measure,

$$m^*\left(E \sim \bigcup_{j=1}^{\infty} I'_j\right) \leq m^*\left(E \sim \bigcup_{i=1}^{n_p} I_{pi}\right) < \frac{1}{p}.$$

Now if two intervals in $\{I'_j\}_{j=1}^{\infty}$ are not disjoint, discard the one associated with the smaller p index to form a countable *disjoint* collection $\{I_k\}_{k=1}^{\infty}$ of intervals in $\mathcal{F}$. The above inequality still holds with the I'_js replaced by the I_ks because as we step through the interval-discarding procedure, the collection of I_ks contains collection $\{I_{pi}\}_{i=1}^{n_p}$ at step p. Therefore,

$$m^*\left(E \sim \bigcup_{k=1}^{\infty} I_k\right) = 0.$$

14. Give the name E to the union. Let $\mathcal{F}$ be the collection of closed, bounded, nondegenerate intervals contained in E. Collection $\mathcal{F}$ covers E in the sense of Vitali because $\mathcal{F}$ equals E and contains arbitrarily small intervals. By the Vitali covering lemma, for each positive ϵ, there is a finite disjoint subcollection $\{I_k\}_{k=1}^{n}$ of $\mathcal{F}$ for which

$$m^*\left(E \sim \bigcup_{k=1}^{n} I_k\right) < \epsilon.$$

Thus, E is measurable by Theorem 11(iii) of Section 2.4 because $\bigcup_{k=1}^{n} I_k$ is a closed set contained in E.

Another way to end the proof is to use the Exercise 13 variation of the Vitali covering lemma and then Theorem 11(iv) of Section 2.4.

15. Note that $(f(0+t)-f(0))/t = \sin(1/t)$. Furthermore, $\sin(1/t)$ takes all values in its range $[-1,1]$ as $|t|$ runs through the points in interval $(0,h]$ where $h>0$ because there is a natural number n such that $0 < 1/(2\pi(n+1)) < 1/(2\pi n) \le h$. Thus,

$$\overline{D}f(0) = \lim_{h\to 0}\left(\sup_{0<|t|\le h} \sin\frac{1}{t}\right) = 1,$$

$$\underline{D}f(0) = \lim_{h\to 0}\left(\inf_{0<|t|\le h} \sin\frac{1}{t}\right) = -1.$$

16. Use linearity of integration to express f as

$$\begin{aligned} f(x) &= \int_a^x (|g| - |g| + g) \\ &= \int_a^x |g| - \int_a^x (|g| - g) \\ &= f_1(x) - f_2(x) \end{aligned}$$

where we abbreviated $\int_a^x |g|$ to $f_1(x)$ and $\int_a^x (|g|-g)$ to $f_2(x)$. Because $|g|$ and $|g|-g$ are nonnegative, f_1, f_2 are increasing on (a,b). By Lebesgue's theorem, f_1, f_2 are differentiable a.e. on (a,b) so that by definition, $\overline{D}f_1$ equals $\underline{D}f_1$ and is finite a.e. on (a,b) and likewise for f_2. Now Problem 20 tells us that $\underline{D}f_1 - \underline{D}f_2 \le \underline{D}(f_1 - f_2) \le \overline{D}(f_1 - f_2) \le \underline{D}f_1 - \underline{D}f_2$, from which it follows that $\underline{D}(f_1 - f_2)$ equals $\overline{D}(f_1 - f_2)$ and is finite a.e. on (a,b). Thus, $f_1 - f_2$, which is f, is differentiable a.e. on (a,b).

17. We flesh out the remark in which (18) is located. The integral in (15) is independent of the values taken by f at the endpoints by Proposition 15 of Section 4.4 with E_0 equal to $\{a,b\}$ and therefore of measure zero by the example on page 31. We can therefore assign arbitrary values to $f(a), f(b)$ as long as f is increasing on $[a,b]$. Putting $f(a)$ to $\inf_{x\in(a,b)} f(x)$ and $f(b)$ to $\sup_{x\in(a,b)} f(x)$ will do because for x in (a,b), we have that $a<x<b$, and

$$f(a) = \inf_{x\in(a,b)} f(x) \le f(x) \le \sup_{x\in(a,b)} f(x) = f(b).$$

Thus,

$$\int_a^b f' \le f(b) - f(a) = \sup_{x\in(a,b)} f(x) - \inf_{x\in(a,b)} f(x).$$

18. Because c is a local minimizer for f, there is a positive h such that $f(c+t) - f(c) \ge 0$ for all t for which $0 < |t| \le h$. Then

$$\underline{D}f(c) = \lim_{h\to 0}\left(\inf_{0<|t|\le h} \frac{f(c+t)-f(c)}{t}\right) \le 0$$

because $(f(c+t)-f(c))/t \le 0$ when t takes negative values. Analogously, $\overline{D}f(c) \ge 0$. Thus, $\underline{D}f(c) \le 0 \le \overline{D}f(c)$.

19. Take the hint. Because $\underline{D}g > 0$ on (a,b), there is an h (which depends on x) such that $(g(x+t)-g(x))/t > 0$ for all t for which $0<|t|\le h$. In other words, g is increasing on $[x-h, x+h]$. Because that holds for each x in $[a,b]$, evidently g is increasing on all of $[a,b]$.

Now for each positive ϵ, we have by Problem 20 that $\underline{D}(f(x)+\epsilon x) \ge \underline{D}f(x) + \underline{D}(\epsilon x) \ge 0+\epsilon$, which shows that $f(x)+\epsilon x$ is increasing on $[a,b]$ by the previous. Because that holds for each ϵ, it holds also if $\epsilon = 0$. Thus, f is increasing on $[a,b]$.

20. We have that $\underline{D}(f+g) \leq \overline{D}(f+g)$ by the first sentence on page 111 or by Theorem 18 of Section 1.5, which we also use in the following:

$$\begin{aligned}\underline{D}(f+g)(x) &= \lim_{h\to 0}\left(\inf_{0<|t|\leq h}\frac{f(x+t)+g(x+t)-f(x)-g(x)}{t}\right)\\ &\geq \lim_{h\to 0}\left(\inf_{0<|t|\leq h}\frac{f(x+t)-f(x)}{t}+\inf_{0<|t|\leq h}\frac{g(x+t)-g(x)}{t}\right)\\ &= \underline{D}f(x)+\underline{D}g(x).\end{aligned}$$

Analogously, $\overline{D}(f+g) \leq \overline{D}f+\overline{D}g$. Altogether,

$$\underline{D}f+\underline{D}g \leq \underline{D}(f+g) \leq \overline{D}(f+g) \leq \overline{D}f+\overline{D}g.$$

21. (i) $\overline{D}(f\circ g)(\gamma) = \overline{D}f(g(\gamma)) = \overline{D}f(c)$ so that if $\overline{D}f(c) = \pm\infty$, then $\overline{D}(f\circ g)(\gamma)$ and $\overline{D}f(c)g'(\gamma)$ each equals $\pm\infty$. Now assume that $\overline{D}f(c)$ is finite. Take a positive number ϵ. Let ϵ_1 equal $\min\{1, \epsilon/(g'(\gamma)+1+\overline{D}f(c))\}$. Because g is differentiable at γ, there is a positive h_1 such that

$$\left|\frac{g(\gamma+t)-g(\gamma)}{t}-g'(\gamma)\right| < \epsilon_1 \text{ where } 0<t\leq h_1.$$

By definition of an upper derivative, there is a positive h_2 such that

$$\frac{f(c+t_2)-f(c)}{t_2}-\overline{D}f(c) < \epsilon_1 \quad \text{where } 0<t_2\leq h_2, \text{ or} \tag{6.1}$$

$$f(c+t_2)-f(c)-\overline{D}f(c)t_2 \leq \epsilon_1 t_2 \text{ where } 0\leq t_2\leq h_2. \tag{6.2}$$

Because g is continuous and $g'(\gamma)>0$, there is a positive h_3 such that

$$0\leq g(\gamma+t)-g(\gamma)<h_2 \text{ where } 0\leq t<h_3,$$

in which case we can substitute $g(\gamma+t)-g(\gamma)$ or $g(\gamma+t)-c$ for t_2 in (6.2) to get that

$$f(g(\gamma+t))-f(g(\gamma))-\overline{D}f(c)(g(\gamma+t)-g(\gamma)) \leq \epsilon_1(g(\gamma+t)-g(\gamma)) \text{ where } 0\leq t<h_3.$$

Then if $0<t<\min\{h_1,h_3\}$,

$$\begin{aligned}&\frac{f(g(\gamma+t))-f(g(\gamma))}{t}-\overline{D}f(c)g'(\gamma) \qquad (6.3)\\ &\quad= \frac{f(g(\gamma+t))-f(g(\gamma))-\overline{D}f(c)(g(\gamma+t)-g(\gamma))}{t}\\ &\quad\quad+\frac{\overline{D}f(c)(g(\gamma+t)-g(\gamma))-\overline{D}f(c)g'(\gamma)t}{t}\\ &\quad\leq \epsilon_1\frac{g(\gamma+t)-g(\gamma)}{t}+\overline{D}f(c)\left(\frac{g(\gamma+t)-g(\gamma)}{t}-g'(\gamma)\right)\\ &\quad< \epsilon_1(g'(\gamma)+1)+\overline{D}f(c)\epsilon_1\\ &\quad\leq \epsilon.\end{aligned}$$

Hence, $\overline{D}(f\circ g)(\gamma) \leq \overline{D}f(c)g'(\gamma)$. By repeating the above argument with the differences reversed in (6.1) and (6.3), we get $\overline{D}(f\circ g)(\gamma) \geq \overline{D}f(c)g'(\gamma)$. Thus, we conclude that $\overline{D}(f\circ g)(\gamma) = \overline{D}f(c)g'(\gamma)$.

(ii) Take a positive ϵ. Choose a positive ϵ_1 such that $\left(|\overline{D}f(c)| + \epsilon_1\right)\epsilon_1 \le \epsilon$. Because $g'(\gamma) = 0$, there is a positive h_1 such that

$$\left|\frac{g(\gamma+t) - g(\gamma)}{t}\right| < \epsilon_1 \text{ where } 0 < |t| \le h_1.$$

Because the upper and lower derivatives of f at c are finite, there is a positive h_2 such that

$$\left|\frac{f(c+t_2) - f(c)}{t_2}\right| < |\overline{D}f(c)| + \epsilon_1 \text{ where } 0 < |t_2| \le h_2. \tag{6.4}$$

Because g is continuous, there is a positive h_3 such that

$$|g(\gamma+t) - g(\gamma)| < h_2 \text{ where } |t| < h_3$$

in which case we can substitute into (6.4) to get that

$$\left|\frac{f(g(\gamma+t) - f(g(\gamma))}{g(\gamma+t) - g(\gamma)}\right| < |\overline{D}f(c)| + \epsilon_1 \text{ where } 0 < |t| < h_3.$$

Then if $0 < |t| < \min\{h_1, h_3\}$,

$$\begin{aligned}
\left|\frac{f(g(\gamma+t)) - f(g(\gamma))}{t}\right| &< \left(|\overline{D}f(c)| + \epsilon_1\right)\left|\frac{g(\gamma+t) - g(\gamma)}{t}\right| \\
&< \left(|\overline{D}f(c)| + \epsilon_1\right)\epsilon_1 \\
&\le \epsilon.
\end{aligned}$$

Because that holds for each ϵ, it holds also if $\epsilon = 0$. Thus, $\overline{D}(f \circ g)(\gamma) = 0$.

22. We can use the same reasoning as in our solution to the first part of Problem 24 of Section 3.2 for a monotone function, strict or not, defined on an interval.

    Here is a solution that uses the result for a strictly increasing function to show that an increasing function f on an interval I is measurable. Define a strictly increasing function g by $g(x) = f(x) + x$, that is, $g = f + id$ where id is the identity mapping. Function g is strictly increasing—therefore measurable—because it is the sum of an increasing and strictly increasing function. Mapping id is measurable by Proposition 3 of Section 3.1. By Theorem 6 of Section 3.1, $g - id$, hence f, is measurable. We can apply that theorem even if f is not finite a.e. on I because the purpose of that condition is to preclude application of the theorem to the case where the two functions take infinite values of opposite sign, which is not the case here since id is finite. The solution for a decreasing function is analogous.

23. By hypothesis, there is a nonnegative M such that $-M \le \underline{D}f \le \overline{D}f \le M$ on (a, b). Define functions g, h on $[a, b]$ by

$$\begin{aligned}
g(x) &= f(x) + Mx \\
h(x) &= f(x) - Mx.
\end{aligned}$$

Functions g, h are continuous by Exercise 49(i) of Section 1.6. Function g is increasing on $[a, b]$ by Problem 19 because for each x in (a, b),

$$\begin{aligned}
\underline{D}g(x) &= \underline{D}(f(x) + Mx) \\
&\ge \underline{D}f(x) + \underline{D}(Mx) && \text{Problem 20} \\
&\ge -M + M = 0.
\end{aligned}$$

Analogously, h is decreasing. Then if $a \le u \le v \le b$,

$$\begin{aligned} g(v) - g(u) &\ge 0 \\ f(v) + Mv - f(u) - Mu &\ge 0 \\ f(u) - f(v) &\le M(v-u), \end{aligned} \qquad \begin{aligned} h(v) - h(u) &\le 0 \\ f(v) - Mv - f(u) + Mu &\le 0 \\ f(v) - f(u) &\le M(v-u). \end{aligned}$$

Those two results can be written as the single inequality $|f(v) - f(u)| \le M(v-u)$. That suffices to prove that f is Lipschitz on $[a,b]$ because we assumed that $u \le v$. If we want to match the definition on page 25, where the order of u, v is arbitrary, swapping u, v in the above leads to $|f(v) - f(u)| \le M|v-u|$.

We can solve this problem without the explicit statement that $\underline{D}f$ be bounded because for continuous f, bounded $\overline{D}f$ implies bounded $\underline{D}f$. In fact, they share the same bounds, which we now proceed to prove. The proof uses the lemma that $\overline{D}(f(x) + s_1 x + s_2) = \overline{D}f(x) + s_1$ where s_1, s_2 are real numbers. Indeed,

$$\begin{aligned} \overline{D}(f(x) + s_1 x + s_2) &= \lim_{h \to 0} \left(\sup_{0<|t|\le h} \frac{f(x+t) + s_1 x + s_1 t + s_2 - f(x) - s_1 x - s_2}{t} \right) \\ &= \lim_{h \to 0} \left(\sup_{0<|t|\le h} \left(\frac{f(x+t) - f(x)}{t} + s_1 \right) \right) \\ &= \lim_{h \to 0} \left(\sup_{0<|t|\le h} \frac{f(x+t) - f(x)}{t} + s_1 \right) \\ &= \lim_{h \to 0} \left(\sup_{0<|t|\le h} \frac{f(x+t) - f(x)}{t} \right) + s_1 \\ &= \overline{D}f(x) + s_1, \end{aligned}$$

which proves the lemma. Now for the main result. Define set S by

$$S = \left\{ \frac{f(d) - f(c)}{d - c} \,\middle|\, c, d \in [a,b],\ c < d \right\}.$$

Let s be a value in S so that $s = (f(d) - f(c))/(d-c)$ for some c, d in $[a,b]$ where $c < d$. Define function g on $[c,d]$ by $g(x) = f(x) - f(c) - s(x-c)$. Because g is continuous on $[c,d]$ and $g(d) = g(c)$, function g takes a minimum value m_1 and a maximum value m_2 in $[c,d)$ by the extreme value theorem (Section 1.6). By Problem 18, its analogous result for a maximizer, and the lemma,

$$\begin{aligned} 0 &\le \overline{D}g(m_1) = \overline{D}f(m_1) - s, \\ 0 &\ge \overline{D}g(m_2) = \overline{D}f(m_2) - s \end{aligned}$$

so that

$$\inf \overline{D}f \le \overline{D}f(m_2) \le s \le \overline{D}f(m_1) \le \sup \overline{D}f.$$

Because s is arbitrary, $\inf \overline{D}f \le \inf S$, and $\sup S \le \sup \overline{D}f$. We next prove the reverse inequalities.

Let M_1 be a real number such that $\inf \overline{D}f < M_1$. Then a point x_1 in $[a,b]$ and a number t_1 exist such that

$$\overline{D}f(x_1) < M_1 \text{ where } x_1 + t_1 \in [a,b], \text{ and } \frac{f(x_1 + t_1) - f(x_1)}{t_1} < M_1.$$

That yields that $\inf S < M_1$. Because that argument holds for all M_1 greater than $\inf \overline{D}f$, we have that $\inf S \le \inf \overline{D}f$. Similarly, let M_2 be a real number such that $\sup \overline{D}f > M_2$. Then a point x_2 in $[a,b]$ and a number t_2 exist such that

$$\overline{D}f(x_2) > M_2 \text{ where } x_2 + t_2 \in [a,b], \text{ and } \frac{f(x_2+t_2) - f(x_2)}{t_2} > M_2.$$

That yields that $\sup S > M_2$. Because that argument holds for all M_2 less than $\sup \overline{D}f$, we have that $\sup S \ge \sup \overline{D}f$.

Putting the inequalities together, $\inf \overline{D}f = \inf S$ and $\sup \overline{D}f = \sup S$. Analogously, $\inf \underline{D}f = \inf S$ and $\sup \underline{D}f = \sup S$, which shows that $\overline{D}f, \underline{D}f$ share the same bounds. Thus, if $\overline{D}f$ is bounded, then so is $\underline{D}f$.

24. On $(0,1]$, routine calculation gives that $f'(x) = 2x\sin(1/x^2) - 2(\cos(1/x^2))/x$. If $x = 0$,

$$\overline{D}f(0) = \lim_{h\to 0}\left(\sup_{0<t\le h} \frac{t^2\sin(1/t^2) - 0}{t}\right) = \lim_{h\to 0}\left(\sup_{0<t\le h} t\sin(1/t^2)\right) = 0 = \underline{D}f(0),$$

so $f'(0) = 0$.

Now $2x\sin(1/x^2)$ is continuous and bounded and therefore integrable on $[0,1]$ by Proposition 3 of Section 3.1 and Theorem 4 of Section 4.2, so if $(-2\cos(1/x^2))/x$ is not integrable over $[0,1]$, then neither is f' because, by the triangle inequality,

$$\left|-\frac{2}{x}\cos\frac{1}{x^2}\right| \le \left|2x\sin\frac{1}{x^2}\right| + \left|2x\sin\frac{1}{x^2} - \frac{2}{x}\cos\frac{1}{x^2}\right|.$$

Indeed, note that $|\cos(1/x^2)| \ge 1/\sqrt{2}$ when $1/x^2$ is in interval $[\pi n - \pi/4, \pi n + \pi/4]$ where n is an integer, or equivalently, when x is in $\left[1/\sqrt{\pi n + \pi/4}, 1/\sqrt{\pi n - \pi/4}\right] = I_n$. Then

$$\begin{aligned}
\int_0^1 \left|-\frac{2}{x}\cos\frac{1}{x^2}\right| dx &\ge \sum_{n=1}^{\infty}\int_{I_n}\left|\frac{2}{x}\cos\frac{1}{x^2}\right| dx \\
&\ge \sum_{n=1}^{\infty} \frac{2}{1/\sqrt{\pi n + \pi/4}}\frac{1}{\sqrt{2}}\left(\frac{1}{\sqrt{\pi n - \pi/4}} - \frac{1}{\sqrt{\pi n + \pi/4}}\right) \\
&= \sum_{n=1}^{\infty}\sqrt{2}\left(\sqrt{\frac{4n+1}{4n-1}} - 1\right) \\
&> \sum_{n=1}^{\infty}\frac{\sqrt{2}}{4n} \\
&= \infty,
\end{aligned}$$

the last inequality justified by a Laurent series. Thus, f' is not integrable over $[0,1]$.

6.3 Functions of Bounded Variation: Jordan's Theorem

25. Put f to be the restriction to $[0,1]$ of Weierstrass's function mentioned in the remark on page 113. Then f is continuous on $[0,1]$ but fails to be differentiable at any point. By the contrapositive of Corollary 6, the restriction of f to $[a,b]$ is not of bounded variation.

26. Follow the example on page 117. For a natural number n, consider partition P_n of $[0,1]$ given by $P_n = \{0, y_1, q_2, y_2, q_3, \ldots, y_{n-1}, q_n, y_n, 1\}$ where the y_i are irrational and the q_i are rational. Such a partition exists for each n by Section 1.2's Theorem 2 and Problem 12. Then $V(f, P_n) = 2n$. Thus, f is not of bounded variation on $[0,1]$ because $2n$ diverges.

27. Work through the details of (23) to get that

$$h(x) = \begin{cases} 2\sin x \\ 2 \\ 4 + 2\sin x \end{cases} \qquad g(x) = \begin{cases} \sin x & x \in [0, \pi/2), \\ 2 - \sin x & x \in [\pi/2, 3\pi/2), \\ 4 + \sin x & x \in [3\pi/2, 2\pi]. \end{cases}$$

A simpler approach is to give h all of f when f is increasing, and when f is decreasing, set h to a constant. Then put g to $h - f$. That approach gives that

$$h(x) = \begin{cases} \sin x \\ 1 \\ 2 + \sin x \end{cases} \qquad g(x) = \begin{cases} 0 & x \in [0, \pi/2), \\ 1 - \sin x & x \in [\pi/2, 3\pi/2), \\ 2 & x \in [3\pi/2, 2\pi]. \end{cases}$$

28. By the definition of a step function on page 69, $f(x) = c_i$ if $x_{i-1} < x < x_i$ for partition P given by $P = \{x_0, \ldots, x_k\}$. For this problem, we also need f at the partition points themselves, so define $f(x_i) = C_i$. To calculate the total variation, we wish to find $\sup V(f, Q)$ where Q is a partition of $[a, b]$. The paragraph after the example on page 117 allows us to consider just those Q that contain P, in which case we have induced partitions Q_i of the subintervals $[x_{i-1}, x_i]$, say $Q_i = \{y_{i0}, y_{i1}, \ldots, y_{ik_i}\}$ where $y_{i0} = x_{i-1}$, $y_{ik_i} = x_i$. Then,

$$V(f_{Q_i}, Q_i) = \sum_{j=1}^{k_i} |f(y_{ij}) - f(y_{i,j-1})| = \begin{cases} |C_i - C_{i-1}| & Q_i = \{x_{i-1}, x_i\}, \\ |c_i - C_{i-1}| + |C_i - c_i| & \text{otherwise} \end{cases}$$

where we keep only the first and last summands in the latter case because for terms corresponding to two interior points of Q_i, we have that $x_{i-1} < y_{i,j-1} < y_{ij} < x_i$ so that $|f(y_{ij}) - f(y_{i,j-1})| = |c_i - c_i| = 0$. Moreover, for those two expressions for $V(f_{Q_i}, Q_i)$, the triangle inequality tells us that $|C_i - C_{i-1}| \le |c_i - C_{i-1}| + |C_i - c_i|$ so consequently $TV(f_{Q_i}) = |c_i - C_{i-1}| + |C_i - c_i|$. The additivity formula (20) extends to finite sums. Thus,

$$TV(f) = \sum_{i=1}^{k} (|c_i - C_{i-1}| + |C_i - c_i|).$$

29. (i) The second remark on page 114 holds with $[0,1]$ replaced by $[-1,1]$ because f is an even function and also with sin replaced by cos. Then f' is not integrable over $[-1,1]$. By the contrapositive of Corollary 6, f is not of bounded variation on $[-1,1]$.
Here is another solution. For a natural number n, consider this partition P_n of $[-1,1]$:

$$P_n = \left\{-1, 0, \frac{1}{\sqrt{n\pi}}, \frac{1}{\sqrt{(n-1/2)\pi}}, \frac{1}{\sqrt{(n-1)\pi}}, \ldots, \frac{1}{\sqrt{2\pi}}, \frac{1}{\sqrt{(3/2)\pi}}, \frac{1}{\sqrt{\pi}}, \frac{1}{\sqrt{(1/2)\pi}}, 1\right\}.$$

Function f is zero at partition points 0 and $1/\sqrt{(i-1/2)\pi}$ where $i = 1, \ldots, n$. At partition points $1/\sqrt{i\pi}$, we have that $\left|f(1/\sqrt{i\pi})\right| = 1/(i\pi)$. Then in the defining sum for V (page 116), the first and last terms are $\cos 1$, and the other terms are $1/(i\pi)$ each repeated twice so that $V(f, P_n) = 2\cos 1 + 2(1 + 1/2 + \cdots + 1/n)/\pi$. Thus, f is not of bounded variation on $[-1,1]$ because the harmonic series diverges.

(ii) Note that

$$g'(x) = \begin{cases} 2x\cos(1/x) + \sin(1/x) & x \neq 0, \\ 0 & x = 0 \end{cases}$$

so that $|g'| < 3$ on $[-1, 1]$. Hence, g is Lipschitz. By the second example on page 116, g is of bounded variation on $[-1, 1]$. (Use the definition of a derivative to verify that $g'(0) = 0$.)

30. By Problem 34,

$$TV(\alpha f + \beta g) \leq TV(\alpha f) + TV(\beta g) = |\alpha| TV(f) + |\beta| TV(g) < \infty$$

because $TV(f), TV(g) < \infty$. Thus, $\alpha f + \beta g$ is also of bounded variation.

As in the second part of the proof of Theorem 6 of Section 3.1, it suffices to show that f^2 is of bounded variation. Because f is of bounded variation, f is bounded; otherwise, in the definition of TV, a partition that contains a point at which f is not bounded would make a summand in the definition of V either ∞ or undefined—in either case, f would not be of bounded variation. There is therefore a nonnegative number M such that $|f| \leq M$. Now for a partition $\{x_0, \ldots, x_k\}$, say P, of the interval on which f is defined,

$$\begin{aligned}
V(f^2, P) &= \sum_{i=1}^{k} |f^2(x_i) - f^2(x_{i-1})| && \text{definition of } V \\
&= \sum_{i=1}^{k} |f(x_i) + f(x_{i-1})||f(x_i) - f(x_{i-1})| \\
&\leq 2M \sum_{i=1}^{k} |f(x_i) - f(x_{i-1})| && |f(x_i) + f(x_{i-1})| \leq |f(x_i)| + |f(x_{i-1})| \\
&= 2M\, V(f, P) && \text{definition of } V, \\
TV(f^2) &\leq 2M\, TV(f) && \text{take supremum among the } P \\
&< \infty && TV(f) < \infty.
\end{aligned}$$

31. If $P = P'$, the claim is trivially true. Otherwise, there is a point x in P between two consecutive points x'_{i-1}, x'_i of P'. By the triangle inequality,

$$|f(x'_i) - f(x'_{i-1})| \leq |f(x) - f(x'_{i-1})| + |f(x'_i) - f(x)|.$$

Therefore, if P has just one more partition point than P', we see from the definition of V that $V(f, P') \leq V(f, P)$. Repeating our reasoning for the other points in P but not in P' does not change the conclusion that $V(f, P') \leq V(f, P)$.

32. By definition of *supremum*, there is a partition P'_n of $[a, b]$ such that

$$TV(f) - V(f, P'_n) < 1/n.$$

Let P_n contain each partition point of $P'_1, P'_2, \ldots, P'_n$. Then P_n is a refinement of both P_{n-1} where $n \neq 1$ and P'_n so that by Problem 31, $\{V(f, P_n)\}$ is increasing, and

$$TV(f) - V(f, P_n) \leq TV(f) - V(f, P'_n) < 1/n.$$

Now for a positive ϵ, choose index N large enough so that $1/N \leq \epsilon$. Then if $n \geq N$, we have that $TV(f) - V(f, P_n) < 1/N \leq \epsilon$. Thus, $\{V(f, P_n)\} \to TV(f)$.

33. For each partition P of $[a,b]$ given by $P = \{x_0, \ldots, x_k\}$,

$$
\begin{aligned}
V(f,P) &= \sum_{i=1}^{k} |f(x_i) - f(x_{i-1})| && \text{definition of } V \\
&= \lim_{n\to\infty} \sum_{i=1}^{k} |f_n(x_i) - f_n(x_{i-1})| && \{f_n\} \to f \text{ pointwise} \\
&= \lim_{n\to\infty} V(f_n, P) && \text{definition of } V \\
&\leq \liminf TV(f_n) && V(f_n,P) \leq TV(f_n) \text{ by definition of } TV, \\
TV(f) &\leq \liminf TV(f_n) && \text{take supremum among the } P.
\end{aligned}
$$

In taking the supremum, the right side does not change because it does not depend on P.

34. Let P be a partition of $[a,b]$ given by $P = \{x_0, \ldots, x_k\}$. Then

$$
\begin{aligned}
V(f+g,P) &= \sum_{i=1}^{k} |(f+g)(x_i) - (f+g)(x_{i-1})| && \text{definition of } V \\
&\leq \sum_{i=1}^{k} |f(x_i) - f(x_{i-1})| + \sum_{i=1}^{k} |g(x_i) - g(x_{i-1})| && \text{triangle inequality} \\
&= V(f,P) + V(g,P) && \text{definition of } V, \\
TV(f+g) &\leq TV(f) + TV(g) && \text{see (19), (20).}
\end{aligned}
$$

Next,

$$
\begin{aligned}
V(\alpha f,P) &= \sum_{i=1}^{k} |\alpha f(x_i) - \alpha f(x_{i-1})| && \text{definition of } V \\
&= |\alpha| \sum_{i=1}^{k} |f(x_i) - f(x_{i-1})| \\
&= |\alpha| V(f,P) && \text{definition of } V, \\
TV(\alpha f) &= |\alpha| TV(f) && \text{take supremum among the } P.
\end{aligned}
$$

35. If $\alpha > \beta$, then f' is integrable over $[0,1]$ because for the Riemann integral,

$$
\begin{aligned}
\int_0^1 |f'| &= \int_0^1 \left|\alpha x^{\alpha-1} \sin(1/x^\beta) - \beta x^{\alpha-\beta-1} \cos(1/x^\beta)\right| dx \\
&\leq \int_0^1 \left|\alpha x^{\alpha-1} \sin(1/x^\beta)\right| dx + \int_0^1 \left|\beta x^{\alpha-\beta-1} \cos(1/x^\beta)\right| dx && \text{triangle inequality} \\
&< \int_0^1 \alpha x^{\alpha-1}\, dx + \int_0^1 \beta x^{\alpha-\beta-1}\, dx && |\sin|, |\cos| \leq 1 \\
&= 1 + \beta/(\alpha-\beta) \\
&< \infty && \alpha > \beta.
\end{aligned}
$$

Now f is bounded on $[0,1]$ and continuous on $(0,1]$. (Actually, f is continuous at zero also, but we do not need to prove it for this exercise.) By Theorem 8 of Section 5.3, f is Riemann integrable, so we can use the fundamental theorem of calculus along with certain properties of Lebesgue-integrable functions. For each partition P of $[0,1]$ given by $P = \{x_0, \dots, x_k\}$, we have that

$$
\begin{aligned}
V(f,P) &= \sum_{i=1}^{k} |f(x_i) - f(x_{i-1})| && \text{definition of } V \\
&= \sum_{i=1}^{k} \left| \int_{x_{i-1}}^{x_i} f' \right| && \text{(i) on page 107} \\
&\le \sum_{i=1}^{k} \int_{x_{i-1}}^{x_i} |f'| && \text{Proposition 16 of Section 4.4} \\
&= \int_0^1 |f'| && \text{Corollary 18 of Section 4.4} \\
&< \infty && \text{previous result.}
\end{aligned}
$$

Because that holds for all partitions, $TV(f) < \infty$. Thus, f is of bounded variation on $[0,1]$.

If $\alpha \le \beta$, put α, β each to 2 so that we have the second remark on page 114. Because f' is not integrable over $[0,1]$, f is not of bounded variation on $[0,1]$ by the contrapositive of Corollary 6. We can also solve this problem for general α, β as in our second solution to Problem 29(i).

36. Put x_0 to $\sup\{x \mid f_{[0,x]} \text{ is of bounded variation}\}$. Consider subinterval $[x_0 - \delta_1, x_0 + \delta_2]$ in $[0,1]$ where $\delta_1 \ge 0$, and $\delta_2 > 0$ if $x_0 < 1$, or $\delta_2 = 0$ if $x_0 = 1$. Then

$$
\begin{aligned}
TV\big(f_{[x_0-\delta_1, x_0+\delta_2]}\big) &= TV\big(f_{[0,x_0+\delta_2]}\big) - TV\big(f_{[0,x_0-\delta_1]}\big) && \text{formula (20)} \\
&= \infty - TV\big(f_{[0,x_0-\delta_1]}\big) && f_{[0,x_0+\delta_2]} \text{ is not of bounded var.} \\
&= \infty && f_{[0,x_0-\delta_1]} \text{ is of bounded var.}
\end{aligned}
$$

Thus, $[x_0 - \delta_1, x_0 + \delta_2]$ is a subinterval that contains x_0 and has arbitrarily small length and on which f fails to be of bounded variation.

6.4 Absolutely Continuous Functions

37. (i) Consider the function, continuous on $[0,1]$, in the last remark on page 114. Function f is Lipschitz on $[\epsilon, 1]$ since $|f'(x)| = \big|2x\sin(1/x^2) - 2\cos(1/x^2)/x\big| < 2 + 2/\epsilon$, hence, absolutely continuous on $[\epsilon, 1]$ by Proposition 7. But by Problem 35, f is not of bounded variation on $[0,1]$, thus, not absolutely continuous on $[0,1]$ by the contrapositive of Theorem 8.

    (ii) Because f is continuous at the point 0, for each $\eta > 0$, there is a number ϵ such that $f(\epsilon) - f(0) < \eta/2$. Then for every finite disjoint collection $\{(a_k, b_k)\}_{k=1}^m$ of open intervals in $(0, \epsilon)$,

$$\sum_{k=1}^{m} |f(b_k) - f(a_k)| \le f(\epsilon) - f(0) < \frac{\eta}{2}$$

because f is increasing and continuous on $[0, \epsilon]$.

Choose δ' as a response to the $\eta/2$ challenge regarding the criterion for the absolute continuity of f on $[\epsilon, 1]$. Put δ to $\min\{\epsilon, \delta'\}$. Let $\{(a_k, b_k)\}_{k=1}^n$ be a disjoint collection of open subintervals of $(0,1)$ for which $\sum_{k=1}^{n}(b_k - a_k) < \delta$. Index those intervals so that the first m intervals are to the left of ϵ and the others are to the right. By the displayed inequality above and choice of δ in relation to the absolute continuity of f on $[\epsilon, 1]$,

$$\sum_{k=1}^{n} |f(b_k) - f(a_k)| = \sum_{k=1}^{m} |f(b_k) - f(a_k)| + \sum_{k=m+1}^{n} |f(b_k) - f(a_k)| < \frac{\eta}{2} + \frac{\eta}{2} = \eta.$$

If one of the intervals (a_k, b_k) contains ϵ, we can write $|f(b_k) - f(a_k)|$ as the sum $|f(b_k) - f(\epsilon)| + |f(\epsilon) - f(a_k)|$ because f is increasing. Allocate the first term to the sum above associated with $(0, \epsilon)$ and the second to the sum associated with $(\epsilon, 1)$. The inequality remains valid. Thus, f is absolutely continuous on $[0, 1]$.

(iii) By our solution to Problem 50 of Section 1.6, f is not Lipschitz on $[0, 1]$, but f is uniformly continuous, hence continuous, on $[0, 1]$. Function f is Lipschitz on $[\epsilon, 1]$ because $|f'(x)| = 1/(2\sqrt{x}) \leq 1/(2\epsilon)$. Hence, f is absolutely continuous on $[\epsilon, 1]$. Finally, f is increasing, so it is absolutely continuous on $[0, 1]$ by part (ii).

38. First assume the *if* statement. By the definition on page 13, the family of all countable disjoint collections of open intervals in (a, b) includes the family of all such finite collections. Thus, f is absolutely continuous.

Conversely, assume that f is absolutely continuous. Let δ respond to the $\epsilon/2$ challenge regarding the criterion for the absolute continuity of f. Let $\{(a_k, b_k)\}_{k=1}^{n}$ be a disjoint collection of open subintervals of (a, b) for which $\sum_{k=1}^{\infty}(b_k - a_k) < \delta$. Then we have that $\sum_{k=1}^{n}(b_k - a_k) \leq \sum_{k=1}^{\infty}(b_k - a_k) < \delta$ for each natural number n so that

$$\sum_{k=1}^{n} |f(b_k) - f(a_k)| < \frac{\epsilon}{2}.$$

Take the limit as $n \to \infty$ to obtain that

$$\sum_{k=1}^{\infty} |f(b_k) - f(a_k)| \leq \frac{\epsilon}{2} < \epsilon.$$

39. Assume first that f is absolutely continuous. Let δ respond to the ϵ challenge of Problem 38. If $m(E) < \delta/2$, then by Theorem 11(i) of Section 2.4, Proposition 9 of Section 1.4, and the excision property applied to measurable sets, there is a countable, disjoint collection $\{(a_k, b_k)\}_{k=1}^{\infty}$ of open intervals containing E for which

$$\sum_{k=1}^{\infty} (b_k - a_k) < \frac{\delta}{2} + m(E) < \delta$$

so that

$$\sum_{k=1}^{\infty} |f(b_k) - f(a_k)| < \epsilon.$$

By Problem 55 of Section 1.6, the image of $\{(a_k, b_k)\}_{k=1}^{\infty}$ is a collection $\{(f(a_k), f(b_k))\}_{k=1}^{\infty}$ of intervals. The $(f(a_k), f(b_k))$ are disjoint because f is increasing; moreover, they cover $f(E)$ because $E \subseteq \bigcup_{k=1}^{\infty}(a_k, b_k)$. Thus, by definition of outer measure,

$$m^*(f(E)) \leq m\left(f\left(\bigcup_{k=1}^{\infty}(a_k, b_k)\right)\right) = m\left(\bigcup_{k=1}^{\infty} f((a_k, b_k))\right) = \sum_{k=1}^{\infty} |f(b_k) - f(a_k)| < \epsilon.$$

Conversely, assume the *if* statement. Let the measurable subset E be a finite disjoint collection $\{(a_k, b_k)\}_{k=1}^{n}$ of open intervals in (a, b). Subset E is measurable by Propositions 8 and 5 of Section 2.3. We have that

$$m^*(f(E)) = \sum_{k=1}^{n} |f(b_k) - f(a_k)|, \qquad m(E) = m\left(\bigcup_{k=1}^{n} (a_k, b_k)\right) = \sum_{k=1}^{n} (b_k - a_k).$$

Substitution into the *if* statement yields the definition of an absolutely continuous function.

40. Let E be a zero-measure subset of $[a, b]$. Subset E is measurable by Proposition 4 of Section 2.3. Because $m(E) < \delta$ for all positive δ, we have that $m^*(f(E)) < \epsilon$ for each positive ϵ and therefore for ϵ equal to 0. Thus, f maps sets of measure zero onto sets of measure zero.

    Next, ψ is increasing and continuous on $[0, 1]$ and maps the Cantor set, which has measure zero, onto a set of measure 1. By the contrapositive the previous, ψ is not absolutely continuous. Then neither is φ; otherwise, ψ, which is $\varphi + id$, would be absolutely continuous by Problem 42 and the fact that the identity mapping id is absolutely continuous.

41. We proved (i) in the second paragraph of our solution to Problem 37 of Section 2.7. Let E be a measurable subset of $[a, b]$. Set E therefore has finite measure. By the second part of Problem 18 of Section 2.4, there is an F_σ set F contained in E such that $m(F) = m(E)$. That implies that $E = F \cup E_0$ where E_0 is some set of measure zero. Then $f(E) = f(F) \cup f(E_0)$. Thus, $f(E)$ is measurable by (i), (ii), and Proposition 5 of Section 2.3.

42. We prove more generally that linear combinations of absolutely continuous functions are absolutely continuous. Let functions f, g be absolutely continuous on $[a, b]$, and α, β real numbers. Let δ_f respond to the $\epsilon/|2\alpha|$ challenge regarding the criterion for the absolute continuity of f. Let δ_g respond to the $\epsilon/|2\beta|$ challenge regarding the criterion for the absolute continuity of g. Put δ to $\min\{\delta_f, \delta_g\}$. Let $\{(a_k, b_k)\}_{k=1}^{n}$ be a disjoint collection of open subintervals of (a, b) for which $\sum_{k=1}^{n}(b_k - a_k) < \delta$. By the triangle inequality and choice of δ in relation to the absolute continuity of f, g on $[a, b]$,

$$\begin{aligned}\sum_{k=1}^{n} |(\alpha f + \beta g)(b_k) - (\alpha f + \beta g)(a_k)| &\leq |\alpha| \sum_{k=1}^{n} |f(b_k) - f(a_k)| + |\beta| \sum_{k=1}^{n} |g(b_k) - g(a_k)| \\ &< |\alpha| \frac{\epsilon}{2|\alpha|} + |\beta| \frac{\epsilon}{2|\beta|} \\ &= \epsilon,\end{aligned}$$

    which shows that $\alpha f + \beta g$ is absolutely continuous on $[a, b]$.

    As in the second part of the proof of Theorem 6 of Section 3.1, it suffices to show that f^2 is absolutely continuous. By the extreme value theorem (Section 1.6), there is a nonnegative number M such that $|f| \leq M$. Choose δ now as a response to the $\epsilon/(2M)$ challenge regarding the criterion for the absolute continuity of f. Then because $|f(b_k) + f(a_k)| \leq M + M$,

$$\begin{aligned}\sum_{k=1}^{n} |f^2(b_k) - f^2(a_k)| &= \sum_{k=1}^{n} |f(b_k) + f(a_k)||f(b_k) - f(a_k)| \\ &\leq 2M \sum_{k=1}^{n} |f(b_k) - f(a_k)| \\ &< 2M \frac{\epsilon}{2M} = \epsilon,\end{aligned}$$

which shows that f^2 is absolutely continuous on $[a, b]$. By repeated application of those results, linear combinations and products of a finite number of absolutely continuous functions are absolutely continuous.

43. (i) Use Problem 37(ii). Function f is continuous and increasing on $[0, 1]$ and Lipschitz on $[\epsilon, 1]$ for each ϵ in $(0, 1)$ because $|f'(x)| = 1/(3x^{2/3}) \le 1/(3\epsilon^{2/3})$. Hence, f is absolutely continuous on $[0, 1]$. By symmetry, f is absolutely continuous on $[-1, 0]$. Thus, f is absolutely continuous on $[-1, 1]$.

    Function g is absolutely continuous by our solution to Problem 29(ii) and Proposition 7.

    (ii) Function g, hence $f \circ g$, is zero at partition points $-1, 0$, and $1/(2i-1)$ where $i = 1, \ldots, n$. At the other partition points $1/(2i)$, we have that $|(f \circ g)(1/(2i))| = 1/(2i)^{2/3}$. Then in the defining sum for V (page 116), the nonzero terms are $1/(2i)^{2/3}$ each repeated twice so that

$$\begin{aligned} V(f \circ g, P_n) &= 2(1/2^{2/3} + 1/4^{2/3} + \cdots + 1/(2n)^{2/3}) \\ &= 2^{1/3}(1 + 1/2^{2/3} + \cdots + 1/n^{2/3}). \end{aligned}$$

    (iii) $f \circ g$ fails to be of bounded variation on $[-1, 1]$ because the above p-series diverges since $p \le 1$. Composition $f \circ g$ thus fails to be absolutely continuous by the contrapositive of Theorem 8.

44. Let c be a Lipschitz constant for f on $\mathbf{R}$. Choose δ as a response to the ϵ/c challenge regarding the absolutely continuity of g on $[a, b]$. Let $\{(a_k, b_k)\}_{k=1}^n$ be a disjoint collection of open subintervals of (a, b) for which $\sum_{k=1}^n (b_k - a_k) < \delta$. By definition of a composition of functions, the Lipschitz property of f, and the choice of δ in relation to the absolute continuity of g on $[a, b]$,

$$\begin{aligned} \sum_{k=1}^n |(f \circ g)(b_k) - (f \circ g)(a_k)| &= \sum_{k=1}^n |f(g(b_k)) - f(g(a_k))| \\ &\le c \sum_{k=1}^n |g(b_k) - g(a_k)| \\ &< c(\epsilon/c) \\ &= \epsilon. \end{aligned}$$

    Thus, $f \circ g$ is absolutely continuous on $[a, b]$.

45. Choose η as a response to the ϵ challenge regarding the absolutely continuity of f on $[g(a), g(b)]$. Note that $g(a) < g(b)$ because $a < b$. (If $a = b$, there is nothing to prove.) Choose δ as a response to the η challenge regarding the absolutely continuity of g on $[a, b]$. Let $\{(a_k, b_k)\}_{k=1}^n$ be a disjoint collection of subintervals of (a, b). Then $\{(g(a_k), g(b_k))\}_{k=1}^n$ is a disjoint collection of open subintervals of $(g(a), g(b))$ because g is strictly monotone.

$$\text{If } \sum_{k=1}^n (b_k - a_k) < \delta, \text{ then } \sum_{k=1}^n (g(b_k) - g(a_k)) < \eta, \text{ and then } \sum_{k=1}^n |f(g(b_k)) - f(g(a_k))| < \epsilon.$$

    Thus, $f \circ g$ is absolutely continuous on $[a, b]$.

46. The first inclusion is strict by Problem 37(iii). The second inclusion is strict by Problem 40 and the fact that $\varphi \in \mathcal{F}_{BV}$ by Jordan's theorem and Proposition 20 of Section 2.7.

$\mathcal{F}_{BV}, \mathcal{F}_{AC}$ are closed with respect to the formation of a finite number of linear combinations by Problem 30 and our solution to Problem 42. As for $\mathcal{F}_{Lip}$, let functions f, g be Lipschitz on $[a, b]$ with some Lipschitz constants c, d. Let α, β be real numbers. For all x, x' in $[a, b]$,

$$\begin{aligned} |(\alpha f + \beta g)(x') - (\alpha f + \beta g)(x)| &\le |\alpha||f(x') - f(x)| + |\beta||g(x') - g(x)| && \text{triangle inequality} \\ &\le (|\alpha|c + |\beta|d)|x' - x| && f, g \text{ Lipschitz,} \end{aligned}$$

which shows that $|\alpha|c + |\beta|d$ is a Lipschitz constant for $\alpha f + \beta g$. Thus, $\mathcal{F}_{Lip}$ is closed with respect to the formation of a finite number of linear combinations.

Next, consider a function f in $\mathcal{F}_{BV}$. The total variation function for f is increasing by the sentence after formula (21) and thus also in $\mathcal{F}_{BV}$ by Jordan's theorem. If f is in $\mathcal{F}_{AC}$, then the total variation function for f is also in $\mathcal{F}_{AC}$ by the proof of Theorem 8. Lastly, let f be in $\mathcal{F}_{Lip}$, and c a Lipschitz constant for f. Let $[x, x']$ be a closed interval in $[a, b]$. c is also a Lipschitz constant for f on $[x, x']$. Then by formula (21) and the example spanning pages 116–117, $\left|TV(f_{[a,x']}) - TV(f_{[a,x]})\right| = TV(f_{[x,x']}) \le c(x' - x)$. Thus, the total variation function for f is also in $\mathcal{F}_{Lip}$.

47. Assume first that f is absolutely continuous on $[a, b]$. By the triangle inequality,

$$\left|\sum_{k=1}^{n}(f(b_k) - f(a_k))\right| \le \sum_{k=1}^{n}|f(b_k) - f(a_k)| < \epsilon \text{ if } \sum_{k=1}^{n}(b_k - a_k) < \delta.$$

Conversely, assume the *if* statement. Choose δ as a response to the $\epsilon/2$ challenge. Then if $\sum_{k=1}^{n}(b_k - a_k) < \delta$, we have also that

$$\sum_{\{k \mid f(b_k) \ge f(a_k)\}} (b_k - a_k) < \delta,$$
$$\sum_{\{k \mid f(b_k) < f(a_k)\}} (b_k - a_k) < \delta,$$

which we use to justify the third step of

$$\begin{aligned} \sum_{k=1}^{n}|f(b_k) - f(a_k)| &= \sum_{\{k \mid f(b_k) \ge f(a_k)\}} (f(b_k) - f(a_k)) + \sum_{\{k \mid f(b_k) < f(a_k)\}} |f(b_k) - f(a_k)| \\ &= \left|\sum_{\{k \mid f(b_k) \ge f(a_k)\}} (f(b_k) - f(a_k))\right| + \left|\sum_{\{k \mid f(b_k) < f(a_k)\}} (f(b_k) - f(a_k))\right| \\ &< \epsilon/2 + \epsilon/2 \\ &= \epsilon. \end{aligned}$$

Thus, f is absolutely continuous on $[a, b]$.

6.5 Integrating Derivatives: Differentiating Indefinite Integrals

In the proof of Corollary 12, the "two nonnegative numbers" are actually nonpositive.

48. By Proposition 20 of Section 2.7 and its proof, $\varphi(0) = 0$, $\varphi(1) = 1$, and $\varphi' = 0$ on $\mathcal{O}$ where $\mathcal{O}$ is in $[0, 1]$ and $m(\mathcal{O}) = 1$. Evidently, φ is differentiable a.e. on $[0, 1]$. Then φ is not absolutely continuous by the contradiction from Theorem 10 that $0 = \int_0^1 \varphi' = \varphi(1) - \varphi(0) = 1$.

Problem 40 is more closely linked to the definition of absolute continuity (see lead-up Problems 38 and 39) than Theorem 10, which takes more work to prove. Both reasonings take advantage of the fact that even though the Cantor set **C** has measure zero, **C** makes some significant contribution in the sense that $m(\psi(\mathbf{C})) = 1$, and that Theorem 10 to fails due to **C**'s inclusion in $[a, b]$.

49. Because f is continuous, we have (29) by the first paragraph of the section. Because f is differentiable a.e. on (a, b), we have the second equality of (30) by the third sentence in the proof of Theorem 10. The desired result follows immediately from those two equations.

50. The solution is the proof of Theorem 10 except that the predicates of the second and fourth sentences are given rather than deduced from the absolute continuity of f. Note also that formula (29) holds for f continuous.

51. This is Exercise 50 except that Lebesgue's dominated convergence theorem justifies passage of the limit under the integral sign rather than Theorem 9 and Vitali's convergence theorem.

52. By the proof of Theorem 10, $\{\mathrm{Diff}_{1/n}\, g\}$ is uniformly integrable over $[a, b]$. We claim that $\{f\, \mathrm{Diff}_{1/n}\, g\}$ is also uniformly integrable over $[a, b]$. Indeed, by the extreme value theorem, there is a nonnegative M such that $|f| \leq M$. Referring now to the definition on page 93, choose δ as a response to the ϵ/M challenge regarding the criterion for the uniform integrability of $\{\mathrm{Diff}_{1/n}\, g\}$. If subset A of $[a, b]$ is measurable, and $m(A) < \delta$, then $\int_A |f\, \mathrm{Diff}_{1/n}\, g| \leq M \int_A |\mathrm{Diff}_{1/n}\, g| < M(\epsilon/M) = \epsilon$. That proves the claim. Then by equation (30) and by symmetry,

$$\lim_{n\to\infty} \int_a^b f\, \mathrm{Diff}_{1/n}\, g = \int_a^b fg',$$

$$\lim_{n\to\infty} \int_a^b g\, \mathrm{Diff}_{1/n}\, f = \int_a^b gf'.$$

On the other hand,

$$\begin{aligned}
\int_a^b f\, \mathrm{Diff}_{1/n}\, g &= \int_a^b f(x) \frac{g(x+1/n) - g(x)}{1/n}\, dx \\
&= n \int_a^b f(x)g(x+1/n)\, dx - n \int_{a-1/n}^{b-1/n} f(x+1/n)g(x+1/n)\, dx \\
&= \int_a^b \frac{f(x) - f(x+1/n)}{1/n} g(x+1/n)\, dx \\
&\quad + n \int_{b-1/n}^{b} f(x+1/n)g(x+1/n)\, dx - n \int_{a-1/n}^{a} f(x+1/n)g(x+1/n)\, dx \\
&= -\int_a^b g(x+1/n)\, \mathrm{Diff}_{1/n}\, f(x)\, dx + \mathrm{Av}_{1/n}(fg)(b) - \mathrm{Av}_{1/n}(fg)(a).
\end{aligned}$$

Now take the limit as $n \to \infty$, and make two substitutions using the first two displayed equations above. We can make another substitution using equality

$$\lim_{n\to\infty} \big(\mathrm{Av}_{1/n}(fg)(b) - \mathrm{Av}_{1/n}(fg)(a)\big) = (fg)(b) - (fg)(a) = f(b)g(b) - f(a)g(a)$$

by the first paragraph of the section and Problem 49(i) of Section 1.6. Those substitutions give us the result we seek:

$$\int_a^b fg' = f(b)g(b) - f(a)g(a) - \int_a^b f'g.$$

53. Assume first that f has a Lipschitz constant c on $[a, b]$. Then a.e. on $[a, b]$,

$$
\begin{aligned}
|f'| &= \left|\lim_{n\to\infty} \text{Diff}_{1/n} f\right| && \text{proof of Theorem 10} \\
&= \left|\lim_{n\to\infty} \frac{f(x+1/n) - f(x)}{1/n}\right| && \text{definition of Diff} \\
&\le c && |f(x+1/n) - f(x)| \le c(1/n) \text{ for all } [x, x+1/n] \text{ in } [a, b].
\end{aligned}
$$

Conversely, assume that $|f'| \le c$ a.e. on $[a, b]$. Let $[x_1, x_2]$ be a nondegenerate closed interval in $[a, b]$. Then

$$
\begin{aligned}
\left|\frac{f(x_2) - f(x_1)}{x_2 - x_1}\right| &= \frac{1}{x_2 - x_1}\left|\int_{x_1}^{x_2} f'\right| && \text{Theorem 10} \\
&\le \frac{1}{x_2 - x_1}\int_{x_1}^{x_2} |f'| && \text{Proposition 16 of Section 4.4} \\
&\le \frac{1}{x_2 - x_1}\int_{x_1}^{x_2} c && \text{hypothesis; Section 4.4's Prop. 15 and Theorem 17} \\
&= c.
\end{aligned}
$$

Thus, c is a Lipschitz constant for f on $[a, b]$.

54. (i) Let E be the subset of $[a, b]$ on which f' vanishes. By definition of a singular function, $m(E) = b - a$. On E, sequence $\{\text{Diff}_{1/n} f\} \to 0$ pointwise. By Egoroff's theorem (Section 3.3), there is a closed set F contained in E for which $\{\text{Diff}_{1/n} f\} \to 0$ uniformly on F, and $m(E \sim F) = b - a - m(F) < \delta/2$. We can assume that F does not contain isolated points because E does not contain isolated points (otherwise, for such a point x, there would be a positive r for which $(x - r, x + r) \cap E = \{x\}$, which shows that f' would not vanish on a set with positive measure $2r$). Choose an index N for which $|\text{Diff}_{1/n} f| = \text{Diff}_{1/n} f < \epsilon/(b - a)$ on F if $n \ge N$ where we drop the absolute value symbol because f is increasing. Let $\mathcal{F}$ be the collection of closed intervals in F that have measures $1/n$ where $n \ge N$. Collection $\mathcal{F}$ covers F in the sense of Vitali. By the Vitali covering lemma, there is a finite disjoint subcollection $\{[c_k, d_k]\}_{k=1}^p$ of $\mathcal{F}$ for which

$$m\left(F \sim \bigcup_{k=1}^p [c_k, d_k]\right) = m(F) - m\left(\bigcup_{k=1}^p [c_k, d_k]\right) < \frac{\delta}{2}.$$

Put $\{(a_k, b_k)\}_{k=1}^n$ to be the intervals complementary to those of $\{[c_k, d_k]\}_{k=1}^p$ in (a, b). Then,

$$
\begin{aligned}
\sum_{k=1}^n (b_k - a_k) &= m\left(\bigcup_{k=1}^n (a_k, b_k)\right) \\
&= m\left((a, b) \sim \bigcup_{k=1}^p [c_k, d_k]\right) \\
&= b - a - m\left(\bigcup_{k=1}^p [c_k, d_k]\right) \\
&< b - a - m(F) + \delta/2 \\
&< \delta/2 + \delta/2 \\
&= \delta,
\end{aligned}
$$

and

$$\begin{aligned}\sum_{k=1}^{n}(f(b_k)-f(a_k)) &= f(b)-f(a)-\sum_{k=1}^{p}(f(d_k)-f(c_k)) \quad f \text{ increasing}\\ &> f(b)-f(a)-\frac{\epsilon}{b-a}\sum_{k=1}^{p}(d_k-c_k) \quad d_k-c_k=\frac{1}{n} \text{ for some } n\\ &> f(b)-f(a)-\epsilon.\end{aligned}$$

(ii) Assume, to get a contradiction, that f were not singular. Then in $[a, b]$ there is a subset X such that $m(X) > 0$, and $f' \geq \eta > 0$ on X. Let δ equal $m(X)/2$. Then for each finite disjoint collection $\{(a_k, b_k)\}_{k=1}^{n}$ of open intervals in (a, b), if $\sum_{k=1}^{n}(b_k - a_k) < \delta$, then $(a, b) \sim \bigcup_{k=1}^{n}(a_k, b_k)$, which is $\bigcup_{k=1}^{p}[c_k, d_k]$, contains a set of measure greater than $\delta/2$ on which $f' \geq \eta$. Then if $\epsilon \leq \eta\delta/2$,

$$\begin{aligned}\sum_{k=1}^{n}(f(b_k)-f(a_k)) &= f(b)-f(a)-\sum_{k=1}^{p}(f(d_k)-f(c_k))\\ &< f(b)-f(a)-\eta(\delta/2)\\ &\leq f(b)-f(a)-\epsilon,\end{aligned}$$

which contradicts the property described in part (i). Thus, f is singular.

(iii) Let E_{0n} be the subset of $[a, b]$ on which f_n' does not vanish. Let E denote the set $[a, b] \sim \bigcup_{n=1}^{\infty} E_{0n}$. Then each f_n' vanishes on E. By extension of Problem 20 to countable sums and the fact that $\underline{D}f_n = f_n' = 0 = \overline{D}f_n$ on E, we have that $f' = \sum_{n=1}^{\infty} f_n' = 0$ on E. By definition of a singular function, $m(E_{0n}) = 0$ for each n. By Proposition 3 of Section 2.2, $m\left(\bigcup_{n=1}^{\infty} E_{0n}\right) \leq \sum_{n=1}^{\infty} m(E_{0n}) = 0$. Thus, f is singular because f' vanishes on $[a, b]$ except possibly on a subset of measure zero.

55. (i) Because f is of bounded variation, f, v are differentiable a.e. on (a, b) by Corollary 6. Therefore, $\{\text{Diff}_{1/n} f\}$ is a sequence of functions that converges to f' pointwise a.e. on $[a, b]$. Then a.e. on $[a, b]$,

$$\begin{aligned}|f'(x)| &= \left|\lim_{n\to\infty} \text{Diff}_{1/n} f(x)\right|\\ &= \lim_{n\to\infty}\frac{|f(x+1/n)-f(x)|}{1/n}\\ &= \lim_{n\to\infty}\frac{V\left(f_{[x,x+1/n]}, \{x, x+1/n\}\right)}{1/n} && \text{definition of } V \text{ (page 116; } k=1)\\ &\leq \lim_{n\to\infty}\frac{TV\left(f_{[x,x+1/n]}\right)}{1/n} && \text{definition of } TV\\ &= \lim_{n\to\infty}\frac{v(x+1/n)-v(x)}{1/n} && \text{formula (21) and definition of } v\\ &= v'(x).\end{aligned}$$

Corollary 6 also tells us that v', f', hence $|f'|$, are integrable over $[a, b]$ so that

$$\int_a^b |f'| \leq \int_a^b v' \leq v(b) - v(a) = TV\left(f_{[a,b]}\right) - TV\left(f_{[a,a]}\right) = TV(f) - 0 = TV(f)$$

where in the second step, we used Corollary 4 and the fact that v is increasing by the sentence after (21).

(ii) Assume first that f is absolutely continuous. Then f is of bounded variation by Theorem 8, so $f', |f'|$ are integrable. Let P be a partition of $[a, b]$ given by $P = \{x_0, \ldots, x_k\}$. Then

$$
\begin{aligned}
\int_a^b |f'| &= \sum_{i=1}^{k} \int_{x_{i-1}}^{x_i} |f'| && \text{Corollary 18 of Section 4.4} \\
&\geq \sum_{i=1}^{k} \left| \int_{x_{i-1}}^{x_i} f' \right| && \text{Proposition 16 of Section 4.4} \\
&= \sum_{i=1}^{k} |f(x_i) - f(x_{i-1})| && \text{Theorem 10} \\
&= V(f, P) && \text{definition of } V.
\end{aligned}
$$

Because that holds for all partitions, $\int_a^b |f'| \geq TV(f)$, which, in view of part (i), is thus an equality.

Conversely, assume that $\int_a^b |f'| = TV(f)$. Then the last displayed inequality in our solution to part (i) is an equality. Furthermore, that equality holds with b replaced by each x in $[a, b]$ so that $v(x) = v(a) + \int_a^x v' = \int_a^x v'$, which shows that v is absolutely continuous by Theorem 11. Let δ respond to the ϵ challenge regarding the criterion for the absolute continuity of v. Let $\{(a_k, b_k)\}_{k=1}^n$ be a disjoint collection of open subintervals of (a, b) for which $\sum_{k=1}^{\infty}(b_k - a_k) < \delta$. Then

$$
\begin{aligned}
\sum_{k=1}^{n} |f(b_k) - f(a_k)| &= \sum_{k=1}^{n} V\left(f_{[a_k, b_k]}, \{a_k, b_k\}\right) && \text{definition of } V \text{ (page 116)} \\
&\leq \sum_{k=1}^{n} TV\left(f_{[a_k, b_k]}\right) && \text{definition of } TV \\
&= \sum_{k=1}^{n} (v(b_k) - v(a_k)) && \text{formula (21) and definition of } v \\
&< \epsilon && \text{choice of } \delta.
\end{aligned}
$$

Thus, f is absolutely continuous on $[a, b]$.

(iii) Corollary 4 has f increasing. Then using part (i) and the first example on page 116,

$$\int_a^b |f'| = \int_a^b f' \leq TV(f) = f(b) - f(a),$$

which is the consequent of Corollary 4.

Corollary 12 has f monotone. If f is increasing, the reasoning above with part (ii) gives us equality $\int_a^b f' = f(b) - f(a)$. If f is decreasing, we get the same result:

$$-\int_a^b |f'| = \int_a^b f' = -TV(f) = f(b) - f(a).$$

We therefore have Corollary 12, that is, f is absolutely continuous on $[a, b]$ if and only if $\int_a^b f' = f(b) - f(a)$.

56. (i) By Proposition 9 of Section 1.4, $\mathcal{O}$ is the union $\bigcup_{k=1}^{\infty}(a_k, b_k)$ of a countable, disjoint collection of open intervals. By Problem 55 of Section 1.6, $g(\mathcal{O})$ is a union of a countable

collection of intervals with endpoints $g(a_k), g(b_k)$. Thus, using Theorem 10 in the second step,

$$m(g(\mathcal{O})) = \sum_{k=1}^{\infty}(g(b_k) - g(a_k)) = \sum_{k=1}^{\infty}\int_{a_k}^{b_k} g'(x)\,dx = \int_{\mathcal{O}} g'(x)\,dx.$$

(ii) By definition, E is the intersection $\bigcap_{k=1}^{\infty} G_k$ of a countable collection of open sets. For each natural number n, define E_n by $E_n = \bigcap_{k=1}^{n} G_k$. Because each E_n is an open set (Proposition 8 of Section 1.4), part (i) gives us that

$$m(g(E_n)) = \int_{E_n} g'(x)\,dx.$$

Now take the limit as $n \to \infty$, and apply Theorem 15(ii) of Section 2.5 to the left side. We can apply that theorem because $\{g(E_k)\}_{k=1}^{\infty}$ is descending since $\{E_k\}_{k=1}^{\infty}$ is descending, and because $m(g(E_1)) < \infty$ by the extreme value theorem. To the right side apply Theorem 21(ii) of Section 4.5 where $\bigcap_{n=1}^{\infty} E_n = \bigcap_{n=1}^{\infty}\bigcap_{k=1}^{n} G_k = \bigcap_{k=1}^{\infty} G_k = E$. The result is that

$$m(g(E)) = \int_{E} g'(x)\,dx.$$

(iii) The equalities follow from Problem 40 and Problem 9 of Section 4.2.

(iv) There is a G_δ set E containing A for which $m(E \sim A) = 0$ by Theorem 11(ii) of Section 2.4. Then

$$\begin{aligned}
m(g(E)) &= \int_{E} g'(x)\,dx && \text{part (ii)},\\
m(g(A \cup (E \sim A))) &= \int_{A\cup(E\sim A)} g'(x)\,dx && E = (E \sim A) \cup A,\\
m(g(A)) + m(g(E \sim A)) &= \int_{A} g'(x)\,dx + \int_{E\sim A} g'(x)\,dx && g \text{ inc.; } A, E \sim A \text{ disjoint},\\
m(g(A)) &= \int_{A} g'(x)\,dx && \text{part (iii)}, m(E \sim A) = 0.
\end{aligned}$$

(v) Refer to Problem 8 of Section 3.1. Function g is Borel measurable because its domain $[a, b]$ is a Borel set, and for each r, set $\{x \in [a, b] \mid g(x) > r\}$ is an interval or the empty set since g is increasing and continuous, and both of those sets are Borel. Let G be an open (Borel measurable) subset of $[c, d]$. Then $g^{-1}(G)$ is a Borel set, hence measurable. Characteristic function χ_G is Borel measurable because its domain $[c, d]$ is a Borel set, and for each s, set $\{y \in [c, d] \mid \chi_G(y) > s\}$ is either G or $\emptyset$, both Borel sets. Hence, $\chi_G \circ g$, that is, $\chi_{g^{-1}(G)}$, is measurable. By part (iv) above,

$$\begin{aligned}
m(g(g^{-1}(G))) &= \int_{g^{-1}(G)} g'(x)\,dx,\\
m(G) &= \int_a^b \chi_{g^{-1}(G)}(x)g'(x)\,dx\\
&= \int_a^b \chi_G(g(x))g'(x)\,dx.
\end{aligned}$$

Analogously, for a closed subset F of $[c, d]$,

$$m(F) = \int_a^b \chi_F(g(x))g'(x)\,dx.$$

Let E be a measurable subset of $[c, d]$. By Theorem 11(i),(iii) of Section 2.4, we can choose a descending sequence $\{G_n\}$ of open sets containing E and an ascending sequence $\{F_n\}$ of closed sets contained in E such that $\{m(G_n)\} \to m(E)$, $\{m(F_n)\} \to m(E)$. Define functions h, u_n, v_n by

$$\begin{aligned} h(x) &= \chi_E(g(x))g'(x), \\ u_n(x) &= \chi_{F_n}(g(x))g'(x), \\ v_n(x) &= \chi_{G_n}(g(x))g'(x). \end{aligned}$$

Then for each n, we have that $0 \leq u_1 \leq \cdots \leq u_n \leq h \leq v_n \leq \cdots \leq v_1$ because $g' \geq 0$ since g is increasing. Furthermore, monotone sequences $\{u_n\}, \{v_n\}$ are bounded by g' because χ is either 0 or 1. By Theorem 15 of Section 1.5, $\{u_n\}, \{v_n\}$ converge pointwise. By Theorem 10, g' is integrable. Then by Lebesgue's dominated convergence theorem (Section 4.4),

$$\begin{aligned} \int_a^b \lim_{n\to\infty} (v_n(x) - u_n(x))\, dx &= \lim_{n\to\infty} \int_a^b (v_n(x) - u_n(x))\, dx \\ &= \lim_{n\to\infty} \int_a^b v_n(x)\, dx - \lim_{n\to\infty} \int_a^b u_n(x)\, dx \\ &= \lim_{n\to\infty} \int_a^b \chi_{G_n}(g(x))g'(x)\, dx - \lim_{n\to\infty} \int_a^b \chi_{F_n}(g(x))g'(x)\, dx \\ &= \lim_{n\to\infty} m(G_n) - \lim_{n\to\infty} m(F_n) \\ &= m(E) - m(E) \\ &= 0. \end{aligned}$$

Because $v_n(x) - u_n(x)$ is a nonnegative measurable function, Proposition 9 of Section 4.3 tells us that

$$\lim_{n\to\infty} (v_n - u_n) = 0 \text{ a.e. on } [a, b]$$

so that

$$\lim_{n\to\infty} (h - u_n) \leq \lim_{n\to\infty} (v_n(x) - u_n(x)) = 0 \text{ a.e. on } [a, b],$$

which shows that $\{u_n\}$ is an (increasing) sequence of measurable functions that converges pointwise a.e. on $[a, b]$ to h, and that h is measurable by Proposition 9 of Section 3.2. By the monotone convergence theorem (Section 4.3),

$$\begin{aligned} \lim_{n\to\infty} \int_a^b u_n(x)\, dx &= \int_a^b h(x)\, dx \\ m(E) &= \int_a^b \chi_E(g(x))g'(x)\, dx. \end{aligned}$$

Let φ have canonical representation $\sum_{i=1}^m a_i \chi_{E_i}$. Then

$$\int_c^d \varphi(y)\, dy = \sum_{i=1}^m a_i m(E_i) = \sum_{i=1}^m a_i \int_a^b \chi_{E_i}(g(x))g'(x)\, dx = \int_a^b \varphi(g(x))g'(x)\, dx.$$

(vi) Function f is measurable because f is integrable. Then by Problem 24(i) of Section 4.3, there is an increasing sequence $\{\varphi_n\}$ of nonnegative simple functions on $[c, d]$ that converges pointwise on $[c, d]$ to f. By part (v),

$$\int_c^d \varphi_n(y)\,dy = \int_a^b \varphi_n(g(x))g'(x)\,dx,$$

so by the monotone convergence theorem,

$$\int_c^d f(y)\,dy = \int_a^b f(g(x))g'(x)\,dx.$$

(vii) We have that

$$\begin{aligned}\int_c^d \chi_{g(\mathcal{O})}(y)\,dy &= \int_a^b \chi_{g(\mathcal{O})}(g(x))g'(x)\,dx,\\ m(g(\mathcal{O})) &= \int_a^b \chi_{\mathcal{O}}(x)g'(x)\,dx\\ &= \int_{\mathcal{O}} g'(x)\,dx.\end{aligned}$$

57. Yes, the formula remains true. Nowhere in our derivation of the formula did we require that g increase *strictly.*

58. Refer to Problems 38 and 39 of Section 2.7. Taking the hint, put E to $[0,1] \sim F$, and put f to $\int_0^x \chi_E$. Function f is absolutely continuous by Theorem 11. Function f is strictly increasing because if $u < v$, then $f(v) - f(u) = \int_0^v \chi_E - \int_0^u \chi_E = \int_u^v \chi_E = m(E \cap [u, v])$, which is positive because E is open and dense in $[0, 1]$, that is, between u, v, there is an open interval contained in E. Finally, by Theorem 14, $f' = \chi_E$ on $[0, 1]$, so $f' = 0$ on $[0, 1] \sim E$, which is F and so has positive measure $1 - \alpha$.

59. Yes, it is possible. Function $f(s)$ is integrable over $[c, d]$ by hypothesis. Furthermore, our solution to Problem 56(v) and (vi) shows that $f(g(t))g'(t)$ is integrable over $[a, b]$. Note that we are not assuming the final result of Problem 56(vi)—that would be begging the question with regard to this problem—but just the parts of our solution that show that $f(g(t))g'(t)$ is integrable. Define functions F, G by

$$F(x) = \int_{g(a)}^x f(s)\,ds, \quad G(x) = \int_a^x f(g(t))g'(t)\,dt.$$

Then $F, G, F \circ g$ are absolutely continuous by Theorem 11 and Problem 45 and therefore differentiable a.e. by Theorem 10. By Theorem 14, $F'(x) = f(x)$, and $G'(x) = f(g(x))g'(x)$ a.e. By Problem 21(i), $(F \circ g)'(x) = F'(g(x))g'(x) = f(g(x))g'(x) = G'(x)$ a.e., which shows that $F \circ g - G$ is singular and proves the last clause in this problem's statement. Furthermore, $F \circ g - G$ is absolutely continuous by Problem 42 and so is constant on $[a, b]$ by Problem 60. Thus, $\int_{g(a)}^{g(b)} f(y)\,dy - \int_a^b f(g(x))g'(x)\,dx = (F \circ g)(b) - G(b) = (F \circ g)(a) - G(a) = 0$.

60. For each $x \in [a, b]$,

$f(x) = f(a) + \int_a^x f'$	first two sentences in proof of Theorem 11
$= f(a)$	Lemma 13; $f' = 0$ a.e. on $[a, b]$ because f is singular.

Thus, f is constant (and equal to $f(a)$).

Next, redefine the Lebesgue decomposition of a function f of bounded variation as

$$g(x) = \int_a^x f' + f(a), \text{ and } h(x) = f(x) - \int_a^x f' - f(a) \text{ for all } x \in [a,b]$$

so that g, h have the same properties described in the last paragraph of the section along with the requirement that $h(a) = 0$. Suppose that there were another decomposition $f = g_2 + h_2$ such that g_2 is absolutely continuous, h_2 is singular, and $h_2(a) = 0$. Then $g_2 - g = h - h_2$. The right side $h - h_2$ is singular and vanishes at x equal to a. Moreover, $h - h_2$ is absolutely continuous because the left side $g_2 - g$ is absolutely continuous by Problem 42. Then by the first part of this problem, $h - h_2$ is constant and equal to zero. Hence, $h_2 = h$. It follows now from equality $g_2 - g = h - h_2 = 0$ that $g_2 = g$. Thus, the Lebesgue decomposition of a function of bounded variation is unique in this case.

6.6 Convex Functions

In the first displayed line in the proof of Theorem 18, replace f with φ.

In the last sentence of the proof of Jensen's inequality, replace $[a,b]$ with $[0,1]$.

61. Assume first that φ is convex. Proceed by induction. The claim is (trivially) true if $n = 1$ and also if $n = 2$ by definition, so assume that the claim is true if $n = m$. If $n = m+1$, choose a λ_k less than one, and name it λ_1 by re-indexing if necessary the λ_ks. Then

$$\begin{aligned}
\varphi\left(\sum_{k=1}^{m+1} \lambda_k x_k\right) &= \varphi\left(\lambda_1 x_1 + (1-\lambda_1)\sum_{k=2}^{m+1} \frac{\lambda_k}{1-\lambda_1} x_k\right) \\
&\le \lambda_1 \varphi(x_1) + (1-\lambda_1)\varphi\left(\sum_{k=2}^{m+1} \frac{\lambda_k}{1-\lambda_1} x_k\right) && \varphi \text{ convex; } \sum_{k=2}^{m+1} \frac{\lambda_k}{1-\lambda_1} x_k \in (a,b) \\
&\le \lambda_1 \varphi(x_1) + (1-\lambda_1)\sum_{k=2}^{m+1} \frac{\lambda_k}{1-\lambda_1} \varphi(x_k) && \text{induct. hypoth.; } \sum_{k=2}^{m+1} \frac{\lambda_k}{1-\lambda_1} = 1 \\
&= \sum_{k=1}^{m+1} \lambda_k \varphi(x_k).
\end{aligned}$$

Hence, the claim is true if $n = m+1$ whenever it is true for $n = m$. Thus, by the induction principle, the claim holds if $n = 1, 2, \ldots$.

Conversely, assume the *if* part of the statement. Put n to 2. Then φ is convex by definition.

For f a simple function,

$$\begin{aligned}
\varphi\left(\int_0^1 f(x)\,dx\right) &= \varphi\left(\sum_{k=1}^{n} a_k m(E_k)\right) && \text{definition on page 71 where } \psi = f,\ E = [0,1] \\
&\le \sum_{k=1}^{n} \varphi(a_k) m(E_k) && \text{first part of this problem where } \lambda_k = m(E_k) \\
&= \int_0^1 (\varphi \circ f)(x)\,dx.
\end{aligned}$$

62. If φ is convex, put λ to $1/2$ in inequality (38) to get that $\varphi((x_1+x_2)/2) \le (\varphi(x_1)+\varphi(x_2))/2$ for all x_1, x_2 in (a,b).

Conversely, assume that $\varphi((x_1+x_2)/2) \le (\varphi(x_1)+\varphi(x_2))/2$ for all x_1, x_2 in (a,b). We first show by induction that

$$\varphi\left(\frac{q}{2^n}x_1 + \left(1-\frac{q}{2^n}\right)x_2\right) \le \frac{q}{2^n}\varphi(x_1) + \left(1-\frac{q}{2^n}\right)\varphi(x_2) \text{ for each } q \text{ in } \{0,1,\dots,2^n\}.$$

The claim is true if $n=1$, trivially if $q=0$ or 2, and by assumption if $q=1$. So presume that the claim is true if $n=m$. If $n=m+1$, and with q equal to $2d+r$ where $r=0$ or 1,

$$\begin{aligned}
\varphi\left(\frac{q}{2^{m+1}}x_1 + \left(1-\frac{q}{2^{m+1}}\right)x_2\right) &= \varphi\left(\frac{2d+r}{2^{m+1}}x_1 + \left(1-\frac{2d+r}{2^{m+1}}\right)x_2\right)\\
&= \varphi\left(\frac{1}{2}\left(\frac{d}{2^m}x_1 + \left(1-\frac{d}{2^m}\right)x_2\right) + \frac{1}{2}\left(\frac{d+r}{2^m}x_1 + \left(1-\frac{d+r}{2^m}\right)x_2\right)\right)\\
&\le \frac{1}{2}\varphi\left(\frac{d}{2^m}x_1 + \left(1-\frac{d}{2^m}\right)x_2\right) + \frac{1}{2}\varphi\left(\frac{d+r}{2^m}x_1 + \left(1-\frac{d+r}{2^m}\right)x_2\right) && \text{assump.}\\
&\le \frac{1}{2}\left(\frac{d}{2^m}\varphi(x_1) + \left(1-\frac{d}{2^m}\right)\varphi(x_2)\right) + \frac{1}{2}\left(\frac{d+r}{2^m}\varphi(x_1) + \left(1-\frac{d+r}{2^m}\right)\varphi(x_2)\right) && \text{ind. hyp.}\\
&= \frac{q}{2^{m+1}}\varphi(x_1) + \left(1-\frac{q}{2^{m+1}}\right)\varphi(x_2).
\end{aligned}$$

Hence, the claim is true if $n=m+1$ whenever it is true if $n=m$. Thus, by the induction principle, the claim holds if $n=1,2,\dots$.

By Problem 44 of Section 1.5, there is a sequence $\{q_n\}$ of 0s and 1s such that

$$\lambda = \sum_{n=1}^{\infty}\frac{q_n}{2^n} = \lim_{m\to\infty}\sum_{n=1}^{m}\frac{q_n}{2^n} = \lim_{m\to\infty}\frac{q(m)}{2^m}$$

for some $q(m)$ in $\{0,1,\dots,2^m\}$ where $q(m)$ signifies that the numerator is a function of m. Then starting with the left side of inequality (38),

$$\begin{aligned}
\varphi(\lambda x_1 + (1-\lambda)x_2) &= \lim_{m\to\infty}\varphi\left(\frac{q(m)}{2^m}x_1 + \left(1-\frac{q(m)}{2^m}\right)x_2\right) && \text{Proposition 21 of Sec. 1.6}\\
&\le \lim_{m\to\infty}\left(\frac{q(m)}{2^m}\varphi(x_1) + \left(1-\frac{q(m)}{2^m}\right)\varphi(x_2)\right) && \text{claim; Th. 18 of Sec. 1.5}\\
&= \lambda\varphi(x_1) + (1-\lambda)\varphi(x_2).
\end{aligned}$$

Thus, φ is convex.

Here is another proof of the converse. Suppose, to get a contradiction, that φ were not convex. Then there is a pair x_1, x_2 for which inequality (38) is false. Define function f on $[0,1]$ by

$$f(\lambda) = \varphi(\lambda x_1 + (1-\lambda)x_2) - \lambda\varphi(x_1) - (1-\lambda)\varphi(x_2).$$

By our supposition, there is a λ in $(0,1)$ such that $f(\lambda) > 0$. Denote $\max\{f(\lambda)\}$ by positive number M, which exists by the extreme value theorem (f is continuous by Problem 49(i) of Section 1.6 and the hypothesis that φ is continuous). Let λ_{m} be the smallest λ such that $f(\lambda) = M$. Let positive δ be small enough so that open interval $(\lambda_{\mathrm{m}}-\delta, \lambda_{\mathrm{m}}+\delta)$ is contained in $(0,1)$. Define a pair of points x_1', x_2' by

$$\begin{aligned}
x_1' &= (\lambda_{\mathrm{m}}+\delta)x_1 + (1-\lambda_{\mathrm{m}}-\delta)x_2\\
x_2' &= (\lambda_{\mathrm{m}}-\delta)x_1 + (1-\lambda_{\mathrm{m}}+\delta)x_2.
\end{aligned}$$

Then

$$\begin{aligned} f(\lambda_{\mathrm{m}}+\delta)+f(\lambda_{\mathrm{m}}-\delta) &= \varphi(x_1')+\varphi(x_2')-2(\lambda_{\mathrm{m}}\varphi(x_1)+(1-\lambda_{\mathrm{m}})\varphi(x_2)) \\ &= \varphi(x_1')+\varphi(x_2')+2(f(\lambda_{\mathrm{m}})-\varphi(\lambda_{\mathrm{m}}x_1+(1-\lambda_{\mathrm{m}})x_2)) \\ &= 2f(\lambda_{\mathrm{m}})+\varphi(x_1')+\varphi(x_2')-2\varphi\Big(\frac{x_1'+x_2'}{2}\Big) \\ &\geq 2f(\lambda_{\mathrm{m}}) \qquad \text{hypothesis.} \end{aligned}$$

Hence,

$$f(\lambda_{\mathrm{m}}) \leq \frac{f(\lambda_{\mathrm{m}}+\delta)+f(\lambda_{\mathrm{m}}-\delta)}{2} < \frac{M+M}{2} < M,$$

which contradicts that $f(\lambda_{\mathrm{m}}) = M$. Thus, φ is convex.

63. No. For example, function $x \mapsto -\sqrt{x}$ is convex on $[0,1]$ but not Lipschitz by the last sentence of the first paragraph of Section 1.6 (or our solution to Problem 50 there).

64. If φ'' is nonnegative, then φ is convex by Proposition 15. Conversely, if φ is convex, then its derivative φ' is an increasing function by Theorem 18, that is, φ'' is nonnegative.

65. Because $[0,\infty)$ is not open, we use the definition of Problem 63. We know that φ is continuous on $[0,\infty)$. Consider φ just on $(0,\infty)$ for now. Second derivative $t \mapsto p(p-1)b^2(a+bt)^{p-2}$ is nonnegative, so φ is convex on $(0,\infty)$ by Proposition 15. Now consider φ on $[0,\infty)$. If $x_1, x_2 = 0$, then inequality (38) holds (it is the trivial equality $a^p = a^p$). It remains to show that inequality (38) holds when one of the points, say x_1, is zero and the other is in $(0,\infty)$. Let $\{x_{1n}\}$ be a sequence of numbers in $(0,\infty)$ that converges to 0 (such as $\{1/n\}$). Then

$$\begin{aligned} \varphi(\lambda x_{1n}+(1-\lambda)x_2) &\leq \lambda\varphi(x_{1n})+(1-\lambda)\varphi(x_2) \text{ for all } n && \varphi \text{ convex on } (0,\infty), \\ \lim_{n\to\infty}\varphi(\lambda x_{1n}+(1-\lambda)x_2) &\leq \lim_{n\to\infty}(\lambda\varphi(x_{1n})+(1-\lambda)\varphi(x_2)) && \text{Theorem 18 of Section 1.5,} \\ \varphi(\lambda 0+(1-\lambda)x_2) &\leq \lambda\varphi(0)+(1-\lambda)\varphi(x_2) && \text{Proposition 21 of Sec. 1.6} \end{aligned}$$

so that inequality (38) holds for this final case. Thus, φ is convex on $[0,\infty)$.

66. We claim that Jensen's inequality is always an equality if and only if φ is linear. Assume first that φ is linear. Then φ equals its supporting line, that is, $\varphi = y$ in the proof of Jensen's inequality. We can therefore change the inequalities there to equalities and obtain Jensen's "equality."

    Conversely, assume that Jensen's inequality is always an equality. Suppose, to get a contradiction, that φ were not linear. Then there are x_1, x_2 and λ in $[0,1]$ such that inequality (38) is strict inequality

$$\varphi(\lambda x_1+(1-\lambda)x_2) < \lambda\varphi(x_2)+(1-\lambda)\varphi(x_2).$$

    Put f to $x_1\chi_{[0,\lambda]}+x_2\chi_{(\lambda,1]}$. Both f and $\varphi\circ f$, that is, $\varphi(x_1)\chi_{[0,\lambda]}+\varphi(x_2)\chi_{(\lambda,1]}$ are integrable over $[0,1]$ by Lemma 1 of Section 4.1. Then

$$\begin{aligned} \varphi\Big(\int_0^1 f(x)\,dx\Big) &= \varphi\Big(\int_0^1 \big(x_1\chi_{[0,\lambda]}+x_2\chi_{(\lambda,1]}\big)(x)\,dx\Big) \\ &= \varphi(\lambda x_1+(1-\lambda)x_2) \\ &< \lambda\varphi(x_2)+(1-\lambda)\varphi(x_2) \\ &= \int_0^1 (\varphi\circ f)(x)\,dx, \end{aligned}$$

    which contradicts that Jensen's inequality is always an equality. Thus, φ is linear.

67. **Jensen's inequality** Let φ be a convex function on $(-\infty, \infty)$, f an integrable function over $[a, b]$, and $\varphi \circ f$ also integrable over $[a, b]$. Then

$$\varphi\left(\frac{1}{b-a}\int_a^b f(x)\,dx\right) \le \frac{1}{b-a}\int_a^b (\varphi \circ f)(x)\,dx.$$

Proof Define α by $\alpha = \dfrac{1}{b-a}\displaystyle\int_a^b f(x)\,dx$. Continue as in the text's proof with $[0, 1]$ replaced by $[a, b]$ and the end of the proof replaced by

$$\begin{aligned}\int_a^b \varphi(f(x))\,dx &\ge \int_a^b (m(f(x)-\alpha)+\varphi(\alpha))\,dx \\ &= m\left(\int_a^b f(x)\,dx-(b-a)\alpha\right)+(b-a)\varphi(\alpha) \\ &= (b-a)\varphi(\alpha).\end{aligned}$$

68. Function exp is convex on $(-\infty, \infty)$ by the example. Function f is integrable over $[0, 1]$ by hypothesis. Finally, because $\exp \circ f$ is nonnegative, Jensen's inequality holds by the words after its proof. Then the claim follows directly from Jensen's inequality.

69. Define semisimple function ψ on $[0, 1]$ by

$$\psi = \sum_{k=1}^{\infty} \log \zeta_k \cdot \chi_{E_k} \text{ where } E_k = \left[\sum_{n=1}^{k-1} \alpha_n, \sum_{n=1}^{k} \alpha_n\right),$$

in other words, $\psi(x) = \log \zeta_k$ when $\sum_{n=1}^{k-1} \alpha_n \le x < \sum_{n=1}^{k} \alpha_n$. Then

$$\begin{aligned}\prod_{n=1}^{\infty} \zeta_n^{\alpha_n} &= \exp\left(\sum_{n=1}^{\infty} \alpha_n \log \zeta_n\right) \\ &= \exp\left(\int_0^{\sum_{n=1}^{\infty} \alpha_n} \psi(x)\,dx\right) \\ &\le \int_0^{\sum_{n=1}^{\infty} \alpha_n} \exp(\psi(x))\,dx \qquad \text{Problem 68; sum of } \{\alpha_n\} \text{ is } 1 \\ &= \sum_{n=1}^{\infty} \alpha_n \zeta_n.\end{aligned}$$

70. By the example, $-\log$ is convex on $(0, \infty)$, which suffices for the first condition in the statement of Jensen's inequality because we need consider $\log\left(\int_0^1 g(x)\,dx\right) \ge \int_0^1 \log(g(x))\,dx$ only when each side is defined, and log is defined on $(0, \infty)$ only. If $\int_0^1 g(x)\,dx = \infty$, then the inequality holds and we are done. Otherwise, continue with g integrable (definition on page 84). The third condition to check in the statement of Jensen's inequality is that $-\log \circ g$ be integrable over $[0, 1]$, which it is because g is finite a.e. on $[0, 1]$ by Proposition 13 of Section 4.3, so $-\log \circ g$ is finite a.e. on $[0, 1]$ so that we can use Proposition 15 of Section 4.4 to write $-\int_0^1 \log(g(x))\,dx < \infty$. Then Jensen's inequality tells us that

$$-\log\left(\int_0^1 g(x)\,dx\right) \le -\int_0^1 \log(g(x))\,dx.$$

71. Proposition 7 of Section 3.1 tells us that $\varphi \circ f$ is measurable whenever f is. In Proposition 16 of Section 4.4, put g to be $c_1 + c_2|x_1|$, a nonnegative integrable function that dominates $\varphi \circ f$ by (43) and is integrable over $[0, 1]$. Then $\varphi \circ f$ is integrable over $[0, 1]$.

Conversely, assume that $\varphi \circ f$ is integrable over $[0, 1]$ whenever f is. Suppose, to get a contradiction, that for all c_1, c_2, there were an x such that $|\varphi(x)| > c_1 + c_2|x|$. Then for each natural number n, there is an x_n such that $|\varphi(x_n)| > n + n|x_n| = n(1 + |x_n|)$. Define f on $[0, 1]$ to be semisimple function $\sum_{n=0}^{\infty} x_n \chi_{E_n}$ where $x_0 = 0$, and sets E_n are given by

$$E_1 = \left[0, \frac{c\min\{1, 1/|x_1|\}}{1^2}\right) \text{ where } c^{-1} = \sum_{n=1}^{\infty} \frac{1}{n^2} = \frac{\pi^2}{6},$$

$$E_n = \left[\frac{c}{(n-1)^2}, \frac{c}{(n-1)^2} + \frac{c\min\{1, 1/|x_n|\}}{n^2}\right) \text{where } n = 2, 3, 4, \ldots,$$

$$E_0 = [0, 1] \sim \bigcup_{n=1}^{\infty} E_n.$$

The purpose of constant c is to ensure that the E_n are contained in $[0, 1]$. Note that the E_n are disjoint and that $m(E_n) = c\min\{1, 1/|x_n|\}/n^2$ where $n = 1, 2, \ldots$. Then f is integrable by the beginning of Section 4.4 because

$$\begin{aligned}
\int_0^1 |f| &= \sum_{n=1}^{\infty} |x_n| m(E_n) \\
&= c\sum_{n=1}^{\infty} \frac{|x_n| \min\{1, 1/|x_n|\}}{n^2} \\
&= c\sum_{n=1}^{\infty} \frac{\min\{|x_n|, 1\}}{n^2} \\
&\leq 1 \\
&< \infty.
\end{aligned}$$

But $\varphi \circ f$ is not integrable, contradicting our assumption, because

$$\begin{aligned}
\int_0^1 |\varphi \circ f| &= \sum_{n=1}^{\infty} |\varphi(x_n)| m(E_n) \\
&> c\sum_{n=1}^{\infty} \frac{n(1 + |x_n|)\min\{1, 1/|x_n|\}}{n^2} \\
&= c\sum_{n=1}^{\infty} \frac{\min\{1 + |x_n|, 1/|x_n| + 1\}}{n} \\
&\geq c\sum_{n=1}^{\infty} \frac{1}{n} \\
&= \infty.
\end{aligned}$$

Thus, there are constants c_1, c_2 for which (43) holds.

Chapter 7

The L^p Spaces: Completeness and Approximation

7.1 Normed Linear Spaces

In the last example, the third sentence should begin "Since each continuous function on $[a, b]$ takes a minimum and maximum value, ..."

1. Functions in $C[a, b]$ are bounded by the extreme value theorem and measurable by Proposition 3 of Section 3.1 so therefore integrable over $[a, b]$ by Theorem 4 of Section 4.2. Then $C[a, b] \subseteq L^1[a, b]$. Thus, $\|\cdot\|_1$ is a norm on $C[a, b]$ by the first example with E equal to $[a, b]$. Note that nonnegativity in that example is justified by Chapter 4's Theorem 17 (monotonicity) and Proposition 9.

   Next, suppose, to get a contradiction, that there were a positive number c for which $\|f\|_{\max} \leq c\|f\|_1$ for all f in $C[a, b]$. For each natural number n, define piecewise-linear function f_n on $[0, 1]$ by

$$f_n(x) = \begin{cases} n - n^2x & 0 \leq x \leq 1/n, \\ 0 & 1/n < x \leq 1. \end{cases}$$

   Functions f_n are in $C[0, 1]$ because $f_n(1/n) = 0$. We calculate that

$$\begin{aligned} \|f_n\|_1 &= \int_0^1 |f_n| = \int_0^{1/n} (n - n^2x)\, dx = 1/2, \\ \|f_n\|_{\max} &= \max_{x\in[0,1]} |f_n(x)| = \max_{x\in[0,1]} \{n - n^2x, 0\} = n. \end{aligned}$$

   Because $\|f\|_{\max} \leq c\|f\|_1$ for all f in $C[0, 1]$, we have that $n \leq c/2$ for all n, a contradiction. Thus, there is no such real number c.

   On the other hand, because $|f| \leq \|f\|_{\max}$ for all f in $C[a, b]$,

$$\|f\|_1 = \int_a^b |f| \leq \int_a^b \|f\|_{\max} = (b - a)\|f\|_{\max}$$

   by Theorem 17 (monotonicity) of Section 4.4. Thus, $b - a$ will do for c in this case.

2. By definition, a polynomial $a_1 + a_2x + a_3x^2 + \cdots + a_{n+1}x^n$ has only a finite number of nonzero coefficients a_k. Now refer to the third example. We can represent p by sequence $(a_1, a_2, a_3, \ldots, a_{n+1}, 0, \ldots)$, which is in ℓ^1 because $\sum_{k=1}^{\infty}|a_k| = \sum_{k=1}^{n+1}|a_k| < \infty$. Hence, X is a subset of ℓ^1. It follows that X is a linear space because linear combinations of polynomials are polynomials, and that $\|p\| = \|\{a_k\}\|_1$ is a norm as claimed in the example and proved in Problem 5.

3. Follow the first example:

$$\|f + g\| = \int_a^b x^2|(f + g)(x)|\, dx \leq \int_a^b x^2(|f(x)| + |g(x)|)\, dx = \|f\| + \|g\|.$$

   Clearly, $\|\cdot\|$ is positively homogeneous. Finally, $\|\cdot\|$ is nonnegative by Chapter 4's Theorem 17 (monotonicity) and Proposition 9. Thus, $\|\cdot\|$ defines a norm on $L^1[a, b]$.

4. By the second example, $\|f\|_\infty$ is the smallest essential upper bound for f. Hence, we can replace *infimum* in the definition of $\|f\|_\infty$ with *minimum*. Putting E to $[a, b]$ gives us therefore that

$$\begin{aligned} \|f\|_\infty &= \min\{M \mid |f(x)| \leq M \text{ for almost all } x \text{ in } [a, b]\} \\ &= \min\{M \mid m\{x \in [a, b] \mid |f(x)| > M\} = 0\} \qquad \text{definition of } \textit{almost all}. \end{aligned}$$

Now let f be continuous on $[a, b]$. Starting with the first definition above,

$$\|f\|_\infty = \min\{M \mid |f(x)| \leq M \text{ for almost all } x \text{ in } [a, b]\} \leq \max_{x \in [a,b]} |f(x)| = \|f\|_{\max}.$$

That inequality does not actually require that f be continuous, but the reverse inequality does: By the extreme value theorem, there is a point x' in $[a, b]$ such that $f(x') = \|f\|_{\max}$. Let δ respond to the ϵ challenge regarding the criterion for the continuity of f so that if x is in $[a, b] \cap (x' - \delta, x' + \delta)$, then $f(x) > \|f\|_{\max} - \epsilon$. Now consider set G defined by $G = \{x \in [a, b] \mid |f(x)| > \|f\|_{\max} - \epsilon\}$. Looking at our statement of the continuity of f, our definition of G, and the first part of this problem, we see that $m(G) \geq \delta \neq 0$ so that $\|f\|_\infty > \|f\|_{\max} - \epsilon$. Because that holds for ϵ arbitrarily small, $\|f\|_\infty = \|f\|_{\max}$.

5. By the third example, ℓ^1 is a linear space. By similar reasoning, ℓ^∞ is a linear space: the sum of two real bounded sequences is a real bounded sequence, and also a real multiple of a real bounded sequence is a real bounded sequence. To see that $\|\{a_k\}\|_\infty$ is a norm on ℓ^∞, follow the first example:

$$\begin{aligned} \|\{a_k\} + \{b_k\}\|_\infty &= \sup_{1 \leq k < \infty} |a_k + b_k| \\ &\leq \sup_{1 \leq k < \infty} (|a_k| + |b_k|) \leq \sup_{1 \leq k < \infty} |a_k| + \sup_{1 \leq k < \infty} |b_k| = \|\{a_k\}\|_\infty + \|\{b_k\}\|_\infty. \end{aligned}$$

Clearly, $\|\cdot\|_\infty$ satisfies the positive homogeneity and nonnegativity criteria. Thus, ℓ^∞ is a normed linear space. Likewise, for ℓ^1,

$$\|\{a_k\} + \{b_k\}\|_1 = \sum_{k=1}^{\infty} |a_k + b_k| \leq \sum_{k=1}^{\infty} (|a_k| + |b_k|) = \sum_{k=1}^{\infty} |a_k| + \sum_{k=1}^{\infty} |b_k| = \|\{a_k\}\|_1 + \|\{b_k\}\|_1.$$

Clearly, $\|\cdot\|_1$ satisfies the positive homogeneity and nonnegativity criteria. Thus, ℓ^1 is a normed linear space.

7.2 The Inequalities of Young, Hölder, and Minkowski

6. Let f, g be functions whose normalizations $f/\|f\|_p$, $g/\|g\|_q$ satisfy the conditions of Theorem 1. Then by Hölder's inequality,

$$\int_E \left| \frac{f}{\|f\|_p} \frac{g}{\|g\|_q} \right| \leq \left\| \frac{f}{\|f\|_p} \right\|_p \left\| \frac{g}{\|g\|_q} \right\|_q.$$

Multiply both sides by nonnegative constant $\|f\|_p \|g\|_q$ to get that $\int_E |fg| \leq \|f\|_p \|g\|_q$. Thus, Hölder's inequality is true for f, g.

7. For the first example,

$$\int_0^1 |x^\alpha|^p \, dx = \begin{cases} 1/(\alpha p + 1) < \infty & \alpha > -1/p = -1/p_1, \\ \infty & \alpha = -1/p = -1/p_2. \end{cases}$$

Thus, $f \in L^{p_1}(0,1]$, and $f \notin L^{p_2}(0,1]$ so that $f \in L^{p_1}(0,1] \sim L^{p_2}(0,1]$.

For the second example, substitute u for $\ln x$ to get that

$$\int_0^\infty \left| \frac{x^{-1/2}}{1+|\ln x|} \right|^p dx = \int_{-\infty}^\infty \frac{e^{(1-p/2)u}}{(1+|u|)^p}\, du.$$

If $p = 2$, the value of the integral is 2. If $p < 2$, the integrand is not bounded as $u \to \infty$ by an analysis similar to that in our second solution to Exercise 14. Likewise, if $p > 2$, the integrand is not bounded as $u \to -\infty$. Thus, f belongs to $L^p(0,\infty)$ if and only if $p = 2$.

8. Starting with algebraic identity $\lambda^2 f^2 + 2\lambda fg + g^2 = (\lambda f + g)^2 \geq 0$, then integrating over E and applying Chapter 4's Theorem 17 and Proposition 9 gives us that

$$\lambda^2 \int_E f^2 + 2\lambda \int_E fg + \int_E g^2 = \int_E (\lambda f + g)^2 \geq 0$$

where fg is integrable over E by Theorem 1, and $\lambda f + g$ belongs to $L^2(E)$ because it is a linear space. We also have the following because $|f|$ and $|g|$ belong to $L^2(E)$:

$$\lambda^2 \int_E |f|^2 + 2\lambda \int_E |f||g| + \int_E |g|^2 = \int_E (\lambda|f| + |g|)^2 \geq 0.$$

The left side is a quadratic polynomial in λ that is greater than or equal to zero. In that case, the quadratic formula tells us that the discriminant is less than or equal to zero. Thus,

$$\left(2\int_E |f||g|\right)^2 - 4\int_E |f|^2 \cdot \int_E |g|^2 \leq 0,$$

which is equivalent to the Cauchy-Schwarz inequality

$$\int_E |fg| \leq \sqrt{\int_E f^2} \cdot \sqrt{\int_E g^2}.$$

9. From the proof of Young's inequality, we have the equivalent inequality

$$e^{\lambda u + (1-\lambda)v} \leq \lambda e^u + (1-\lambda)e^v, \text{ where } \lambda = 1/p,\ u = \ln a^p, \text{ and } v = \ln b^q.$$

Consider the geometric formulation of that inequality as described after inequality (38) of Section 6.6. Because exp is strictly convex in the sense that it is nowhere linear, if $u \neq v$, then a point on the chord between (u, e^u) and (v, e^v) is strictly above the graph of exp, i.e., the inequality above is strict, and vice versa. Thus, in Young's inequality, there is equality if and only if $u = v$, that is, $a^p = b^q$.

There are algebraic solutions to this problem, but they are not as enlightening as the one above.

10. The claim is false if $p = 1$. For example, put E to $[0,1]$, f to 1, and g to 2. There is equality $\int_0^1 |fg| = \|f\|_1 \|g\|_\infty$ in Hölder's inequality, but there are no α, β such that $\alpha|f|^1 = \beta|g|^\infty$, that is, $\alpha = \beta 2^\infty = \beta\infty$, a.e. on E.

For the sake of argument, assume that $p > 1$. Assume that $f \neq 0$ and $g \neq 0$ for otherwise there is nothing to prove. From the proof of Hölder's inequality, we see that there is equality in Hölder's inequality for normalized f, g if

$$|f||g| = \frac{|f|^p}{p} + \frac{|g|^q}{q} \text{ a.e. on } E,$$

which by Problem 9 is true if and only if $|f|^p = |g|^q$ a.e. on E. For unnormalized f, g, that becomes "if and only if $|f/\|f\|_p|^p = |g/\|g\|_q|^q$ a.e. on E." Put α to $\|g\|_q^q$ and β to $\|f\|_p^p$. Then in Hölder's inequality, there is equality if and only if there are constants α, β not both zero, for which $\alpha|f|^p = \beta|g|^q$ a.e. on E.

11. Sets $\mathbf{R}^n$ and $\mathcal{T}$ defined by $\mathcal{T} = \{T_x \,|\, x \in \mathbf{R}^n\}$ are linear spaces because they satisfy the conditions of a linear space listed on page 253. Moreover, there is a (natural) one-to-one correspondence between $\mathbf{R}^n$ and $\mathcal{T}$ given by $x \mapsto T_x$. Thus, $\|x\|_p$ is a norm on $\mathbf{R}^n$ because $\|x\|_p$ is norm $\|T_x\|_p$ on $\mathcal{T}$, which is a subset of $L^p[1, n+1)$.

    Hölder's inequality for this norm is $\sum_{k=1}^{n} x_k y_k \le \|x\|_p \|y\|_q$ for all x, y in $\mathbf{R}^n$. The proof follows from Hölder's inequality:

$$\sum_{k=1}^{n} x_k y_k = \int_1^{n+1} T_x T_y \le \int_1^{n+1} |T_x T_y| \le \|T_x\|_p \|T_y\|_q = \|x\|_p \|y\|_q.$$

    Minkowski's inequality for this norm is $\|x + y\|_p \le \|x\|_p + \|y\|_p$. The proof follows from Minkowski's inequality for functions:

$$\|x + y\|_p = \|T_{x+y}\|_p = \|T_x + T_y\|_p \le \|T_x\|_p + \|T_y\|_p = \|x\|_p + \|y\|_p.$$

12. If $p = \infty$, then definitions of $\|\cdot\|_\infty$ for ℓ^∞ and $L^\infty[1, \infty)$ tell us that T_a belongs to $L^\infty[1, \infty)$ and that $\|a\|_\infty = \|T_a\|_\infty$. If $p < \infty$,

$$\int_1^{n+1} |T_a|^p = \sum_{k=1}^{n} |a_k|^p \qquad \text{for each natural number } n,$$

$$\lim_{n\to\infty} \int_1^{n+1} |T_a|^p = \sum_{k=1}^{\infty} |a_k|^p \qquad \text{take limit of both sides,}$$

$$\int_1^{\infty} |T_a|^p = \sum_{k=1}^{\infty} |a_k|^p < \infty \qquad \text{monotone convergence theorem; } a \in \ell^p.$$

    Thus, T_a belongs to $L^p[1, \infty)$, and $\|a\|_p = \left(\sum_{k=1}^{\infty} |a_k|^p\right)^{1/p} = \left(\int_1^{\infty} |T_a|^p\right)^{1/p} = \|T_a\|_p$.

    Theorem 1 in ℓ^p Let p be in $[1, \infty)$, and let q be the conjugate of p. If a belongs to ℓ^p and b belongs ℓ^q, then $\sum_{k=1}^{\infty} a_k b_k$ is summable, and Hölder's inequality is $\sum_{k=1}^{\infty} |a_k b_k| \le \|a\|_p \|b\|_q$. Moreover, if $a \ne 0$, a's *conjugate sequence* a^* defined by

$$a_k^* = \begin{cases} \|a\|_p^{1-p} \operatorname{sgn}(a_k) |a_k|^{p-1} & a_k \ne 0, \\ 0 & a_k = 0 \end{cases}$$

    belongs to ℓ^q, $\sum_{k=1}^{\infty} a_k a_k^* = \|a\|_p$, and $\|a^*\|_q = 1$.

    Proof Hölder's inequality in ℓ^p follows from inequality (3):

$$\sum_{k=1}^{\infty} |a_k b_k| = \int_1^{\infty} |T_a T_b| \le \|T_a\|_p \|T_b\|_q = \|a\|_p \|b\|_q.$$

Next, observe that $a_k a_k^* = \|a\|_p^{1-p}|a_k|^p$. Therefore,

$$\sum_{k=1}^{\infty} a_k a_k^* = \|a\|_p^{1-p} \sum_{k=1}^{\infty} |a_k|^p = \|a\|_p^{1-p}\|a\|_p^p = \|a\|_p.$$

Because $q(p-1) = p$, we have that $\|a^*\|_q = 1$. □

Minkowski's Inequality in ℓ^p Let p be in $[1, \infty]$. If sequences a, b belong to ℓ^p, then so does their sum $a + b$, and $\|a + b\|_p \leq \|a\|_p + \|b\|_p$.

Proof The proof follows from Minkowski's inequality for functions:

$$\|a + b\|_p = \|T_{a+b}\|_p = \|T_a + T_b\|_p \leq \|T_a\|_p + \|T_b\|_p = \|a\|_p + \|b\|_p.$$

13. Because f is bounded, there is a nonnegative M such that $|f| \leq M$ on E. Then using Theorem 17 (monotonicity) of Section 4.4 in the second step below,

$$\int_E |f|^{p_2} = \int_E |f|^{p_2 - p_1}|f|^{p_1} \leq M^{p_2 - p_1} \int_E |f|^{p_1} < \infty.$$

Thus, f belongs to $L^{p_2}(E)$.

14. By a Maclaurin series for the exponential function, $\exp(x^{-1/(2p)}) > x^{-1/(2p)}$ if $x \in (0, 1]$ and $1 \leq p < \infty$ so that $\ln(1/x) < 2px^{-1/(2p)}$. Then

$$\int_0^1 |\ln(1/x)|^p \, dx < \int_0^1 \left(2px^{-1/(2p)}\right)^p dx = 2(2p)^p < \infty.$$

Here is another solution to show that f belongs to $L^p(0, 1]$. In the second step below, we choose nonnegative M large enough so that if $u \geq M$, then $e^{u/2} > u^p$:

$$\begin{aligned}
\int_0^1 |\ln(1/x)|^p \, dx &= \int_0^{\infty} u^p e^{-u} \, du && u = \ln(1/x) \\
&= \int_0^M u^p e^{-u} \, du + \int_M^{\infty} u^p e^{-u} \, du \\
&< \int_0^M u^p \, du + \int_M^{\infty} e^{u/2} e^{-u} \, du && e^{-u} \leq 1 \text{ on } [0, M];\ u^p < e^{u/2} \text{ on } [M, \infty) \\
&= \frac{M^{p+1}}{p+1} + 2e^{-M/2} \\
&< \infty.
\end{aligned}$$

We show that such an M exists using a Maclaurin series:

$$\begin{aligned}
\frac{e^{u/2}}{u^p} &= \frac{1}{u^p}\left(1 + \frac{u/2}{1!} + \frac{(u/2)^2}{2!} + \cdots\right) \\
&= u^{-p} + \frac{u^{1-p}}{1! \cdot 2} + \frac{u^{2-p}}{2! \cdot 2^2} + \cdots + \frac{u^{n-p}}{n! \cdot 2^n} + \cdots \\
&> \frac{u^{\lceil p+1 \rceil - p}}{\lceil p+1 \rceil ! \cdot 2^{\lceil p+1 \rceil}} && \lceil \cdot \rceil \text{ denotes the ceiling function.}
\end{aligned}$$

The last expression is not bounded because it equals a constant times u raised to a power at least 1. Hence, there is a u (which we called M above) such that $e^{u/2}/u^p > 1$. We can shorten our analysis above if we are allowed to use the facts that $\int_0^\infty u^p e^{-u}\,du$ is the gamma function and that it converges. In any case, we have shown that f belongs to $L^p(0,1]$.

But f is not in $L^\infty(0,1]$ because f has no essential upper bound. Indeed, for all nonnegative numbers M, we have that $|f| > M$ in interval $(0, \exp(-M))$, which has positive measure.

15. **Hölder's inequality** Let E be a measurable set, let p be in $[1,\infty)$, and let q,r be in $(1,\infty]$ such that $1/p + 1/q + 1/r = 1$ unless $p = 1$, in which case define q,r each to be ∞. If f belongs to $L^p(E)$, g belongs to $L^q(E)$, and h belongs to $L^r(E)$, then their product fgh is integrable over E and

$$\int_E |fgh| \le \|f\|_p \|g\|_q \|h\|_r.$$

Proof 1 If $p = 1$, start with Hölder's inequality applied to f, gh:

$$\begin{aligned}
\int_E |fgh| &= \|f\|_1 \|gh\|_\infty \\
&= \|f\|_1 \inf\{M \mid m\{x \in E \mid |(gh)(x)| > M\} = 0\} \\
&\le \|f\|_1 \inf\{M_1 \mid m\{x \in E \mid |g(x)| > M_1\} = 0\} \inf\{M_2 \mid m\{x \in E \mid |h(x)| > M_2\} = 0\} \\
&= \|f\|_p \|g\|_\infty \|h\|_\infty.
\end{aligned}$$

If $p > 1$, apply Hölder's inequality to f, gh and then again to $|g|^{p/(p-1)}, |h|^{p/(p-1)}$; simplify; and use the fact that $r = pq/(pq - p - q)$:

$$\begin{aligned}
\int_E |fgh| &\le \|f\|_p \|gh\|_{p/(p-1)} \\
&= \|f\|_p \left(\int_E |gh|^{p/(p-1)} \right)^{(p-1)/p} \\
&\le \|f\|_p \left(\left\| |g|^{p/(p-1)} \right\|_{q(p-1)/p} \cdot \left\| |h|^{p/(p-1)} \right\|_{q(p-1)/p/(q(p-1)/p-1)} \right)^{(p-1)/p} \\
&\le \|f\|_p \|g\|_q \left(\int_E |h|^{pq/(pq-p-q)} \right)^{(pq-p-q)/pq} = \|f\|_p \|g\|_q \|h\|_r. \qquad \square
\end{aligned}$$

Proof 2 We first prove that if $1 < p < \infty$, and $1 < q < \infty$, then for three nonnegative numbers a, b, c,

$$abc \le \frac{a^p}{p} + \frac{b^q}{q} + \frac{c^r}{r}.$$

Apply Young's inequality to a, bc and then again to $b^{p/(p-1)}, c^{p/(p-1)}$:

$$\begin{aligned}
abc &\le \frac{a^p}{p} + \frac{(bc)^{p/(p-1)}}{p/(p-1)} \\
&\le \frac{a^p}{p} + \frac{p-1}{p} \left(\frac{\left(b^{p/(p-1)}\right)^{q(p-1)/p}}{q(p-1)/p} + \frac{\left(c^{p/(p-1)}\right)^{q(p-1)/p/(q(p-1)/p-1)}}{q(p-1)/p/(q(p-1)/p-1)} \right) \\
&= \frac{a^p}{p} + \frac{b^q}{q} + \frac{c^{pq/(pq-p-q)}}{pq/(pq-p-q)} \\
&= \frac{a^p}{p} + \frac{b^q}{q} + \frac{c^r}{r}.
\end{aligned}$$

First consider the case where $p = 1$. Then Hölder's inequality follows from the monotonicity of integration and the observation (2) that $\|g\|_\infty, \|h\|_\infty$ are essential upper bounds for g, h

on E. Now consider the case where $p > 1$. Assume that $f, g, h \neq 0$ for otherwise there is nothing to prove. It is clear that if Hölder's inequality is true when f, g, h are replaced by their normalizations $f/\|f\|_p$, $g/\|g\|_q$, $h/\|h\|_r$, then it is true for f, g, h. We therefore assume that $\|f\|_p, \|g\|_q, \|h\|_r = 1$, that is,

$$\int_E |f|^p = 1, \int_E |g|^q = 1, \text{ and } \int_E |h|^r = 1,$$

in which case Hölder's inequality becomes

$$\int_E |fgh| \leq 1.$$

Because $|f|^p$, $|g|^q$, $|h|^r$ are integrable over E, functions f, g, h are finite a.e. on E. Hence, by Young's inequality,

$$|fgh| = |f||g||h| \leq \frac{|f|^p}{p} + \frac{|g|^q}{q} + \frac{|h|^r}{r} \text{ a.e. on } E.$$

We infer from the linearity of integration and the integral comparison test that fgh is integrable over E and, by the monotonicity and linearity of integration, that

$$\int_E |fgh| \leq \frac{1}{p}\int_E |f|^p + \frac{1}{q}\int_E |g|^q + \frac{1}{r}\int_E |h|^r = \frac{1}{p} + \frac{1}{q} + \frac{1}{r} = 1.$$

16. Sequence $\{f_n\}$ is not necessarily uniformly integrable over $[0,1]$. For example, put f_n to $n\chi_{[0,1/n]}$ (as in the example on page 83). Then $\{f_n\}$ is bounded in $L^1[0,1]$ because

$$\|f_n\|_1 = \int_0^1 f_n = \int_0^1 n\chi_{[0,1/n]} = n \cdot m([0,1/n]) = n(1/n) = 1 \text{ for all } f_n,$$

but $\{f_n\}$ is not uniformly integrable (see the definition on page 93). Indeed, for ϵ equal to 1 and all positive δ, choose n large enough so that $1/n < \delta$. Then we have a measurable set $[0,1/n]$ in $[0,1]$ such that $m([0,1/n]) = 1/n < \delta$, and a function f_n that gives us that

$$\int_0^{1/n} |f_n| = \int_0^1 f_n = 1 \not< \epsilon.$$

17. Sequence $\{\chi_{[n,n+1]}\}$ is not tight by the beginning of Section 5.1. The sequence however has bound 1 and is in $L^p(\mathbf{R})$ because $\int_{\mathbf{R}} |\chi_{[n,n+1]}|^p = 1 < \infty$.

18. Because $|f| \leq \|f\|_\infty$ a.e. on E,

$$\|f\|_p = \left(\int_E |f|^p\right)^{1/p} \leq \left(\int_E \|f\|_\infty^p\right)^{1/p} = \|f\|_\infty m(E)^{1/p}.$$

Hence, $\limsup\{\|f\|_p\} \leq \|f\|_\infty$. Now get the reverse inequality with the limit inferior. For an ϵ in $(0, \|f\|_\infty)$, define set S by $S = \{x \in E \mid |f(x)| \geq \|f\|_\infty - \epsilon\}$. Then $0 < m(S) \leq m(E)$, and

$$\|f\|_p = \left(\int_E |f|^p\right)^{1/p} \geq \left(\int_S (\|f\|_\infty - \epsilon)^p\right)^{1/p} = (\|f\|_\infty - \epsilon)m(S)^{1/p}.$$

Because that holds for all ϵ in $(0, \|f\|_\infty)$, it holds also if $\epsilon = 0$. Then, $\liminf\{\|f\|_p\} \geq \|f\|_\infty$. Thus, $\lim_{p\to\infty}\|f\|_p = \|f\|_\infty$ by Chapter 1's Theorem 19(iv) and Problem 41.

19. If $f = 0$, the claim is (trivially) true. Assume now that $f \neq 0$. By Theorem 17 (monotonicity) of Section 4.4 and Hölder's inequality, we have inequality

$$\max_{g \in L^q(E), \|g\|_q \leq 1} \int_E fg \leq \max_{g \in L^q(E), \|g\|_q \leq 1} \int_E |fg| \leq \max_{g \in L^q(E), \|g\|_q \leq 1} \|f\|_p \|g\|_q = \|f\|_p.$$

But we have equality $\int_E fg = \|f\|_p$ by equation (4) with $g = f^*$ where $g \in L^q(E)$ and $\|g\|_q = 1$. Thus,

$$\max_{g \in L^q(E), \|g\|_q \leq 1} \int_E fg = \|f\|_p.$$

20. If $f = 0$, then $\int_E fg = 0$ trivially. Conversely, assume that $\int_E fg = 0$ for all g in $L^q(E)$. Then starting with equation (4), $\|f\|_p = \int_E ff^* = 0$ by assumption because $f^* \in L^q(E)$. Thus, $f = 0$ by the nonnegativity property of a norm.

21. If the claim holds for $|f|$ for all f in $L^p[0,1]$, then the claim holds for all f in $L^p[0,1]$. Hence, it suffices to consider nonnegative fs only. Assume first that $p = \infty$. If $\lambda < 1$, then

$$\lim_{\epsilon \to 0^+} \frac{1}{\epsilon^\lambda} \int_0^\epsilon f \leq \lim_{\epsilon \to 0^+} \epsilon^{1-\lambda} \|f\|_\infty = 0.$$

If on the other hand $\lambda \geq 1$, put f to λ. Then

$$\lim_{\epsilon \to 0^+} \frac{1}{\epsilon^\lambda} \int_0^\epsilon \lambda = \lim_{\epsilon \to 0^+} \frac{\lambda}{\epsilon^{\lambda-1}} = \infty \neq 0.$$

Hence, if $p = \infty$, the claim holds precisely when $\lambda < 1$.

Now assume that $p < \infty$. If $\lambda \leq 1 - 1/p = 1/q$, we have by Hölder's inequality that

$$\lim_{\epsilon \to 0^+} \frac{1}{\epsilon^\lambda} \int_0^\epsilon f \leq \lim_{\epsilon \to 0^+} \left\| \frac{1}{\epsilon^\lambda} \right\|_q \|f\|_p = \lim_{\epsilon \to 0^+} \epsilon^{1/q - \lambda} \left(\int_0^\epsilon f^p \right)^{1/p} = 0$$

because $1/q - \lambda \geq 0$ and because $\lim_{\epsilon \to 0^+} \int_0^\epsilon f^p = 0$ since f belongs to $L^p[0,1]$. Hence, if $1 \leq p < \infty$, the claim holds when $\lambda \leq 1/q$.

Combining all the results above, we have shown that for all p such that $1 \leq p \leq \infty$, the given statement holds precisely when $\lambda < 1/q$.

(Note that we do not have to prove that the statement is false for the case where $p < \infty$ and $\lambda > 1/q$ because we are to find values of λ for which the statement holds for all p, and when $p = \infty$, the conjugate q is 1, and we proved that the statement is false if $p = \infty$, and $\lambda \geq 1 = 1/q$.)

22. First consider the case where $p = 1$:

$$\begin{aligned}
\sum_{k=1}^n |F(x_k) - F(x_{k-1})| &= \sum_{k=1}^n \left| \int_{x_k}^{x_{k-1}} F' \right| && \text{Theorem 10 of Section 6.5} \\
&\leq \sum_{k=1}^n \int_{x_k}^{x_{k-1}} |F'| && \text{Proposition 16 of Section 4.4} \\
&= \int_a^b |F'| && \text{Corollary 18 of Section 4.4} \\
&< \infty && F' \text{ is an } L^1[a,b] \text{ function by hypothesis.}
\end{aligned}$$

Thus, there is such a constant M. Our proof for when $p > 1$ includes ideas from that of Corollary 2:

$$\begin{aligned}
\sum_{k=1}^{n} \frac{|F(x_k) - F(x_{k-1})|^p}{(x_k - x_{k-1})^{p-1}} &\le \sum_{k=1}^{n} \frac{\left(\int_{x_{k-1}}^{x_k} |F' \cdot 1|\right)^p}{(x_k - x_{k-1})^{p-1}} \\
&\le \sum_{k=1}^{n} \frac{\left(\left(\int_{x_{k-1}}^{x_k} |F'|^p\right)^{1/p} \left(\int_{x_{k-1}}^{x_k} 1^q\right)^{1/q}\right)^p}{(x_k - x_{k-1})^{p-1}} && \text{Hölder's inequality} \\
&= \sum_{k=1}^{n} \frac{\left(\int_{x_{k-1}}^{x_k} |F'|^p\right)(x_k - x_{k-1})^{p-1}}{(x_k - x_{k-1})^{p-1}} && 1/q = (p-1)/p \\
&= \int_a^b |F'|^p \\
&< \infty.
\end{aligned}$$

7.3 L^p Is Complete: The Riesz-Fischer Theorem

In the proof of Proposition 4, symbol k has two meanings. For clarity, replace the second use of k (the choice) with K. There is also a typo in the proof: the last $=$ should be $\le$.

23. Sequence $\{(-1)^n/n\}$ is Cauchy by Theorem 17 of Section 1.5 because $\{(-1)^n/n\}$ converges (to zero). The criterion for the sequence to be rapidly Cauchy is that

$$\left| \frac{(-1)^{k+1}}{k+1} - \frac{(-1)^k}{k} \right| = \frac{2k+1}{k(k+1)} \le \epsilon_k^2 \text{ for all } k.$$

Now $(2k+1)/(k(k+1)) > 1/k$, so $\epsilon_k > \sqrt{1/k}$. Then $\sum_{k=1}^{\infty} \epsilon_k > \sum_{k=1}^{\infty} \sqrt{1/k}$, which does not converge by the p-series test. Thus, $\{(-1)^n/n\}$ is not rapidly Cauchy.

A second counterexample is $\left\{\sum_{j=1}^{n} 1/j^2\right\}$, a sequence of partial sums which converges by the p-series test and is therefore Cauchy. The criterion for the sequence to be rapidly Cauchy is that

$$\left| \sum_{j=1}^{k+1} \frac{1}{j^2} - \sum_{j=1}^{k} \frac{1}{j^2} \right| = \frac{1}{(k+1)^2} \le \epsilon_k^2 \text{ for all } k.$$

Then $\sum_{k=1}^{\infty} \epsilon_k \ge \sum_{k=1}^{\infty} 1/(k+1) = \sum_{k=2}^{\infty} 1/k$, which does not converge by the p-series test.

24. By the definition on page 144, each f_n, g_n and f, g belong to X, so each $\alpha f_n + \beta g_n$ and $\alpha f + \beta g$ are also in X because X is a linear space. Now mimic the proof of Theorem 18 (linearity) of Section 1.5. Observe that

$$\|(\alpha f_n + \beta g_n) - (\alpha f + \beta g)\| \le |\alpha| \|f_n - f\| + |\beta| \|g_n - g\| \text{ for all } n.$$

Take a positive ϵ. Choose a natural number N such that

$$\|f_n - f\| < \epsilon/(2 + 2|\alpha|), \text{ and } \|g_n - g\| < \epsilon/(2 + 2|\beta|) \text{ if } n \ge N.$$

We infer from our first inequality that

$$\|(\alpha f_n + \beta g_n) - (\alpha f + \beta g)\| < \epsilon \text{ if } n \ge N.$$

Thus $\{\alpha f_n + \beta g_n\} \to \alpha f + \beta g$ in X.

25. By the definition on page 144 and Corollary 3, each f_n and f belong also to $L^{p_1}(E)$, so each $f - f_n$ is in $L^{p_1}(E)$ because $L^{p_1}(E)$ is a linear space. Then

$$\begin{aligned} \lim_{n\to\infty} \|f - f_n\|_{p_1} &\le \lim_{n\to\infty} c\|f - f_n\|_{p_2} && \text{Theorem 18 of Section 1.5, Corollary 3} \\ &= c \lim_{n\to\infty} \|f - f_n\|_{p_2} && \text{Theorem 18 of Section 1.5} \\ &= 0 && \{f_n\} \to f \text{ in } L^{p_2}(E). \end{aligned}$$

Thus, $\{f_n\} \to f$ in $L^{p_1}(E)$.

26. As in the proof of Theorem 7, we can assume that on all of E, convergence is pointwise, $\{|f_n|^p\} \to |f|^p$ pointwise, and $|f_n| \le g$ for all n. By the reasoning in the beginning of the second sentence after inequality (10), $|f_n|^p \le g^p$ for all n on E. Each $|f_n|^p$ in $\{|f_n|^p\}$ is measurable by Proposition 7 of Section 3.1. Function g^p is integrable because g is in $L^p(E)$. By Lebesgue's dominated convergence theorem, $|f|^p$ is integrable over E, and $\lim_{n\to\infty} \int_E |f_n|^p = \int_E |f|^p$. Sequence $\{f_n\}$ is in $L^p(E)$ because for all n, $\int_E |f_n|^p \le \int_E g^p < \infty$. Function f is also in $L^p(E)$ because $|f|^p$ is integrable over E. Thus, $\{f_n\} \to f$ in $L^p(E)$ by Theorem 7.

27. The Riesz-Fischer theorem gives us a subsequence $\{f_{n_j}\}$ that converges pointwise a.e. on E to f. As in the proof of Proposition 5, we can put $\{f_{n_k}\}$ to be a subsequence of $\{f_{n_j}\}$ for which $|f - f_{n_k}| \le (1/2)^k$ a.e. on E for all k. Next, we define sequence $\{g_m\}$ and also give it an upper bound: $g_m = |f| + \sum_{k=1}^{m} |f - f_{n_k}| \le |f| + \sum_{k=1}^{m} \frac{1}{2^k} \le |f| + 1$. Theorem 15 of Section 1.5 allows us to put g to be the function to which $\{g_m\}$ converges pointwise a.e. on E because $\{g_m\}$ is an increasing sequence bounded by $|f| + 1$. Then $g \in L^p(E)$ because

$$\begin{aligned} \|g\|_p &\le \||f| + 1\|_p && \text{Theorem 17 (monotonicity) of Section 4.4} \\ &< \infty && |f| + 1 \in L^p(E) \text{ because } \{f_n\} \to f \text{ in } L^p(E) \text{ so that } |f| \in L^p(E). \end{aligned}$$

Furthermore, $|f_{n_k}| \le |f| + |f - f_{n_k}| \le g$ a.e. on E.

28. By hypothesis and Corollary 3, $\{f_n\} \subseteq L^{p+\theta}(E) \subseteq L^p(E)$. Function f is also in $L^p(E)$ because as in the proof of Theorem 7, since $\{|f_n|^p\} \to |f|^p$ pointwise a.e. on E, we infer from Fatou's lemma that $\int_E |f|^p \le \liminf\{\int_E |f_n|^p\} < \infty$. Likewise, f is in $L^{p+\theta}(E)$. Then $\{f_n - f\}$ belongs to and is bounded as a subset of $L^{p+\theta}(E)$ because it is a linear space. So there is a nonnegative number M such that $\|f_n - f\|_{p+\theta} \le M$, which also holds true if the norm is over all measurable subsets A of E by Theorem 11 of Section 4.3. Now $\{|f_n - f|^p\} \to 0$ pointwise a.e. on E as in the proof of Theorem 8. Put δ to $(\epsilon/M^p)^{(p+\theta)/\theta}$. That δ responds to the ϵ challenge regarding the criterion for $\{|f_n - f|^p\}$ to be uniformly integrable over E because

$$\begin{aligned} \int_A |f_n - f|^p &= \|f_n - f\|_p^p && \text{definition on page 139} \\ &\le \left(m(A)^{\theta/(p(p+\theta))} \|f_n - f\|_{p+\theta}\right)^p && \text{Corollary 3} \\ &< \left((\epsilon/M^p)^{(p+\theta)/\theta}\right)^{\theta/(p+\theta)} M^p && m(A) < \delta \\ &= \epsilon. \end{aligned}$$

By the Vitali convergence theorem of Section 4.6, $\lim_{n\to\infty} \int_E |f_n - f|^p = 0$. Thus, by the sentence before the definition on page 145, $\{f_n\} \to f$ in $L^p(E)$.

29. The space is not a Banach space. For example, define a sequence $\{p_n\}$ of polynomials on $[0,1]$ by

$$p_n(x) = 1 + \frac{x}{2} + \frac{x^2}{4} + \cdots + \frac{x^n}{2^n} = \sum_{k=0}^{n} \frac{x^k}{2^k}.$$

The sequence is Cauchy because for each positive ϵ, there is an index N where $N \geq -\log_2 \epsilon$ such that if $n \geq m \geq N$, then

$$\|p_n - p_m\|_{\max} = \left\| \sum_{k=0}^{n} \frac{x^k}{2^k} - \sum_{k=0}^{m} \frac{x^k}{2^k} \right\|_{\max} = \left\| \sum_{k=m+1}^{n} \frac{x^k}{2^k} \right\|_{\max} = \sum_{k=m+1}^{n} \frac{1}{2^k} < \frac{1}{2^m} \leq \frac{1}{2^N} \leq \epsilon.$$

By the formula for a geometric series, $\sum_{k=0}^{\infty}(x/2)^k = 2/(2-x)$, so $\{p_n\}$ converges to $2/(2-x)$. But $2/(2-x)$ is not a polynomial because none of its kth derivatives vanishes.

30. This and Problem 31 are Proposition 10 of Section 9.4. Following is a more detailed version of the proof.

We first need to show that $\|\cdot\|_{\max}$ is a norm on $C[a,b]$ (i.e., complete the last example on page 138). Indeed, if $f, g \in C[a,b]$, then we infer from the triangle inequality for real numbers that

$$\begin{aligned}
\|f+g\|_{\max} &= \max_{x\in[a,b]} |(f+g)(x)| \\
&\leq \max_{x\in[a,b]} (|f(x)| + |g(x)|) \\
&\leq \max_{x\in[a,b]} |f(x)| + \max_{x\in[a,b]} |g(x)| \\
&= \|f\|_{\max} + \|g\|_{\max}.
\end{aligned}$$

Norm $\|\cdot\|_{\max}$ is positively homogeneous by Problem 7(i) of Section 1.1 and clearly nonnegative. Hence, $\|\cdot\|_{\max}$ is a norm on $C[a,b]$. Another proof uses the last part of Problem 4 and the fact that $\|\cdot\|_\infty$ is a norm on $L^\infty[a,b]$, which contains $C[a,b]$.

Now for all k, n and all x in $[a,b]$,

$$|f_{n+k}(x) - f_n(x)| \leq \max_{x'\in[a,b]} |f_{n+k}(x') - f_n(x')| = \|f_{n+k} - f_n\|_{\max} \leq \sum_{j=n}^{\infty} a_j$$

where the last inequality comes from inequality (6). As in the proof of Proposition 5, we infer from the above inequality that for each x in $[a,b]$, sequence $\{f_n(x)\}$ is a Cauchy sequence of real numbers. Put f to be the function to which $\{f_n\}$ converges for each x in $[a,b]$ (see Theorem 17 of Section 1.5). To show that $\{f_n\} \to f$ uniformly on $[a,b]$, we verify the conditions of definition (iii) on page 60: Take a positive ϵ. Proposition 20 of Section 1.5 gives us an index N for which $\sum_{j=n}^{\infty} a_j < \epsilon$ if $n \geq N$. Take the limit as $k \to \infty$ in the displayed inequality above to get that

$$|f - f_n| \leq \sum_{j=n}^{\infty} a_j < \epsilon \text{ on } [a,b] \text{ if } n \geq N.$$

Thus, $\{f_n\} \to f$ uniformly on $[a,b]$.

It remains to show that f is continuous at each x in $[a,b]$. Because $\{f_n\} \to f$ uniformly on $[a,b]$, there is an index N for which $|f - f_N| < \epsilon/3$. Because f_N is continuous, there is a positive δ for which if $x' \in [a,b]$ and $|x' - x| < \delta$, then $|f_N(x') - f_N(x)| < \epsilon/3$. In that case,

$$|f(x') - f(x)| \leq |f(x') - f_N(x')| + |f_N(x') - f_N(x)| + |f_N(x) - f(x)| < \epsilon/3 + \epsilon/3 + \epsilon/3 = \epsilon.$$

Thus, we have a function f in $C[a,b]$ such that $\{f_n\} \to f$ uniformly on $[a,b]$.

31. Let $\{f_n\}$ be a Cauchy sequence in $C[a,b]$. We can inductively choose a strictly increasing sequence of natural numbers $\{n_k\}$ such that $\|f_{n_{k+1}} - f_{n_k}\|_{\max} \leq (1/2)^k$ for all k. Subsequence $\{f_{n_k}\}$ satisfies the conditions of Problem 30 because a geometric series with ratio 1/2 converges; hence, there is a function f in $C[a,b]$ such that $\{f_{n_k}\} \to f$ uniformly on $[a,b]$. Then

$$\lim_{k\to\infty} \|f - f_{n_k}\|_{\max} = \lim_{k\to\infty} \max_{x\in[a,b]} |(f - f_{n_k})(x)| = \max_{x\in[a,b]} \lim_{k\to\infty} |(f - f_{n_k})(x)| = 0.$$

By the proof of Proposition 4, $\{f_n\}$ also converges to f in $C[a,b]$. Thus, by the definition on page 145, $C[a,b]$ normed by the maximum norm is a Banach space.

32. Because $L^\infty(E)$ is a linear space, $f_{n+k} - f_n$ is essentially bounded for all k, n. Then for all k, n and using *minimum* in the definition of $\|\cdot\|_\infty$ as explained in our solution to Problem 4,

$$\begin{aligned} |f_{n+k}(x) - f_n(x)| &\leq \min\{M \mid |(f_{n+k} - f_n)(x')| \leq M \text{ for almost all } x' \text{ in } E\} \\ &= \min\{M \mid |(f_{n+k} - f_n)(x')| \leq M \text{ for all } x' \text{ in } E \sim E_0\} \\ &= \|f_{n+k} - f_n\|_\infty \\ &\leq \sum_{j=n}^{\infty} a_j \qquad \text{inequality (6)} \end{aligned}$$

where E_0 is a zero-measure subset of E. The proof that there is an f such that $\{f_n\} \to f$ uniformly on $E \sim E_0$ is the same as that in our solution to Problem 30 with $[a,b]$ replaced by $E \sim E_0$. It remains to show that f belongs to $L^\infty(E)$. Because $\{f_n\} \to f$ uniformly on $E \sim E_0$, for each positive ϵ, there is an index N for which $|f - f_N| < \epsilon$ on $E \sim E_0$. Because f_N is essentially bounded, there is some nonnegative M for which $|f_N(x)| \leq M$ if $x \in E \sim E_0$. In that case,

$$|f(x)| \leq |f(x) - f_N(x)| + |f_N(x)| < \epsilon + M \text{ for almost all } x \text{ in } E.$$

Thus, we have an f in $L^\infty(E)$ such that $\{f_n\} \to f$ uniformly on $E \sim E_0$.

33. Let $\{f_n\}$ be a Cauchy sequence in $L^\infty(E)$. We can inductively choose a strictly increasing sequence of natural numbers $\{n_k\}$ such that $\|f_{n_{k+1}} - f_{n_k}\|_\infty \leq (1/2)^k$ for all k. Subsequence $\{f_{n_k}\}$ satisfies the conditions of Problem 32; hence, there is a function f in $L^\infty(E)$ such that $\{f_{n_k}\} \to f$ uniformly on $E \sim E_0$ where E_0 is a zero-measure subset of E. Then

$$\begin{aligned} \lim_{k\to\infty} \|f - f_{n_k}\|_\infty &= \lim_{k\to\infty} \min\{M \mid |(f - f_{n_k})(x)| \leq M \text{ for almost all } x \text{ in } E\} \\ &= \min\left\{M \,\middle|\, \lim_{k\to\infty} |(f - f_{n_k})(x)| \leq M \text{ on } E \sim E_0\right\} \\ &= 0. \end{aligned}$$

By the proof of Proposition 4, $\{f_n\}$ also converges to f in $L^\infty(E)$. Thus, $L^\infty(E)$ is a Banach space.

34. Assume that $1 \leq p < \infty$. By the first example on page 138, ℓ^p is a linear space. To show that ℓ^p is a normed linear space, we verify that $\|\cdot\|_p$ is a norm on ℓ^p by checking the three properties in the definition on page 137. The triangle inequality holds by our solution to Problem 12. Norm $\|\cdot\|_p$ is positively homogeneous because for a sequence $\{a_k\}$ in ℓ^p,

$$\|\alpha\{a_k\}\|_p = \left(\sum_{k=1}^{\infty} |\alpha a_k|^p\right)^{1/p} = |\alpha| \left(\sum_{k=1}^{\infty} |a_k|^p\right)^{1/p} = \alpha\|\{a_k\}\|_p.$$

Norm $\|\cdot\|_p$ is nonnegative because

$$\|\{a_k\}\|_p = \left(\sum_{k=1}^{\infty}|a_k|^p\right)^{1/p} \geq 0, \text{ and } \left(\sum_{k=1}^{\infty}|a_k|^p\right)^{1/p} = 0 \text{ if and only if } \{a_k\} = 0,$$

that is, each term of $\{a_k\}$ is zero. Hence, ℓ^p is a normed linear space.

In the following proof that every Cauchy sequence in ℓ^p converges to a sequence in ℓ^p, it may be helpful to think of $\{a_k\}$ in ℓ^p as the f in X in the text of this section. Let $\{\{a_k\}_n\}$ be a Cauchy sequence (of sequences) in ℓ^p. Then for each positive ϵ, there is an index N such that

$$\|\{a_k\}_n - \{a_k\}_m\|_p = \left(\sum_{k=1}^{\infty}|a_{nk} - a_{mk}|^p\right)^{1/p} < \epsilon \text{ if } m, n \geq N$$

where a_{nk} denotes the kth term of $\{a_k\}_n$. Then for each k, we have that $|a_{nk} - a_{mk}| < \epsilon$ if $m, n \geq N$, so $\{a_{nk}\}_{n=1}^{\infty}$ is a Cauchy sequence of real numbers for each k and hence converges, to a_k say, by Theorem 17 of Section 1.5. We next show that $\{a_k\}$ is in ℓ^p. Because $\{a_{nk}\}_{n=1}^{\infty} \to a_k$, for each natural number r and k equal to $1, \ldots, r$, there is an index N_{rk} for which if $n \geq N_{rk}$, then $|a_k - a_{nk}| < \epsilon/r^{1/p}$. Then for each r,

$$\sum_{k=1}^{r}|a_k - a_{N_{rk}k}|^p < \sum_{k=1}^{r}\left(\epsilon/r^{1/p}\right)^p = \epsilon^p.$$

Take the limit as $r \to \infty$ noting that the rightmost side is independent of r. Then there is an index m given by $\max\{N_{rk}\}$ for which

$$\sum_{k=1}^{\infty}|a_k - a_{mk}|^p < \epsilon^p < \infty.$$

Hence, $\{a_k\} - \{a_k\}_m$ is in ℓ^p. Because ℓ^p is a linear space, $\{a_k\}$ is also in ℓ^p. Finally, to show that $\{\{a_k\}_n\}$ converges to $\{a_k\}$ in ℓ^p, we have that

$$\lim_{n\to\infty}\|\{a_k\} - \{a_k\}_n\|_p = \lim_{n\to\infty}\left(\sum_{k=1}^{\infty}|a_k - a_{nk}|^p\right)^{1/p} < (\epsilon^p)^{1/p} = \epsilon.$$

Because that holds for ϵ arbitrarily small, $\{\{a_k\}_n\} \to \{a_k\}$ in ℓ^p. Thus, ℓ^p is a Banach space.

Now assume that $p = \infty$. By Problem 5, ℓ^∞ is a normed linear space. Let $\{\{a_k\}_n\}$ be a Cauchy sequence in ℓ^∞. Then there is an index N such that

$$\|\{a_k\}_n - \{a_k\}_m\|_\infty = \sup_{1\leq k<\infty}|a_{nk} - a_{mk}| < \epsilon \text{ if } m, n \geq N.$$

Then for each k, we have that $|a_{nk} - a_{mk}| < \epsilon$ if $m, n \geq N$, so $\{a_{nk}\}_{n=1}^{\infty}$ is a Cauchy sequence of real numbers for each k and hence converges, to a_k say. Because $\{a_{nk}\}_{n=1}^{\infty} \to a_k$, for each k, there is an index N_k for which if $n \geq N_k$, then $|a_k - a_{nk}| < \epsilon$. Then

$$\sup_{1\leq k<\infty}|a_k| \leq \sup_{1\leq k<\infty}|a_k - a_{N_k k}| + \sup_{1\leq k<\infty}|a_{N_k k}| < \epsilon + \sup_{1\leq k<\infty}|a_{N_k k}| < \infty$$

because $\{a_k\}_{N_k}$ is bounded since it is in ℓ^∞. Hence, $\{a_k\}$ is also bounded and therefore in ℓ^∞. Lastly,

$$\lim_{n\to\infty}\|\{a_k\} - \{a_k\}_n\|_\infty = \lim_{n\to\infty}\sup_{1\leq k<\infty}|a_k - a_{nk}| < \epsilon$$

so that $\{\{a_k\}_n\} \to \{a_k\}$ in ℓ^∞. Thus, ℓ^∞ is a Banach space.

35. Sequences in c are bounded by Proposition 14 of Section 1.5, so c is a subset of ℓ^∞. Moreover, c is closed with respect to the formation of linear combinations by Theorem 18 (linearity) of Section 1.5. Consequently, c is a normed linear space.

    Now let $\{\{a_k\}_n\}$ be a Cauchy sequence in c. Then for each positive ϵ, there is an index N such that

$$\|\{a_k\}_n - \{a_k\}_m\|_\infty = \sup_{1\le k<\infty} |a_{nk} - a_{mk}| < \epsilon/3 \text{ if } m, n \ge N$$

    where a_{nk} denotes the kth term of $\{a_k\}_n$. Then for each k, we have that $|a_{nk} - a_{mk}| < \epsilon/3$ if $m, n \ge N$, so $\{a_{nk}\}_{n=1}^\infty$ is a Cauchy sequence of real numbers for each k and hence converges, to a_k say, by Theorem 17 of Section 1.5. We next show that $\{a_k\}$ is in c. Because $\{a_{nk}\}_{n=1}^\infty \to a_k$, there is an index N_k for which if $n \ge N_k$, then $|a_k - a_{nk}| < \epsilon/3$. Then $\{a_k\}$ is a Cauchy sequence of real numbers because there is an index N' given by $\max\{N, N_j, N_k\}$ for which if $j, k \ge N'$, then $|a_j - a_k| \le |a_j - a_{N'j}| + |a_{N'j} - a_{N'k}| + |a_{N'k} - a_k| < \epsilon/3 + \epsilon/3 + \epsilon/3 = \epsilon$. Appealing again to Theorem 17, $\{a_k\}$ converges and therefore is in c. Finally, to show that $\{\{a_k\}_n\}$ converges to $\{a_k\}$ in c, we have that

$$\lim_{n\to\infty} \|\{a_k\} - \{a_k\}_n\|_\infty = \lim_{n\to\infty} \sup_{1\le k<\infty} |a_k - a_{nk}| < \epsilon/3.$$

    Hence, $\{\{a_k\}_n\} \to \{a_k\}$ in c. Thus, c is a Banach space.

    Next, c_0 is a subset of c. We infer from Theorem 18 (linearity) of Section 1.5 that c_0 is closed with respect to the formation of linear combinations. Consequently, c_0 is a normed linear space.

    Now let $\{\{a_k\}_n\}$ be a Cauchy sequence in c_0. The proof that $\{\{a_k\}_n\}$ converges to a sequence in c_0 is the same as ours for c to the point where were we showed that if $n \ge N_k$, then $|a_k - a_{nk}| < \epsilon/3$. We also have that because $\{a_k\}_{N_k} \to 0$, there is an index K for which if $k \ge K$, then $|a_{N_k k}| < \epsilon/3$. Then $\{a_k\}$ converges to zero because if $k \ge K$, then $|a_k| \le |a_k - a_{N_k k}| + |a_{N_k k}| < \epsilon/3 + \epsilon/3 < \epsilon$. Hence, $\{a_k\}$ converges to zero and therefore is in c_0. The proof that $\{\{a_k\}_n\}$ converges to $\{a_k\}$ in c_0 is the same as our proof for c. Thus, c_0 is a Banach space.

7.4 Approximation and Separability

36. Assume first that $\mathcal{S}$ is dense in X. Then for each function g in X and natural number n, there is a function s_n in $\mathcal{S}$ for which $\|s_n - g\| < 1/n$. Then $\lim_{n\to\infty} \|s_n - g\| = 0$. Thus, $\{s_n\}$ is a sequence in $\mathcal{S}$ for which $\lim_{n\to\infty} s_n = g$ in X.

    Conversely, assume that for each g in X, there is a $\{s_n\}$ in $\mathcal{S}$ for which $\lim_{n\to\infty} s_n = g$ in X. Then for each positive ϵ, there is an index N for which $\|s_N - g\| < \epsilon$. Thus, $\mathcal{S}$ is dense in X.

37. By hypothesis, for each function h in $\mathcal{H}$ and positive ϵ, there is a function g in $\mathcal{G}$ for which $\|g - h\| < \epsilon/2$. Again by hypothesis, there is a function f in $\mathcal{F}$ for which $\|f - g\| < \epsilon/2$. Then for each h in $\mathcal{H}$, $\|f - h\| \le \|f - g\| + \|g - h\| < \epsilon/2 + \epsilon/2 = \epsilon$. Thus, $\mathcal{F}$ is dense in $\mathcal{H}$.

38. For each natural number n, subcollection $\mathcal{P}_{n-1}$ of polynomials with rational coefficients and degree $n-1$ is equipotent to Cartesian product $\overbrace{\mathbf{Q} \times \cdots \times \mathbf{Q}}^{n \text{ times}}$, which by Corollary 4(ii) of Section 1.3, is equipotent to $\overbrace{\mathbf{N} \times \cdots \times \mathbf{N}}^{n \text{ times}}$, which is equipotent to $\mathbf{N}$ by Corollary 4(i). Because equipotency is an equivalence relation, $\mathcal{P}_{n-1}$ is countably infinite. Then by Corollary 6 of

Section 1.3, the collection of polynomials with rational coefficients is countable because it is a union $\mathcal{P}_0 \cup \mathcal{P}_1 \cup \mathcal{P}_2 \cup \cdots$ of a countable collection of countable subcollections.

39. Because $g \in L^p(E)$, there is a positive M for which $\|g\|_p \le M$. Because $\mathcal{S}$ is dense in $L^q(E)$, for each function h in $L^q(E)$ and positive ϵ, there is a function f in $\mathcal{S}$ for which $\|h - f\|_q < \epsilon/M$. Then

$$\begin{aligned}
\int_E gh &= \int_E g(h-f) && \textstyle\int_E gf = 0 \text{ because } f \in \mathcal{S} \\
&\le \int_E |g(h-f)| && \text{Theorem 17 (monotonicity) of Section 4.4} \\
&\le \|g\|_p \|h-f\|_q && \text{Hölder's inequality (Section 7.2)} \\
&< M \cdot \epsilon/M = \epsilon && \text{choice of } M \text{ and } f.
\end{aligned}$$

Because that holds for ϵ arbitrarily small, $\int_E gh = 0$. Thus, $g = 0$ by Problem 20. Note that Hölder's inequality is true also if $p = \infty$ because by the symmetry with respect to p, q, we can use compound inequality $1 \le q < \infty$ in the statement of Theorem 1 so that it holds if $q = 1$ and $p = \infty$.

Another solution, only superficially different from the one above, uses the first sentence on page 151 and definition of convergence.

40. **$\mathcal{S}'[a,b]$ is dense in $\mathcal{S}[a,b]$ with respect to $\|\cdot\|_p$.** We use an argument not unlike the proof of Proposition 10. Each function in $\mathcal{S}[a,b]$ is a sum of constant functions on an interval. Consequently, if each such constant function can be arbitrarily closely approximated in the $\|\cdot\|_p$ norm by a function in $\mathcal{S}'[a,b]$, because $\mathcal{S}'[a,b]$ is a linear space, so can each function in $\mathcal{S}[a,b]$. Let φ be a constant function on (c,d). Take a positive ϵ. It suffices to find a rational function ψ constant on (x_1, x_2) and zero on $(x_1, x_2)^C$ with x_1, x_2 also rational for which $\|\psi - \varphi\|_p < \epsilon$. Because $\mathbf{Q}$ is dense in $\mathbf{R}$, we can choose x_1, x_2 such that $x_1 \in [c, c + \epsilon^p/(3|\varphi|^p)]$, $x_2 \in [d - \epsilon^p/(3|\varphi|^p), d]$, $x_1 < x_2$, and $\psi \in [\varphi, \varphi + \epsilon/(3(x_2 - x_1))^{1/p})$ on (x_1, x_2). Then

$$\|\psi - \varphi\|_p = \left(\int_c^d |\psi - \varphi|^p\right)^{1/p} = \left(\int_c^{x_1} |\varphi|^p + \int_{x_1}^{x_2} |\psi - \varphi|^p + \int_{x_2}^d |\varphi|^p\right)^{1/p} < \epsilon.$$

$\mathcal{S}'[a,b]$ is a countable set. For each n, the subcollection $\mathcal{S}'_n$ of functions in $\mathcal{S}'[a,b]$ with $n+1$ partition points is equipotent to a subset of $\overbrace{\mathbf{Q} \times \cdots \times \mathbf{Q}}^{n \text{ times}} \times \overbrace{\mathbf{Q} \times \cdots \times \mathbf{Q}}^{n-1 \text{ times}}$ where the first braced Cartesian product corresponds to the constant rational values on each interval and the second to the choice of interior partition points. (I use the word *subset* because an interior partition point cannot take on an arbitrary rational number but rather one between the partition points on either side of it.) Our Cartesian product is equipotent to $\mathbf{N}$ by Corollary 4 of Section 1.3. Because equipotency is an equivalence relation, $\mathcal{S}'_n$ is countably infinite. Then by Corollary 6 of Section 1.3, $\mathcal{S}'[a,b]$ is countable because it is a union $\mathcal{S}'_1 \cup \mathcal{S}'_2 \cup \cdots$ of a countable collection of countable subcollections.

$\mathcal{F}$ is a countable collection of functions that is dense in $L^p(\mathbf{R})$. $\mathcal{F}$ is countable by Corollary 6 of Section 1.3. As for denseness, take a positive ϵ, and let f be a function in $L^p(\mathbf{R})$. Function f is clearly the limit of sequence $\{f_n\}$ where f_n is f restricted to $[-n, n]$ so that f_n is in $L^p[-n,n]$. Because $\mathcal{S}'[-n,n]$ is dense in $L^p[-n,n]$, there is a function ψ_n in $\mathcal{F}_n$ for which $\|\psi_n - f_n\|_p < \epsilon$. Put ψ to be the limit of $\{\psi_n\}$. Function ψ is in $\mathcal{F}$, and $\psi - f$ is in $L^p(\mathbf{R})$. Then

$$\|\psi - f\|_p = \left(\int_{\mathbf{R}} |\psi - f|^p\right)^{1/p} = \left(\lim_{n\to\infty} \int_{[-n,n]} |\psi_n - f_n|^p\right)^{1/p} = \lim_{n\to\infty} \|\psi_n - f_n\|_p < \epsilon$$

where the first norm is over $\mathbf{R}$ and the other over $[-n, n]$. Thus, $\mathcal{F}$ is dense in $L^p(\mathbf{R})$.

41. $L^{p_2}(E)$ normed by $\|\cdot\|_{p_1}$ is not always a Banach space. For a counterexample, use the first example on page 143 where $p_1 = 1$, $p_2 = 2$ so that $f(x) = 1/\sqrt{x}$ is in $L^1(0,1]$ but not in $L^2(0,1]$. Sequence $\{\chi_{[1/n,1]}/\sqrt{x}\}$ is in $L^2(0,1]$ because

$$\int_0^1 \left|\frac{\chi_{[1/n,1]}}{\sqrt{x}}\right|^2 dx = \int_{1/n}^1 \frac{1}{x}\, dx = \log 1 - \log\frac{1}{n} = 0 + \log n = \log n < \infty.$$

Furthermore, the sequence is Cauchy normed by $\|\cdot\|_1$ because for each positive ϵ, there is an index N given by $\lceil 4/\epsilon^2 \rceil$ such that if $n \geq m \geq N$, then

$$\left\|\frac{\chi_{[1/n,1]}}{\sqrt{x}} - \frac{\chi_{[1/m,1]}}{\sqrt{x}}\right\|_1 = \int_{1/n}^{1/m} \left|\frac{1}{\sqrt{x}}\right| dx = \sqrt{\frac{4}{m}} - \sqrt{\frac{4}{n}} < \sqrt{\frac{4}{m}} \leq \sqrt{\frac{4}{N}} \leq \sqrt{\frac{4\epsilon^2}{4}} = \epsilon.$$

But the function $1/\sqrt{x}$ to which the sequence converges is not in $L^2(0,1]$ and therefore not in $L^2(0,1]$ normed by $\|\cdot\|_1$. Thus, $L^2(0,1]$ normed by $\|\cdot\|_1$ is not a Banach space.

42. Put E to be a set E_0 of measure zero. Let f be a function in $L^\infty(E_0)$ so that there is some nonnegative M for which $|f(x)| \leq M$ for almost all x in E_0. Now because "almost all x in E_0" means on $E_0 \sim E_0$, which is $\emptyset$, number M can (vacuously) take on an arbitrary nonnegative value. Then $\|f\|_\infty = \inf\{M \mid M \geq 0\} = 0$ so that by the nonnegativity property of a norm, $f = 0$, that is, there is only one function (the zero function) in $L^\infty(E_0)$. Then $L^\infty(E_0)$ itself is a countable subset that is dense in $L^\infty(E_0)$ because for each function f in $L^\infty(E_0)$ and positive ϵ, we have that $\|f - f\|_\infty = 0 < \epsilon$. Thus, $L^\infty(E)$ is separable if E has measure zero.

We answer the second part by verifying details in the example on pages 152–153:

η is a one-to-one mapping of $[a, b]$ onto a set of natural numbers. The reason is that $\eta(x_2) \neq \eta(x_1)$. Otherwise $f_{\eta(x_1)}, f_{\eta(x_2)}$ would be identical, giving the contradiction that

$$\begin{aligned} 1 = \left\|\chi_{[a,x_1]} - \chi_{[a,x_2]}\right\|_\infty &= \left\|\chi_{[a,x_1]} - f_{\eta(x_1)} - \chi_{[a,x_2]} + f_{\eta(x_2)}\right\|_\infty \\ &\leq \left\|\chi_{[a,x_1]} - f_{\eta(x_1)}\right\|_\infty + \left\|\chi_{[a,x_2]} - f_{\eta(x_2)}\right\|_\infty < 1/2 + 1/2 = 1. \end{aligned}$$

Now $\eta(x_2) \neq \eta(x_1)$ simply means that η maps each $x \in [a, b]$ to a different natural number.

A set of natural numbers is countable, and $[a, b]$ is not countable. See Theorems 3 and 7 of Section 1.3.

43. We are given that $X_0 \subseteq X$. It remains to show that $X \subseteq X_0$. Let f be a function in X. By the top of page 151, there is a sequence $\{f_n\}$ in X_0 that converges to f in X. By Proposition 4, $\{f_n\}$ is Cauchy. Because X_0 is complete, f is in X_0 by definition (page 145). Thus, $X \subseteq X_0$ so that $X = X_0$.

44. Let $\mathcal{Q}$ be the collection of rational sequences $(q_1, \ldots, q_n, 0, \ldots)$ with a finite number of nonzero terms. By reasoning analogous to our solution of Problem 38, $\mathcal{Q}$ is countable. To prove that $\mathcal{Q}$ is dense in ℓ^p, take a positive ϵ and a sequence a in ℓ^p where $a = (a_1, a_2, \ldots)$ so that $\sum_{k=1}^\infty |a_k|^p < \infty$. By Proposition 20(i) of Section 1.5, there is an index N for which $\sum_{k=N+1}^\infty |a_k|^p < \epsilon^p/2$. Because $\mathbf{Q}$ is dense in $\mathbf{R}$, we can find a rational number q_k such that $|a_k - q_k| < \epsilon/(2N)^{1/p}$ for each k equal to $1, \ldots, N$, which defines a sequence $q = (q_1, \ldots, q_N, 0, \ldots)$ in $\mathcal{Q}$. Then

$$\|a - q\|_p = \left(\sum_{k=1}^\infty |a_k - q_k|^p\right)^{1/p} = \left(\sum_{k=1}^N |a_k - q_k|^p + \sum_{k=N+1}^\infty |a_k|^p\right)^{1/p} < \left(\frac{\epsilon^p}{2} + \frac{\epsilon^p}{2}\right)^{1/p} = \epsilon$$

which shows that $\mathcal{Q}$ is dense in ℓ^p. Thus, ℓ^p is separable.

The collection $2^{\mathbf{N}}$ of sets of natural numbers is uncountable by Problem 22 of Section 1.3. To show that ℓ^∞ is not separable, argue as in the example on pages 152–153 with details verified by our solution to the second part of Problem 42. Suppose that there were a countable set $\{b_n\}_{n=1}^\infty$ (each b_n is a sequence) that is dense in ℓ^∞. For two different sequences e, e' of 0s and 1s, select natural numbers $\eta(e), \eta(e')$ for which $\|e - b_{\eta(e)}\|_\infty < 1/2$, and $\|e' - b_{\eta(e')}\|_\infty < 1/2$. Observe that $\|e - e'\|_\infty = 1$. Now by our solution to Problem 22 of Section 1.3, the set of all sequences of 0s and 1s is equipotent to $2^{\mathbf{N}}$ and therefore uncountable. Consequently, η is a one-to-one mapping of an uncountable set onto a set of natural numbers. We conclude from that contradiction that ℓ^∞ is not separable.

45. We answer this problem by verifying details in the paragraph before Theorem 12:

Each f in $\mathcal{F}$ is the limit in $L^p(\mathbf{R})$ of a sequence of continuous, piecewise linear functions, each of which vanishes outside a bounded set. We are given an f that is zero outside $[a, b]$ and a step function on $[a, b]$ with partition $\{a = x_0, x_1, \ldots, x_{m-1}, x_m = b\}$ and numbers $c_1, \ldots, c_m$ such that $f(x) = c_k$ if $x_{k-1} < x < x_k$ where $1 \le k \le m$. Define a sequence $\{f_n\}$ of continuous, piecewise linear functions on $\mathbf{R}$ as follows: Put f_n to 0 outside (a, b). For k equal to $1, \ldots, m-1$ and with δ_{nk} equal to $\min\{1/n, (x_k - x_{k-1})/2\}$ and c_0 equal to 0, define f_n by

$$f_n(x) = \begin{cases} c_{k-1} + \left(\dfrac{c_k - c_{k-1}}{\delta_{nk}}\right)(x - x_{k-1}) & x_{k-1} < x < x_{k-1} + \delta_{nk} \\ c_k & x_{k-1} + \delta_{nk} \le x \le x_k, \end{cases}$$

and for k equal to m, define f_n by

$$f_n(x) = \begin{cases} c_{m-1} + \left(\dfrac{c_m - c_{m-1}}{\delta_{nm}}\right)(x - x_{m-1}) & x_{m-1} < x < x_{m-1} + \delta_{nm} \\ c_m & x_{m-1} + \delta_{nm} \le x \le x_m - \delta_{nm} = b - \delta_{nm} \\ c_m - \dfrac{c_m}{\delta_{nm}}(x - (x_m - \delta_{nm})) & x_m - \delta_{nm} < x < x_m = b. \end{cases}$$

Then

$$\begin{aligned} \lim_{n\to\infty} \|f - f_n\|_p &= \lim_{n\to\infty} \left(\int_{\mathbf{R}} |f - f_n|^p \right)^{1/p} \\ &= \lim_{n\to\infty} \left(\sum_{k=1}^{m} \left| \frac{(c_k - c_{k-1})\delta_{nk}}{2} \right|^p + \left| \frac{c_m \delta_{nm}}{2} \right|^p \right)^{1/p} \\ &= 0 \end{aligned}$$

because $\lim_{n\to\infty} \delta_{nk} = 0$ since $\delta_{nk} = 1/n$ for large enough n.

$\mathcal{F}'$ is dense in $L^p(\mathbf{R})$. We first show that $\mathcal{F}'$ is a subset of $L^p(\mathbf{R})$. By the extreme value theorem (Section 1.6), each function f in $\mathcal{F}'$ is bounded on $[a, b]$ so therefore on all of $\mathbf{R}$ because f vanishes outside $[a, b]$. By Proposition 3 of Section 3.1, f is measurable. By Theorem 7 of Section 5.3, f is integrable over $[a, b]$ so therefore over all $\mathbf{R}$ because f vanishes outside $[a, b]$. Consequently, f belongs to $L^1(\mathbf{R})$. Then f belongs to $L^p(\mathbf{R})$ by Problem 13. Hence, $\mathcal{F}'$ is a subset of $L^p(\mathbf{R})$.

It remains to show, by the top of page 151, that for each g in $L^p(\mathbf{R})$ there is a sequence $\{f_n\}$ in $\mathcal{F}'$ for which $\lim_{n\to\infty} f_n = g$ with respect to $\|\cdot\|_p$, that is, $\lim_{n\to\infty} \|g - f_n\|_p = 0$. Indeed, because $\mathcal{F}$ is dense in $L^p(\mathbf{R})$, for each positive ϵ, there is function f in $\mathcal{F}$ for which $\|g - f\|_p < \epsilon$. We showed above that for each f, there is sequence $\{f_n\}$ in $\mathcal{F}'$ for which $\lim_{n\to\infty} \|f - f_n\|_p = 0$. Then starting with Minkowski's inequality, $\lim_{n\to\infty} \|g - f_n\|_p \le \lim_{n\to\infty} (\|g - f\|_p + \|f - f_n\|_p) < \epsilon$. Because that holds for ϵ arbitrarily small, $\lim_{n\to\infty} \|g - f_n\|_p = 0$. Thus, $\mathcal{F}'$ is dense in $L^p(\mathbf{R})$. (Note that we cannot use statement (13) in this case because $\mathcal{F}'$ is not dense in $\mathcal{F}$ since $\mathcal{F}'$ is not a subset of $\mathcal{F}$.)

46. If a or b is zero, or $b = a$, the inequality is true. Now assume that neither a nor b is zero, and that $b \ne a$. In the following cases, the first inequality comes from the fact that $x \mapsto x^p$ is a convex function on $(0, \infty)$ if $p \ge 1$ by the example on page 131.

- $\operatorname{sgn}(b) = \operatorname{sgn}(a)$: If $|b| < |a|$, then

$$\begin{aligned}\left|\operatorname{sgn}(a)|a|^{1/p} - \operatorname{sgn}(b)|b|^{1/p}\right|^p &= \left||a|^{1/p} - |b|^{1/p}\right|^p \\ &= \left||a|^{1/p} - |b|^{1/p}\right|^p + \left(|b|^{1/p}\right)^p - |b| \\ &\le \left||a|^{1/p} - |b|^{1/p} + |b|^{1/p}\right|^p - |b| \\ &= |a| - |b| \\ &= |a - b| \\ &< 2^p|a - b|.\end{aligned}$$

  That result also holds if $|b| > |a|$ because we can interchange a and b in the first and last expressions above and get an equivalent inequality.
- $\operatorname{sgn}(b) = -\operatorname{sgn}(a)$: The claim holds also in this case because

$$\begin{aligned}\left|\operatorname{sgn}(a)|a|^{1/p} - \operatorname{sgn}(b)|b|^{1/p}\right|^p &= \left(|a|^{1/p} + |b|^{1/p}\right)^p \\ &= 2^p\left(|a|^{1/p}/2 + |b|^{1/p}/2\right)^p \\ &\le 2^{p-1}(|a| + |b|) \\ &< 2^p|a - b|.\end{aligned}$$

47. If a or b is zero, or $b = a$, the inequality is true. Now assume that neither a nor b is zero, and $b \ne a$.

- $\operatorname{sgn}(a) = \operatorname{sgn}(b)$: By the reasoning in our solution to Problem 46, we can assume without loss of generality that $|b| < |a|$ so that $|\operatorname{sgn}(a)|a|^p - \operatorname{sgn}(b)|b|^p| = |a|^p - |b|^p$. The generalized binomial theorem $(x + y)^p = x^p + px^{p-1}y + \cdots$ leads to inequality $(x + y)^p - x^p \le py(x + y)^{p-1}$, which holds even if the infinite series does not converge. Into that inequality substitute $|b|$ for x and $|a| - |b|$ for y to get that

$$\begin{aligned}(|b| + |a| - |b|)^p - |b|^p &= |a|^p - |b|^p \\ &< p(|a| - |b|)(|b| + |a| - |b|)^{p-1} \\ &< p|a - b|(|a| + |b|)^{p-1}.\end{aligned}$$

- Otherwise, assume without loss of generalization that $a > 0$ and $b < 0$. Then

$$\begin{aligned}|\operatorname{sgn}(a)|a|^p - \operatorname{sgn}(b)|b|^p| &= a^p + |b|^p \\ &\le (a + |b|)^p && x \mapsto x^p \text{ is convex} \\ &= |a - b|(|a| + |b|)^{p-1} \\ &< p|a - b|(|a| + |b|)^{p-1}.\end{aligned}$$

48. Function $\Phi(f)$ belongs to $L^p(E)$ because

$$\int_E |\Phi(f)|^p = \int_E \left|\text{sgn}(f)|f|^{1/p}\right|^p = \int_E |f| < \infty$$

since f is in $L^1(E)$. Moreover, for all f, g in $L^1(E)$,

$$\begin{aligned}\|\Phi(f) - \Phi(g)\|_p^p &= \int_E |\Phi(f) - \Phi(g)|^p \\ &= \int_E \left|\text{sgn}(f)|f|^{1/p} - \text{sgn}(g)|g|^{1/p}\right|^p \\ &\le 2^p \int_E |f - g| \\ &= 2^p \|f - g\|_1.\end{aligned}$$

Next, if $\{f_n\} \to f$ in $L^1(E)$, then each f_n and f belong to $L^1(E)$, and $\lim_{n\to\infty} \|f - f_n\|_1 = 0$. By the previous, each $\Phi(f_n)$ and $\Phi(f)$ belong to $L^p(E)$, and

$$\lim_{n\to\infty} \|\Phi(f) - \Phi(f_n)\|_p = \left(\lim_{n\to\infty} \|\Phi(f) - \Phi(f_n)\|_p^p\right)^{1/p} \le 2\left(\lim_{n\to\infty} \|f - f_n\|_1\right)^{1/p} = 0,$$

where we justify the first step by Proposition 21 of Section 1.6 and the fact that function $x \mapsto x^{1/p}$ is continuous on $[0, \infty)$. Thus, Φ is a continuous mapping of $L^1(E)$ into $L^p(E)$.

Next, we show that Φ is one-to-one by showing that if $\Phi(f) = \Phi(g)$, then $f = g$. Indeed, if $\text{sgn}(f(x))|f(x)|^{1/p} = \text{sgn}(g(x))|g(x)|^{1/p}$ for each x in E, then $\text{sgn}(f) = \text{sgn}(g)$ so that $|f|^{1/p} = |g|^{1/p}$. Consequently, $|f| = |g|$ because $x \mapsto x^{1/p}$ is one-to-one on $[0, \infty)$. Hence, $f = g$ because $\text{sgn}(f) = \text{sgn}(g)$.

We show that the image of Φ is $L^p(E)$ by showing that each function h in $L^p(E)$ is $\Phi(f)$ for some f in $L^1(E)$. Define f on E by $f(x) = \text{sgn}(h(x))|h(x)|^p$. Function f is in $L^1(E)$ because

$$\int_E |f| = \int_E |\text{sgn}(h)|h|^p| = \int_E |h|^p < \infty$$

since h is in $L^p(E)$. Moreover, $\Phi(f) = \text{sgn}(\text{sgn}(h)|h|^p)|\text{sgn}(h)|h|^p|^{1/p} = h$. Thus, the image of Φ is $L^p(E)$. Along the way, we found a formula $\text{sgn}(h)|h|^p$ for the inverse mapping Φ^{-1}.

To see that Φ^{-1} is a continuous mapping from $L^p(E)$ to $L^1(E)$, consider a sequence $\{h_n\}$ of functions such that $\{h_n\} \to h$ in $L^p(E)$. Then each h_n and h belong to $L^p(E)$ so that by the previous, each $\Phi^{-1}(h_n)$ and $\Phi^{-1}(h)$ belong to $L^1(E)$. Furthermore,

$$\begin{aligned}\lim_{n\to\infty} \int_E \left|\Phi^{-1}(h) - \Phi^{-1}(h_n)\right| &= \lim_{n\to\infty} \int_E |\text{sgn}(h)|h|^p - \text{sgn}(h_n)|h_n|^p| \\ &\le p \lim_{n\to\infty} \int_E |h - h_n|(|h| + |h_n|)^{p-1} && \text{Problem 47} \\ &\le p \lim_{n\to\infty} \|h - h_n\|_p \cdot \left\|(|h| + |h_n|)^{p-1}\right\|_{p/(p-1)} && \text{Theorem 1} \\ &\le p \cdot 0 \cdot \lim_{n\to\infty} \||h| + |h_n|\|_p^{p-1} \\ &= 0.\end{aligned}$$

In order to apply Hölder's inequality, function $(|h|+|h_n|)^{p-1}$ needs to belong to $L^{p/(p-1)}(E)$, which is true by the second part of Theorem 1 because each $|h_n|$ and $|h|$ belong to $L^p(E)$, and $L^p(E), L^{p/(p-1)}(E)$ are linear spaces. That $\lim_{n\to\infty} \|h - h_n\|_p = 0$ is just our initial assumption that $\{h_n\} \to h$ in $L^p(E)$. Finally, we are justified in setting the next-to-last expression to zero because $\lim_{n\to\infty} \| |h| + |h_n| \|_p^{p-1}$ is bounded since, once again, each $|h_n|$ and $|h|$ belong to $L^p(E)$. Consequently, $\{\Phi^{-1}(h_n)\} \to \Phi^{-1}(h)$ in $L^1(E)$ by the sentence before the definition on page 145. Thus, Φ^{-1} is a continuous mapping from $L^p(E)$ to $L^1(E)$. (This problem is an example in Section 11.4.)

49. Separability of $L^1(E)$ implies that there is a countable subset $\mathcal{F}$ such that for each g in $L^1(E)$ and positive ϵ, there is an f in $\mathcal{F}$ for which $\|f - g\|_1 < (\epsilon/2)^p$. Because Φ is a function whose image is $L^p(E)$, image $\Phi(\mathcal{F})$ is a countable subset of $\Phi(L^1(E))$, which is equal to $L^p(E)$, and each function in $L^p(E)$ can be expressed as $\Phi(g)$ for some g in $L^1(E)$. Then for each function $\Phi(g)$ in $L^p(E)$, we have a function $\Phi(f)$ in $\Phi(\mathcal{F})$ for which

$$\|\Phi(f) - \Phi(g)\|_p \leq 2\|f - g\|_1^{1/p} < 2((\epsilon/2)^p)^{1/p} = \epsilon.$$

Thus, separability of $L^1(E)$ implies separability of $L^p(E)$ for p in $(1, \infty)$.

50. Suppose, to get a contradiction, that there were such a mapping. Because Φ is onto, we can express each function in $L^\infty[a, b]$ by $\Phi(g)$ for some g in $L^1[a, b]$. Because Φ is continuous, for each positive ϵ, there is a positive δ for which if $f \in L^1[a, b]$ and $\|f - g\|_1 < \delta$, then $\|\Phi(f) - \Phi(g)\|_\infty < \epsilon$. Indeed, separability of $L^1[a, b]$ (Theorem 11) guarantees existence of such an f and that we can find f in a countable subset $\mathcal{F}$. We deduce from all the previous that there is a function $\Phi(f)$ in countable set $\Phi(\mathcal{F})$ contained in $L^\infty[a, b]$ for which $\|\Phi(f) - \Phi(g)\|_\infty < \epsilon$. Hence, $L^\infty[a, b]$ is separable, a contradiction by Problem 42. Thus, there is no continuous mapping from $L^1[a, b]$ onto $L^\infty[a, b]$.

51. $C_c(E)$ is a subset of $L^p(E)$ by an argument similar to that which we used in our solution to Problem 45 to show that $\mathcal{F}'$ is a subset of $L^p(\mathbf{R})$. Now by the proof of Theorem 11 (see the last sentence in particular), the collection, say $\mathcal{F}''$, consisting of restrictions to E of functions in $\mathcal{F}$ is dense in $L^p(E)$. Then by definition, for each function g in $L^p(E)$ and positive ϵ, there is a function ψ in $\mathcal{F}''$ for which $\|\psi - g\|_p < \epsilon/2$. Let f be a function in $C_c(E)$. Function $|f - \psi|^p$ is integrable over E because f, ψ belong to $L^p(E)$, and $L^p(E)$ is a linear space. By Proposition 23 of Section 4.6, there is a positive δ for which

$$\text{if subset } E \sim F \text{ of } E \text{ is measurable and } m(E \sim F) < \delta, \text{ then } \int_{E\sim F} |f - \psi|^p < (\epsilon/2)^p.$$

By Lusin's theorem, there is a continuous function f on $\mathbf{R}$ and a closed set F contained in E for which $\psi = f$ on F and $m(E \sim F) < \delta$. We can take for f a function from $C_c(\mathbf{R})$ and then restrict f to E so that f is in $C_c(E)$ because f is continuous on E since it is continuous on $\mathbf{R}$, and because, as described in the proof of Theorem 11, ψ vanishes outside a bounded set. Then

$$
\begin{aligned}
\|f - g\|_p &\leq \|f - \psi\|_p + \|\psi - g\|_p && \text{Minkowski's inequality} \\
&< \left(\int_E |f - \psi|^p\right)^{1/p} + \epsilon/2 && \text{definition of } \|\cdot\|_p;\ \mathcal{F}'' \text{ dense in } L^p(E) \\
&= \left(\int_{E\sim F} |f - \psi|^p\right)^{1/p} + \epsilon/2 && f - \psi = 0 \text{ on } F \\
&< ((\epsilon/2)^p)^{1/p} + \epsilon/2 && E \sim F \text{ measurable, } m(E \sim F) < \delta \\
&= \epsilon.
\end{aligned}
$$

Thus, $C_c(E)$ is dense in $L^p(E)$.

Chapter 8

The L^p Spaces: Duality and Weak Convergence

8.1 The Riesz Representation for the Dual of L^p, $1 \le p < \infty$

1. Denote $\sup\{T(f) \mid f \in X, \|f\| \le 1\}$ by $\|T\|'_*$. Take a nonzero g in X. We have that $T(g/\|g\|) \le \|T\|'_*$ because $g/\|g\|$ has norm 1. Then $T(g) \le \|T\|'_*\|g\|$ by linearity of T and positive homogeneity of a norm. Now $-g$ is also in X because X is a linear space, so $T(-g) \le \|T\|'_*\|g\|$, or $-T(g) \le \|T\|'_*\|g\|$, so that $|T(g)| \le \|T\|'_*\|g\|$ for all g in X including the case where $g = 0$ since $T(0) = 0$. Hence, $\|T\|_* \le \|T\|'_*$ by definition of $\|\cdot\|_*$.

   We now get the opposite inequality. As stated in the text, it is easy to see that inequality (5) holds if $M = \|T\|_*$, that is, $T(g) \le \|T\|_*\|g\|$ for all g in X. In particular, if $\|f\| \le 1$, then $\|T(f)\| \le \|T\|_*\|f\| \le \|T\|_*$. Hence, $\|T\|'_* \le \|T\|_*$. Thus, $\|T\|_* = \|T\|'_*$.

2. Denote the collection of bounded linear functionals on X by $\mathcal{B}$. Refer to the definition of a linear space on page 253. Let α, β be real numbers; T, S in $\mathcal{B}$; and f a function in X. By the next-to-last sentence on page 155, we define linear combinations of linear functionals pointwise, i.e., addition in $\mathcal{B}$ is defined by $(T+S)(f) = T(f) + S(f)$ and scalar multiplication by $(\alpha T)(f) = \alpha T(f)$. Then $T+S$ and αT are in $\mathcal{B}$ because they are bounded linear functionals on X; in particular, 0 and $-T$ are in $\mathcal{B}$ because $0 = 0T$ and $-T = -1T$. By the properties of real numbers (recall T is real-valued), addition in $\mathcal{B}$ is associative and commutative, the identity is the zero linear functional 0, and the inverse of T is $-T$. So far, we have shown that $\mathcal{B}$ is an abelian (commutative) group with a scalar product. Furthermore, again using properties of real numbers,

$$
\begin{aligned}
((\alpha+\beta)T)(f) &= (\alpha+\beta)T(f) = \alpha T(f) + \beta T(f), \\
(\alpha(T+S))(f) &= \alpha(T+S)(f) = \alpha(T(f)+S(f)) = \alpha T(f) + \alpha S(f), \\
((\alpha\beta)T)(f) &= (\alpha\beta)T(f) = \alpha(\beta T(f)) = \alpha(\beta T)(f) = (\alpha(\beta T))(f), \text{ and} \\
(1T)(f) &= 1T(f) = T(f).
\end{aligned}
$$

   Hence, $\mathcal{B}$ is a linear space. To show that $\|\cdot\|_*$ is a norm on $\mathcal{B}$, verify the conditions listed on page 137. The triangle inequality holds because

$$
\begin{aligned}
\|T+S\|_* &= \sup\{T(f)+S(f) \mid \|f\| \le 1\} && \text{formula (8)} \\
&\le \sup\{T(f) \mid \|f\| \le 1\} + \sup\{S(f) \mid \|f\| \le 1\} \\
&= \|T\|_* + \|S\|_*.
\end{aligned}
$$

   Norm $\|\cdot\|_*$ is nonnegative because by definition, $\|T\|_*$ is the infimum of all nonnegative M for which $|T(f)| \le M\|f\|$ for all f. That definition also implies that if $T = 0$, then $\|T\|_* = 0$. Conversely, if $\|T\|_* = 0$, then $|T(f)| = 0$ for all f; consequently, $T = 0$. To show positive homogeneity, first note that

$$
\begin{aligned}
\|-\alpha T\|_* = \|\alpha(-T)\|_* &= \sup\{\alpha(-T(f)) \mid \|f\| \le 1\} \\
&= \sup\{\alpha T(-f) \mid \|f\| \le 1\} \\
&= \sup\{\alpha T(f) \mid \|-f\| \le 1\} \\
&= \sup\{\alpha T(f) \mid \|f\| \le 1\} = \|\alpha T\|_*,
\end{aligned}
$$

   which shows that $\|\alpha T\|_* = \||\alpha|T\|$. Then

$$\|\alpha T\|_* = \||\alpha|T\|_* = \sup\{|\alpha|T(f) \mid \|f\| \le 1\} = |\alpha|\sup\{T(f) \mid \|f\| \le 1\} = |\alpha|\|T\|_*.$$

where we applied the nonnegativity property in the third step. Thus, $\mathcal{B}$ is a linear space on which $\|\cdot\|_*$ is a norm.

3. Assume first that T is bounded and $\{f_n\} \to f$ in X. Then for every positive ϵ, there is an index N for which if $n \ge N$, then $\|f - f_n\| < \epsilon/\|T\|_*$. Then by inequality (6), $|T(f) - T(f_n)| \le \|T\|_*\|f - f_n\| < \|T\|_*\epsilon/\|T\|_* = \epsilon$ if $n \ge N$. Thus, $\{T(f_n)\} \to T(f)$.

   Conversely, assume that if for every positive δ, there is an N_δ for which if $n \ge N_\delta$, then $\|f - f_n\| < \delta$, then there is an index N_ϵ for which if $n \ge N_\epsilon$, then $|T(f) - T(f_n)| < \epsilon$. The preceding holds if we replace each of N_δ, N_ϵ with $\max\{N_\delta, N_\epsilon\} = N$. Now let g be a function in X. Put f to $\delta g/\|g\| + f_N$ for some $\{f_n\}$ that converges to f in X. Then by linearity of T,

$$|T(g)| = |T((f - f_N)\|g\|/\delta)| = |T(f) - T(f_N)|\|g\|/\delta < (\epsilon/\delta)\|g\|.$$

   Thus, T is bounded.

4. Assume first that T is bounded. Then T is Lipschitz by inequality (6) with Lipschitz constant $\|T\|_*$ by definitions of a Lipschitz constant and $\|T\|_*$.

   Conversely, assume that T is Lipschitz. Because the requirement for being Lipschitz holds for all h in X, we can put h to be the zero function 0. Then by linearity of T, we have that $T(h) = T(0) = T(0 \cdot 0) = 0T(0) = 0$ where the first 0 in $T(0 \cdot 0)$ is the zero scalar and the second is the zero function. Then $|T(g)| = |T(g) - T(h)| \le c\|g - h\| = c\|g\|$ for all g in X. Thus, T is bounded with $\|T\|_*$ equal to the Lipschitz constant.

5. By Theorem 12 of Section 7.4, the linear space $C_c(E)$ of continuous real-valued functions on E that vanish outside a bounded set is dense in $L^p(E)$. By definition of denseness, $C_c(E)$ is a subset of $L^p(E)$; consequently, $C_c(E)$ is also a subset of the subset $\mathcal{F}(E)$ of functions in $L^p(E)$ that vanish outside a bounded set. Thus, because $\mathcal{F}(E)$ contains $C_c(E)$, $\mathcal{F}(E)$ is dense in $L^p(E)$.

   Next, consider constant function g defined by $g = 2$. Function g is in $L^\infty(\mathbf{R})$ because g has an essential upper bound, namely, 2. Referring now to the definition on page 150, put ϵ to 1. Any function f in $\mathcal{F}(\mathbf{R})$ vanishes somewhere on $\mathbf{R}$ so that $\|f - g\|_\infty \ge 2 \not< 1 = \epsilon$. Thus, $\mathcal{F}(\mathbf{R})$ is not dense in $L^\infty(\mathbf{R})$.

6. This problem refers to the second sentence of the proof of Theorem 5. Function Φ is Lipschitz (definition on page 25) because for all x', x in $[a, b]$,

$$\begin{aligned} |\Phi(x') - \Phi(x)| &= \left|T\left(\chi_{[a,x')}\right) - T\left(\chi_{[a,x)}\right)\right| && \text{definition of } \Phi \\ &\le \|T\|_*\left\|\chi_{[a,x')} - \chi_{[a,x)}\right\|_1 && \text{inequality (6) (} T \text{ a bounded linear functional)} \\ &= \|T\|_* \int_a^b \left|\chi_{[a,x')} - \chi_{[a,x)}\right| && \text{definition of } \|\cdot\|_1 \\ &= \|T\|_*|x' - x|. \end{aligned}$$

   Then by Proposition 7 of Section 6.4, Φ is absolutely continuous on $[a, b]$. To establish the Riesz Representation theorem in this case where $p = 1$, skip to the second paragraph of the proof of Theorem 5 and follow from there to the end of the proof of the Riesz Representation theorem.

7. **Riesz Representation Theorem for the Dual of ℓ^p** Let p be in $[1, \infty)$, and let q be the conjugate of p. For each sequence $(b_1, b_2, \dots)$ denoted by b in ℓ^q, define functional $\mathcal{R}_b$ on ℓ^p by

$$\mathcal{R}_b(a) = \sum_{k=1}^{\infty} b_k a_k \text{ for all } (a_1, a_2, \dots) \text{ denoted by } a \text{ in } \ell^p.$$

Then $\mathcal{R}_b$ is a bounded linear functional on ℓ^p; $\|\mathcal{R}_b\|_* = \|b\|_q$; and for each bounded linear functional T on ℓ^p, there is a unique sequence b in ℓ^q for which $T = \mathcal{R}_b$.

Proof By linearity of summation, $\mathcal{R}_b$ is linear. By the triangle inequality and our solution to Problem 12 of Section 7.2,

$$|\mathcal{R}_b(a)| = \left|\sum_{k=1}^{\infty} b_k a_k\right| \le \sum_{k=1}^{\infty} |b_k a_k| \le \|b\|_q \|a\|_p \text{ for all } a \text{ in } \ell^p$$

from which we infer that $\mathcal{R}_b$ is a bounded linear functional on ℓ^p and $\|\mathcal{R}_b\|_* \le \|b\|_q$. If $p > 1$, according to that same problem (with p, q interchanged), the conjugate sequence b^* defined by

$$b_k^* = \begin{cases} \|b\|_q^{1-q} \operatorname{sgn}(b_k)|b_k|^{q-1} & b_k \neq 0, \\ 0 & b_k = 0 \end{cases}$$

belongs to ℓ^p, $\mathcal{R}_b(b^*) = \|b\|_q$, and $\|b^*\|_p = 1$. It follows from formula (8) that $\|\mathcal{R}_b\|_* = \|b\|_q$. If $p = 1$, we argue by contradiction. If $\|b\|_\infty > \|\mathcal{R}_b\|_*$, there is an index j such that $|b_j| > \|\mathcal{R}_b\|_*$. Define a to be the sequence with jth term equal to $\operatorname{sgn}(b_j)$ and all other terms zero. Then $\|a\|_1 = 1$ and yet $\mathcal{R}_b(a) > \|\mathcal{R}_b\|_*$, which is a contradiction.

By linearity of summation, for each b, b' in ℓ^q, we have that $\mathcal{R}_{b'} - \mathcal{R}_b = \mathcal{R}_{b'-b}$. Hence if $\mathcal{R}_{b'} = \mathcal{R}_b$, then $\mathcal{R}_{b'-b} = 0$ and therefore $\|b' - b\|_q = 0$ so that $b' = b$. Therefore, for a bounded linear functional T on ℓ^p, there is at most one sequence b in ℓ^q for which $T = \mathcal{R}_b$. It remains to show that for each bounded linear functional T on ℓ^p, there is a sequence b in ℓ^q for which $T = \mathcal{R}_b$. Such a sequence does exist—it is $(T(e_1), T(e_2), \dots)$ denoted by b where e_j is the sequence (not the jth term of a sequence e) with jth term 1 and all other terms zero; for example, $e_2 = (0, 1, 0, 0, \dots)$. Sequence b is well defined because each e_j belongs to ℓ^p since $\sum_{k=1}^{\infty} |e_j^k|^p = 1 < \infty$ where e_j^k is the kth term of sequence e_j, and

$$\begin{aligned}
T(a) &= T((a_1, a_2, \dots)) \\
&= T(a_1(1, 0, \dots) + a_2(0, 1, \dots) + \cdots) \\
&= T(a_1 e_1 + a_2 e_2 + \cdots) \\
&= T\left(\sum_{k=1}^{\infty} a_k e_k\right) \\
&= \sum_{k=1}^{\infty} a_k T(e_k) && \text{linearity of } T \\
&= \sum_{k=1}^{\infty} a_k b_k \\
&= \mathcal{R}_b(a).
\end{aligned}$$

Lastly, we show that b belongs to ℓ^q. If $p = 1$ and $q = \infty$, then

$$\begin{aligned}
\|b\|_\infty &= \sup_{1 \le k < \infty} |b_k| \\
&= \sup_{1 \le k < \infty} |T(e_k)| \\
&\le \sup_{1 \le k < \infty} \|T\|_* \|e_k\|_\infty && \text{inequality (5)} \\
&= \|T\|_* \cdot 1 \\
&< \infty.
\end{aligned}$$

If $p > 1$, define sequence d given by $(d_1, d_2, \dots)$ where $d_k = |b_k|^{q-1} \operatorname{sgn}(b_k)$. Any truncation $(d_1, \dots, d_K, 0, 0, \dots)$ of d belongs to ℓ^p. Then

$$\begin{aligned}
\sum_{k=1}^{K} |b_k|^q &= \sum_{k=1}^{K} |b_k|^{q-1} \operatorname{sgn}(b_k) b_k \\
&= \sum_{k=1}^{K} d_k T(e_k) \\
&= T\left(\sum_{k=1}^{K} d_k e_k\right) \\
&\leq \|T\|_* \left\| \{d_k\}_{k=1}^{K} \right\|_p && \text{inequality (5)} \\
&= \|T\|_* \left(\sum_{k=1}^{K} |d_k|^p\right)^{1/p} \\
&= \|T\|_* \left(\sum_{k=1}^{K} |b_k|^{(q-1)p}\right)^{1/p} \\
&= \|T\|_* \left(\sum_{k=1}^{K} |b_k|^q\right)^{1/p},
\end{aligned}$$

and solving for $\|T\|_*$ yields that

$$\left(\sum_{k=1}^{K} |b_k|^q\right)^{1-1/p} = \left(\sum_{k=1}^{K} |b_k|^q\right)^{1/q} \leq \|T\|_*.$$

The right side is independent of K. Therefore, $\sum_{k=1}^{\infty} |b_k|^q \leq \|T\|_*^q < \infty$. Thus, b belongs to ℓ^q.

8. Although it is more efficient to determine the dual space c^* of c first and then specialize the result to c_0, we determine c_0^* first to facilitate understanding our subsequent representation of c^* and also to provide a simpler solution for the c_0 case. For each sequence $(b_1, b_2, \dots)$, denoted by b, in ℓ^1, define functional $\mathcal{R}_b$ on c_0 by $\mathcal{R}_b(a) = \sum_{k=1}^{\infty} b_k a_k$ for all $(a_1, a_2, \dots)$, denoted by a, in c_0. By linearity of summation, $\mathcal{R}_b$ is linear. By the triangle inequality, the fact that c_0 is a subset of ℓ^∞ by Theorem 19(iv) of Section 1.6, and our solution to Problem 12 of Section 7.2,

$$|\mathcal{R}_b(a)| = \left|\sum_{k=1}^{\infty} b_k a_k\right| \leq \sum_{k=1}^{\infty} |b_k a_k| \leq \|b\|_1 \|a\|_\infty \text{ for all } a \text{ in } c_0,$$

from which we infer that $\mathcal{R}_b$ is a bounded linear functional on c_0.

By linearity of summation, for each b, b' in ℓ^1, we have that $\mathcal{R}_{b'} - \mathcal{R}_b = \mathcal{R}_{b'-b}$. Hence if $\mathcal{R}_{b'} = \mathcal{R}_b$, then $\mathcal{R}_{b'-b} = 0$ and therefore $\|b' - b\|_1 = 0$ so that $b' = b$. Therefore, for a bounded linear functional T on c_0, there is at most one sequence b in ℓ^1 for which $T = \mathcal{R}_b$. As in our solution to Problem 7, such a sequence b does exist—it is $(T(e_1), T(e_2), \dots)$. Sequence b is well defined because each e_j belongs to c_0 since $e_j^k = 0$ if $k > j$. To show that b belongs to ℓ^1, define sequence s by $s_k = \operatorname{sgn}(b_k)$. Any truncation $(s_1, \dots, s_K, 0, 0, \dots)$ of

s belongs to c_0 for the same reason that each e_j belongs to c_0. Then

$$\begin{aligned}\sum_{k=1}^{K}|b_k| &= \sum_{k=1}^{K} s_k b_k \\ &= \sum_{k=1}^{K} s_k T(e_k) \\ &= T\left(\sum_{k=1}^{K} s_k e_k\right) \\ &\leq \|T\|_* \left\|\{s_k\}_{k=1}^{K}\right\|_\infty \\ &= \|T\|_* \sup_{1\leq k<\infty} |s_k| \\ &= \|T\|_* \cdot 1.\end{aligned}$$

The right side is independent of K. Therefore, $\sum_{k=1}^{\infty}|b_k| \leq \|T\|_* < \infty$. Hence, b belongs to ℓ^1.

To summarize the foregoing, each b in ℓ^1 defines a bounded linear functional $\mathcal{R}_b$ in c_0^*, and there is exactly one b in ℓ^1 for which a bounded linear functional in c_0^* is an $\mathcal{R}_b$. We have thus determined that c_0^* is exactly the space of R_b for which b belongs to ℓ^1.

For c^*, we get a similar result by modifying the above argument. For each b in ℓ^1, extend $\mathcal{R}_b$ to be on all of c and defined by $\mathcal{R}_b(a) = b_0 a_\infty + \sum_{k=1}^{\infty} b_k a_k$ for all a in c where a_∞ denotes $\lim_{k\to\infty} a_k$ and indexing for the terms of b starts at 0 (to simplify notation). By linearity of summation and convergence of real sequences (Theorem 18 of Section 1.5), $\mathcal{R}_b$ is linear. We infer from inequality

$$|\mathcal{R}_b(a)| = \left|b_0 a_\infty + \sum_{k=1}^{\infty} b_k a_k\right| \leq |b_0 a_\infty| + \sum_{k=1}^{\infty}|b_k a_k| \leq \|b\|_1 \|a\|_\infty \text{ for all } a \text{ in } c$$

that $\mathcal{R}_b$ is a bounded linear functional on c.

By the same reasoning as before, there is exactly one sequence b in ℓ^1 for which a bounded linear functional T on c is an $\mathcal{R}_b$, and that b is $\left(T(e) - \sum_{j=1}^{\infty} T(e_j), T(e_1), T(e_2), \ldots\right)$ where e_1, e_2, ... are as before, and e is the sequence each term of which is 1, that is, $e = (1, 1, \ldots)$. Sequence b is well defined because e, e_1, e_2, ... belong to c, and

$$\begin{aligned}T(a) &= T(a_\infty \cdot 0 + a_1 e_1 + a_2 e_2 + \cdots) \\ &= T\left(a_\infty\left(e - \sum_{k=1}^{\infty} e_k\right) + \sum_{k=1}^{\infty} a_k e_k\right) \\ &= a_\infty\left(T(e) - \sum_{k=1}^{\infty} T(e_k)\right) + \sum_{k=1}^{\infty} a_k T(e_k) \\ &= \mathcal{R}_b(a).\end{aligned}$$

Sequence b belongs to ℓ^1 because

$$\sum_{k=0}^{\infty}|b_k| = \left|T(e) - \sum_{k=1}^{\infty} T(e_k)\right| + \sum_{k=1}^{\infty}|b_k| < \infty$$

since the first absolute value expression to the right of the equal sign is a real number because T is bounded on c, and the last summation is less than ∞ as before. Thus, c^* is exactly the space of our extended R_b for which b belongs to ℓ^1.

9. T is bounded because $|T(f)| = |f(x_0)| \le \max_{x \in [a,b]} |f(x)| = \|f\|_{\max}$ for all f in $C[a,b]$ where the inequality comes from the extreme value theorem (Section 1.6).

   Next, put the desired function of bounded variation to be $\chi_{[x_0,b]}$ on $[a,b]$. Function $\chi_{[x_0,b]}$ is of bounded variation (page 116) because $TV(\chi_{[x_0,b]}) = 1 < \infty$. The Riemann-Stieltjes integral
$$\int_a^b f(x)\, d\chi_{[x_0,b]}(x) \text{ for all } f \text{ in } C[a,b]$$
   is properly defined and linear by the second example and equal to $f(x_0)$. Thus, definition $T(f) = f(x_0)$ is given by Riemann-Stieltjes integration against a function of bounded variation. A partial check on our answer is its agreement with inequality (4).

10. See our solution to Problem 9. By the extreme value theorem, there is an x_0 in $[a,b]$ such that $f(x_0) = \|f\|_{\max}$. Put g to $\chi_{[x_0,b]}$. Then $TV(\chi_{[x_0,b]}) = 1$, and
$$\int_a^b f\, d\chi_{[x_0,b]} = f(x_0) = \|f\|_{\max}.$$

11. Function Φ is Lipschitz on $[a,b]$ because for all x', x in $[a,b]$,
$$\begin{aligned} |\Phi(x') - \Phi(x)| &= |T(g_{x'}) - T(g_x)| \\ &\le \|T\|_* \|g_{x'} - g_x\|_{\max} && \text{inequality (6) because } T \text{ is bounded} \\ &= \|T\|_* |(x' - a) - (x - a)| \\ &= \|T\|_* |x' - x|. \end{aligned}$$

8.2 Weak Sequential Convergence in L^p

In the definition of ϵ_{n+1} on page 164, change f_n to f_k.

12. The claim is true by the contrapositive of the third sentence on page 145 because
$$\lim_{n\to\infty} \int_0^1 |f_n - 0|^p = \int_0^1 1^p = 1 \ne 0.$$

13. f_n is the same step function of the first example except that f_n here alternates between β and α instead of 1 and -1. We can use Theorem 11 because $\{f_n\}$ is bounded in $L^p(I)$ since
$$\|f_n\|_p = \left(\int_0^1 |f_n|^p\right)^{1/p} = \left(\frac{|\beta|^p}{2} + \frac{|\alpha|^p}{2}\right)^{1/p} < \infty \text{ for all } n,$$
   and of course the constant function $(\alpha+\beta)/2$ belongs to $L^p(I)$. Then for all x in I,
$$\begin{aligned} \lim_{n\to\infty} \int_0^x f_n &= \lim_{n\to\infty} \left(\sum_{\substack{k=0 \\ k \text{ even}}}^{\lfloor 2^n x - 1 \rfloor} \frac{\beta}{2^n} + \sum_{\substack{k=1 \\ k \text{ odd}}}^{\lfloor 2^n x - 1 \rfloor} \frac{\alpha}{2^n} + \begin{cases} \beta(x - \lfloor 2^n x \rfloor / 2^n) & \lfloor 2^n x \rfloor \text{ even} \\ \alpha(x - \lfloor 2^n x \rfloor / 2^n) & \lfloor 2^n x \rfloor \text{ odd} \end{cases} \right) \\ &= \frac{\beta x}{2} + \frac{\alpha x}{2} + 0 = \frac{\beta + \alpha}{2} x = \int_0^x \frac{\alpha + \beta}{2} \end{aligned}$$

where $\lfloor \cdot \rfloor$ denotes the floor function. Thus, $\{f_n\}$ converges weakly in $L^p(I)$ to constant function $(\alpha+\beta)/2$.

Next, because for $n \neq m$, absolute difference $|f_n - f_m|$ takes the value $|\beta - \alpha|$ on a set of measure $1/2$, we have that $\|f_n - f_m\|_p = |\beta - \alpha|/2^{1/p}$. For $\alpha \neq \beta$ therefore, no subsequence of $\{f_n\}$ is Cauchy in $L^p(I)$, and thus no subsequence of $\{f_n\}$ converges strongly in $L^p(I)$.

14. This generalizes the Riemann-Lebesgue-lemma example. Observe that for each n, we have that $|f_n| \leq M$ on $[a,b]$ by inequality (10) of Section 1.6. By the Archimedean property, there is an n large enough for which $nx - na > T$ and $x \in [a,b]$. Let j be the smallest index such that $na \leq jT$, and k the largest such that $kT \leq nx$. Then for all such n and all x in $[a,b]$,

$$\begin{aligned}
\lim_{n\to\infty} \int_a^x h(nt)\,dt &= \lim_{n\to\infty} \frac{1}{n} \int_{na}^{nx} h(u)\,du && u = nt \\
&= \lim_{n\to\infty} \frac{1}{n} \left(\int_{na}^{jT} h(u)\,du + \int_{jT}^{kT} h(u)\,du + \int_{kT}^{nx} h(u)\,du \right) \\
&\leq \lim_{n\to\infty} (MT + 0 + MT)/n \\
&= 0.
\end{aligned}$$

Therefore, if $1 < p < \infty$, we infer from Theorem 11 that $\{f_n\} \rightharpoonup f$ in $L^p[a,b]$.

If $p = 1$, let A be a measurable subset of $[a,b]$. Then A has finite outer measure. By Theorem 12 of Section 2.4, for each positive ϵ, there is a finite disjoint collection of open intervals $\{I_k\}_{k=1}^m$ for which if $G = \bigcup_{k=1}^m I_k$, then $m(A \sim G) + m(G \sim A) < \epsilon/M$ so that

$$\begin{aligned}
\lim_{n\to\infty} \int_A f_n &= \lim_{n\to\infty} \left(\int_G f_n + \int_{A\sim G} f_n - \int_{G\sim A} f_n \right) \\
&\leq 0 + Mm(A \sim G) + Mm(G \sim A) \\
&\leq M \cdot \epsilon/M \\
&= \epsilon
\end{aligned}$$

where $\lim_{n\to\infty} \int_G f_n = 0$ because $\lim_{n\to\infty} \int_{I_k} f_n = 0$ for each k by the same reasoning as in the case where $1 < p < \infty$. Because $\lim_{n\to\infty} \int_A f_n \leq \epsilon$ holds for all ϵ, it holds also if $\epsilon = 0$. Therefore, we infer from Theorem 10 that $\{f_n\} \rightharpoonup f$ in $L^p[a,b]$. (With only superficial changes to our solution, we can use Theorem 10 to solve this problem without breaking it into two cases.)

15. This generalizes the example spanning pages 167–168. Let A be a measurable subset of $\mathbf{R}$ of finite measure. Because A has finite measure, it is contained in some interval $[a,b]$. Next, consider sequence $\{f_0\chi_{[-n,n]}\}$, which is in $L^p(\mathbf{R})$ and converges to f_0 pointwise on $\mathbf{R}$. Moreover,

$$\begin{aligned}
\lim_{n\to\infty} \int_{\mathbf{R}} \left|f_0\chi_{[-n,n]}\right|^p &= \lim_{n\to\infty} \int_{-n}^{n} |f_0|^p \\
&= \int_{\mathbf{R}} |f_0|^p.
\end{aligned}$$

Hence, $\{f_0\chi_{[-n,n]}\} \to f_0$ in $L^p(\mathbf{R})$ by Theorem 7 of Section 7.3. Then for every positive ϵ, there is an index N greater than $b - a$ for which $\left\|f_0 - f_0\chi_{[-N,N]}\right\|_p < \epsilon/m(A)^{1/q}$. That shows that if A is outside $[-N,N]$, then the norm on $L^p(A)$ of f_0 is less than $\epsilon/m(A)^{1/q}$. Now $f_0(x - N)$ shifts $f_0(x)$ far enough to the right along the x-axis so that A is outside

$[-N, N]$ with respect to $f_0(x)$ so that on A, we have that $\|f_0(x-N)\|_p < \epsilon/m(A)^{1/q}$. Next, the constant function 1 on A belongs to $L^q(A)$, and $\|1\|_q = \left(\int_A 1^q\right)^{1/q} = m(A)^{1/q}$. Then

$$\begin{aligned}
\lim_{n\to\infty} \int_A f_n &\le \lim_{n\to\infty} \int_A |1 f_0(x-n)|\, dx \\
&\le \lim_{n\to\infty} \|1\|_q \|f_0(x-n)\|_p \qquad \text{Hölder's inequality, Theorem 18 of Section 1.5} \\
&< m(A)^{1/q} \cdot \epsilon/m(A)^{1/q} \\
&= \epsilon.
\end{aligned}$$

Because that holds for arbitrarily small ϵ, we thereby infer from Theorem 10 that $\{f_n\} \rightharpoonup f$ in $L^p(\mathbf{R})$.

That is not true if $p = 1$ by the counterexample in the example spanning pages 167–168: $\int_{\mathbf{R}} f_n = 1$ for all n, but $\int_{\mathbf{R}} f = 0$.

16. Verify the condition in the third sentence on page 145. By the first chain of equations on page 169 and Theorem 18 (linearity) of Section 1.5,

$$\begin{aligned}
\lim_{n\to\infty} \int_E |f_n - f|^2 &= \lim_{n\to\infty} \int_E f_n^2 - 2 \lim_{n\to\infty} \int_E f_n f + \int_E f^2 \\
&= \int_E f^2 - 2\int_E f^2 + \int_E f^2 \\
&= 0.
\end{aligned}$$

17. The answer is "yes" if $\{f_n\}$ possesses properties (i) and (iii), properties (ii) and (iii), or property (iv). The answer is "no" in the other cases. Following are details.

By the reasoning in our solution to Problem 6 of Section 5.2, if a sequence converges, every subsequence converges to the same real number or function as the case may be. As a result, we reduce redundancy in the following cases in which $\{f_n\}$ possesses two properties by just determining whether a subsequence possesses the other two properties.

If (i) and (ii) are true, (but (i) already implies (ii) by Theorem 12), it is not necessary that a subsequence possess properties (iii) or (iv). Consider sequence $\{f_n\}$ given by $\{n^{1/p}\chi_{(0,1/n]}\}$ from the example of Section 7.3. That sequence is bounded in $L^p[0,1]$ because $\|f_n\|_p = 1$ for all n and converges pointwise on $[0,1]$ to $f = 0$, which of course is also in $L^p[0,1]$. But for every subsequence $\{f_k\}$ of $\{f_n\}$, we have that $\lim_{k\to\infty} \|f_k\|_p = 1 \neq 0 = \|f\|_p$. Thus, there is no subsequence such that $\{\|f_k\|_p\}$ converges to $\|f\|_p$. Then neither is there a subsequence that possesses property (iv) by the contrapositives of Theorem 7 of Section 7.3 or the Radon-Riesz theorem. Here is a direct way to show that no subsequence possesses property (iv) using that same counterexample sequence. For all positive ϵ and indices m, n such that $n > m$ and $n \ge m/(1-\epsilon^p)$,

$$\begin{aligned}
\|f_n - f_m\|_p &= \left(\int_0^1 \left|n^{1/p}\chi_{(0,1/n]} - m^{1/p}\chi_{(0,1/m]}\right|^p\right)^{1/p} \\
&= \left(\left(n^{1/p} - m^{1/p}\right)^p/n + m(1/m - 1/n)\right)^{1/p} \\
&> (1 - m/n)^{1/p} \\
&\ge \epsilon.
\end{aligned}$$

Therefore, no subsequence is Cauchy in $L^p[0,1]$ and hence no subsequence can converge strongly in $L^p[0,1]$. Another counterexample is to put E to $[1,\infty)$ and f_n to $\chi_{[n,n+1)}$.

If (i) and (iii) are true, all subsequences possess property (ii), and all subsequences possess property (iv) by Theorem 7 of Section 7.3 or the Radon-Riesz theorem.

If (i) and (iv) are true, all subsequences possess property (ii) by Theorem 12 or the second sentence after the definition on page 163, and all subsequences possess property (iii) by Theorem 7 of Section 7.3 or the Radon-Riesz theorem.

If (ii) and (iii) are true, all subsequences possess property (iv), and a subsequence possesses property (i) by the Riesz-Fischer theorem. It is not necessary, however, that the full sequence possess property (i). Consider the $\{f_n\}$ in the example on page 100. That sequence is bounded by 1 and in $L^p[0,1]$. As shown in the example, $\{f_n\}$ does not converge to the zero function pointwise, but $\{f_n\} \to 0$ in $L^p[0,1]$ by the third sentence on page 145 because

$$\lim_{n\to\infty} \int_0^1 |f_n - 0|^p \leq \lim_{n\to\infty} 1^p \ell(I_n) = \lim_{n\to\infty} \ell(I_n) = 0.$$

A subsequence that converges to the zero function pointwise is $\chi_{[0,1]}, \chi_{[0,1/2]}, \chi_{[0,1/3]}, \ldots$.

If (ii) and (iv) are true (but (iv) already implies (ii)), every subsequence possesses property (iii), and a subsequence possesses property (i).

If (iii) and (iv) are true (but (iv) already implies (iii) by the first two sentences in the proof of the Radon-Riesz theorem), every subsequence possesses property (ii), and a subsequence possesses property (i).

18. Because $\|f_n\|$ is unbounded, there is an index n_1 such that $\|f_{n_1}\| \geq 1 \cdot 3^1 = 3$. Likewise, there is an index n_2 greater than n_1 such that $\|f_{n_2}\| \geq 2 \cdot 3^2 = 18$. Continuing in that fashion and relabeling the indices $n_1 \mapsto 1$, $n_2 \mapsto 2$, ..., we can assume that $\|f_n\| \geq \alpha_n = n3^n$ for all n.

    Sequence $\{\alpha_n/\|f_n\|\}$ of real numbers is bounded below by 0 and above by 1 so that by the Bolzano-Weierstrass theorem (Section 1.5), $\{\alpha_n/\|f_n\|\}$ has a convergent subsequence. Say that that subsequence converges to the real number $1/\alpha$ in $[0,1]$. Then by relabeling the subsequence, we can assume that $\{\|f_n\|/\alpha_n\} \to \alpha$ where $\alpha \in [1,\infty]$.

    Next, $\{g_n\}$ converges weakly to the function $1/\alpha \cdot f$ because for all T in X^*,

$$\begin{aligned}
\lim_{n\to\infty} T(g_n) &= \lim_{n\to\infty} T(\alpha_n/\|f_n\| \cdot f_n) && \text{definition of } g_n \\
&= \lim_{n\to\infty} (\alpha_n/\|f_n\| \cdot T(f_n)) && \text{linearity of } T \\
&= \lim_{n\to\infty} \alpha_n/\|f_n\| \cdot \lim_{n\to\infty} T(f_n) && \text{our solution to Problem 42 of Section 1.5} \\
&= 1/\alpha \cdot T(f) && \text{previous; } \{f_n\} \rightharpoonup f \text{ in } X \\
&= T(1/\alpha \cdot f) && \text{linearity of } T,
\end{aligned}$$

    and $\|g_n\| = \|\alpha_n/\|f_n\| \cdot f_n\| = \alpha_n/\|f_n\| \cdot \|f_n\| = \alpha_n = n3^n$ for all n where we used the positive homogeneity property of a norm (page 137) in the second step.

19. Assume that $\{\zeta_n\} \rightharpoonup \zeta$ in ℓ^p. Our solution to Problem 7 tells us that every bounded linear functional on ℓ^p is given by summation against a sequence in ℓ^q where q is the conjugate of p; in other words, we have the ℓ^p analogue of Proposition 6. For each index k, take sequence $\{e_k\}$ in ℓ^q where $\{e_k\}$ is defined as in Problem 21 so that $\lim_{n\to\infty} \sum_{j=1}^{\infty} e_k^j \zeta_n^j = \sum_{j=1}^{\infty} e_k^j \zeta^j$, which simplifies to $\lim_{n\to\infty} \zeta_n^k = \zeta^k$. Thus, $\{\zeta_n\}$ converges componentwise. That proof is valid if $1 \leq p < \infty$.

Conversely, assume that $\{\zeta_n\}$ converges componentwise. Define functions T_{ζ_n} and T_ζ as in Problem 12 of Section 7.2. Then $\{T_{\zeta_n}\}$ is a bounded sequence in $L^p[1,\infty)$ that converges to T_ζ pointwise on $[1,\infty)$. By Theorem 12, $\{T_{\zeta_n}\} \rightharpoonup T_\zeta$ in $L^p[1,\infty)$. By Proposition 6,

$$\lim_{n\to\infty}\int_1^\infty T_b T_{\zeta_n} = \int_1^\infty T_b T_\zeta \text{ for all } T_b \text{ in } L^q[1,\infty).$$

We infer from our solution to Problem 12 of Section 7.2 that b is in ℓ^q if and only if T_b is in $L^q[1,\infty)$ so that the previous equation is equivalent to

$$\lim_{n\to\infty}\sum_{k=1}^\infty b^k\zeta_n^k = \sum_{k=1}^\infty b^k\zeta^k \text{ for all } b \text{ in } \ell^q.$$

Thus, $\{\zeta_n\} \rightharpoonup \zeta$ in ℓ^p by the ℓ^p analogue of Proposition 6.

20. Let q_1, q_2 be the conjugates of p_1, p_2. Then $1 < q_2 < q_1 \le \infty$, so $L^{q_1}[0,1] \subseteq L^{q_2}[0,1]$ by Corollary 3 of Section 7.2. Hence, if $\{f_n\} \rightharpoonup f$ in $L^{p_2}[0,1]$, then $\{f_n\} \rightharpoonup f$ in $L^{p_1}[0,1]$ by two applications of Proposition 6.

    Conversely, assume that $\{f_n\} \rightharpoonup f$ in $L^{p_1}[0,1]$. Then f belongs to $L^{p_1}[0,1]$ by definition (page 163), and $\{f_n\}$ is bounded in $L^{p_1}[0,1]$ by Theorem 7. Let φ be a simple function in $L^{q_2}[0,1]$. Simple function φ is bounded by definition (page 61) and so belongs also to $L^{q_1}[0,1]$ by Problem 13 of Section 7.2. Then $\lim_{n\to\infty}\int_0^1 f_n\varphi = \int_0^1 f\varphi$ by Proposition 6. Furthermore, $\{f_n\}$ is bounded in $L^{p_2}[0,1]$ because $\{f_n\}$ is bounded in $L^{p_1}[0,1]$ and $\int_0^1 |f_n|^{p_2} = \int_0^1 |f_n|^{p_1(p_2/p_1)}$. By Proposition 9 of Section 7.4, the subspace of simple functions in $L^{q_2}[0,1]$ is dense in $L^{q_2}[0,1]$. Thus, $\{f_n\} \rightharpoonup f$ in $L^{p_2}[0,1]$ by Proposition 9 of this section.

21. Clearly $\{e_n\}$ is a bounded sequence in ℓ^p, and the zero sequence belongs to ℓ^p. Now refer to Problem 19 and its notation. For each index k, we have that $\lim_{n\to\infty} e_n^k = 0$ because if $n > k$, then $e_n^k = 0$. Hence, $\{e_n\}$ converges componentwise to 0 and therefore weakly to 0 in ℓ^p if $p > 1$. On the other hand, because if $n \ne m$, then $|e_n^k - e_m^k|$ takes the value 1 when $k = m$ or n, we have that

$$\|e_n - e_m\|_p = \left(\sum_{k=1}^\infty |e_n^k - e_m^k|^p\right)^{1/p} = (1^p + 1^p)^{1/p} = 2^{1/p}.$$

    Therefore, no subsequence of $\{e_n\}$ is Cauchy in ℓ^p, and thus no subsequence converges strongly in ℓ^p if $1 \le p < \infty$.

    Next, our solution to Problem 7 tells us that every bounded linear functional on ℓ^1 is given by summation against a sequence in ℓ^∞. Associate sequence $(1,1,\dots)$ denoted by e in ℓ^∞ with bounded linear functional T in ℓ^{1*} so that $T(e_n) = \sum_{k=1}^\infty e^k e_n^k = 1$, and consequently, $\lim_{n\to\infty} T(e_n) = 1$. Now by the same reasoning as in the paragraph spanning pages 163–164 along with our solution to Problem 12 (Theorem 1 in ℓ^p) of Section 7.2, $\{e_n\}$ can converge weakly to at most one sequence in ℓ^1. But $\lim_{n\to\infty} T(e_n) = 1 = T(e_1) = T(e_2)$, and $e_2 \ne e_1$. Thus, sequence $\{e_n\}$ does not converge weakly in ℓ^1.

22. **Radon-Riesz Theorem in ℓ^2** Let p be in $(0,\infty)$. In the notation of Problem 19, assume that $\{\zeta_n\} \rightharpoonup \zeta$ in ℓ^2. Then $\{\zeta_n\} \to \zeta$ in ℓ^2 if and only if $\lim_{n\to\infty}\|\zeta_n\|_2 = \|\zeta\|_2$.

Proof First assume that $\{\zeta_n\} \to \zeta$ in ℓ^2. Then $\lim_{n\to\infty} \|\zeta_n\|_2 = \|\zeta\|_2$ by the reverse triangle inequality: $\lim_{n\to\infty} |\|\zeta_n\|_2 - \|\zeta\|_2| \le \lim_{n\to\infty} \|\zeta_n - \zeta\|_2 = 0$.

Conversely, let $\{\zeta_n\}$ be a sequence for which $\{\zeta_n\} \rightharpoonup \zeta$ in ℓ^2 and $\lim_{n\to\infty} \sum_{k=1}^{\infty} (\zeta_n^k)^2 = \sum_{k=1}^{\infty} (\zeta^k)^2$. Observe that for each n,

$$\|\zeta_n - \zeta\|_2^2 = \sum_{k=1}^{\infty} |\zeta_n^k - \zeta^k|^2 = \sum_{k=1}^{\infty} (\zeta_n^k - \zeta^k)^2 = \sum_{k=1}^{\infty} (\zeta_n^k)^2 - 2\sum_{k=1}^{\infty} \zeta_n^k \zeta^k + \sum_{k=1}^{\infty} (\zeta^k)^2.$$

Because ζ belongs to $\ell^q = \ell^2$, we have that $\lim_{n\to\infty} \sum_{k=1}^{\infty} \zeta_n^k \zeta^k = \sum_{k=1}^{\infty} (\zeta^k)^2$. So $\{\zeta_n\} \to \zeta$ in ℓ^2.□

23. By the sixth and seventh sentences on page 135 and the extreme value theorem (Section 1.6), $C[a,b]$ is a subset of $L^2[a,b]$ so that $\{f_n\} \rightharpoonup f$ in $L^2[a,b]$. By Theorem 7, $\{f_n\}$ is bounded in $L^2[a,b]$. By definition of *weak convergence*, f belongs to $L^2[a,b]$. Then we can use Theorem 11. Take the derivative of each side of equation (24): $\frac{d}{dx} \lim_{n\to\infty} \int_a^x f_n = \frac{d}{dx} \int_a^x f$. Because each f_n and f are continuous on $[a,b]$, we can apply the fundamental theorem of calculus and interchange the derivative and limit to get that $\lim_{n\to\infty} f_n(x) = f(x)$ for all x in $[a,b]$. Thus, $\{f_n\}$ converges pointwise on $[a,b]$ to f.

24. For all x in $[a,b]$, functional T_x defined on $L^\infty[a,b]$ by $T_x(g) = \int_a^x g$ is linear by linearity of integration and bounded because $|T_x(g)| = \left|\int_a^x g\right| \le \int_a^x |g| \le (b-a)\|g\|_\infty$ for all g in $L^\infty[a,b]$. Hence, T_x belongs to $L^\infty[a,b]^*$. The result now follows from definition of *weak convergence* (page 163).

25. (i) Because X is a linear space, $f_2 - f_1 \in X$. Then

$$\|f_2 - f_1\| = T(f_2 - f_1) = T(f_2) - T(f_1) = \lim_{n\to\infty} T(f_n) - \lim_{n\to\infty} T(f_n) = 0.$$

Thus, $f_2 = f_1$ by the nonnegativity property of a norm (page 137).

(ii) Similar to the argument in the first paragraph of the proof of Theorem 7,

$$\|f\| = T(f) = \lim_{n\to\infty} T(f_n) \le \lim_{n\to\infty} \|T\|_* \|f_n\| \le \liminf\{\|f_n\|\}.$$

26. The proof is the same as given in the second paragraph of the proof of Theorem 7 with the next-to-last sentence replaced by "This is a contradiction because $\left\{\int_E g f_n\right\}$ is bounded."

8.3 Weak Sequential Compactness

27. Put the sequence to be $\{\sin nx\}$ for x in $[-\pi,\pi]$. Sequence $\{\sin nx\}$ is bounded and in $C[-\pi,\pi]$. We claim that $\|\sin nx - \sin mx\|_{\max} \ge 1/2$. By a trigonometric identity, that holds if $\sin((n-m)x/2)\cos((n+m)x/2) \ge 1/4$ which in turn holds if $\sin((n-m)x/2)$ and $\cos((n+m)x/2)$ are each at least $1/2$. That happens when $(n-m)x/2 \in [\pi/6, 5\pi/6]$ and $(n+m)x/2 \in [-\pi/3, \pi/3]$; "solving" for x gives $x \in [\pi/(3(n-m)), 5\pi/(3(n-m))]$ and $x \in [-2\pi/(3(n+m)), 2\pi/(3(n+m))]$. Those two intervals overlap when $n > m$ and $\pi/(3(n-m) \le 2\pi/(3(n+m))$, that is, when $n \ge 3m$. Any subsequence has indices m, n such that $n \ge 3m$. As a result, $\|\sin nx - \sin mx\|_{\max} \ge 1/2$. Therefore, no subsequence of $\{\sin nx\}$ is Cauchy in $C[-\pi,\pi]$, and hence no subsequence can converge strongly in $C[-\pi,\pi]$.

28. In our solution to Problem 21, we show that $\{e_n\}$, a bounded sequence in ℓ^p, fails to have a strongly convergent subsequence when $1 \leq p < \infty$. Our reasoning holds also when $p = \infty$ with the calculation of $\|e_n - e_m\|_p$ replaced by $\|e_n - e_m\|_\infty = \sup_{1 \leq k < \infty} \left|e_n^k - e_m^k\right| = 1$.

29. Let the nondegenerate interval be $[a, b]$. If $x \in [a, b]$, then $(x - a)/(b - a) \in [0, 1]$, which shows that by rescaling, we can assume that the nondegenerate interval is $[0, 1]$. Then the sequence given in the example will do.

    Next, for E the empty set, the claim is vacuously true. For a slightly less trivial solution, put E to be a set of outer measure zero. Set E is measurable by Proposition 4 of Section 2.3. Let $\{f_n\}$ be a bounded sequence in $L^1(E)$. Then because $m(E) = m^*(E) = 0$, all of the f_n are the same function as explained in the first and third paragraphs of Section 7.1. Thus, *every* subsequence of $\{f_n\}$ converges weakly.

30. Assume first that $\{T_n\} \to T$ with respect to $\|\cdot\|_*$. On $\{f \in X \mid \|f\| \leq 1\}$,

$$\begin{aligned}
\lim_{n\to\infty}(T_n(f) - T(f)) &\leq \lim_{n\to\infty}|T_n(f) - T(f)| \\
&= \lim_{n\to\infty}|(T_n - T)(f)| && \text{next-to-last sentence on page 155} \\
&\leq \lim_{n\to\infty}\|T_n - T\|_*\|f\| && \text{inequality (5); } T_n - T \text{ is in linear space } X^* \\
&\leq \lim_{n\to\infty}\|T_n - T\|_* && \|f\| \leq 1 \\
&= 0 && \{T_n\} \to T \text{ with respect to } \|\cdot\|_*.
\end{aligned}$$

    That is, for each positive ϵ, there is an index N for which $|T_n - T| < \epsilon$ on $\{f \in X \mid \|f\| \leq 1\}$ if $n \geq N$. Thus, $\{T_n\} \to T$ uniformly on $\{f \in X \mid \|f\| \leq 1\}$ by definition (iii) on page 60.

    Conversely, assume that $\{T_n\} \to T$ uniformly on $\{f \in X \mid \|f\| \leq 1\}$. Then $\{T_n\} \to T$ with respect to $\|\cdot\|_*$ because by formula (8) and our assumption,

$$\lim_{n\to\infty}\|T_n - T\|_* = \lim_{n\to\infty}\sup\{T_n(f) - T(f) \mid f \in X,\ \|f\| \leq 1\} = 0.$$

31. No, the sequence is not uniformly integrable by the example in Section 4.3 and the contrapositive of the Vitali convergence theorem of Section 4.6.

    Another explanation is the contrapositive of the first sentence of the remark.

    A third solution is to show that there is no δ that satisfies the definition on page 93. Put ϵ to $1/2$ and A to $[0, \min\{1, \delta\}/2]$. Then subset A of $[0, 1]$ is measurable and $m(A) < \delta$, but $\int_A |f_n| = \int_A |n\chi_{I_n}| \geq \min\{1, n\delta\}/2$, which is greater than or equal to ϵ if $n \geq 1/\delta$.

32. It fails at the invocation of Theorem 11 of Chapter 7 because $L^\infty(E)$ is not separable by the example after that theorem. Then Helly's theorem cannot be invoked. See Problem 36. The proof also fails at the invocation of the Riesz Representation theorem by the last remark of Section 8.1. Note that the first two sentences in that remark with $[a, b]$ replaced by E explain why the use of Proposition 2 does not cause a problem in the proof of Theorem 14 if $p = 1$: T_n is a bounded linear functional on $L^q(E)$, and $\|T_n\|_* \leq \|f_n\|_p$.

33. Mimic the proof of Theorem 14. Let $\{\zeta_n\}$ be a bounded sequence of sequences in ℓ^p. Note that each ζ_n is itself a sequence $(\zeta_n^1, \zeta_n^2, \ldots)$. Define functional T_n on ℓ^q where q is the conjugate of p by

$$T_n(a) = \sum_{j=1}^{\infty} \zeta_n^j a^j \text{ where } a \text{ is a sequence } \left(a^1, a^2, \ldots\right) \text{ in } \ell^q.$$

Our solution to Problem 7 with p, q interchanged tells us that each T_n is a bounded linear functional on ℓ^q and $\|T_n\|_* = \|\zeta_n\|_p$. Because $\{\zeta_n\}$ is a bounded sequence in ℓ^p, sequence $\{T_n\}$ is bounded in ℓ^{q*}. Moreover, according to Problem 44 of Section 7.4, ℓ^q is separable. Therefore, by Helly's theorem, there is a subsequence $\{T_{n_k}\}$ and T in ℓ^{q*} such that

$$\lim_{k\to\infty} T_{n_k}(a) = T(a) \text{ for all } a \text{ in } \ell^q.$$

Our solution to Problem 7 with p, q interchanged tells us that there is a sequence $(\zeta^1, \zeta^2, \ldots)$ denoted by ζ in ℓ^p for which

$$T(a) = \sum_{j=1}^{\infty} \zeta^j a^j \text{ for all } a \text{ in } \ell^q.$$

But the three previous displayed equations mean that

$$\lim_{k\to\infty} \sum_{j=1}^{\infty} \zeta_{n_k}^j a^j = \sum_{j=1}^{\infty} \zeta^j a^j \text{ for all } a \text{ in } \ell^q.$$

Our solution to Problem 7 implies the ℓ^p analogue of Proposition 6. Accordingly, $\{\zeta_{n_k}\}$ converges weakly to ζ in ℓ^p.

34. By the definition and subsequent paragraph on page 152, $C[0,1]$ is a separable normed linear space. By the second example of Section 8.1, $T_n(g) = \int_0^1 g(x)\,df_n(x)\,dx$ defines a bounded sequence $\{T_n\}$ of bounded linear functionals on $C[0,1]$. By Helly's theorem, there is a subsequence $\{T_{n_k}\}$ of $\{T_n\}$ for which $T_{n_k}(g)$ converges to some real number $T(g)$. Thus, by Theorem 17 of Section 1.5, $\{T_{n_k}(g)\}$, or $\left\{\int_0^1 g(x)\,df_{n_k}(x)\,dx\right\}$, is Cauchy.

35. (i) By definition of denseness, for each g in X and positive ϵ, there is an f in $\mathcal{S}$ for which $\|f-g\| < \epsilon/(3M)$ so that $|T_n(f) - T_n(g)| \le \|T_n\|_* \|f-g\| \le M\epsilon/(3M) = \epsilon/3$ for all n by inequality (6). The Cauchy hypothesis tells us that there is an index N for which if $m, n \ge N$, then $|T_n(f) - T_m(f)| < \epsilon/3$. Then for each g in X and if $m, n \ge N$,

$$\begin{aligned} |T_n(g) - T_m(g)| &\le |T_n(g) - T_n(f)| + |T_n(f) - T_m(f)| + |T_m(f) - T_m(g)| && \text{tri. inequal.} \\ &< \epsilon/3 + \epsilon/3 + \epsilon/3 \\ &= \epsilon. \end{aligned}$$

Thus, $\{T_n(g)\}$ is Cauchy for all g in X.

(ii) T is linear by linearity of T_n and linearity of convergence of real numbers. To show explicitly that T is linear, check the definition on page 155:

$$\begin{aligned} T(\alpha g + \beta h) &= \lim_{n\to\infty} T_n(\alpha g + \beta h) && \text{definition of } T \\ &= \lim_{n\to\infty} (\alpha T_n(g) + \beta T_n(h)) && T_n \text{ is linear because } T_n \text{ is in } X^* \\ &= \alpha \lim_{n\to\infty} T_n(g) + \beta \lim_{n\to\infty} T_n(h) && \text{Theorem 18 (linearity) of Section 1.5} \\ &= \alpha T(g) + \beta T(h) && \text{definition of } T. \end{aligned}$$

That T is bounded is shown at the end of the proof of Helly's theorem with n_k replaced by n and f replaced by g.

36. The point of this problem is to show that X must be separable—$L^\infty[0,1]$ is a normed linear space by the second example in Section 7.1 but is not separable by the example in Section 7.4. Define a sequence $\{g_n\}$ in $L^1[0,1]$ by $g_n = n\chi_{(0,1/n]}$ so that $\|g_n\|_1 = 1$—see the second example on page 167 with f, g interchanged. Define T_n on $L^\infty[0,1]$ by $T_n(f) = \int_0^1 fg_n$. Then $\{T_n\}$ is a bounded sequence in dual space $L^\infty[0,1]^*$ by the last remark of Section 8.1 with f, g interchanged. Suppose, to get a contradiction, that there were a subsequence $\{T_{n_k}\}$ of $\{T_n\}$ that converges for all f in $L^\infty[0,1]$. Consider the f in $L^\infty[0,1]$ defined by $f = (-1)^k$. Then

$$\begin{aligned}
\lim_{k\to\infty} T_{n_k}(f) &= \lim_{k\to\infty} \int_0^1 fg_{n_k} \\
&= \lim_{k\to\infty} (-1)^k \int_0^1 g_{n_k} \\
&= \lim_{k\to\infty} (-1)^k \|g_{n_k}\|_1 \\
&= \lim_{k\to\infty} (-1)^k,
\end{aligned}$$

which does not converge, a contradiction. Thus, the conclusion of Helly's theorem is not true if $X = L^\infty[0,1]$.

37. In the following, we separate cases $p = 1$ and $p > 1$ only if the answers differ.

(i) holds. (ii) All subsequences converge weakly to f in $L^p(E)$ by the second sentence after the definition on page 163. (iii) A subsequence converges to f pointwise a.e. on E by the Riesz-Fischer theorem (Section 7.3). (iv) Because each f_n and f belong to $L^p(E)$, the f_n are measurable, and f is finite a.e. on E. Then the subsequence of (iii) above converges in measure to f by Proposition 3 of Section 5.2.

(ii) holds. (i) It is not necessary that a subsequence converge strongly in $L^p(E)$ by the Riemann-Lebesgue-lemma example (Section 8.2). But when $p > 1$, if $\lim_{n\to\infty} \|f_n\|_p = \|f\|_p$, then all subsequences converge strongly to f in $L^p(E)$ by the Radon-Riesz theorem, and if $\|f\|_p = \liminf\{\|f_n\|_p\}$, then a subsequence converges strongly to f in $L^p(E)$ by Corollary 13. (iii) It is not necessary that a subsequence converge pointwise a.e. on E by the Riemann-Lebesgue-lemma example along with our Problem 27 reasoning with the maximum norm replaced by absolute value. (iv) Consequently, it is not necessary that a subsequence converge in measure to f by the contrapositive of Problem 11 of Section 5.2.

(iii) holds. (i) It is not necessary that a subsequence converge strongly to f in $L^p(E)$ by our solution to the (i)-and-(ii) case of Problem 17, which holds also if $p = 1$. (ii) If $p > 1$, then all subsequences converge weakly to f in $L^p(E)$ by Theorem 12. But it is not necessary that a subsequence converge weakly to f in $L^1(E)$ by the second example on page 167 and the example of this section. (iv) All subsequences converge in measure to f.

(iv) holds. (i) It is not necessary that a subsequence converge strongly to f in $L^p(E)$ by the (i)-and-(ii) case of our solution to Problem 17. The sequence in that example converges in measure to f because it converges to f pointwise. (iii) A subsequence converges to f pointwise a.e. on E by Theorem 4 of Section 5.2. (ii) If $p > 1$, then that subsequence also converges weakly to f in $L^p(E)$. But it is not necessary that a subsequence converge weakly to f in $L^1(E)$.

8.4 The Minimization of Convex Functionals

38. Our solution to Problem 21 takes care of the second sentence. Next, put the subsequence to be sequence $\{e_n\}$ itself. Then

$$\begin{aligned}\left\|\frac{e_1+e_2+\cdots+e_n}{n}\right\|_p &= \left\|\left(\frac{1}{n},\frac{1}{n},\ldots,\frac{1}{n},0,0,\ldots\right)\right\|_p && \text{first } n \text{ terms are } \frac{1}{n}\\ &= \left(\sum_{k=1}^{n}\left|\frac{1}{n}\right|^p\right)^{1/p}\\ &= \left(\frac{n}{n^p}\right)^{1/p}\\ &= \frac{1}{n^{(p-1)/p}}\end{aligned}$$

which tends to 0 as $n \to \infty$ because $(p-1)/p > 0$. Thus, the arithmetic means converge strongly to 0 in ℓ^p. Our solution shows that each subsequence of $\{e_n\}$ will do.

39. By replacing each a_n with $a_n - a$, we can assume that $\{a_n\} \to 0$. Then for every positive ϵ, there is an index N for which if $n > N$, then $|a_n| < \epsilon/2$. By Proposition 14 of Section 1.5, there is a nonnegative M for which $|a_n| \le M$ for all n. By the Archimedean property, there is another index N' greater than or equal to N such that $NM/N' < \epsilon/2$. Then if $n > N'$,

$$\begin{aligned}\left|\frac{a_1+\cdots+a_n}{n}\right| &\le \frac{|a_1|+\cdots+|a_N|}{n}+\frac{|a_{N+1}|+\cdots+|a_n|}{n} && \text{triangle inequality}\\ &< \frac{NM}{n}+\frac{(n-N)\cdot\epsilon/2}{n}\\ &< \epsilon/2+\epsilon/2\\ &= \epsilon.\end{aligned}$$

Thus, the sequence of arithmetic means also converges to 0.

40. **Banach-Saks theorem in ℓ^2** Assume that $\{\zeta_n\} \rightharpoonup \zeta$ in ℓ^2. Then there is a subsequence $\{\zeta_{n_k}\}$ for which the sequence of arithmetic means converges strongly to ζ in ℓ^2.

Proof By replacing each ζ_n with $\zeta_n - \zeta$, we can assume that $\{\zeta_n\} \rightharpoonup 0$ in ℓ^2. Let a be a sequence in ℓ^2. Define functions T_{ζ_n} and T_a as in Problem 12 of Section 7.2. Then $\|\zeta_n\|_2 = \|T_{\zeta_n}\|_2$ and

$$\sum_{i=1}^{\infty}\zeta_n^i a^i = \int_1^{\infty} T_{\zeta_n}T_a$$

where ζ_n^i, a^i are the ith terms of sequences ζ_n, a. Then by Proposition 6 and the ℓ^p analogue of Proposition 6 (which our solution to Problem 7 gives us), $T_{\zeta_n} \rightharpoonup 0$ in $L^2[1,\infty)$. Theorem 7 tells us that $\{T_{\zeta_n}\}$ is bounded in $L^2[1,\infty)$. Then $\{\zeta_n\}$ is bounded in ℓ^2. (Another way to show that $\{\zeta_n\}$ is bounded in ℓ^2 is to adapt the second paragraph of the proof of Theorem 7 for ℓ^2.) Choose a nonnegative M for which $\sum_{i=1}^{\infty}(\zeta_n^i)^2 \le M$ for all n. We will choose a subsequence $\{\zeta_{n_j}\}$ such that

$$\sum_{i=1}^{\infty}\left(\zeta_{n_1}^i+\cdots+\zeta_{n_j}^i\right)^2 < 2j + Mj \text{ for all } j.$$

Indeed, define $n_1 = 1$. Assume that we have chosen $n_1, n_2, \ldots, n_k$ such that $n_1 < n_2 < \cdots < n_k$ and

$$\sum_{i=1}^{\infty}\left(\zeta_{n_1}^i + \cdots + \zeta_{n_j}^i\right)^2 < 2j + Mj \text{ where } j = 1, \ldots, k.$$

Because $\zeta_{n_1} + \cdots + \zeta_{n_k}$ belongs to ℓ^2 and $\{\zeta_n\}$ converges weakly in ℓ^2 to 0, we can choose n_{k+1} for which $n_{k+1} > n_k$ and

$$\sum_{i=1}^{\infty}\left(\zeta_{n_1}^i + \cdots + \zeta_{n_k}^i\right)\zeta_{n_{k+1}}^i < 1$$

by the ℓ^p analogue of Proposition 6. However,

$$\sum_{i=1}^{\infty}\left(\zeta_{n_1}^i + \cdots + \zeta_{n_{k+1}}^i\right)^2 = \sum_{i=1}^{\infty}\left(\zeta_{n_1}^i + \cdots + \zeta_{n_k}^i\right)^2 + 2\sum_{i=1}^{\infty}\left(\zeta_{n_1}^i + \cdots + \zeta_{n_k}^i\right)\zeta_{n_{k+1}}^i + \sum_{i=1}^{\infty}\left(\zeta_{n_{k+1}}^i\right)^2,$$

and therefore

$$\begin{aligned}\sum_{i=1}^{\infty}\left(\zeta_{n_1}^i + \cdots + \zeta_{n_{k+1}}^i\right)^2 &< 2k + Mk + 2 + M \\ &= 2(k+1) + M(k+1).\end{aligned}$$

Subsequence $\{\zeta_{n_k}\}$ has been inductively chosen so that

$$\sum_{i=1}^{\infty}\left(\frac{\zeta_{n_1}^i + \cdots + \zeta_{n_k}^i}{k}\right)^2 < \frac{2+M}{k} \text{ for all } k.$$

Therefore, the sequence of arithmetic means of $\{\zeta_{n_k}\}$ converges strongly to 0 in ℓ^2. $\square$

41. Assume that $E \subseteq [a, b]$ so that by $T(f)$, we mean that T is restricted to $L^p(E)$. Set K is a closed, bounded, convex subset of $L^p(E)$ by definition and the second example. Functional T is convex because it is linear—for f, g in K and λ in $[0, 1]$, we have that

$$T(\lambda f + (1-\lambda)g) = \lambda T(f) + (1-\lambda)T(g).$$

Then the proof of Theorem 17 tells us that there is a sequence $\{f_n\}$ in K such that $\{f_n\} \rightharpoonup f_0$ in $L^p(E)$, and

$$T(f_0) = \inf\{T(f) \mid f \in K\} = c.$$

Functional T is also bounded by the sentence after the definition of *continuous* (page 176). By Riesz's representation theorem, there is a unique function h in $L^q(E)$, where q is the conjugate of p, for which $\int_E hf = T(f)$ for all f in $L^p(E)$, and $\|h\|_q = \|T\|_*$. If $h = 0$, then $T = 0$, and each f in K will do for f_0. Otherwise, continue with what Proposition 6 tells us, namely, that

$$\begin{aligned}\lim_{n\to\infty}\int_E hf_n &= \int_E hf_0 \\ &= T(f_0) \\ &= c.\end{aligned}$$

We deduce from equations (4) of Section 7.2 with p, q interchanged and f replaced by h that a suitable f_0 is defined by $f_0 = ch^*/\|h\|_q$ where h^* is the conjugate function of h.

42. We can assume that $c_1, c_2 \geq 0$ because

$$c_1 + c_2|s|^{p_1/p_2} \leq |c_1| + |c_2||s|^{p_1/p_2}.$$

Sequence $\{\varphi \circ f_n\}$ is in $L^{p_2}(E)$ because for each n,

$$\begin{aligned} \int_E |\varphi \circ f_n|^{p_2} &\leq \int_E \left(c_1 + c_2|f_n|^{p_1/p_2}\right)^{p_2} \\ &\leq 2^{p_2} \int_E (c_1^{p_2} + c_2^{p_2}|f_n|^{p_1}) && \text{inequality (1) of Sec. 7.1, Th. 17 of Sec. 4.4} \\ &\leq 2^{p_2}\left(m(E)c_1^{p_2} + c_2^{p_2}\int_E |f_n|^{p_1}\right) \\ &< \infty && \{f_n\} \subseteq L^{p_1}(E). \end{aligned}$$

Likewise, $\varphi \circ f$ belongs to $L^{p_2}(E)$. Furthermore, as in the proof of Corollary 18, we can assume that $\{\varphi \circ f_n\}$ converges pointwise a.e. on E to $\varphi \circ f$, and because exponentiation of an absolute value is a continuous (composed) function, $\{|\varphi \circ f_n|^{p_2}\}$ converges pointwise a.e. on E to $|\varphi \circ f|^{p_2}$. Using the same reasoning and function g as in the proof of Corollary 18,

$$\begin{aligned} |\varphi \circ f_n|^{p_2} &\leq \left(c_1 + c_2|f_n|^{p_1/p_2}\right)^{p_2} \\ &\leq \left(c_1 + c_2 g^{p_1/p_2}\right)^{p_2} \end{aligned}$$

so that

$$\lim_{n\to\infty} \int_E |\varphi \circ f_n|^{p_2} = \int_E |\varphi \circ f|^{p_2}.$$

The parts of the proof of Corollary 18 that we used above are valid if $p = 1$ (Corollary 18 has that $p > 1$ as a condition because the argument uses Theorem 17 later on). Thus, by Theorem 7 of Section 7.3 with f_n replaced by $\varphi \circ f_n$ and p by p_2, we have that $\{\varphi \circ f_n\} \to \varphi \circ f$ in $L^{p_2}(E)$.

According to the third example, this problem is to show that the T in the example is properly defined, continuous, and convex. That is mostly shown in the proof of Corollary 18. The contributions of this problem are that we can assume that $a, b \geq 0$ and that the relevant parts of the proof of Corollary 18 are valid if $p = 1$

43. Norm $\|g - f_0\|_p$ for all g in C defines a functional T on C. To see that T is continuous, observe that if a sequence $\{g_n\}$ in C converges strongly to g in C, then $\{g_n - f_0\} \to g - f_0$ in $L^p(E)$. Hence, by the first two sentences of the proof of the Radon-Riesz theorem, $\{\|g_n - f_0\|_p\} \to \|g - f_0\|_p$. Functional T is also convex by the first sentence on page 177. To see that explicitly, observe that if g, h belong to C, and λ is in $[0, 1]$, then

$$\begin{aligned} \|\lambda g + (1-\lambda)h - f_0\|_p &= \|\lambda(g - f_0) + (1-\lambda)(h - f_0)\|_p \\ &\leq \lambda\|g - f_0\|_p + (1-\lambda)\|h - f_0\|_p. \end{aligned}$$

Thus, by Theorem 17, there is a g_0 in C for which $\|g_0 - f_0\|_p \leq \|g - f_0\|_p$ for all g in C.

44. Each f_n is a periodic step function with period $1/2^n$. Each period consists of exactly two steps, the first one of length $1/2^{2n+1}$ and value $1 - 2^{n+1}$ and the second one of length $1/2^n - 1/2^{2n+1}$ and value 1.

(i) The integral of f_n over one period is zero—

$$\begin{aligned}\int_0^{1/2^n} f_n &= \int_0^{1/2^{2n+1}} (1 - 2^{n+1}) + \int_{1/2^{2n+1}}^{1/2^n} 1 \\ &= \frac{1 - 2^{n+1}}{2^{2n+1}} + \frac{1}{2^n} - \frac{1}{2^{2n+1}} \\ &= 0\end{aligned}$$

—which implies that to prove the requested inequality, it suffices to consider x in $[0, 1/2^n)$. Furthermore, because the values of the two steps in $[0, 1/2^n]$ have opposite signs, $\left|\int_0^x f_n\right|$ is greatest when x is at the end of the first step, i.e., when $x = 1/2^{2n+1}$. Hence,

$$\begin{aligned}\left|\int_0^x f_n\right| &\leq \int_0^{1/2^{2n+1}} (2^{n+1} - 1) \\ &= \frac{2^{n+1} - 1}{2^{2n+1}} \\ &< \frac{2^{n+1}}{2^{2n+1}} \\ &= \frac{1}{2^n}\end{aligned}$$

for all x in $[0, 1]$ and all n, and therefore

$$\lim_{n\to\infty} \int_0^x f_n = 0 = \int_0^x 0 = \int_0^x f \text{ for all } x \text{ in } [0, 1].$$

Another way to get that result is by an analysis like that of the first part of our solution to Problem 14.

(ii) The measure of the subset of $[0, 1]$ on which $f_n = 1$ is $2^n(1/2^n - 1/2^{2n+1})$, or $1 - 1/2^{n+1}$, which increases with n, so

$$\begin{aligned}\int_E f_n &= m(E) \\ &> 1 - 1/2^{n+1} \\ &> 0 \text{ for all } n.\end{aligned}$$

(iii) Using our description above of step function f_n,

$$\begin{aligned}\|f_n\|_1 &= \int_0^1 |f_n| \\ &= 2^n\left(\int_0^{1/2^{2n+1}} (2^{n+1} - 1) + \int_{1/2^{2n+1}}^{1/2^n} 1\right) \\ &= 2 - 1/2^n \\ &< 2 \text{ for all } n,\end{aligned}$$

which shows that $\{f_n\}$ is bounded in $L^1[0, 1]$. By part (ii),

$$\lim_{n\to\infty} \int_E f_n > \frac{1}{2} > 0 = \int_E 0 = \int_E f.$$

E is measurable by Proposition 7 of Section 2.3 because E is the union of a countable collection of intervals. Then by the contrapositive of Theorem 10, $\{f_n\}$ does not converge weakly in $L^1[0,1]$ to f. That and part (i) do not contradict Theorem 11 because it does not hold if $p = 1$ as explained after that theorem.

(iv) $\{f_n\}$ is in $L^p[0,1]$ because for each n,

$$\begin{aligned}\int_0^1 |f_n|^p &= 2^n\left(\int_0^{1/2^{2n+1}} \left(2^{n+1}-1\right)^p + \int_{1/2^{2n+1}}^{1/2^n} 1^p\right)\\ &= \frac{\left(2^{n+1}-1\right)^p}{2^{n+1}} + 1 - \frac{1}{2^{n+1}}\\ &< \infty.\end{aligned}$$

But $\{f_n\}$ does not converge weakly in $L^p[0,1]$ to f for the same reason we gave in part (iii). That and part (i) do not contradict Theorem 11 because $\{f_n\}$ is not bounded in $L^p[0,1]$ as we see from our expression for $\int_0^1 |f_n|^p$ above—since $p > 1$, we can make $\|f_n\|_p$, which is $\left(\int_0^1 |f_n|^p\right)^{1/p}$, greater than each real number by choosing n sufficiently large.

45. I think the second sentence should begin "Use this sequence ..." Put $\{g_n\}$ to be the sequence of the first example of Section 8.2. As shown there, $\{g_n\}$ is in $L^2[0,1]$. Moreover, $\|g_n - g_m\|_2 = \sqrt{2}$ if $m \neq n$; therefore, $\{g_n\}$ has no Cauchy subsequence in $L^2[0,1]$. By the denseness of the continuous functions in $L^2[0,1]$ (Theorem 12 of Section 7.4), for each g_n, there is a continuous function h_n on $[0,1]$ for which $\|h_n - g_n\|_2 < \sqrt{2}/3$. Then

$$\begin{aligned}\|g_n - g_m\|_2 &\leq \|g_n - h_n\|_2 + \|h_n - h_m\|_2 + \|h_m - g_m\|_2 && \text{triangle inequality,}\\ \|h_n - h_m\|_2 &\geq \|g_n - g_m\|_2 - \|g_n - h_n\|_2 - \|h_m - g_m\|_2 && \text{rearrangement}\\ &> \sqrt{2} - \sqrt{2}/3 - \sqrt{2}/3\\ &= \sqrt{2}/3.\end{aligned}$$

Therefore, no subsequence of $\{h_n\}$ is Cauchy in $L^2[0,1]$, and hence no subsequence of $\{h_n\}$ converges in $L^2[0,1]$.

In this next solution, $\{g_n\}$ itself is a sequence of continuous functions, in which case we automatically satisfy the problem's request to use the denseness of continuous functions in L^2 spaces. If $y \in [0,1]$, then $x = 2\pi y - \pi \in [-\pi,\pi]$, which shows that by rescaling, we can work in $L^2[-\pi,\pi]$. Put $\{g_n\}$ to be the sequence of the Riemann-Lebesgue lemma example (Section 8.2). As shown there, $\{g_n\}$ is in $L^2[-\pi,\pi]$. Moreover, if $m \neq n$,

$$\begin{aligned}\|g_n - g_m\|_2 &= \left(\int_{-\pi}^{\pi} |\sin nx - \sin mx|^2\, dx\right)^{1/2}\\ &= \sqrt{2\pi};\end{aligned}$$

therefore, $\{g_n\}$ has no Cauchy subsequence in $L^2[-\pi,\pi]$. Then $\{g_n\}$ is a sequence of continuous functions on $[-\pi,\pi]$ for which no subsequence converges in $L^2[-\pi,\pi]$.

Chapter 9

Metric Spaces: General Properties

9.1 Examples of Metric Spaces

1. If there is such a c, then ρ and τ are equivalent by definition. Conversely, assume that ρ and τ are equivalent. Then there exist positive numbers c_1 and c_2 such that $c_1 \leq c_2$, and $c_1\tau(u,v) \leq \rho(u,v) \leq c_2\tau(u,v)$. Put c to $\max\{c_2, 1/c_1\}$. Then $1/c \leq c_1$ so that

$$\frac{1}{c}\tau(u,v) \leq c_1\tau(u,v) \leq \rho(u,v) \leq c_2\tau(u,v) \leq c\tau(u,v).$$

2. First verify that ρ^*, ρ^+ are metrics on $\mathbf{R}^n$ using the definition on page 182. Euclidean space $\mathbf{R}^n$ is nonempty because it contains the point $(0,\ldots,0)$. By hypothesis, ρ^*, ρ^+ are functions from $\mathbf{R}^n \times \mathbf{R}^n$ to $\mathbf{R}$. Functions ρ^*, ρ^+ satisfy property—

   (i) because each $|x_k - y_k|$ is at least zero,

   (ii) because each ρ^*, ρ^+ equals zero if and only if $y_k = x_k$ for each k, that is, if and only if $y = x$,

   (iii) because $|y_k - x_k| = |x_k - y_k|$ for each k, and

   (iv) because $|x_k - y_k| \leq |x_k - z_k| + |z_k - y_k|$ for each k by the triangle inequality for absolute value.

   To see that ρ^*, ρ^+ are equivalent, observe that $\rho^+(x,y) \leq \rho^*(x,y)$ because $\rho^+(x,y)$ is equal to just one of the terms $|x_k - y_k|$ whereas $\rho^*(x,y)$ is the sum of that term and other nonzero terms. On the other hand, $\rho^*(x,y) \leq n\rho^+(x,y)$ because each of the n terms $|x_k - y_k|$ in the formula for $\rho^*(x,y)$ is less than or equal to the maximum of those terms. Thus, ρ^*, ρ^+ define equivalent metrics on $\mathbf{R}^n$ because, altogether, $(1/n)\rho^*(x,y) \leq \rho^+(x,y) \leq \rho^*(x,y)$.

3. Put the metric to be the discrete metric ρ defined on page 183. Suppose, to get a contradiction, that ρ and ρ^* were equivalent. Then there is a positive number c such that $\rho(x,y) \leq c\rho^*(x,y)$ for all x, y in $\mathbf{R}^n$. Choose x to be $(1/(2c), 0, 0, \ldots, 0)$ and y to be $(0,\ldots,0)$. Then, $\rho(x,y) = 1 > 1/2 = c/(2c) = c\rho^*(x,y)$, a contradiction. Thus, ρ fails to be equivalent to ρ^*. The same argument with the same x, y shows that ρ also fails to be equivalent to ρ^+.

4. Suppose, to get a contradiction, that the two norms were equivalent. Then there is a positive number c such that $\max\{|(f-g)(x)| \mid x \in [a,b]\} \leq c\int_a^b |f-g|$ for all f, g in $C[a,b]$. Choose g to be 0. Let δ equal $\min\{1/(2c), (b-a)/2\}$. Choose f to be the tent function (see the example spanning pages 167–168) that vanishes outside $(a, a+2\delta)$, is linear on intervals $[a, a+\delta]$ and $[a+\delta, a+2\delta]$, and has peak value $f(a+\delta)$ equal to 1. In symbols,

$$f(x) = \begin{cases} (x-a)/\delta & x \in (a, a+\delta] \\ 1 - (x-(a+\delta))/\delta & x \in (a+\delta, a+2\delta) \\ 0 & \text{otherwise.} \end{cases}$$

   Then $\max\{|(f-g)(x)| \mid x \in [a,b]\} = 1 > 1/2 = c/(2c) \geq c\delta = c\int_a^b |f-g|$, a contradiction. Thus, the two norms are not equivalent.

   Intuitively, the norms cannot be equivalent because geometrically speaking, the maximum norm is a length whereas the $L^1[a,b]$ norm is an area.

5. Check that ρ satisfies the definition on page 182 excepting the *only-if* direction of property (ii). Set X is nonempty because it contains E and $\emptyset$. By hypothesis, ρ is a function from $X \times X$ to $\mathbf{R}$. Function ρ satisfies property—

   (i) because m is equal to m^* by the definition on page 43 and therefore nonnegative since m^* is a sum of lengths—see the definition of *outer measure* on page 31,
   (ii) because if $A = B$, then $m((A \sim B) \cup (B \sim A)) = m(\emptyset \cup \emptyset) = m(\emptyset) = 0$,
   (iii) because $(A \sim B) \cup (B \sim A) = (B \sim A) \cup (A \sim B)$, and
   (iv) because for C in X and with the notation of the Venn diagram below,

$$\begin{aligned}
\rho(A,C) + \rho(C,B) &= m((A \sim C) \cup (C \sim A)) + m((C \sim B) \cup (B \sim C)) \\
&= m(a) + m(d) + m(c) + m(e) + m(c) + m(f) + m(b) + m(d) \\
&\geq m(a) + m(f) + m(b) + m(e) \\
&= m((A \sim B) \cup (B \sim A)) \\
&= \rho(A, B).
\end{aligned}$$

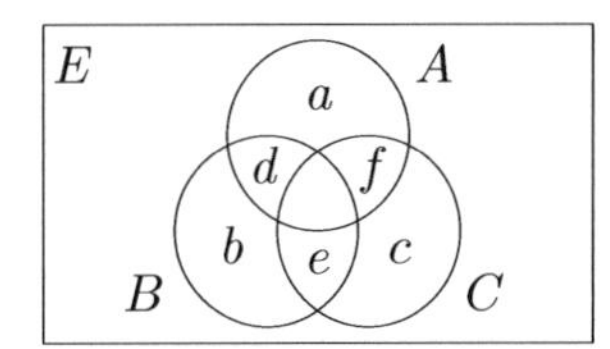

$a = A \sim (B \cup C)$ $\quad$ $d = A \cap B \sim C$
$b = B \sim (C \cup A)$ $\quad$ $e = B \cap C \sim A$
$c = C \sim (A \cup B)$ $\quad$ $f = C \cap A \sim B$

The first step is just the definition of ρ. Proposition 13 of Section 2.5 justifies the second step because the subsets involved are disjoint. In the third step, we discard four of the measures. Reasoning for the remaining steps is reverse that of the previous steps.

We next show that $\widetilde{\rho}$ is a metric on $X/\cong$. Collection $X/\cong$ is nonempty because it contains $[E]$. By hypothesis, $\widetilde{\rho}$ is a function from $X/\cong \times X/\cong$ to $\mathbf{R}$. Function $\widetilde{\rho}$ satisfies metric properties (i), (iii), and (iv) because $\widetilde{\rho}$ inherits those properties from ρ due to definition $\widetilde{\rho}([A],[B]) = \rho(A,B)$ and property (ii) because $\widetilde{\rho}([A],[B]) = 0$ if and only if $\rho(A,B) = 0$ if and only if $A \cong B$ if and only if $[A] = [B]$.

Finally,

$$\begin{aligned}
&\int_E |\chi_A - \chi_B| \\
&= \int_{E \sim (A \cup B)} |\chi_A - \chi_B| + \int_{A \sim B} |\chi_A - \chi_B| + \int_{B \sim A} |\chi_A - \chi_B| + \int_{A \cap B} |\chi_A - \chi_B| \\
&= m(E \sim (A \cup B))|0 - 0| + m(A \sim B)|1 - 0| + m(B \sim A)|0 - 1| + m(A \cap B)|1 - 1| \\
&= 0 + m(A \sim B) + m(B \sim A) + 0 \\
&= m((A \sim B) \cup (B \sim A)) \\
&= \rho(A, B).
\end{aligned}$$

This formula for ρ gives us a second, easier way to solve the first part of this problem: Function ρ satisfies properties (i) and (iii) by the properties of absolute value and property (iv) by the triangle inequality for absolute value because

$$\begin{aligned}
\rho(A,B) &= \int_E |\chi_A - \chi_B| \leq \int_E (|\chi_A - \chi_C| + |\chi_C - \chi_B|) && \text{triangle inequality; (12) of Sec. 4.3} \\
&= \int_E |\chi_A - \chi_C| + \int_E |\chi_C - \chi_B| && \text{(11) of Section 4.3} \\
&= \rho(A,C) + \rho(C,B).
\end{aligned}$$

6. Upon multiplying both sides of the latter inequality by $(1+a)(1+b)(1+c)$ and simplifying, $a \le b+c+2bc+abc$, which is true because $a \le b+c$ and $2bc+abc$ is nonnegative.

7. Check that ρ satisfies the definition on page 182 excepting the *only-if* direction of property (ii). Set X is nonempty because it contains the zero function. By hypothesis, ρ is a function from $X \times X$ to $\mathbf{R}$. Function ρ satisfies property—

   (i) because $|f-g|/(1+|f-g|)$ is nonnegative, so $\int_E |f-g|/(1+|f-g|)$ is nonnegative,

   (ii) because if $f=g$, then $|f-g|=0$,

   (iii) because $|f-g|=|g-f|$, and

   (iv) because for h in X, we have that $|f-g| \le |f-h|+|h-g|$ by the triangle inequality for absolute value so that by Problem 6 and Theorem 10 of Section 4.3,

$$\rho(f,g) = \int_E \frac{|f-g|}{1+|f-g|} \le \int_E \frac{|f-h|}{1+|f-h|} + \int_E \frac{|h-g|}{1+|h-g|} = \rho(f,h)+\rho(h,g).$$

   We next show that $\widetilde{\rho}$ is a metric on $X/\cong$. The last paragraph of the section applies here because $f \cong g$ if and only if f, g are equal a.e. on E if and only if $|f-g|/(1+|f-g|)=0$ a.e. on E if and only if $\rho(f,g)=\int_E |f-g|/(1+|f-g|)=0$ by Proposition 9 of Section 4.3. Check that $\widetilde{\rho}$ satisfies the definition on page 182. Collection $X/\cong$ is nonempty because it contains $[0]$. By hypothesis, $\widetilde{\rho}$ is a function from $X/\cong \times X/\cong$ to $\mathbf{R}$. Function $\widetilde{\rho}$ satisfies metric properties (i), (iii), and (iv) because $\widetilde{\rho}$ inherits those properties from ρ due to definition $\widetilde{\rho}([f],[g])=\rho(f,g)$ and property (ii) because $\widetilde{\rho}([f],[g])=0$ if and only if $\rho(f,g)=0$ if and only if $f \cong g$ if and only if $[f]=[g]$.

8. We have four solutions:

   - $(a+b)^p = \dfrac{a+b}{(a+b)^{1-p}} = \dfrac{a}{(a+b)^{1-p}} + \dfrac{b}{(a+b)^{1-p}} \le \dfrac{a}{a^{1-p}} + \dfrac{b}{b^{1-p}} = a^p+b^p.$

   - Multiply both sides of $\dfrac{a}{a+b}+\dfrac{b}{a+b} \le \left(\dfrac{a}{a+b}\right)^p + \left(\dfrac{b}{a+b}\right)^p$ by $(a+b)^p$ and simplify.

   - Function φ on $[0,\infty)$ defined by $\varphi(x)=-x^p$ is convex by Proposition 15 of Section 6.6 because φ's second derivative $x \mapsto -p(p-1)x^{p-2}$ is nonnegative on $[0,\infty)$. Then

$$\begin{aligned}
\varphi(\lambda x_1+(1-\lambda)x_2) &\le \lambda\varphi(x_1)+(1-\lambda)\varphi(x_2) && \text{(38) of Section 6.6;}\\
-a^p &\le -\frac{a}{a+b}(a+b)^p && \text{put } \lambda \text{ to } \frac{a}{a+b},\ x_1 \text{ to } a+b,\ x_2 \text{ to } 0,\\
-b^p &\le -\frac{b}{a+b}(a+b)^p && \text{put } \lambda \text{ to } \frac{b}{a+b},\ x_1 \text{ to } a+b,\ x_2 \text{ to } 0;\\
a^p+b^p &\ge (a+b)^p && \text{add the two inequalities.}
\end{aligned}$$

   - If $b=0$, the claim is true. Otherwise, we proceed to show that function f on $[0,\infty)$ defined by $f(x)=(1+x)^p-x^p$ is less than or equal to 1. Observe that $f(0)=1$ and the first derivative $x \mapsto p\big((1+x)^{p-1}-x^{p-1}\big)$ is negative if $x>0$, that is, $f(x)$ is decreasing from $f(0)$ if $x>0$. It follows that $f(x)=(1+x)^p-x^p \le 1$ if $x \ge 0$. Now substitute a/b for x into that inequality, and multiply both sides by b^p to get that $(a+b)^p \le a^p+b^p$.

9. Let X be the collection defined in this problem. Check that ρ_p satisfies the definition on page 182 excepting the *only-if* direction of property (ii). Set X is nonempty because it contains the zero function. By hypothesis, ρ_p is a function from $X \times X$ to $\mathbf{R}$. Function ρ_p satisfies property—

(i) because $|g-h|^p$ is nonnegative, so $\int_E |g-h|^p$ is nonnegative,

(ii) because if $h = g$, then $|g-h|^p = 0$,

(iii) because $|g-h| = |h-g|$, and

(iv) because for f in X,

$$\begin{aligned}
\rho_p(h,g) &= \int_E |g-h|^p && \text{definition of } \rho_p \\
&\le \int_E (|g-f| + |f-h|)^p && \text{triangle inequality; statement (12) of Section 4.3} \\
&\le \int_E |g-f|^p + \int_E |f-h|^p && \text{Problem 8; inequality (11) of Section 4.3} \\
&= \rho_p(f,g) + \rho_p(h,f) && \text{definition of } \rho_p.
\end{aligned}$$

We next show that $\widetilde{\rho_p}$ is a metric on $X/\cong$. By the same reasoning we give in the corresponding part of our solution to Problem 7, $h \cong g$ if and only if $\rho(h,g) = 0$ so that the last paragraph of the section applies here. Check that $\widetilde{\rho}$ satisfies the definition on page 182. Collection $X/\cong$ is nonempty because it contains $[0]$. By hypothesis, $\widetilde{\rho_p}$ is a function from $X/\cong \times X/\cong$ to $\mathbf{R}$. Function $\widetilde{\rho_p}$ satisfies metric properties (i), (iii), and (iv) because $\widetilde{\rho_p}$ inherits those properties from ρ_p due to definition $\widetilde{\rho_p}([h],[g]) = \rho_p(h,g)$ and property (ii) because $\widetilde{\rho_p}([h],[g]) = 0$ if and only if $\rho_p(h,g) = 0$ if and only if $h \cong g$ if and only if $[h] = [g]$.

10. Check that ρ_* satisfies the definition on page 182: ρ_* is a function from $\prod_{n=1}^{\infty} X_n$ to $\mathbf{R}$ by hypothesis and because

$$\rho_*(x,y) = \sum_{n=1}^{\infty} \frac{1}{2^n} \frac{\rho_n(x,y)}{1+\rho_n(x,y)} \le \sum_{n=1}^{\infty} \frac{1}{2^n} = 1 < \infty.$$

In showing that ρ_* satisfies properties (i), (ii), (iii), and (iv), we use the fact that each ρ_n satisfies those properties. Function ρ_* satisfies property—

(i) because $(1/2^n)(\rho_n(x,y)/(1+\rho_n(x,y))$ is nonnegative for each n, so the sum is nonnegative,

(ii) because $\rho_* = 0$ if and only if $(1/2^n)(\rho_n(x,y)/(1+\rho_n(x,y)) = 0$ for each n if and only if $\rho_n(x,y) = 0$ for each n if and only if $x = y$,

(iii) because $\rho_n(x,y) = \rho_n(y,x)$ for each n, and

(iv) because for point z in $\prod_{n=1}^{\infty} X_n$,

$$\begin{aligned}
\rho_*(x,y) &= \sum_{n=1}^{\infty} \frac{1}{2^n} \frac{\rho_n(x,y)}{1+\rho_n(x,y)} && \text{definition of } \rho_* \\
&\le \sum_{n=1}^{\infty} \frac{1}{2^n} \frac{\rho_n(x,z)}{1+\rho_n(x,z)} + \sum_{n=1}^{\infty} \frac{1}{2^n} \frac{\rho_n(z,y)}{1+\rho_n(z,y)} && \text{Prob. 6; mono., linear. of sums} \\
&= \rho_*(x,z) + \rho_*(z,y) && \text{definition of } \rho_p.
\end{aligned}$$

11. To find the desired metric σ, rewrite the second definition on page 184 in the notation of this problem: A mapping f from a metric space (A,σ) to a metric space (X,ρ) is said to be an *isometry* provided that f maps A onto X and $\sigma(a_1,a_2) = \rho(f(a_1), f(a_2))$ for all a_1, a_2 in A. To show that σ is unique, let ς be another such metric. Then by the same reasoning, $\varsigma(a_1,a_2) = \rho(f(a_1), f(a_2))$ for all a_1, a_2 in A. Thus, $\varsigma = \sigma$.

12. This is like Problem 11 of Section 7.2 with p equal to 2 and $[1, n+1)$ scaled to $[0,1]$. For a point $(x_1, x_2, \dots, x_n)$, denoted by x, in $\mathbf{R}^n$, define T_x to be the step function on $[0,1)$ that takes the value $\sqrt{n}x_k$ on $[(k-1)/n, k/n)$ where $k = 1, \dots, n$. Then T_x belongs to $L^2[0,1]$, and

$$\|T_x\|_2 = \left(\int_0^1 |T_x|^2\right)^{1/2} = \left(\sum_{k=1}^{n} \frac{1}{n}\left|\sqrt{n}x_k\right|^2\right)^{1/2} = \left(x_1^2 + \cdots + x_n^2\right)^{1/2} = \|x\|.$$

Thus, for a point y in $\mathbf{R}^n$, $\|x+y\| = \|T_{x+y}\|_2 = \|T_x + T_y\|_2 \le \|T_x\|_2 + \|T_y\|_2 = \|x\| + \|y\|$. We also get the triangle inequality for the induced metric by the second and third sentences after definition (1).

9.2 Open Sets, Closed Sets, and Convergent Sequences

In the fifth sentence after the definition on page 188, replace X with x.

13. In a metric space X, it is possible for $B(u,r)$ to equal $B(v,r)$ if, for example, X has the discrete metric ρ (page 183) and $r = 2$. Then for all u' in X, we have that $\rho(u', u) < 2$, so $B(u,2) = X$. Likewise, $B(v,2) = X$. Thus, $B(u,2) = B(v,2)$. Our solution below for a normed linear space is another situation in which $B(u,r) = B(v,r)$.

    In $\mathbf{R}^n$ with the Euclidean norm, it is not possible for $B(u,r)$ to equal $B(v,r)$. Without loss of generality, we can translate and rotate the coordinates so that u is the origin and v is point $(v_1, 0, \dots, 0)$ where $v_1 > 0$. If $v_1 \ge r$, point v itself is in $B(v,r)$ because $\|v-v\| = 0 < r$ but not in $B(u,r)$ because $\|v-u\| = \left((v_1-0)^2 + (0-0)^2 + \cdots + (0-0)^2\right)^{1/2} = v_1 \ge r$. If on the other hand $v_1 < r$, consider point $v' = (v_1 + (r - v_1/2), 0, \dots, 0)$. Point v' is in $B(v,r)$ because $\|v'-v\| = \left((v_1 + (r-v_1/2) - v_1)^2 + (0-0)^2 + \cdots + (0-0)^2\right)^{1/2} = r - v_1/2 < r$ but not in $B(u,r)$ because $\|v'-u\| = \left((v_1 + (r-v_1/2) - 0)^2 + (0-0)^2 + \cdots + (0-0)^2\right)^{1/2} = r + v_1/2 > r$. In either case, we found a point in $B(v,r)$ that is not in $B(u,r)$. Thus, $B(u,r) \ne B(v,r)$.

    In a normed linear space, it is possible for $B(u,r)$ to equal $B(v,r)$. For example, in the normed linear subspace of Euclidean space $\mathbf{R}^n$ that contains only the point $(0, \dots, 0)$, $B(u,r) = B(v,r)$ vacuously because there are no distinct points.

14. By hypothesis, for each positive ϵ, there are indices N_u, N_v such that if $n \ge N_u$, then $\rho(u_n, u) < \epsilon/2$, and if $n \ge N_v$, then $\rho(v_n, v) < \epsilon/2$. Consequently, there is an index N given by $\max\{N_u, N_v\}$ for which if $n \ge N$, then

$$\begin{aligned}
\rho(u_n, v_n) \le \rho(u_n, u) + \rho(u, v_n) &\le \rho(u_n, u) + \rho(u, v) + \rho(v, v_n) && \text{triangle inequality} \\
&< \epsilon/2 + \rho(u,v) + \epsilon/2 && n \ge N \ge N_u, N_v \\
&= \rho(u,v) + \epsilon.
\end{aligned}$$

    Likewise, $\rho(u,v) < \rho(u_n, v_n) + \epsilon$. The two inequalities imply that $|\rho(u_n, v_n) - \rho(u,v)| < \epsilon$. Thus, $\{\rho(u_n, v_n)\} \to \rho(u,v)$ by the definition on page 21.

15. (i) We have three solutions to show that $\overline{B}(x,r)$ is closed. This first one mimics the paragraph after the definition on page 186. It suffices by Proposition 4 to show that $X \sim \overline{B}(x,r)$, that is, $\{x' \in X \mid \rho(x', x) > r\}$, which we name set G, is open. In turn, it suffices to show that if $x' \in G$ and $r' = \rho(x', x) - r$, then $B(x', r') \subseteq G$. To verify that, let y be in $B(x', r')$. Then $\rho(y, x') < r'$ so that by the triangle inequality, $\rho(y, x) \ge \rho(x', x) - \rho(y, x') > \rho(x', x) - r' = r$. Therefore, $B(x', r') \subseteq G$.

    This second solution uses reasoning like that in the proof of Proposition 3. Set $\overline{B}(x,r)$ is closed if it contains all its points of closure. Let y be a point of closure of $\overline{B}(x,r)$.

Consider neighborhood $B(y,\delta)$ of y. There is a point y' in $\overline{B}(x,r) \cap B(y,\delta)$. Then $\rho(y',x) \le r$, and $\rho(y',y) < \delta$ so that $\rho(x,y) \le \rho(x,y') + \rho(y',y) < r + \delta$. Because that holds for δ arbitrarily small, $\rho(x,y) \le r$; hence, $y \in \overline{B}(x,r)$. So set $\overline{B}(x,r)$ is closed.

This third solution uses Proposition 6. Let a sequence $\{y_n\}$ in $\overline{B}(x,r)$ converge to a limit y in X. Then $\rho(y,x) \le \rho(y,y_n) + \rho(y_n,x) \le \rho(y_n,y) + r$. Now take the limit as $n \to \infty$ to get that $\rho(y,x) \le r$ by the definition on page 188. Then the limit y belongs to $\overline{B}(x,r)$. Thus, $\overline{B}(x,r)$ is closed.

Next, let x' be a point in $B(x,r)$. Then $\rho(x',x) < r \le r$ so that x' is also in $\overline{B}(x,r)$. In other words, $\{x' \in X \,|\, \rho(x',x) < r\} \subseteq \{x' \in X \,|\, \rho(x',x) \le r\}$. Thus, $\overline{B}(x,r)$ contains $B(x,r)$.

(ii) We want to prove that $\overline{B}(x,r) = \overline{B(x,r)}$. By part (i), $\overline{B}(x,r)$ is closed, and by definition, $B(x,r) \subseteq \overline{B}(x,r)$, so by Proposition 3, $\overline{B(x,r)} \subseteq \overline{B}(x,r)$. That inclusion holds in a general metric space. We now show that $\overline{B}(x,r) \subseteq \overline{B(x,r)}$ in a normed linear space. Let y be a point outside of $\overline{B(x,r)}$. By Proposition 3, $\overline{B(x,r)}$ is closed. So there is an open ball $B(y,\delta)$ that does not contain a point in $\overline{B(x,r)}$ or in $B(x,r)$. We claim that $\|y-x\| \ge r+\delta$. Indeed, let Φ be a mapping that makes a linear combination of x, y defined by $\Phi(\lambda) = \lambda x + (1-\lambda)y$ where λ is a real number. Observe that for λ in $[0, r/\|y-x\|)$, point $\Phi(\lambda)$ is in $B(x,r)$, and that for λ in $(1-\delta/\|y-x\|, 1]$, point $\Phi(\lambda)$ is in $B(y,\delta)$. Those two intervals $[0, r/\|y-x\|)$ and $(1-\delta/\|y-x\|, 1]$ are disjoint because $B(x,r)$ and $B(y,\delta)$ are disjoint, so $r/\|y-x\| \le 1-\delta/\|y-x\|$. Solving for $\|y-x\|$ proves our claim. Then because $\|y-x\| \ge r+\delta$, point y is not in $\overline{B}(x,r)$ either. In other words, there are no points outside of $\overline{B(x,r)}$ that are in $\overline{B}(x,r)$. Hence, $\overline{B}(x,r) \subseteq \overline{B(x,r)}$. Thus, $\overline{B}(x,r) = \overline{B(x,r)}$.

Now let X be a general metric space. Consider the case in which X contains more than one point and has the discrete metric (page 183). Then $\overline{B}(x,1) = X$. As for $\overline{B(x,1)}$, the next-to-last sentence on page 188 tells us that $\overline{B(x,1)} = B(x,1) = \{x\}$. Explicitly, note that $B(x,1)$ has no points of closure other than its only member x because any other point y in X has a neighborhood $B(y,1) = \{y\}$ that does not contain a (the only) point in $B(x,1)$. Consequently, $\overline{B(x,1)} = \{x\}$. Therefore, $\overline{B}(x,r) \ne \overline{B(x,r)}$. Thus, $\overline{B}(x,r)$ is not always the closure of $B(x,r)$ in a general metric space.

16. Assume first that E is open in X. Then for every point x in E, there is an open ball $\{x' \in X \,|\, \rho(x',x) < r_x\}$ that is contained in E. Define corresponding open balls in Y for each x in E by $Y_x = \{x' \in Y \,|\, \rho(x',x) < r_x\}$, and put $\mathcal{O}$ equal to $\bigcup_{x\in E} Y_x$, which is open in Y by Proposition 1. Because each x in E is in Y_x which is in $\mathcal{O}$, and $E \subseteq X$ by hypothesis, $E \subseteq \mathcal{O} \cap X$. On the other hand, because $Y_x \cap X$ is a subset of the first open ball $\{x' \in X \,|\, \rho(x',x) < r_x\}$ we constructed, which is contained in E, we have that $Y_x \cap X \subseteq E$ for each x in E so that $\mathcal{O} \cap X \subseteq E$. Thus, $E = X \cap \mathcal{O}$.

    Conversely, assume that $E = X \cap \mathcal{O}$ where $\mathcal{O}$ is open in Y. If $E = \emptyset$, it is open. Otherwise, let x belong to E so that $x \in \mathcal{O} \cap X$. Because $\mathcal{O}$ is an open set containing x, there is an open ball Y_x centered at x that is contained in $\mathcal{O}$. Then $Y_x \cap X \subseteq E$. Thus, E is open in X.

17. By the symmetry of the proposition, it suffices to prove just one direction of the *if-and-only-if* statement. So let a subset G of X be open in (X,ρ). Then for a point x in G, there is an open ball $\{x' \in X \,|\, \rho(x',x) < r\}$ contained in G. Because ρ, σ are equivalent metrics, there is a positive number c such that $\rho(x',x) \le c\sigma(x',x)$. Consequently, if $c\sigma(x',x) < r$, then $\rho(x',x) < r$. Then altogether, $\{x' \in X \,|\, c\sigma(x',x) < r\} \subseteq \{x' \in X \,|\, \rho(x',x) < r\} \subseteq G$. Hence, open ball $\{x' \in X \,|\, \sigma(x',x) < r/c\}$ is contained in G. Thus, G is open in (X,σ).

18. By Proposition 4, we can write an equivalent statement which is true by Proposition 2: "$X \sim A$ is open in X if and only if $X \sim A = X \cap (Y \sim F)$ where $Y \sim F$ is open in Y."

19. (i) Subset $\mathcal{O}$ is not necessarily open in Y. For example, put Y to be $\mathbf{R}$ and both X and $\mathcal{O}$ to be singleton set $\{0\}$. Set X is a subspace of Y by the last paragraph on page 183. Set $\mathcal{O}$ is open in X by Proposition 1 because $\mathcal{O} = X$. But $\mathcal{O}$ is not open in Y because no open ball $\{y \in Y \mid |y - 0| < r\}$ is contained in $\{0\}$.

    Now assume that X is an open subset of Y. Because $\mathcal{O}$ is open in X, Proposition 2 tells us that $\mathcal{O} = X \cap G$ where G is open in Y. Thus, because $\mathcal{O}$ is the intersection of two open subsets of Y, subset $\mathcal{O}$ is open in Y by Proposition 1.

    (ii) Subset F is not necessarily closed in Y. For example, put Y to be $\mathbf{R}$ and both X and F to be the interval $(0, 1)$. Subset F is closed in X by Proposition 5 because $F = X$. But F is not closed in Y because every neighborhood of point 0 contains a point in $(0, 1)$, so 0 is a point of closure of $(0, 1)$, but 0 is not in $(0, 1)$.

    Now assume that X is a closed subset of Y. Because F is closed in X, Problem 18 tells us that $F = X \cap F'$ where F' is closed in Y. Thus, because F is the intersection of closed subsets of Y, subset F is closed in Y by Proposition 5.

20. If $x \in \operatorname{int} E$, then there is an open ball $B(x, r)$ contained in E. By the paragraph after the definition on page 186, if $x' \in B(x, r)$, then there is a $B(x', r')$ such that $B(x', r') \subseteq B(x, r)$. Then $B(x', r') \subseteq E$, so $x' \in \operatorname{int} E$. So because each point x' of $B(x, r)$ is an interior point of E, we have that $B(x, r) \subseteq \operatorname{int} E$. Thus, $\operatorname{int} E$ is open.

    The next part of the problem generalizes Problem 32(i) of Section 1.4. Subset E is open if and only if for every point y in E, there is a $B(y, r)$ that is contained in E if and only if $y \in \operatorname{int} E$ if and only if $E \subseteq \operatorname{int} E$. We can replace that last inclusion with equality $E = \operatorname{int} E$ if we can prove that $\operatorname{int} E \subseteq E$. But that is true for every set E because for each x in $\operatorname{int} E$, there is an open ball centered at x (and hence containing x) that is contained in E.

21. Comparing definitions in this problem and Problem 20, we see that $\operatorname{ext} E = \operatorname{int}(X \sim E)$, which is always open. Thus, $\operatorname{ext} E$ is open.

    By Proposition 4, the last statement of this problem is equivalent to "$X \sim E$ is open if and only if $X \sim E = \operatorname{int}(X \sim E)$," which is true by Problem 20.

22. (i) Set $\operatorname{bd} E$ is closed if it contains all its points of closure. Let x be a point of closure of $\operatorname{bd} E$. Then every neighborhood of x, in particular, every $B(x, r)$, contains a point in $\operatorname{bd} E$. Therefore every $B(x, r)$ contains points in E and points in $X \sim E$, and hence, $x \in \operatorname{bd} E$. Thus, $\operatorname{bd} E$ is closed.

    (ii) Assume first that E is open. Then for every point y in E, there is a $B(y, r)$ that is contained in E. Then $B(y, r)$ contains no points in $X \sim E$ so that $y \notin \operatorname{bd} E$. Thus, $E \cap \operatorname{bd} E = \emptyset$.

    Conversely, assume that $E \cap \operatorname{bd} E = \emptyset$. Then for every y in E, there is a $B(y, r)$ that does not contain points in E and points in $X \sim E$. Because $y \in E$, we can sharpen that statement by saying that there is a $B(y, r)$ that does not contain points in $X \sim E$. Consequently, all the points that $B(y, r)$ contains are in E so that $B(y, r) \subseteq E$. Thus, E is open.

    (iii) Assume first that E is closed, that is, E contains all of its points of closure. Then for each y in E, every $B(y, r)$ contains a point in E. Thus, $\operatorname{bd} E \subseteq E$ because every point in $\operatorname{bd} E$ contains a point in E by definition.

    Conversely, assume that $\operatorname{bd} E \subseteq E$. Subset E is closed if it contains all its points of closure. Let y be a point of closure of E. Then every $B(y, r)$ contains a point in E. If all of the points that $B(y, r)$ contains are in E, then $y \in E$. Otherwise, $B(y, r)$ contains not only a point in E but also points in $X \sim E$ so that $y \in \operatorname{bd} E$. Therefore $y \in E$ because $\operatorname{bd} E \subseteq E$. In either case, $y \in E$. Thus, E is closed.

23. By hypothesis and by the first sentence after the definition page 187, $A \subseteq B \subseteq \overline{B}$. Thus, $\overline{A} \subseteq \overline{B}$ by Proposition 3.

    Next, because $A \subseteq A \cup B$, we have by the previous that $\overline{A} \subseteq \overline{A \cup B}$. Likewise, $\overline{B} \subseteq \overline{A \cup B}$. Hence, $\overline{A} \cup \overline{B} \subseteq \overline{A \cup B}$. To get the inclusion in the other direction, note that inclusions $A \subseteq \overline{A}$ and $B \subseteq \overline{B}$ together imply that $A \cup B \subseteq \overline{A} \cup \overline{B}$, which is closed by Proposition 5. Then $\overline{A \cup B} \subseteq \overline{A} \cup \overline{B}$ by Proposition 3. Thus, $\overline{A \cup B} = \overline{A} \cup \overline{B}$.

    Inclusions $A \subseteq \overline{A}$ and $B \subseteq \overline{B}$ imply also that $A \cap B \subseteq \overline{A} \cap \overline{B}$, which is closed by Proposition 5. Thus, $\overline{A \cap B} \subseteq \overline{A} \cap \overline{B}$ by Proposition 3.

24. Let $\mathcal{F}$ be the collection of all closed subsets of X that contain E. By Proposition 3, $\overline{E}$ is the smallest set in $\mathcal{F}$, and $\overline{E} \subseteq F$ for every F in $\mathcal{F}$. It follows that $\overline{E} = \bigcap_{F \in \mathcal{F}} F$, which is the symbolic form of the desired result. As a check, note that both $\overline{E}$ and $\bigcap_{F \in \mathcal{F}} F$ are closed—the former by Proposition 3 and the latter by Proposition 5.

25. Function τ is a metric with an upper bound of 1 by Problem 10 with ρ_* replaced by $\tau/2$ and the countable collection of metric spaces consisting of just the one metric space (X, ρ).

    Next, let $\{x_n\}$ in (X, ρ) converge to x. By our solution to Problem 42 of Section 1.5, $\lim_{n \to \infty} \tau(x_n, x) = \lim_{n \to \infty} \rho(x_n, x) \cdot \lim_{n \to \infty} 1/(1 + \rho(x_n, x))$ so that $\lim_{n \to \infty} \tau(x_n, x) = 0$ if and only if $\lim_{n \to \infty} \rho(x_n, x) = 0$. Thus, convergence of sequences with respect to ρ, τ is the same.

    We conclude that sets that are closed with respect to ρ are closed with respect to τ by Proposition 6 and then deduce from Proposition 4 that sets that are open with respect to ρ are open with respect to τ.

    Metrics ρ, τ are not equivalent if, for example, (X, ρ) is a normed linear space. If ρ, τ were equivalent, then there would be a positive number c such that $\|u\| = \rho(u, 0) \leq c\tau(u, 0) \leq c$ by the definition on page 184, by definition (1), and by the upper bound 1 on τ. But by property (iii) on page 183, we can make $\|\cdot\|$ greater than c by choosing α large enough. Thus, ρ, τ are not equivalent.

9.3 Continuous Mappings Between Metric Spaces

26. Define mapping $f : \mathbf{R} \to \mathbf{R}$ by $f(x) = x^2$. To show that f is continuous, it suffices by the product property of Problem 49(i) of Section 1.6 to show that identity mapping $id_{\mathbf{R}}$ is continuous. Indeed, for every point x in $\mathbf{R}$ and sequence $\{x_n\}$ in $\mathbf{R}$, if $\{x_n\} \to x$, then $\{id_{\mathbf{R}}(x_n)\} \to id_{\mathbf{R}}(x)$ by definition of $id_{\mathbf{R}}$. So $id_{\mathbf{R}}$, and therefore f, is continuous.

    Now suppose, to get a contradiction, that f were also uniformly continuous. Then there would be a positive δ such that for x, y in $\mathbf{R}$, if $|y - x| < \delta$, then $|y^2 - x^2| < 2$. But if $x > 1/\delta$ and $y = x + 1/x$, then $|y - x| < \delta$ and $|y^2 - x^2| = 2 + 1/x^2$, a contradiction. Thus, f is not uniformly continuous.

    We exhibit a uniformly continuous mapping that is not Lipschitz in our solution to Problem 50 of Section 1.6.

27. By the next-to-last sentence on page 188, every subset G of X is open. That is because for every point x in G, open ball $B(x, 1)$ is contained in G since $B(x, 1) = \{x\}$. If $G = \emptyset$, then G is open by Proposition 1. In either case, every subset of X is open. Then the inverse image under f of each (open) subset of Y is an open subset of X. Thus, by Proposition 8, f is continuous.

28. Let f be such a mapping. Let G be a subset of X. Image $f(G)$ is a subset of Y and therefore open as explained in our solution to Problem 27. Because f is one-to-one, $f^{-1}(f(G)) = G$. Because f is continuous, G is an open subset of X by Proposition 8. Thus, every subset of X is open.

29. See the first paragraph on page 184 for definition of a product metric. Let (x, y) be a point in $X \times X$. If a sequence $\{(x_n, y_n)\}$ in $X \times X$ converges to (x, y), then we have that $\lim_{n\to\infty} \left(\rho(x_n, x)^2 + \rho(y_n, y)^2\right)^{1/2} = 0$. It follows that $\lim_{n\to\infty} \rho(x_n, x) = 0$ and $\lim_{n\to\infty} \rho(y_n, y) = 0$ by Theorem 18 of Section 1.5 and Proposition 21 of Section 1.6 along with the facts that function $u \mapsto u^2$ is continuous on $\mathbf{R}$ and function $v \mapsto v^{1/2}$ is continuous on $[0, \infty)$. Then starting with the triangle inequality,

$$\lim_{n\to\infty} \rho(x_n, y_n) \leq \lim_{n\to\infty} (\rho(x_n, x) + \rho(x, y) + \rho(y, y_n)) = \rho(x, y).$$

Likewise,

$$\rho(x, y) \leq \lim_{n\to\infty} (\rho(x, x_n) + \rho(x_n, y_n) + \rho(y_n, y)) = \lim_{n\to\infty} \rho(x_n, y_n).$$

Thus, $\lim_{n\to\infty} \rho(x_n, y_n) = \rho(x, y)$, that is, $\{\rho(x_n, y_n)\} \to \rho(x, y)$, so metric $\rho : X \times X \to \mathbf{R}$ is continuous by the definition on page 190.

30. Take points x and y in X. By the triangle inequality, $\rho(x, z) \leq \rho(x, y) + \rho(y, z)$ so that $\rho(x, z) - \rho(y, z) \leq \rho(x, y)$. Likewise, $\rho(y, z) - \rho(x, z) \leq \rho(x, y)$. Those two results give us that $|\rho(x, z) - \rho(y, z)| \leq \rho(x, y)$. Function f is uniformly continuous because δ equal to ϵ responds to all ϵ challenges: If $\rho(x, y) < \delta$, then $|\rho(x, z) - \rho(y, z)| \leq \rho(x, y) < \delta = \epsilon$.

31. This is like Problem 49(ii) of Section 1.6. Let $f : X \to Y$ and $g : Y \to Z$ be uniformly continuous where (X, ρ), (Y, σ), (Z, τ) are metric spaces. Because g is uniformly continuous, for every positive ϵ, there is a positive η such that for u, v in X, if $\sigma(f(u), f(v)) < \eta$, then $\tau(g(f(u)), g(f(v))) < \epsilon$. Because f is uniformly continuous, for that η, there is a positive δ such that if $\rho(u, v) < \delta$, then $\sigma(f(u), f(v)) < \eta$. Those two statements together say that for every positive ϵ, there is a positive δ such that for u, v in X, if $\rho(u, v) < \delta$, then $\tau((g \circ f)(u), (g \circ f)(v)) < \epsilon$. Thus, $g \circ f$ is uniformly continuous.

32. This follows directly from Propositions 8 and 7.

33. (i) This generalizes Problem 30 including the stronger result that f is uniformly continuous. Let x, z be points in X. For each point y' in E,

$$f(x) = \inf\{\rho(x, y) \mid y \in E\} \leq \rho(x, y') \leq \rho(x, z) + \rho(z, y').$$

Because that holds for each y' in E,

$$f(x) \leq \rho(x, z) + \inf\{\rho(z, y) \mid y \in E\} = \rho(x, z) + f(z).$$

Likewise,

$$f(z) \leq \rho(x, z) + f(x).$$

Together, $|f(z) - f(x)| \leq \rho(x, z)$. Function f is uniformly continuous because δ equal to ϵ responds to all ϵ challenges: If $\rho(x, z) < \delta$, then $|f(z) - f(x)| \leq \rho(x, z) < \delta = \epsilon$.

(ii) A point x is in $\{x \in X \mid \text{dist}(x, E) = 0\}$ if and only if $\inf\{\rho(x, y) \mid y \in E\} = 0$ if and only if every open ball $\{x' \in X \mid \rho(x, x') < r\}$ centered at x, hence every neighborhood of x, contains a point in E if and only if $x \in \overline{E}$. Thus, $\{x \in X \mid \text{dist}(x, E) = 0\} = \overline{E}$.

34. See Section 3.1's Proposition 2 and proof. Assume first that there is such an f. Note that $\{x \in X \mid f(x) > 0\} = f^{-1}((0, \infty))$, and that $(0, \infty)$ is an open subset of $\mathbf{R}$. Hence, E is an open subset of X by Proposition 8.

    Conversely, assume that E is open. Put f to be the function $f(x) = \operatorname{dist}(x, X \sim E)$. Function f is real-valued by definition and continuous by Problem 33(i). Now

$$
\begin{aligned}
\{x \in X \mid f(x) > 0\} &= \{x \in X \mid \operatorname{dist}(x, X \sim E) > 0\} && \text{definition of } f \\
&= X \sim \{x \in X \mid \operatorname{dist}(x, X \sim E) = 0\} \\
&= X \sim \overline{X \sim E} && \text{Problem 33(ii)} \\
&= X \sim (X \sim E) && \text{explanation below} \\
&= E && \text{next-to-last sentence on p. 187.}
\end{aligned}
$$

    In the fourth step, $\overline{X \sim E} = X \sim E$ by Proposition 3 because $X \sim E$ is closed by Proposition 4 since E is open. Thus, there is such an f.

35. Assume first that there is such an f. Then by Problem 36 with Y equal to $\mathbf{R}$ and C equal to (closed) singleton set $\{0\}$, inverse image $f^{-1}(0)$ is closed. Thus, E is closed.

    Conversely, assume that E is closed. Put f to be the function defined by $f(x) = \operatorname{dist}(x, E)$. Function f is real-valued by definition and continuous by Problem 33(i). Observe that $f^{-1}(0) = \{x \in X \mid f(x) = 0\} = \{x \in X \mid \operatorname{dist}(x, E) = 0\}$, so by Problem 33(ii), our assumption, and Proposition 3, $f^{-1}(0) = \overline{E} = E$. Thus, there is such an f.

36. Note that for every subset C of Y, we have that $f^{-1}(Y \sim C) = X \sim f^{-1}(C)$. Otherwise, some points in $Y \sim C$ would have preimages in $f^{-1}(C)$ and therefore be in C, a contradiction. Then by Proposition 4, the problem statement is equivalent to "$f : X \to Y$ is continuous if and only if $f^{-1}(Y \sim C)$ is open in X whenever $Y \sim C$ is open in Y," which is true by Proposition 8.

37. This is true by Problem 4 of Section 8.1 and the second example on page 156 with T replaced by ψ and $g(x)$ equal to x. In that example, inequality (4) applied to this exercise is $|\psi(f)| \leq (b - a)\|f\|_{\max}$, which shows that ψ is bounded by the subsequent definition.

9.4 Complete Metric Spaces

38. (i) Mimic the proof of Proposition 4 of Section 7.3. Let $\{x_n\}$ converge to x in (X, ρ). By the triangle inequality for the metric, $\rho(x_m, x_n) \leq \rho(x_m, x) + \rho(x_n, x)$ for all m, n. Therefore, $\{x_n\}$ is Cauchy. (ii) Let $\{x_n\}$ be a Cauchy sequence. Then there is an index N for which if $m, n \geq N$, then $\rho(x_m, x_n) < 1$. Let M be equal to $\max\{\rho(x_n, x_N) \mid n < N\}$. Then for all m, n, we have that $\rho(x_m, x_n) \leq \rho(x_m, x_N) + \rho(x_n, x_N) < 2M + 2$ by considering all the cases in which each of m, n is less than, equal to, or greater than N. By the definition on page 194, $\{x_n\}$ has a finite diameter and therefore is bounded.

39. If a Cauchy sequence converges, then that sequence itself is a convergent subsequence. For the converse, mimic the proof of the second part of Proposition 4 of Section 7.3. Let $\{x_n\}$ be a Cauchy sequence in (X, ρ) that has a subsequence $\{x_{n_k}\}$ that converges to x in (X, ρ). Take a positive ϵ. Because $\{x_n\}$ is Cauchy, we can choose N such that $\rho(x_m, x_n) < \epsilon/2$ if $m, n \geq N$. Because $\{x_{n_k}\}$ converges to x, we can choose K such that $n_K \geq N$ and $\rho(x_{n_K}, x) < \epsilon/2$. Then

$$
\begin{aligned}
\rho(x_n, x) &\leq \rho(x_n, x_{n_K}) + \rho(x_{n_K}, x) && \text{triangle inequality for the metric} \\
&< \epsilon && \text{if } n \geq N.
\end{aligned}
$$

Therefore, $\{x_n\} \to x$ in (X, ρ).

40. By the two definitions on page 192, it suffices to show that an index N such that $1/2^{N-1} \leq \epsilon$ responds to the ϵ challenge for a Cauchy sequence. Indeed, if $m, n \geq N$ and, say, $m < n$,

$$\begin{aligned}
\rho(x_m, x_n) &\leq \rho(x_m, x_{m+1}) + \rho(x_{m+1}, x_{m+2}) + \cdots + \rho(x_{n-1}, x_n) && \text{triangle inequality}\\
&< 1/2^m + 1/2^{m+1} + \cdots + 1/2^{n-1} && \text{hypothesis}\\
&< 1/2^m + 1/2^{m+1} + \cdots + 1/2^{n-1} + \cdots &&\\
&= 1/2^{m-1} && \text{sum of geometric series}\\
&\leq 1/2^{N-1} && N \leq m\\
&\leq \epsilon. &&
\end{aligned}$$

Thus, $\{x_n\}$ is Cauchy and therefore converges because (X, ρ) is a complete metric space.

In the next case, $\{x_n\}$ does not necessarily converge. For example, put X to $\mathbf{R}$, which is complete by the sentence after the definition of completeness on page 192, and put x_n equal to $\sum_{k=1}^{n} \frac{1}{k}$, for which inequality $|x_{n+1} - x_n| = 1/(n+1) < 1/n$ holds. Sequence $\{x_n\}$ does not converge by the p-series test with p equal to 1.

41. Interval $(0, 1]$ with the metric induced by the absolute-value norm is a metric space by the last sentence on page 183. Define a descending sequence $\{F_n\}$ of closed, nonempty sets in $(0, 1]$ by $F_n = (0, 1/n]$. Then $\bigcap_{n=1}^{\infty} F_n = \emptyset$. That does not contradict Cantor's intersection theorem because $(0, 1]$ is not complete by the contrapositive of Proposition 11 since $(0, 1]$ is not a closed subset of $\mathbf{R}$, which *is* complete by the sentence after the definition of completeness on page 192.

42. By the symmetry of the statement, it suffices to prove just one direction of the *if-and-only-if* statement. So assume that (X, ρ) is complete. Let $\{x_n\}$ be a Cauchy sequence in (X, σ). Because ρ, σ are equivalent, there is a positive number c such that for all x', x'' in X, we have that $\sigma(x', x'')/c \leq \rho(x', x'') \leq c\sigma(x', x'')$ by Problem 1. Because $\{x_n\}$ is Cauchy in (X, σ), for each positive ϵ, there is an index N for which if $m, n \geq N$, then $\sigma(x_m, x_n) < \epsilon/c$. Then $\{x_n\}$ is also Cauchy in (X, ρ) because $\rho(x_m, x_n) \leq c\sigma(x_m, x_n) < \epsilon$ if $m, n \geq N$. Because (X, ρ) is complete, $\{x_n\}$ converges in (X, ρ) to some x in X.

By the sentence before Proposition 7, $\{x_n\}$ also converges in (X, σ) to x. To see why, note that by Proposition 6, x is a point of closure of (X, ρ) and therefore a point of closure of (X, σ) because the closed sets are the same in (X, ρ) and (X, σ) by Propositions 7 and 4. Hence, by Proposition 6 in the other direction and the paragraph before that proposition, $\{x_n\}$ converges in (X, σ) to x. Here is another way to show that $\{x_n\} \to x$ in (X, σ): Because $\{x_n\} \to x$ in (X, ρ), there is an N such that $\rho(x_n, x) < \epsilon/c$ if $n \geq N$. Then $\sigma(x_n, x) \leq c\rho(x_n, x) < \epsilon$ if $n \geq N$.

Thus, (X, σ) is complete because each Cauchy sequence in (X, σ) converges to a point in X.

43. Let (X, ρ) and (Y, σ) be two complete metric spaces. Take a Cauchy sequence $\{(x_n, y_n)\}$ in $X \times Y$. For each positive ϵ, there is an index N such that if $m, n \geq N$, then

$$\begin{gathered}
\left(\rho(x_m, x_n)^2 + \sigma(y_m, y_n)^2\right)^{1/2} < \epsilon, \text{ or}\\
\rho(x_m, x_n)^2 + \sigma(y_m, y_n)^2 < \epsilon^2
\end{gathered}$$

where we use the product metric (page 184). Consequently, $\rho(x_m, x_n) < \epsilon$ and $\sigma(y_m, y_n) < \epsilon$ if $m, n \geq N$, which show that $\{x_n\}$ and $\{y_n\}$ are Cauchy sequences. Because (X, ρ) and

(Y, σ) are complete, $\{x_n\}$ converges to a point x in X, and $\{y_n\}$ converges to a point y in Y. Then there is an index K for which if $n \geq K$, then $\rho(x_n, x) < \epsilon/\sqrt{2}$ and $\sigma(y_n, y) < \epsilon/\sqrt{2}$. So if $n \geq K$,

$$\left(\rho(x_n, x)^2 + \sigma(y_n, y)^2\right)^{1/2} < \left(\left(\epsilon/\sqrt{2}\right)^2 + \left(\epsilon/\sqrt{2}\right)^2\right)^{1/2} = \epsilon,$$

which shows that $\{(x_n, y_n)\}$ converges to point (x, y) in $X \times Y$. Thus, $X \times Y$ is complete.

44. Assume first that f is uniformly continuous. Then for every positive ϵ, there is a positive δ such that if $\rho(u_n, v_n) < \delta$, then $\sigma(f(u_n), f(v_n)) < \epsilon$. Now if $\lim_{n\to\infty} \rho(u_n, v_n) = 0$, there is an index N for which $\rho(u_n, v_n) < \delta$ if $n \geq N$. It follows that $\sigma(f(u_n), f(v_n)) < \epsilon$ if $n \geq N$, that is, $\lim_{n\to\infty} \sigma(f(u_n), f(v_n)) = 0$.

    Conversely, assume the *if* part of the statement. Suppose, to get a contradiction, that f were not uniformly continuous. Then there would be an ϵ such that for each n, there existed u_n, v_n for which $\rho(u_n, v_n) < 1/n$ but $\sigma(f(u_n), f(v_n)) \geq \epsilon$. Then $\lim_{n\to\infty} \rho(u_n, v_n) = 0$ but $\lim_{n\to\infty} \sigma(f(u_n), f(v_n)) \geq \epsilon \neq 0$, a contradiction. Thus, f is uniformly continuous.

45. (i) Let ρ be the metric on X, and σ the metric on Y. Let $\{x_n\}$ be a Cauchy sequence in E. Because f is uniformly continuous, for every positive ϵ, there is a positive δ such that if $\rho(x_m, x_n) < \delta$, then $\sigma(f(x_m), f(x_n)) < \epsilon$. Because $\{x_n\}$ is Cauchy, there is an index N for which $\rho(x_m, x_n) < \delta$ if $m, n \geq N$. It follows that $\sigma(f(x_m), f(x_n)) < \epsilon$ if $m, n \geq N$, so $\{f(x_n)\}$ is a Cauchy sequence in Y. Thus, f maps Cauchy sequences in E to Cauchy sequences in Y.

    (ii) We first show that $\overline{f}$ exists. Sequence $\{x_n\}$ exists by Proposition 6. By Problem 38(i), $\{x_n\}$ is Cauchy. By part (i), $\{f(x_n)\}$ is a Cauchy sequence in Y. Because Y is complete, $\{f(x_n)\}$ converges and therefore has a limit. Hence, $\overline{f}$ exists.

    We next show that for each x in $\overline{E}$, function $\overline{f}(x)$ returns the same value no matter what sequence in E we choose to converge to x. So let sequences $\{u_n\}, \{v_n\}$ in E both converge to x, that is, $\lim_{n\to\infty} \rho(u_n, x) = 0$, and $\lim_{n\to\infty} \rho(v_n, x) = 0$. By the triangle inequality and Theorem 18 of Section 1.5,

    $$\lim_{n\to\infty} \rho(u_n, v_n) \leq \lim_{n\to\infty} \rho(u_n, x) + \lim_{n\to\infty} \rho(x, v_n) = 0 + 0 = 0.$$

    By Problem 44, $\lim_{n\to\infty} \sigma(f(u_n), f(v_n)) = 0$. We showed above that $\{f(u_n)\}$ converges, say to y, and $\{f(v_n)\}$, say to y'. By the triangle inequality,

    $$\sigma(y, y') \leq \sigma(y, f(u_n)) + \sigma(f(u_n), f(v_n)) + \sigma(f(v_n), y') \text{ for all } n.$$

    Take the limit as $n \to \infty$ to see that $\sigma(y, y') = 0$. By property (ii) of a metric (page 182), $y = y'$, that is, $\lim_{n\to\infty} f(u_n) = \lim_{n\to\infty} f(v_n)$. Each of those two limits is $\overline{f}(x)$ by definition of $\overline{f}$. That they are equal shows that $\overline{f}$ is properly defined.

    (iii) Take a positive ϵ. Because f is uniformly continuous on E, there is positive δ such that for u, v in E, if $\rho(u, v) < \delta$, then $\sigma(f(u), f(v)) < \epsilon/3$. Assume for $\bar{u}, \bar{v}$ in $\overline{E}$ that $\rho(\bar{u}, \bar{v}) < \delta/3$. Let $\{u_n\}$ be a sequence in E that converges to $\bar{u}$, and $\{v_n\}$ another that converges to $\bar{v}$. Then there is an index N such that $\rho(u_n, \bar{u}) < \delta/3$ and $\rho(v_n, \bar{v}) < \delta/3$ if $n \geq N$. By the triangle inequality,

    $$\rho(u_n, v_n) \leq \rho(u_n, \bar{u}) + \rho(\bar{u}, \bar{v}) + \rho(\bar{v}, v_n) < \delta/3 + \delta/3 + \delta/3 = \delta \text{ if } n \geq N.$$

Then $\sigma(f(u_n), f(v_n)) < \epsilon/3$ if $n \geq N$. By $\overline{f}$'s definition, $\{f(u_n)\} \to \overline{f}(\bar{u})$ and $\{f(v_n)\} \to \overline{f}(\bar{v})$ so that by increasing N if necessary, $\sigma\big(f(u_N), \overline{f}(\bar{u})\big) < \epsilon/3$ and $\sigma\big(f(v_N), \overline{f}(\bar{v})\big) < \epsilon/3$. Then along with the triangle inequality,

$$\begin{aligned}\sigma\big(\overline{f}(\bar{u}), \overline{f}(\bar{v})\big) &\leq \sigma\big(\overline{f}(\bar{u}), f(u_N)\big) + \sigma(f(u_N), f(v_N)) + \sigma\big(f(v_N), \overline{f}(\bar{v})\big) \\ &< \epsilon/3 + \epsilon/3 + \epsilon/3 \\ &= \epsilon.\end{aligned}$$

Thus, $\overline{f}$ is uniformly continuous on $\overline{E}$. Another solution, not simpler, uses Problem 44.

(iv) Function $\overline{f}$ is an extension of f because $E \subseteq \overline{E}$ (by the sentence after the definition on page 187) and because $\overline{f} = f$ on E. Indeed, for x in E, we can choose the constant sequence $\{x\}$, which of course converges to x, so that $\overline{f}(x) = f(x)$.
To show that $\overline{f}$ is unique, let $\overline{g}$ be a continuous function from $\overline{E}$ to Y such that $\overline{g} = f$ on E. For x in $\overline{E}$, choose a sequence $\{x_n\}$ in E such that $\{x_n\} \to x$. By the definition on page 190, $\{\overline{g}(x_n)\} \to \overline{g}(x)$, that is, $\lim_{n\to\infty} \overline{g}(x_n) = \overline{g}(x)$. Because $x_n \in E$ for all n and because $\overline{g} = f$ on E, we have that $\overline{g}(x) = \lim_{n\to\infty} f(x_n) = \overline{f}(x)$. Thus, $\overline{f}$ is unique.

46. (i) This is Problem 10.

(ii) We first establish that a sequence $\{x^k\}$ in (Z, σ) converges if and only if for each n, sequence $(x_n^1, x_n^2, \dots)$ converges in (X_n, ρ_n) (note that superscripts are term indices). So assume that $\{x^k\} \to x$ in (Z, σ). Take a positive number ϵ. Put ϵ' equal to $\min\{\epsilon, 1/2\}/2^{n+1}$. There is an index K such that $\sigma\big(x^k, x\big) < \epsilon'$ if $k \geq K$, that is,

$$\sum_{n=1}^{\infty} 2^{-n} \frac{\rho_n\big(x_n^k, x_n\big)}{1 + \rho_n\big(x_n^k, x_n\big)} < \epsilon' \text{ if } k \geq K.$$

Each term in that sum is also less than ϵ' if $k \geq K$; hence, for each n, if $k \geq K$, then

$$\begin{aligned} 2^{-n} \frac{\rho_n\big(x_n^k, x_n\big)}{1 + \rho_n\big(x_n^k, x_n\big)} &< \epsilon', \\ \rho_n\big(x_n^k, x_n\big) &< \frac{1}{2^{-n}/\epsilon' - 1} && \text{solve for } \rho_n\big(x_n^k, x_n\big) \\ &= \frac{1}{2/\min\{\epsilon, 1/2\} - 1} && \epsilon' = \min\{\epsilon, 1/2\}/2^{n+1} \\ &= \begin{cases} 1/3 & \epsilon \geq 1/2 \\ \epsilon/(2-\epsilon) & \epsilon < 1/2 \end{cases} \\ &< \epsilon. \end{aligned}$$

Thus, $\big(x_n^1, x_n^2, \dots\big) \to x_n$ in (X_n, ρ_n) for each n.
Conversely, assume that $\big(x_n^1, x_n^2, \dots\big) \to x_n$ in (X_n, ρ_n) for each n. Choose an index N large enough such that $1/2^N \leq \epsilon/2$. Then

$$\begin{aligned}\sum_{n=N+1}^{\infty} 2^{-n} \frac{\rho_n\big(x_n^k, x_n\big)}{1 + \rho_n\big(x_n^k, x_n\big)} &< \sum_{n=N+1}^{\infty} 2^{-n} \\ &= 1/2^N \\ &\leq \epsilon/2.\end{aligned}$$

For all n from 1 to N, there is an index K such that $\rho_n(x_n^k, x_n) < \epsilon/2$ if $k \geq K$. Then

$$\begin{aligned}\sigma(x^k, x) &= \sum_{n=1}^{N} 2^{-n} \frac{\rho_n(x_n^k, x_n)}{1+\rho_n(x_n^k, x_n)} + \sum_{n=N+1}^{\infty} 2^{-n} \frac{\rho_n(x_n^k, x_n)}{1+\rho_n(x_n^k, x_n)} \\ &< \sum_{n=1}^{N} 2^{-n}(\epsilon/2) + \epsilon/2 \qquad \text{if } k \geq K \\ &< \epsilon.\end{aligned}$$

Thus, $\{x^k\} \to x$ in (Z, σ).

We are now ready to solve the problem. Assume first that (Z, σ) is complete. Consider a sequence $\{x^k\}$ in Z whose kth term is $(0, \ldots, 0, x_m^k, 0, \ldots)$ for each k and such that sequence $(x_m^1, x_m^2, \ldots)$ is Cauchy in (X_m, ρ_m). Then there is an index K for which $\rho_m(x_m^j, x_m^k) < \epsilon$ if $j, k \geq K$. Moreover,

$$\sigma(x^j, x^k) = \sum_{n=1}^{\infty} 2^{-n} \frac{\rho_n(x_n^j, x_n^k)}{1+\rho_n(x_n^j, x_n^k)} = 2^{-m} \frac{\rho_m(x_m^j, x_m^k)}{1+\rho_m(x_m^j, x_m^k)} < \epsilon \text{ if } j, k \geq K$$

so that $\{x^k\}$ is Cauchy in Z. Because Z is complete, $\{x^k\}$ converges in Z, say to $(0, \ldots, 0, \ x_m, 0, \ldots)$. Then $(x_m^1, x_m^2, \ldots) \to x_m$ in (X_m, ρ_m). Hence, (X_m, ρ_m) is complete. Our argument holds for m equal to each n. Thus, each (X_n, ρ_n) is complete.

Conversely, assume that each (X_n, ρ_n) is complete. Let $\{x^k\}$ be a Cauchy sequence in (Z, σ). Put ϵ' equal to $\min\{\epsilon, 1/2\}/2^{n+1}$. Then there is an index K for which $\sigma(x^j, x^k) < \epsilon'$ if $j, k \geq K$. Now proceed as in the first paragraph above in our solution to this part (ii) to conclude that $(x_n^1, x_n^2, \ldots)$ is a Cauchy sequence in (X_n, ρ_n) for each n. Each of those sequences converges because each (X_n, ρ_n) is complete. Therefore, $\{x^k\}$ converges in (Z, σ). Thus, (Z, σ) is complete.

47. Because x^n and $\cos(x/n)$ for each n as well as α, β considered as constant functions are continuous, $\{f_n\}$ is in $C[0,1]$ by Problem 49(i) of Section 1.6. By Theorem 12(iii), the second definition on page 192, and Problem 38(i), $\{f_n\}$ is Cauchy if and only if it converges to a function in $C[0,1]$. Consequently, it suffices to find values of α, β for which $\{f_n\}$ converges to a function in $C[0,1]$. Observe that

$$\{x^n\} \to \begin{cases} 0 & 0 \leq x < 1 \\ 1 & x = 1 \end{cases} \qquad \text{and} \qquad \{\cos(x/n)\} \to 1$$

so that by Problem 24 of Section 7.3 with X a normed linear space such as $L^\infty[0,1]$ that includes $C[0,1]$ as well as functions that are not continuous,

$$\{f_n(x)\} = \{\alpha x^n + \beta \cos(x/n)\} \to \begin{cases} \beta & 0 \leq x < 1 \\ \alpha + \beta & x = 1. \end{cases}$$

We see that f_n does not converge to a function in $C[0,1]$ if $\alpha \neq 0$ because in that case, $\lim_{n\to\infty} f_n(x)$ fails to be continuous at point x equal to 1. So put α equal to 0. Then for all x such that $0 \leq x \leq 1$, we have that $\lim_{n\to\infty} f_n(x) = \beta$, which, being a constant function, is continuous and therefore in $C[0,1]$. Thus, $\{f_n\}$ is a Cauchy sequence in $C[0,1]$ precisely when α is 0, and β is a real number.

48. Subspace $\mathcal{D}$ is not complete. Consider subspace $\mathcal{P}$ of $\mathcal{D}$ consisting of polynomials with rational coefficients. By the sentence before Theorem 11 of Section 7.4, $\mathcal{P}$ is dense in $C[0,1]$. Choose a function f in $C[0,1]$ that is not differentiable. For example, define f by $f(x) = 1 - |1 - 2x|$, a tent function, which is not differentiable at x equal to $1/2$. Another example is the function mentioned in the remark on page 113. By definition of denseness (page 150), for each natural number n, there is a polynomial p_n in $\mathcal{P}$ for which $\|p_n - f\|_{\max} < 1/n$. Sequence $\{p_n\}$ therefore converges to f. By Problem 38(i), $\{p_n\}$ is a Cauchy sequence in $\mathcal{D}$. But f is not in $\mathcal{D}$. Thus, $\mathcal{D}$ is not complete.

    Here is another, albeit similar, solution. Define $f_n : [0,1] \to \mathbf{R}$ by $f_n(x) = 1 - |1-2x|^{1+1/n}$. Then $\{f_n\}$ is a sequence in $\mathcal{D}$ that converges to $1 - |1 - 2x|$, which is not in $\mathcal{D}$.

49. Subspace $\mathcal{L}$ is not complete. Our two solutions are similar to our solutions to Problem 48. Consider subspace $\mathcal{P}$ of $\mathcal{L}$ consisting of polynomials with rational coefficients. To see that $\mathcal{P} \subseteq \mathcal{L}$, consider a polynomial p in $\mathcal{P}$. Derivative p' is another polynomial and therefore a continuous, real-valued function on the nonempty, closed, bounded set $[0,1]$. By the extreme value theorem (Section 1.6), there is a real number c such that $|p'(x)| \le c$ for all x in $[0,1]$. By the mean value theorem, for u, v in $[0,1]$ where, say, $u < v$, there is an x in $[u,v]$ such that $p(v) - p(u) = p'(x)(v-u)$. We deduce that $|p(v) - p(u)| \le c|v-u|$ for all u, v in $[0,1]$. Hence, p is Lipschitz. So $\mathcal{P} \subseteq \mathcal{L}$.

    Now choose a function in $C[0,1]$ that is not Lipschitz. For example, function f defined by $f(x) = \sqrt{x}$ is continuous on $[0,1]$ but not Lipschitz by the last sentence of the first paragraph of Section 1.6. For each natural number n, there is a polynomial p_n in $\mathcal{P}$ for which $\|p_n - f\|_{\max} < 1/n$. We therefore have a Cauchy sequence $\{p_n\}$ in $\mathcal{L}$ that converges to a function not in $\mathcal{L}$. Thus, $\mathcal{L}$ is not complete.

    This next solution uses the same reasoning but with piecewise linear functions instead of polynomials. By Problem 51 of Section 1.6, for each n, there is a piecewise linear function φ_n on $[0,1]$ for which $\|\varphi_n - f\|_{\max} < 1/n$. Now $\{\varphi_n\}$ is in $\mathcal{L}$ because for each n, function φ_n is continuous, and because for u, v in $[0,1]$, we have that $|\varphi_n(v) - \varphi_n(u)|/|v-u|$ is less than or equal to the maximum of the absolute values of the slopes of the linear pieces of φ_n. Such a maximum exists because φ_n has only a finite number of linear pieces. We therefore have a Cauchy sequence $\{\varphi_n\}$ in $\mathcal{L}$ that converges to a function not in $\mathcal{L}$.

50. (i) If $\{x_n\}, \{y_n\}$ are Cauchy sequences, for each positive ϵ, there is an index N for which if $m, n \ge N$, then $\rho(x_m, x_n) < \epsilon/2$ and $\rho(y_m, y_n) < \epsilon/2$. By the triangle inequality,

$$\rho(x_m, y_m) \le \rho(x_m, x_n) + \rho(x_n, y_n) + \rho(y_n, y_m), \text{ and}$$
$$\rho(x_n, y_n) \le \rho(x_n, x_m) + \rho(x_m, y_m) + \rho(y_m, y_n)$$

    so that

$$\rho(x_m, y_m) - \rho(x_n, y_n) \le \rho(x_m, x_n) + \rho(y_m, y_n), \text{ and}$$
$$\rho(x_n, y_n) - \rho(x_m, y_m) \le \rho(x_m, x_n) + \rho(y_m, y_n).$$

    Together,

$$|\rho(x_n, y_n) - \rho(x_m, y_m)| \le \rho(x_m, x_n) + \rho(y_m, y_n),$$

    which is less than $\epsilon/2 + \epsilon/2$, or ϵ, if $m, n \ge N$. Hence, $\{\rho(x_n, y_n)\}$ is a Cauchy sequence of real numbers and therefore converges by Theorem 17 of Section 1.5.

    (ii) In this solution, lim denotes $\lim_{n\to\infty}$. Refer to the last paragraph of Section 9.1. Function ρ' satisfies the requirements of a metric (page 182) excepting the *only-if* direction of property (ii) because ρ' inherits those requirements from ρ. We also need Theorem 18 (linearity and monotonicity of convergence of real sequences) of Section 1.5 for properties (i) and (iv). Here are the details: Set X' is nonempty because X is nonempty, so for a point x in X, set X' contains the (constant) Cauchy sequence $\{x\}$. By hypothesis, ρ' is a function from $X' \times X'$ to $\mathbf{R}$. Function ρ' satisfies property—

(i) because $\lim \rho(x_n, y_n) \geq 0$,
(ii) because if $\{x_n\} = \{y_n\}$, then $\lim \rho(x_n, y_n) = 0$,
(iii) because $\lim \rho(x_n, y_n) = \lim \rho(y_n, x_n)$, and
(iv) because $\lim \rho(x_n, y_n) \leq \lim(\rho(x_n, z_n) + \rho(z_n, y_n)) = \lim \rho(x_n, z_n) + \lim \rho(z_n, y_n)$.

Thus, ρ' is a pseudometric on X'.

(iii) Recall the definition of an equivalence relation (page 5). The relation is reflexive because $\lim \rho(x_n, x_n) = 0$ and symmetric because if $\lim \rho(x_n, y_n) = 0$, then $\lim \rho(y_n, x_n) = 0$. To prove transitivity, assume that $\lim \rho(x_n, y_n)$ and $\lim \rho(y_n, z_n)$ both equal 0. Because ρ' is a pseudometric, $\lim \rho(x_n, z_n) \geq 0$; also, $\lim \rho(x_n, z_n) \leq \lim \rho(x_n, y_n) + \lim \rho(y_n, z_n)$ so that $\lim \rho(x_n, z_n) \leq 0$. As a result, $\lim \rho(x_n, z_n) = 0$, which shows that the relation is transitive. Thus, the relation is an equivalence relation in X'.

Function $\widehat{\rho}$ is properly defined provided that it does not depend on which representative we take from an equivalence class. So assume that $\{y_n\}, \{z_n\}$ both represent equivalence class $[\{y_n\}]$ so that $\rho'(\{y_n\}, \{z_n\}) = 0$. Then along with the triangle inequality,

$$\rho'(\{x_n\}, \{y_n\}) \leq \rho'(\{x_n\}, \{z_n\}) + \rho'(\{z_n\}, \{y_n\}) = \rho'(\{x_n\}, \{z_n\}), \text{ and}$$
$$\rho'(\{x_n\}, \{z_n\}) \leq \rho'(\{x_n\}, \{y_n\}) + \rho'(\{y_n\}, \{z_n\}) = \rho'(\{x_n\}, \{y_n\}),$$

so that $\rho'(\{x_n\}, \{z_n\}) = \rho'(\{x_n\}, \{y_n\})$, that is, the ρ' distance is the same if equivalent $\{z_n\}$ represents $[\{y_n\}]$ instead of $\{y_n\}$. Thus, $\widehat{\rho}$ is properly defined.

Function $\widehat{\rho}$ satisfies metric properties (i), (iii), and (iv) because $\widehat{\rho}$ inherits them from ρ' since $\widehat{\rho}([\{x_n\}], [\{y_n\}]) = \rho'(\{x_n\}, \{y_n\})$ and property (ii) because $\widehat{\rho}([\{x_n\}], [\{y_n\}]) = 0$ if and only if $\rho'(\{x_n\}, \{y_n\}) = 0$ if and only if $\{x_n\} \cong \{y_n\}$ if and only if $[\{x_n\}] = [\{y_n\}]$.

(iv) Our proof of the hint's second sentence does not require the step described in the hint's first sentence. Because $\{\{x_{n,m}\}_{n=1}^{\infty}\}_{m=1}^{\infty}$ represents a Cauchy sequence in $\widehat{X}$, there is an index M for which

$$\rho'(\{x_{n,m}\}_{n=1}^{\infty}, \{x_{n,m'}\}_{n=1}^{\infty}) < \epsilon/2 \text{ if } m, m' \geq M. \tag{9.1}$$

Then by definition of ρ', $\lim \rho(x_{n,m}, x_{n,m'}) < \epsilon/2$ if $m, m' \geq M$. By definition of a limit and by increasing M if necessary, $\rho(x_{n,m}, x_{n,m'}) < \epsilon/2$ if $m, m', n \geq M$. In particular, $\rho(x_{m,m}, x_{m,m'}) < \epsilon/2$ if $m, m' \geq M$. Also, $\rho(x_{m,m'}, x_{m',m'}) < \epsilon/2$ if $m, m' \geq M$ because $\{x_{n,m'}\}_{n=1}^{\infty}$ is Cauchy in X. Then

$$\begin{aligned} \rho(x_{m,m}, x_{m',m'}) &\leq \rho(x_{m,m}, x_{m,m'}) + \rho(x_{m,m'}, x_{m',m'}) && \text{triangle inequality} \\ &< \epsilon/2 + \epsilon/2 && \text{if } m, m' \geq M, \end{aligned}$$

which shows that $\{x_{n,n}\}_{n=1}^{\infty}$ is a Cauchy sequence from X. To show that $\{x_{n,n}\}_{n=1}^{\infty}$ represents the limit of the Cauchy sequences from $\widehat{X}$, it suffices to show that for every $m \geq M$, distance $\widehat{\rho}([\{x_{n,m}\}_{n=1}^{\infty}], [\{x_{n,n}\}_{n=1}^{\infty}]) < \epsilon$. Indeed,

$$\begin{aligned} \widehat{\rho}([\{x_{n,m}\}_{n=1}^{\infty}], [\{x_{n,n}\}_{n=1}^{\infty}]) &= \rho'(\{x_{n,m}\}_{n=1}^{\infty}, \{x_{n,n}\}_{n=1}^{\infty}) && \text{definition of } \widehat{\rho} \\ &< \epsilon/2 && \text{if } m \geq M \end{aligned}$$

where the inequality comes from inequality (9.1) and the fact that $n \to \infty$ by definition of ρ' so that, in particular, the second n in $x_{n,n}$ is greater than M.

Thus, $(\widehat{X}, \widehat{\rho})$ is complete because representative Cauchy sequence $\{\{x_{n,m}\}_{n=1}^{\infty}\}_{m=1}^{\infty}$ in $\widehat{X}$ converges to representative Cauchy sequence $\{x_{n,n}\}_{n=1}^{\infty}$ in $\widehat{X}$.

(v) By definition of denseness (page 150), $h(X)$ is dense in $\widehat{X}$ provided that—

i. $h(X) \subseteq \widehat{X}$, and

ii. for each equivalence class $[\{x_n\}]$ in $\widehat{X}$, there is an equivalence class $[\{x\}]$ in $h(X)$ for which $\widehat{\rho}([\{x_n\}], [\{x\}]) < \epsilon$, or by definitions of $\widehat{\rho}, \rho'$, for which $\lim \rho(x_n, x) < \epsilon$.

Condition i is true because constant sequence $\{x\}$ is Cauchy in X since $\rho(x, x) = 0 < \epsilon$. As for condition ii, since $\{x_n\}$ is a Cauchy sequence, there is an index N for which if $n, n' \geq N$, then $\rho(x_n, x_{n'}) < \epsilon$. Put x equal to x_N. Then $x \in X$ because all points in sequence $\{x_n\}$ are in X; furthermore, $\lim \rho(x_n, x) < \epsilon$ because $n \geq N$ since $n \to \infty$. Thus, $h(X)$ is dense in $\widehat{X}$.

Next, $\widehat{\rho}(h(u), h(v)) = \rho'(\{u\}, \{v\}) = \rho(u, v)$ by definitions of h, $\widehat{\rho}$, and ρ' where we dropped the limit notation because sequences $\{u\}, \{v\}$ are constant and therefore do not depend on term index n. Note that h considered as $h : X \to h(X)$ is onto. Then by the definition of an isometry on page 184 and the subsequent paragraph, (X, ρ) and $(h(X), \widehat{\rho})$ are exactly the same from the viewpoint of metric spaces.

(vi) (X, ρ) is a subspace of $(\widetilde{X}, \widetilde{\rho})$ because $X \subseteq X \cup (\widehat{X} \sim h(X))$, and $\widetilde{\rho}(u, v) = \rho(u, v)$ if $u, v \in X$.

Next, by the last sentence of our solution to part (v),

$$X \cup (\widehat{X} \sim h(X)) = X \cup (\widehat{X} \sim X) = \widehat{X},$$

which shows that $\widetilde{X}$ and $\widehat{X}$ are the same. Furthermore, because (X, ρ) and $(h(X), \widehat{\rho})$ are the same and because $h(x) = [\{x\}]$, we can consider h to be the identity mapping on X and therefore identify $h(u)$ with u and $h(v)$ with v. Then definitions for $\widetilde{\rho}$ show that $\widetilde{\rho} = \widehat{\rho}$. Hence, $(\widetilde{X}, \widetilde{\rho})$ and $(\widehat{X}, \widehat{\rho})$ are exactly the same from the viewpoint of metric spaces. Thus, $(\widetilde{X}, \widetilde{\rho})$ is complete by part (iv), and (X, ρ) is dense in $(\widetilde{X}, \widetilde{\rho})$ by part (v).

51. Refer to Theorem 13, the subsequent paragraph, and our solution to Problem 50 for background and notation. We want to show that if $(\widetilde{X}_2, \widetilde{\rho}_2)$ is a second completion of (X, ρ), then there is an isometry $\bar{f} : \widetilde{X} \to \widetilde{X}_2$ such that $\bar{f}|_X = id_X$. Analogous to function h, let $h_2(x)$ for all x in X be the equivalence class of the constant sequence in $\widetilde{X}_2$ all of whose terms are x. Put $\bar{f}$ to be the extended mapping described in Problem 45 where f is the composition $h_2 \circ h^{-1} : h(X) \to h_2(X)$. Inverse mapping $h^{-1} : h(X) \to X$ exists because $h : X \to h(X)$ is one-to-one. We now show that $\bar{f}$ does indeed satisfy the properties above. As noted in our solution to Problem 50, X and $h(X)$ and now also $h_2(X)$ are exactly the same from the viewpoint of metric spaces so that we can identify x with $h(x)$ or $h_2(x)$. Then $\bar{f}|_X = id_X$ because for all x in X,

$$\bar{f}(x) = (h_2 \circ h^{-1})(h(x)) = h_2(h^{-1}(h(x))) = h_2(x) = x.$$

To show that we can extend $h_2 \circ h^{-1}$ to $\bar{f}$, verify Problem 45's hypotheses. Composition $h_2 \circ h^{-1}$ maps into a complete metric space because $h_2(X) \subseteq \widetilde{X}_2$. Next, note that an isometry is uniformly continuous by Problem 44 and definition of an isometry (page 184). Then $h_2 \circ h^{-1}$ is uniformly continuous by Problem 31 along with the fact that because h is an isometry, so is h^{-1} by the symmetry of the definition of an isometry. We next show that $\widetilde{X}$ is the closure $\overline{X}$ of X. If a sequence in X converges, then the sequence is Cauchy by Problem 38(i) and therefore converges to a point in $\widetilde{X}$. Then $\widetilde{X} = \overline{X}$ by Proposition 6 and the definition on page 187. Hence, Problem 45 tells us that $h_2 \circ h^{-1}$ has a (unique, uniformly continuous) extension to a mapping $\bar{f}$ of $\widetilde{X}$ to $\widetilde{X}_2$.

It remains to show that $\bar{f}$ is an isometry. To show that $\bar{f}$ is onto, let $[\{x_n\}_2]$ be a (Cauchy sequence) equivalence class in $\widetilde{X}_2$ (subscript 2 on $[\{x_n\}_2]$ is not a term index but rather an indicator that $[\{x_n\}_2]$ is an equivalence class in the second completion, $\widetilde{X}_2$). Because $\widetilde{X}_2$ is a completion of X, sequence $\{x_n\}_2$ converges to a point $\bar{x}$ in $\overline{X}$. There is a Cauchy sequence $\{x_n\}$ in $\widetilde{X}$ that converges to $\bar{x}$. Then

$$\begin{aligned}\bar{f}([\{x_n\}]) &= \lim_{n\to\infty}\left(h_2 \circ h^{-1}\right)([\{x_n\}]) && \text{definition of } \bar{f} \text{ (Problem 45(ii))}\\ &= h_2(\bar{x})\\ &= [\{x_n\}_2].\end{aligned}$$

Hence, $\bar{f}$ is onto. Next, for all $[\{x_n\}], [\{y_n\}]$ in $\widetilde{X}$,

$$\begin{aligned}\widetilde{\rho}_2(\bar{f}([\{x_n\}]), \bar{f}([\{y_n\}]) &= \widetilde{\rho}_2([\{x_n\}_2], [\{y_n\}_2]) && \bar{f}([\{x_n\}]) = [\{x_n\}_2] \text{ as shown above}\\ &= \lim \rho(x_n, y_n) && \text{Problem 50}\\ &= \widetilde{\rho}(\{x_n\}], [\{y_n\}]).\end{aligned}$$

Thus, $\bar{f}$ is an isometry from $\left(\widetilde{X}, \widetilde{\rho}\right)$ onto $\left(\widetilde{X}_2, \widetilde{\rho}_2\right)$ such that $\bar{f}$ is the identity mapping on X.

9.5 Compact Metric Spaces

The last step in the proof of Proposition 14 could use more detail. I provide it in the first paragraph of the solution to Problem 40 of Section 11.5. Replace "topological" with "metric" and ignore the word *countable.*

52. Let E be a subset of $\mathbf{Q}$. Subset E is also a subset of (complete metric space) $\mathbf{R}$, so by Proposition 11, E is complete in $\mathbf{Q}$ (and $\mathbf{R}$) if and only if E is a closed subset of $\mathbf{R}$.

    We can say a little more if E is complete in $\mathbf{Q}$. In that case, E is nowhere dense in $\mathbf{Q}$. Otherwise—referring to Problem 40 of Section 2.7 for the meaning of *nowhere dense*—there would be a nonempty set open in $\mathbf{Q}$ and contained in E. By Proposition 9 of Section 1.4, every nonempty open set is the union of open intervals, so we can assume that a set of the form $\mathbf{Q} \cap (a, b)$ would be contained in E. Note that $\mathbf{Q} \cap (a, b)$ is open in $\mathbf{Q}$ by Proposition 1 with X replaced by $\mathbf{Q}$. Now because $\mathbf{Q}$ is dense in $\mathbf{R}$ (Theorem 2 of Section 1.2), it follows from the definition of denseness on page 150 applied to numbers in $\mathbf{Q}$ and $\mathbf{R}$ that there would be a sequence in $\mathbf{Q} \cap (a, b)$ that converges to an irrational number. That contradicts Proposition 6 because E is closed and contains only rational numbers. Hence, E is nowhere dense in $\mathbf{Q}$.

    But there are nowhere-dense sets in $\mathbf{Q}$ that are not closed in $\mathbf{R}$. For example, an increasing sequence of rational numbers that converges to an irrational number is nowhere dense in $\mathbf{Q}$ but not closed in $\mathbf{R}$.

    Now consider compact subspaces of $\mathbf{Q}$. Let K be a subset of $\mathbf{Q}$. By Theorem 20(ii)$\leftrightarrow$(i), K is compact in $\mathbf{Q}$ (and $\mathbf{R}$) if and only if K is closed and bounded in $\mathbf{R}$, which is true if and only if K is closed in $\mathbf{R}$ and has a minimum and maximum (rational) number. As before, if K is compact in $\mathbf{Q}$, then K is nowhere dense in $\mathbf{Q}$.

53. Points in $\mathbf{R}^n$ are n-tuples, so write $x = (x_1, x_2, \ldots, x_n)$. (i) Consider sequence $\{x^k\}$ in $\mathbf{R}^n$ defined by $x^k = (x_1 + r - 1/k, x_2, \ldots, x_n)$ where k is the term index (not an exponent). $\{x^k\}$ is also in $B(x, r)$ because $\|x^k - x\| = \|(r - 1/k, 0, \ldots, 0)\| = r - 1/k < r$. But each subsequence converges to $(x_1 + r, x_2, \ldots, x_n)$, say point y, which is *not* in $B(x, r)$ because $\|y - x\| = \|(r, 0, \ldots, 0)\| = r \not< r$. Thus, B fails to be compact by the contrapositive of Theorem 20(ii)$\to$(iii).

    (ii) Such an open cover G is $\{B(x, r - 1/k)\}_{k=\lceil 1/r\rceil+1}^{\infty}$ where $\lceil\cdot\rceil$ denotes the ceiling function, used so that the radius of each $B(x, r - 1/k)$ be positive. Set G is open by Proposition 1 and the fact that each $B(x, r - 1/k)$ is open. To see that G covers $B(x, r)$, note that for a point x' in $B(x, r)$, distance $\|x' - x\|$ is less than r and therefore also less than some number $r - \epsilon$. Choose an index K for which $r - 1/K \geq r - \epsilon$ so that $x' \in B(x, r - 1/K)$. Then $x' \in G$, which shows that $B \subseteq G$, that is, G covers B.

On the other hand, open cover G of B has no finite subcover. Otherwise, there is some largest open ball $B(x, r - 1/K_{\max})$ in the subcover and a point $(x_1 + r - 1/(K_{\max} + 1), x_2, \ldots, x_n)$ in $B(x, r)$ but not in $B(x, r - 1/K_{\max})$ and therefore not in the subcover. Thus, B fails to be compact by definition.

(iii) B is not closed by the contrapositive of Proposition 6 along with the counterexample sequence we exhibited above in part (i). Thus, B fails to be compact by the contrapositive of Theorem 20(ii)→(i).

54. We show that X is compact if and only if X is finite. Assume first that X is compact. Then by Theorem 16(ii)→(i), X can be covered by a finite number of open balls of radius 1. With the discrete metric, such an open ball contains at most one point. Thus, X is finite.

    Conversely, assume that X is finite. Let $\{G_\lambda\}_{\lambda \in \Lambda}$ be an open cover of X. Then each point x in X is contained in some set G_{λ_x} of the cover, so collection $\{G_{\lambda_x}\}_{x \in X}$ is a finite subcover of X. Thus, X is compact. This direction of the claim holds for all metrics.

55. This follows directly from Proposition 7 and the definition of a compact metric space.

56. Both metric spaces are complete and totally bounded by Theorem 16(ii)→(i). Their Cartesian product is complete by Problem 43 and totally bounded by Problem 57 and therefore compact by Theorem 16(i)→(ii).

57. Let (X, ρ), (Y, σ) be two totally bounded metric spaces. Take a positive ϵ. There are natural numbers m, n such that $X \subseteq \{B_X(x_k, \epsilon/\sqrt{2})\}_{k=1}^{m}$ where each x_k is in X, and $Y \subseteq \{B_Y(y_k, \epsilon/\sqrt{2})\}_{k=1}^{n}$ where each y_k is in Y. The subscripts on B indicate the metric space in which the open ball is located. Now take an ordered pair (x, y) in $X \times Y$. Point x is in some $B_X(x_i, \epsilon/\sqrt{2})$, and y in some $B_Y(y_j, \epsilon/\sqrt{2})$. Then (x, y) is in open ball $B_{X \times Y}((x_i, y_j), \epsilon)$ because, referring to the top of page 184,

$$\left(\rho(x, x_i)^2 + \sigma(y, y_j)^2\right)^{1/2} < \left(\left(\epsilon/\sqrt{2}\right)^2 + \left(\epsilon/\sqrt{2}\right)^2\right)^{1/2} = \epsilon.$$

    It follows that $X \times Y$ is covered by at most mn open balls of radius ϵ and is therefore totally bounded.

58. Refer to the definition of *totally bounded* and subsequent paragraph. Take a positive ϵ. Assume first that E is totally bounded. Then E considered as a subspace of (X, ρ) can be covered by a finite number of open balls of radius ϵ with centers x_k in E, that is,

$$E \subseteq \{B(x_k, \epsilon)\}_{k=1}^{n} \text{ where } x_k \in E.$$

    Because E is contained in X, centers x_k are also in X, so we can rewrite the previous inclusion as

$$E \subseteq \{B(x_k, \epsilon)\}_{k=1}^{n} \text{ where } x_k \in X.$$

    Thus, E can be covered by a finite number of open balls (open in X) of radius ϵ which have centers belonging to X.

    Conversely, assume that E can be covered by a finite collection of open balls (open in X) of radius $\epsilon/2$ which have centers x_k belonging to X. Ignore those balls that do not contain a point in E. The remaining collection $\{B(x_k, \epsilon/2)\}_{k=1}^{n}$ still covers E. Now construct a finite collection of open balls of radius ϵ which have centers belonging to E: For each k, take a point x'_k from $B(x_k, \epsilon/2)$ that is also in E. Open ball $B(x'_k, \epsilon)$ covers $B(x_k, \epsilon/2)$ because for all points x in $B(x_k, \epsilon/2)$,

$$\begin{aligned} \rho(x, x'_k) &\leq \rho(x, x_k) + \rho(x_k, x'_k) && \text{triangle inequality} \\ &< \epsilon/2 + \epsilon/2 = \epsilon && x, x'_k \in B(x_k, \epsilon/2). \end{aligned}$$

Then $\{B(x'_k, \epsilon)\}_{k=1}^n$ covers $\{B(x_k, \epsilon/2)\}_{k=1}^n$ and therefore covers E also. If we consider E as a subspace of X, then $\{B(x'_k, \epsilon)\}_{k=1}^n$ is a finite collection of open balls in E that covers E. Thus, E is totally bounded.

59. Metric space X is complete by Theorem 16(ii)$\to$(i). So if E is compact, then it is also complete and therefore a closed subset of X by Proposition 11. A stronger form of this direction of the claim is Problem 1 of Section 10.1—X need not be compact. Also, the proof of the stronger and more general Proposition 16 of Section 11.5 works here because a metric space is a Hausdorff topological space.

    Conversely, assume that E is a closed subset of X. Take a sequence in E. Because that sequence is also in X, the sequence has a subsequence that converges to a point in X by Theorem 16(ii)$\to$(iii). That point is also in E by Proposition 6. Thus, E is sequentially compact and therefore compact by Theorem 16(iii)$\to$(ii). Another proof is that of Proposition 15 of Section 11.5; in that proof, F is a typo for K.

60. Metric space (X, ρ) is complete by Theorem 16(ii)$\to$(i). Now refer to page 194. Because $\{F_n\}_{n=1}^\infty$ is descending, $\{\text{diam}\, F_n\}$ is a decreasing sequence. Then because diam is bounded below by 0, sequence $\{\text{diam}\, F_n\}$ converges by the proof of the monotone convergence criterion for real sequences (Section 1.5). If $\lim_{n\to\infty} \text{diam}\, F_n = 0$, then the claim is true by Cantor's intersection theorem. Otherwise, $\lim_{n\to\infty} \sup\{\rho(x,y) \mid x, y \in F_n\} > 0$. That inequality along with the fact that the F_n are descending shows that each F_n contains at least two points of F_{n-1} so that $\bigcap_{n=1}^\infty F_n \neq \emptyset$.

    Here is another solution that borrows ideas from the first part of the proof of Cantor's intersection theorem: For each index n, select an x_n from F_n. Because X is compact, there is a subsequence of $\{x_n\}$ that converges to some x in X (Theorem 16(ii)$\to$(iii)). By reindexing if necessary, we can assume that $\{x_n\} \to x$. For each index n, set F_n is closed and $x_k \in F_n$ if $k \geq n$ so that x belongs to F_n. Thus, x belongs to $\bigcap_{n=1}^\infty F_n$.

61. Take a positive ϵ. Assume first that $\overline{E}$ is totally bounded. Then by Problem 58, $\overline{E}$ can be covered by a finite number of open balls (open in X) of radius ϵ which have centers in X. Because $E \subseteq \overline{E}$, that open cover of $\overline{E}$ also covers E. Thus, by Problem 58 in the other direction, E is totally bounded.

    Conversely, assume E is totally bounded. Then E can be covered by a finite number of open balls (open in (X, ρ)) of radius $\epsilon/2$ which have centers x_k in X, that is, $E \subseteq \{B(x_k, \epsilon/2)\}_{k=1}^n$. We have two proofs. Here is the first:

$$\begin{aligned} \overline{E} &\subseteq \overline{\{B(x_k, \epsilon/2)\}_{k=1}^n} && \text{first part of Problem 23} \\ &= \{\overline{B(x_k, \epsilon/2)}\}_{k=1}^n && \text{second part of Problem 23} \\ &\subseteq \{\overline{B}(x_k, \epsilon/2)\}_{k=1}^n && \text{our solution to Problem 15(ii)} \\ &\subseteq \{B(x_k, \epsilon)\}_{k=1}^n && \text{definitions of open and closed balls (pages 186, 187),} \end{aligned}$$

    which shows that $\overline{E}$ can be covered by a finite number of open balls (open in X) of radius ϵ which have centers x_k in X. Thus, $\overline{E}$ is totally bounded.

Here is the second proof: Let $\bar{x}$ be a point in $\overline{E}$. Then by definition (page 187), every neighborhood of $\bar{x}$ contains a point in E. In particular, $B(\bar{x}, \epsilon/2)$ contains a point x in E. Because $x \in E$, point x is in some $B(x_k, \epsilon/2)$. Then $\bar{x} \in B(x_k, \epsilon)$ because

$$\begin{aligned} \rho(\bar{x}, x_k) &\leq \rho(\bar{x}, x) + \rho(x, x_k) && \text{triangle inequality} \\ &< \epsilon/2 + \epsilon/2 && x \in B(\bar{x}, \epsilon/2);\ x \in B(x_k, \epsilon/2). \end{aligned}$$

Hence, $\overline{E} \subseteq \{B(x_k, \epsilon)\}_{k=1}^{n}$.

62. Assume first that E is totally bounded. Then $\overline{E}$ is totally bounded by Problem 61. Moreover, $\overline{E}$ is closed by Proposition 3 and therefore complete by Proposition 11. Thus, $\overline{E}$ is compact by Theorem 16(i)→(ii).

    Conversely, assume that $\overline{E}$ is compact. Then $\overline{E}$ is totally bounded by Theorem 16(ii)→(i). Thus, E is totally bounded by Problem 61.

63. Refer to the example on pages 197–198 for background and nomenclature. (i) Consider sequence $\{e_n\}$. If $m \neq n$, then $\|e_n - e_m\|_2 = \sqrt{2}$. Hence, no subsequence of $\{e_n\}$ is Cauchy since $\sqrt{2} \not< \epsilon$ for all positive ϵ. Then by the contrapositive of Problem 38(i), no subsequence of $\{e_n\}$ converges, that is, B is not sequentially compact. Thus, B fails to be compact by the contrapositive of Theorem 16(ii)→(iii).

    (ii) Such an open cover is $\{B(\{x_n\}, 1/3)\}_{\{x_n\} \in B}$, in words, the collection of open balls of radius $1/3$ with centers at each sequence in B. That collection covers B because each $\{x_n\}$ in B is in $B(\{x_n\}, 1/3)$. But B cannot be contained in a finite number of balls of radius $1/3$ because one of those balls would contain two of the e_n, which are distance $\sqrt{2}$ apart, and yet the ball has diameter less than 1. Thus, B fails to be compact by definition.

    (iii) The example already shows that B is not totally bounded. Thus, B fails to be compact by the contrapositive of Theorem 16(ii)→(i).

64. Refer to the Riemann-Lebesgue lemma example (Section 8.2) and our second solution to Problem 45 of Section 8.4 for background, integrations, and nomenclature. (i) Consider the sequence of functions $\{f_n\}$ where $f_n : [a, b] \to \mathbf{R}$ is defined by

$$f_n(x) = \sqrt{\frac{2}{b-a}} \sin\left(n\left(-\pi + 2\pi\frac{x-a}{b-a}\right)\right).$$

    Sequence $\{f_n\}$ is in B because

$$\begin{aligned} \|f_n\|_2^2 &= \int_a^b \left| \sqrt{\frac{2}{b-a}} \sin\left(n\left(-\pi + 2\pi\frac{x-a}{b-a}\right)\right) \right|^2 dx && \text{definition of } \|\cdot\|_2 \\ &= \frac{1}{\pi} \int_{-\pi}^{\pi} \sin^2(nt)\, dt && t = -\pi + 2\pi\frac{x-a}{b-a} \\ &= 1. \end{aligned}$$

    As shown in the example, no subsequence of $\{f_n\}$ converges, that is, B is not sequentially compact. Thus, B fails to be compact by the contrapositive of Theorem 16(ii)→(iii).

    (ii) Such an open cover is $\{B(f, 1/3)\}_{f \in B}$, that is, the collection of open balls of radius $1/3$ with centers at each function in B. That collection covers B because each f in B is in $B(f, 1/3)$. Now observe that if $m \neq n$, then

$$\|f_n - f_m\|_2 = \left(\frac{1}{\pi} \int_{-\pi}^{\pi} (\sin nt - \sin mt)^2\, dt\right)^{1/2} = \sqrt{2}.$$

Hence, B cannot be contained in a finite number of balls of radius $1/3$ because one of those balls would contain two of the f_n, which are distance $\sqrt{2}$ apart, and yet the ball has diameter less than 1. Thus, B fails to be compact by definition.

(iii) By our solution to part (ii), B cannot be contained in a finite number of balls of radius $1/3$, which shows that B is not totally bounded. Thus, B fails to be compact by the contrapositive of Theorem 16(ii)→(i).

In our solution to Problem 13 of Section 10.2, we show that $\{f \in L^p[0,1] \mid \|f\|_p \leq 1\}$ fails to be compact for all p in $[1,\infty]$ using the first example of Section 8.2.

65. (i) Take a positive ϵ. By definition of a uniformly continuous mapping, there is a positive δ such that for all x in X, we have that $f(B(x,\delta)) \subseteq B(f(x),\epsilon)$ (cf. ϵ-δ criterion for continuity on page 190). Because X is totally bounded, we can cover it with a finite collection $\{B(x_k,\delta)\}_{k=1}^n$ of open balls. Now an x in X is in some $B(x_k,\delta)$, in which case $f(x)$ is in $B(f(x_k),\epsilon)$. Therefore, $f(X)$ is contained in $\{B(f(x_k),\epsilon)\}_{k=1}^n$, i.e., $f(X)$ is covered by a finite collection of open balls of radius ϵ. Thus, $f(X)$ is totally bounded.

    (ii) Part (i) need not be true if f is only required to be continuous. For example, put X to interval $(0,1]$. Define $f : X \to \mathbf{R}$ by $f(x) = 1/x$ so that $f(X) = [1,\infty)$. Metric space X is totally bounded by Proposition 15 because $(0,1]$ is a subset of $\mathbf{R}$ and because $[0,1)$ is bounded since it has the finite diameter 1. On the other hand, $f(X)$ is not totally bounded because $[1,\infty)$ does not have a finite diameter. Finally, we show that f is continuous on all of $(0,\infty)$ and therefore on X. Observe that f is convex by Proposition 15 of Section 6.6 because f's second derivative $x \mapsto 2/x^3$ is positive on $(0,\infty)$. By Corollary 17 of Section 6.6, f is continuous on each closed, bounded subinterval of $(0,\infty)$ and therefore continuous at each point in $(0,\infty)$. Hence, f is continuous on $(0,\infty)$.

66. This is Problem 25.

67. Yes. Let f be a continuous real-valued function on E. Then for all sequences $\{x_m\}$ in E, if $\{x_m\} \to x$, then $\{f(x_m)\} \to f(x)$ and also $\{-f(x_m)\} \to -f(x)$. Hence, $-f$ is also a continuous real-valued function on E. By assumption, each of f and $-f$ takes a minimum value. Because $-f$ takes a minimum value, f takes a maximum value also. By the extreme value theorem, E is compact. By Theorem 20(ii)→(i), E is closed and bounded.

68. Suppose, to get a contradiction, that E were not closed. Then there is a point $\bar{x}$ of closure of E that is not in E. Define a real-valued function f on E by $f(x) = 1/\|x-\bar{x}\|$. Function f is continuous on E by Proposition 9 because f is a composition of continuous mapping $x \mapsto \|x-\bar{x}\|$ from E to $(0,\infty)$ and continuous mapping $t \mapsto 1/t$ on $(0,\infty)$. Mapping $x \mapsto \|x-\bar{x}\|$ is continuous by Problem 30. Mapping $t \mapsto 1/t$ is continuous by our solution to Problem 65(ii).

    On the other hand, function f is not uniformly continuous on E, a fact we now prove using the contrapositive of Problem 45(i). By Proposition 6, $\bar{x}$ is the limit of a sequence $\{x_m\}$ in E. Then $\lim_{m\to\infty} \|x_m - \bar{x}\| = 0$. Moreover, $\{x_m\}$ is Cauchy by Problem 38(i). Now f maps sequence $\{x_m\}$ to sequence $\{1/\|x_m-\bar{x}\|\}$, which does *not* converge because $\lim_{m\to\infty} 1/\|x_m-\bar{x}\| = \infty$ where we applied Proposition 21 of Section 1.6. Then because $\mathbf{R}$ is a complete metric space (see the sentence after the definition of completeness on page 192), $\{1/\|x_m-\bar{x}\|\}$ is not a Cauchy sequence in $\mathbf{R}$. Hence, f is not uniformly continuous by the contrapositive of Problem 45(i). But that contradicts the assumption that every continuous real-valued function on E is uniformly continuous. Thus, E is closed.

Here is a second proof of the converse. We use the contrapositive of Problem 44. Take the same sequence $\{x_m\}$ from our first proof. Because $\bar{x}$ is a point of closure of E, we can define another sequence $\{x'_m\}$ in E such that

$$\|x'_m - \bar{x}\| \leq \|x_m - \bar{x}\|/2.$$

An alternative way to get a sequence $\{x'_m\}$ that satisfies that inequality is to relabel an appropriate subsequence of $\{x_m\}$. Now by Theorem 18 of Section 1.5,

$$\begin{aligned} \lim_{m\to\infty} \|x'_m - \bar{x}\| &\leq \lim_{m\to\infty} \|x_m - \bar{x}\|/2 \\ &= 0. \end{aligned}$$

Then by the triangle inequality,

$$\begin{aligned} \lim_{m\to\infty} \|x'_m - x_m\| &\leq \lim_{m\to\infty} \|x'_m - \bar{x}\| + \lim_{m\to\infty} \|\bar{x} - x_m\| \\ &= 0 + 0, \end{aligned}$$

but

$$\begin{aligned} \lim_{m\to\infty} \left| \frac{1}{\|x'_m - \bar{x}\|} - \frac{1}{\|x_m - \bar{x}\|} \right| &> \lim_{m\to\infty} \left(\frac{2}{\|x_m - \bar{x}\|} - \frac{1}{\|x_m - \bar{x}\|} \right) \\ &= \lim_{m\to\infty} \frac{1}{\|x_m - \bar{x}\|} \\ &= \infty. \end{aligned}$$

Hence, f is not uniformly continuous by the contrapositive of Problem 44.

Next, for an example that E need not be bounded, put n to 1, and E to $\mathbf{N}$. Then every (continuous) real-valued function on $\mathbf{N}$ is uniformly continuous because, referring to the definition on page 191, if $|v - u| < 1$, then $v = u$ so that $|f(v) - f(u)| = 0 < \epsilon$. But $\mathbf{N}$ is not bounded because it does not have a finite diameter since for every positive number M, we can choose a natural number N such that $|N - 1| > M$.

69. By Theorem 20(ii)→(i), K is closed in $\mathbf{R}$. By Problem 36, $f^{-1}(K)$ is closed in $\mathbf{R}^n$.

    Theorem 20(ii)→(i) also tells us that K is bounded, i.e., there is a positive M such that $|y| \leq M$ for all y in K. Then for all x in $f^{-1}(K)$, we have that $\|x\| \leq |f(x)|/c \leq M/c$, which shows that $f^{-1}(K)$ is bounded. Thus, by Theorem 20(i)→(ii), $f^{-1}(K)$ is compact.

70. (Real-valued) function ρ is continuous by Problem 29 and therefore takes a maximum value by the extreme value theorem. Then

$$\begin{aligned} \operatorname{diam} X &= \sup\{\rho(u,v) \mid u, v \in X\} && \text{definition of diameter} \\ &= \max\{\rho(u,v) \mid u, v \in X\} && \rho \text{ takes a maximum value,} \end{aligned}$$

    which shows that there are points u, v in X for which $\rho(u,v) = \operatorname{diam} X$.

71. Function $\rho(\,\cdot\,, x_0) : K \to \mathbf{R}$ is continuous by Problem 30. So apply the extreme value theorem by putting z equal to the point in K that minimizes $\rho(\,\cdot\,, x_0)$. Then $\rho(z, x_0) \leq \rho(x, x_0)$ for all x in K.

72. Because x is not in K, for each y in K, the distance from y to x is a positive number r_y. We claim that $B(y, r_y/2) \cap B(x, r_y/2) = \emptyset$. We first show that each point y' in $B(y, r_y/2)$ is not in $B(x, r_y/2)$. By the triangle inequality, $\rho(y,x) \leq \rho(y,y') + \rho(y',x)$ where ρ is the metric on X; rearranging, $\rho(y',x) \geq \rho(y,x) - \rho(y,y') > r_y - r_y/2 = r_y/2$. Because the distance

from y' to x is greater than $r_y/2$, point y' is not in $B(x, r_y/2)$. By the symmetry of that argument, neither are the points of $B(x, r_y/2)$ in $B(y, r_y/2)$. That proves our claim.

Now $\{B(y, r_y/2)\}_{y \in K}$ is an open cover of K, so because K is compact, there is some finite subcover $\{B(y_k, r_{y_k}/2)\}_{k=1}^n$ with centers y_k in K. Put $\mathcal{U}$ to be that subcover. Set $\mathcal{U}$ is open by Proposition 1 and contains K by definition of a subcover. Put $\mathcal{O}$ to be $B(x, r_{\min}/2)$ where $r_{\min} = \min\{r_{y_k}\}_{k=1}^n$. Then $\mathcal{O}$ contains x. Moreover, because $B(x, r_{\min}/2)$ is contained in each $B(x, r_{y_k}/2)$, it follows from our claim above that $B(y, r_{y_k}/2) \cap B(x, r_{\min}/2) = \emptyset$ for each k. Thus, $\mathcal{U} \cap \mathcal{O} = \emptyset$.

73. Assume first that $\text{dist}(A, B) > 0$. Then $A \cap B = \emptyset$; otherwise, there would be a point u in A and B so that $\text{dist}(A, B)$ would be 0 since $\rho(u, u) = 0$. Note that this direction of the claim is true for all subsets A, B of X.

Conversely, assume that $A \cap B = \emptyset$. Suppose, to get a contradiction, that $\text{dist}(A, B) = 0$. Then for each natural number n, there is a u_n in A and a v_n in B such that $\rho(u_n, v_n) \leq 1/n$. By Theorem 16(ii)$\to$(iii), there is a subsequence $\{u_{n_k}\}$ of $\{u_n\}$ that converges to a point u in A, that is, $\lim_{k\to\infty} \rho(u_{n_k}, u) = 0$. Now

$$\begin{aligned} \lim_{k\to\infty} \rho(v_{n_k}, u) &\leq \lim_{k\to\infty} \rho(v_{n_k}, u_{n_k}) + \lim_{k\to\infty} \rho(u_{n_k}, u) \quad \text{triangle inequality; Th. 18 of Sec. 1.5} \\ &\leq \lim_{k\to\infty} 1/n_k + 0 \\ &= 0, \end{aligned}$$

which shows that $\{v_{n_k}\}$ also converges to u. Because B is closed, $u \in B$ by Proposition 6. But that contradicts that u is also in A and our assumption that $A \cap B = \emptyset$. Thus, $\text{dist}(A, B) > 0$.

74. Set $X \sim \mathcal{O}$ is closed by Proposition 4, and $(X \sim \mathcal{O}) \cap K = \emptyset$ because $K \subseteq \mathcal{O}$. Then by Problem 73, $\inf\{\rho(v, x) \mid v \in K,\ x \in X \sim \mathcal{O}\}$ is greater than 0 and therefore also greater than some positive number ϵ. By Theorem 16(ii)$\to$(i), the two sentences after the definition of *totally bounded,* and Problem 58, we can put $\mathcal{U}$ to be an ϵ-net for K. Set $\mathcal{U}$ is open by Proposition 1. Now $K \subseteq \mathcal{U}$ by definition of a cover, and $\mathcal{U} \subseteq \overline{\mathcal{U}}$ by the sentence after the definition on page 187. To show that $\overline{\mathcal{U}} \subseteq \mathcal{O}$, note that every neighborhood of each point $\bar{u}$ in $\overline{\mathcal{U}}$ contains a point in $\mathcal{U}$ so that

$$\inf\{\rho(\bar{u}, u) \mid u \in \mathcal{U}\} = 0.$$

Furthermore, by the way we constructed $\mathcal{U}$,

$$\inf\{\rho(u, x) \mid u \in \mathcal{U},\ x \in X \sim \mathcal{O}\} > 0.$$

We deduce that

$$\inf\{\rho(\bar{u}, x) \mid \bar{u} \in \overline{\mathcal{U}},\ x \in X \sim \mathcal{O}\} > 0.$$

By the first part of our solution to Problem 73, $\overline{\mathcal{U}} \cap (X \sim \mathcal{O}) = \emptyset$, which shows that there are no points in $\overline{\mathcal{U}}$ that are outside of $\mathcal{O}$. Hence, $\overline{\mathcal{U}} \subseteq \mathcal{O}$. Altogether, $K \subseteq \mathcal{U} \subseteq \overline{\mathcal{U}} \subseteq \mathcal{O}$.

9.6 Separable Metric Spaces

We can sharpen Proposition 24 to "A totally bounded metric space is separable."

75. We first show that because D is finite, D is closed. Indeed, let x be a point in D. Then singleton set $\{x\}$ is closed by Proposition 6 because the only possible sequence in set $\{x\}$

is the constant sequence $(x, x, \dots)$, and that sequence converges to x, which belongs to set $\{x\}$. Then D is closed by Proposition 5 because D is the union of a finite number of (closed) singleton sets, each consisting of one of the points in D. Because D is closed, $\overline{D} = D$. That result along with the fact that $\overline{D} = X$ by the first sentence after the definition on page 203 gives us that $X = D$.

76. By the definition on page 187, $\overline{D}$ contains all points of closure of D, that is, every neighborhood of each point in $\overline{D}$ contains a point in D. Then every nonempty open subset of $\overline{D}$ contains a point of D. Thus, D is dense in $\overline{D}$ by definition of denseness.

77. Let f and g be continuous mappings defined on X that take the same values on a dense subset D of X. We have two solutions. (i) Let x be a point in X. By Proposition 6 and the first sentence after the definition on page 203, there is a sequence $\{x_n\}$ in D that converges to x. By definition of a continuous function (page 190), $\{f(x_n)\} \to f(x)$, and $\{g(x_n)\} \to g(x)$. Thus, $g(x) = f(x)$ because $g(x_n) = f(x_n)$ for all n.

    (ii) Define function h on X by $h = g - f$. Function h is continuous. Then $h^{-1}(\{0\})$ is closed in X by Problem 36 because singleton set $\{0\}$ is closed as explained in our solution to Problem 75. Then because $h^{-1}(\{0\})$ contains D, set $h^{-1}(\{0\})$ is also dense in X. By the last two sentences of our solution to Problem 75, a closed, dense subset of X is equal to X so that $X = h^{-1}(\{0\})$. Hence, $h = 0$ on X. Thus, $g = f$ on X.

78. Let $(X, \rho), (Y, \sigma)$ be two separable metric spaces. By definition, there is a countable subset D of X that is dense in X and a countable subset E of Y that is dense in Y. Then $D \times E$ is a subset of $X \times Y$. Furthermore, D, E are equipotent to a subset of $\mathbf{N}$, so Cartesian product $D \times E$ is countable by Section 1.3's Theorem 3 and Corollary 4.

    We now show that every nonempty open subset of $X \times Y$ contains a point of $D \times E$. A nonempty open subset of $X \times Y$ contains some open ball $B_{X\times Y}((x, y), r)$ where center (x, y) is in $X \times Y$. The subscript on B indicates the metric space in which the open ball is located. Cartesian product $B_X(x, r/2) \times B_Y(y, r/2) \subseteq B_{X\times Y}((x, y), r)$ because for a point (x', y') in $B_X(x, r/2) \times B_Y(y, r/2)$,

$$\begin{aligned}\left(\rho(x', x)^2 + \sigma(y', y)^2\right)^{1/2} &< \left((r/2)^2 + (r/2)^2\right)^{1/2} \\ &< r\end{aligned}$$

    where the left side is the product metric (page 184). Abbreviate the three open balls above by $B_{X\times Y}$, B_X, and B_Y. Because D is dense in X, and E is dense in Y, set $(D \cap B_X) \times (E \cap B_Y)$ contains a point of $D \times E$. Moreover,

$$\begin{aligned}(D \cap B_X) \times (E \cap B_Y) &= (D \times E) \cap (B_X \times B_Y) && \text{set-operation identity} \\ &\subseteq (D \times E) \cap B_{X\times Y} && B_X \times B_Y \subseteq B_{X\times Y} \text{ as shown above.}\end{aligned}$$

    Hence, every nonempty open subset of $X \times Y$ contains a point of $D \times E$. Thus, $X \times Y$ is separable because there is a countable subset $D \times E$ of $X \times Y$ that is dense in $X \times Y$.

    Here is a proof of the set-operation identity above:

$$\begin{aligned}(d, e) \in (D \cap B_X) \times (E \cap B_Y) &\text{ if and only if } d \in D \cap B_X \text{ and } e \in E \cap B_Y \\ &\text{ if and only if } d \in D \text{ and } d \in B_X \text{ and } e \in E \text{ and } e \in B_Y \\ &\text{ if and only if } (d, e) \in D \times E \text{ and } (d, e) \in B_X \times B_Y \\ &\text{ if and only if } (d, e) \in (D \times E) \cap (B_X \times B_Y).\end{aligned}$$

Chapter 10

Metric Spaces: Three Fundamental Theorems

10.1 The Arzelà-Ascoli Theorem

From the Arzelà-Ascoli lemma to the Arzelà-Ascoli theorem, the boundedness condition changes from pointwise to uniform. The reason for that is given in Problem 2 and the fact that X is compact in the theorem but not necessarily so in the lemma.

The inference at the end of the proof of Theorem 3 deserves more detail. If $k \geq K$, then starting with a rearrangement of the triangle inequality,

$$\begin{aligned}|f(x_{n_k}) - f(x)| &\geq |f_{n_k}(x_{n_k}) - f_{n_k}(x)| - |f_{n_k}(x_{n_k}) - f(x_{n_k})| - |f(x) - f_{n_k}(x)| \\ &> \epsilon_0 - \epsilon_0/3 - \epsilon_0/3 \\ &= \epsilon_0/3\end{aligned}$$

where the first absolute value on the right is at least ϵ_0 by inequality (3) and the other two absolute values on the right are less than $\epsilon_0/3$ because $\rho_{\max}(f, f_{n_k}) < \epsilon_0/3$ since $k \geq K$.

In the second remark, $\mathcal{S}$ is an equisummable subset of ℓ^p.

1. To show that E is a closed subset of Y, apply Proposition 6 of Section 9.2: Take a sequence in E that converges to a limit in Y. By Problem 38(i) of Section 9.4, the sequence is Cauchy. By Theorem 16(ii)→(i) of Section 9.5, E is complete. By definition of a complete metric space (page 192), the sequence's limit belongs to E. Thus, E is a closed subset of Y.

   Here is a shorter proof that E is closed: By Proposition 24 of Section 9.6, E is separable. By the sentence after the definitions of *dense* and *separable*, E is equal to a closed set.

   A third proof that E is closed is the proof of Proposition 16 of Section 11.5 (a metric space is a Hausdorff topological space).

   Subspace E is totally bounded by Theorem 16(ii)→(i) of Section 9.5. Thus, by the second paragraph after the definition of *totally bounded*, E is bounded.

2. Assume first that the sequence, say $\{f_n\}$, is uniformly bounded. Then there is some positive M for which $|f_n| \leq M$ for all n. Therefore, for each x in the metric space, $|f_n(x)| \leq M$ for all n. Thus, $\{f_n\}$ is pointwise bounded. This direction of the claim does not require that $\{f_n\}$ be equicontinuous nor that the metric space be compact.

   Conversely, assume that $\{f_n\}$ is pointwise bounded. Name the metric space (X, ρ). By the sentence after the definition of *equicontinuous* and by Problem 3, $\{f_n\}$ is uniformly equicontinuous on X. Then there is a positive δ such that for x, x' in X and every f_n in $\{f_n\}$, if $\rho(x, x') < \delta$, then $|f_n(x) - f_n(x')| < 1$. By Theorem 16(ii)→(i) of Section 9.5 and definition of a totally bounded metric space, we can cover X by a finite number m of open balls of radius δ. Name the centers of those balls x_k where $k = 1, \ldots, m$. Then for each x in X, there is an x_k such that $\rho(x, x_k) < \delta$. Now because $\{f_n\}$ is pointwise bounded, there is a positive M_k such that $|f_n(x_k)| < M_k$ for all n where $k = 1, \ldots, m$. Then altogether, for all x in X and for all n,

   $$\begin{aligned}|f_n(x)| &\leq |f_n(x) - f_n(x_k)| + |f_n(x_k)| && \text{triangle inequality} \\ &< 1 + M_k && x_k \text{ chosen such that } \rho(x, x_k) < \delta \\ &\leq 1 + \max\{M_1, \ldots, M_m\}.\end{aligned}$$

   Because $1 + \max\{M_1, \ldots, M_m\}$ is a constant number, say M', we conclude that $\{f_n\}$ is uniformly bounded because $|f_n| < M'$ on X for all n.

We can also prove the converse by contradiction. Suppose that $\{f_n\}$ were not uniformly bounded. Then for each natural number j, there is an f_{n_j} in $\{f_n\}$ and a point x_j in X such that $|f_{n_j}(x_j)| > j$. Because X is compact, a subsequence of $\{x_j\}$ converges, say to x, in X by Theorem 16(ii)→(iii) of Section 9.5. Because $\{f_n\}$ is equicontinuous at x, there is a positive δ such that for every j, if $\rho(x_j, x) < \delta$, then $|f_{n_j}(x_j) - f_{n_j}(x)| < 1$. Take a (large) positive number M. Because $\{x_j\} \to x$, there is an index J such that $J \geq M + 1$ and $\rho(x_J, x) < \delta$. Then starting with a rearrangement of the triangle inequality,

$$\begin{aligned} |f_{n_J}(x)| &\geq |f_{n_J}(x_J)| - |f_{n_J}(x_J) - f_{n_J}(x)| \\ &> J - 1 && |f_{n_J}(x_J)| > J;\ \rho(x_J, x) < \delta \\ &\geq M && J \geq M + 1, \end{aligned}$$

which shows that $\{f_{n_j}(x)\}$ is not bounded, contradicting that $\{f_n\}$ is pointwise bounded.

3. The authors restrict definitions of *equicontinuous* and *uniformly equicontinuous* to real-valued functions, but the problem statement suggests that the functions map into a general metric space. Furthermore, we use that more general interpretation of this problem in our solution to Problem 8. So assume that we have an equicontinuous family $\mathcal{F}$ of (continuous) functions from a compact metric space (X, ρ) into a metric space (Y, σ), and replace the absolute-value metric in the two equicontinuity definitions with metric σ.

   We mimic the proof of Proposition 23 of Section 9.5. Take a positive ϵ. By the criterion for equicontinuity at a point x in X, for every f in $\mathcal{F}$ and x' in X, there is a δ_x for which if $\rho(x', x) < \delta_x$, then $\sigma(f(x'), f(x)) < \epsilon/2$. We claim that δ equal to the Lebesgue number for open cover $\{B(x, \delta_x)\}_{x \in X}$ responds to the ϵ challenge for the uniform equicontinuity of $\mathcal{F}$. Indeed, for u, v in X, if $\rho(u, v) < \delta$, then $u \in B(v, \delta)$. Furthermore, by Lebesgue's covering lemma, some member $B(x, \delta_x)$ of the open cover contains $B(v, \delta)$. Consequently, $u, v \in B(x, \delta_x)$. Then for every f in $\mathcal{F}$,

$$\begin{aligned} \sigma(f(u), f(v)) &\leq \sigma(f(u), f(x)) + \sigma(f(x), f(v)) && \text{triangle inequality for } \sigma \\ &< \epsilon/2 + \epsilon/2 && \rho(u, x), \rho(v, x) < \delta_x \text{ because } u, v \in B(x, \delta_x) \\ &= \epsilon. \end{aligned}$$

   Thus, $\mathcal{F}$ is uniformly equicontinuous.

4. Note that f *must* be in $C(X)$ by Problem 59 of Section 1.6.

   Take a positive ϵ. Because $\{f_n\} \to f$ uniformly on X, there is an index N such that for all x in X, if $n \geq N$, then $|f_n(x) - f(x)| < \epsilon/3$. Now take a point x in X. We show that there is a δ' that responds to the ϵ challenge for the equicontinuity at x of subcollection $\{f_1, \ldots, f_{N-1}\}$ and a δ'' that responds to the ϵ challenge for the equicontinuity at x of subcollection $\{f_N, f_{N+1}, \ldots\}$. Then δ equal to $\min\{\delta', \delta''\}$ responds to the ϵ challenge for the equicontinuity at x of sequence $\{f_n\}$.

   Consider first those f_n for which $n < N$. Because each f_n is continuous, for each n, there is a δ_n such that for every x' in X, if $\rho(x', x) < \delta_n$ where ρ is the metric on X, then $|f_n(x') - f_n(x)| < \epsilon$. Put δ' equal to $\min\{\delta_1, \ldots, \delta_{N-1}\}$. Consequently, if $\rho(x', x) < \delta'$, then $|f_n(x') - f_n(x)| < \epsilon$ for each f_n.

   Consider now the f_n for which $n \geq N$. Because f is continuous, there is a δ'' for which if $\rho(x', x) < \delta''$, then $|f(x') - f(x)| < \epsilon/3$, and

$$\begin{aligned} |f_n(x') - f_n(x)| &\leq |f_n(x') - f(x')| + |f(x') - f(x)| + |f(x) - f_n(x)| && \text{triangle inequality} \\ &< \epsilon/3 + \epsilon/3 + \epsilon/3 = \epsilon && n \geq N,\ \rho(x', x) < \delta''. \end{aligned}$$

Finally, replace δ' and δ'' in the preceding with δ equal to $\min\{\delta', \delta''\}$ so that we can state that for *every* f_n in $\{f_n\}$, if $\rho(x', x) < \delta$, then $|f_n(x') - f_n(x)| < \epsilon$. Hence, $\{f_n\}$ is equicontinuous at x. Because that holds at every point in X, sequence $\{f_n\}$ is equicontinuous.

5. Note that symbol C denotes two different things in the problem statement, the constant in the definition of a Hölder continuous function and the indicator for the linear space of continuous, real-valued functions on $[0, 1]$.

   Let $\mathcal{F}$ be the set of real-valued functions f on $[0, 1]$ for which $\|f\|_\alpha \leq 1$ where $0 < \alpha \leq 1$. We use Theorem 3 to solve this problem, i.e., to show that $\mathcal{F}$ has compact closure as a subset of $C[0, 1]$. Note that $[0, 1]$ is compact by Theorem 20(i)→(ii) of Section 9.5 because $[0, 1]$ is a closed, bounded subset of $\mathbf{R}$.

   In applying Theorem 3 to this problem, it is not necessary that $\mathcal{F}$ be closed because we are asked to show that the *closure* of $\mathcal{F}$ is compact. Nevertheless, we prove that $\mathcal{F}$ does happen to be closed. First confirm that Hölder's norm really is a norm by checking the conditions on page 183. Function $\|\cdot\|_\alpha$ is real-valued and nonnegative. For each pair of real-valued functions f, g on $[0, 1]$ and for all x, y in $[0, 1]$ where $x \neq y$, we have that—

   (i) by inspection, $\|f\|_\alpha = \max\{|f(x)| + |f(x) - f(y)|/|x - y|^\alpha\} = 0$ if and only if $f = 0$,

   (ii) by the fact that $(f + g)(x) = f(x) + g(x)$ and the triangle inequality,

$$\begin{aligned}
\|f + g\|_\alpha &= \max\{|(f + g)(x)| + |(f + g)(x) - (f + g)(y)|/|x - y|^\alpha\} \\
&\leq \max\{|f(x)| + |g(x)| + (|f(x) - f(y)| + |g(x) - g(y)|)/|x - y|^\alpha\} \\
&\leq \max\{|f(x)| + |f(x) - f(y)|/|x - y|^\alpha\} + \max\{|g(x)| + |g(x) - g(y)|/|x - y|^\alpha\} \\
&= \|f\|_\alpha + \|g\|_\alpha, \text{ and}
\end{aligned}$$

   (iii) for each real number c and by the fact that $(cf)(x) = cf(x)$,

$$\begin{aligned}
\|cf\|_\alpha &= \max\{|(cf)(x)| + |(cf)(x) - (cf)(y)|/|x - y|^\alpha\} \\
&= |c| \max\{|f(x)| + |f(x) - f(y)|/|x - y|^\alpha\} \\
&= |c|\|f\|_\alpha.
\end{aligned}$$

   Then $\mathcal{F}$ is closed ball $\{f \text{ a real-valued function on } [0, 1] \mid \|f\|_\alpha \leq 1\}$, which is closed by Problem 15(i) of Section 9.2.

   Set $\mathcal{F}$ is uniformly bounded because for all f in $\mathcal{F}$,

$$|f(x)| \leq \max\{|f(x)| + |f(x) - f(y)|/|x - y|^\alpha \mid x, y \in [0, 1],\ y \neq x\} = \|f\|_\alpha \leq 1.$$

   Next, we show that $\mathcal{F}$ is uniformly equicontinuous. Take a positive ϵ. Put δ equal to $\epsilon^{1/\alpha}$. For x, y in $[0, 1]$ and every f in $\mathcal{F}$, if $|x - y| < \delta$, then

$$\begin{aligned}
|f(x) - f(y)| &\leq (1 - |f(x)|)|x - y|^\alpha && \text{follows from definition of } \|\cdot\|_\alpha \text{ and that } \|f\|_\alpha \leq 1 \\
&\leq |x - y|^\alpha && 0 \leq |f(x)| \leq 1 \\
&< \delta^\alpha && |x - y| < \delta \\
&= \left(\epsilon^{1/\alpha}\right)^\alpha && \delta = \epsilon^{1/\alpha} \\
&= \epsilon.
\end{aligned}$$

   Hence, $\mathcal{F}$ is uniformly equicontinuous. It follows immediately that $\mathcal{F}$ is equicontinuous and that each function in $\mathcal{F}$ is continuous so that $\mathcal{F}$ is a subset of $C[0, 1]$.

   By Theorem 3, $\mathcal{F}$ is a compact subspace of $C[0, 1]$. Thus, $\mathcal{F}$ has compact closure (indeed, $\mathcal{F}$ is already closed as we showed above) as a subset of $C[0, 1]$.

6. If $\overline{\mathcal{F}}$ is equicontinuous, then $\mathcal{F}$ is equicontinuous because $\mathcal{F} \subseteq \overline{\mathcal{F}}$ by the sentence after the definition on page 187.

   Conversely, assume that $\mathcal{F}$ is equicontinuous. Let $\bar{f}$ be a function of closure of $\mathcal{F}$. Then every neighborhood of $\bar{f}$ contains a function in $\mathcal{F}$. So for each positive ϵ, there is an f in $\mathcal{F}$ such that $\max_{x\in X}\left|f(x) - \bar{f}(x)\right| < \epsilon/3$. Moreover, because $\mathcal{F}$ is equicontinuous on X, function f is continuous on X. Let x be a point in X. There is a δ such that for every x' in X, if $\rho(x', x) < \delta$ where ρ is the metric on X, then $|f(x') - f(x)| < \epsilon/3$. Altogether, for every x' in X, if $\rho(x', x) < \delta$, then

$$\begin{aligned} \left|\bar{f}(x') - \bar{f}(x)\right| &\le \left|\bar{f}(x') - f(x')\right| + |f(x') - f(x)| + \left|f(x) - \bar{f}(x)\right| && \text{triangle inequality} \\ &< \epsilon/3 + \epsilon/3 + \epsilon/3 \\ &= \epsilon. \end{aligned}$$

   Hence, $\bar{f}$ is continuous at x. Because that holds for every point in X and every function of closure $\bar{f}$ of $\mathcal{F}$ and because $\overline{\mathcal{F}}$ is the set of all functions of closure of $\mathcal{F}$, the closure $\overline{\mathcal{F}}$ of $\mathcal{F}$ is equicontinuous.

   For the conclusion, assume first that a subset $\mathcal{F}$ of $C(X)$ has compact closure, that is, $\overline{\mathcal{F}}$ is compact. Then $\overline{\mathcal{F}}$ is uniformly bounded and equicontinuous by Theorem 3. Then $\mathcal{F}$ too is equicontinuous and uniformly bounded because $\mathcal{F} \subseteq \overline{\mathcal{F}}$.

   Conversely, assume that $\mathcal{F}$ is equicontinuous and uniformly bounded. Because $\mathcal{F}$ is equicontinuous, its closure $\overline{\mathcal{F}}$ in $C(X)$ is equicontinuous. We next show that $\overline{\mathcal{F}}$ is uniformly bounded. Let $\bar{f}$ be a function of closure of $\mathcal{F}$. Then there is an f in $\mathcal{F}$ such that $\max_{x\in X}\left|f(x) - \bar{f}(x)\right| < 1$. Because $\mathcal{F}$ is uniformly bounded, there is some positive M for which $|f| \le M$ on X for all f in $\mathcal{F}$. Altogether, for all x in X,

$$\begin{aligned} |\bar{f}(x)| &\le \left|\bar{f}(x) - f(x)\right| + |f(x)| && \text{triangle inequality} \\ &\le \max_{x\in X}\left|\bar{f}(x) - f(x)\right| + M && |f| \le M \text{ on } X \\ &< 1 + M. \end{aligned}$$

   Hence, $\bar{f}$ is bounded. Because that holds for every function of closure $\bar{f}$ of $\mathcal{F}$ and because $\overline{\mathcal{F}}$ is the set of all functions of closure of $\mathcal{F}$, the closure $\overline{\mathcal{F}}$ of $\mathcal{F}$ is uniformly bounded. Consequently, $\overline{\mathcal{F}}$ is a compact subspace of $C(X)$ by Theorem 3. Thus, $\mathcal{F}$ has compact closure.

7. No and no. For the first counterexample, put each f_n to the constant function n. Then $\{f_n\}$ is equicontinuous because for every f_n in $\{f_n\}$ and pair of points x, x' in $[a, b]$, we have that $|f_n(x') - f_n(x)| = |n - n| = 0$. On the other hand, if $m \ne n$, we have that $\|f_n - f_m\|_{\max} \ge 1$. Therefore, no subsequence of $\{f_n\}$ is Cauchy in $C[a, b]$, and hence no subsequence can converge in $C[a, b]$.

   The paragraph after the proof of the Arzelà-Ascoli theorem gives us a counterexample to the second question. Sequence $\{f_n\}$ defined by $f_n(x) = x^n$ on $[0, 1]$ for all n converges to discontinuous function

$$x \mapsto \begin{cases} 0 & x \in [0, 1), \\ 1 & x = 1. \end{cases}$$

   Hence, $\{f_n\}$ fails to have a subsequence that converges to a continuous function on $[0, 1]$. But $\{f_n\}$ is in $C[0, 1]$ and is uniformly bounded because $|f_n| \le 1$ on $[0, 1]$ for all n.

8. **Arzelà-Ascoli lemma** Let X be a separable metric space and $\{f_n\}$ an equicontinuous sequence in $C(X, Y)$ such that for each x in X, the closure of set $\{f_n(x) \mid n \text{ a natural number}\}$

is a compact subspace of Y. Then a subsequence of $\{f_n\}$ converges pointwise on all of X to a mapping $f : X \to Y$.

Proof Let $\{x_j\}_{j=1}^{\infty}$ be an enumeration of a dense subset D of X. The closure of set $\{f_n(x_1) \,|\, n \text{ a natural number}\}$ is a compact subset of Y. Therefore, by Theorem 16(ii)→(iii) of Section 9.5, sequence of points in Y defined by $n \mapsto f_n(x_1)$ has a convergent subsequence, that is, there is a strictly increasing sequence of natural numbers $\{s(1,n)\}$ and a point a_1 in Y for which

$$\lim_{n\to\infty} f_{s(1,n)}(x_1) = a_1.$$

The proof continues as in that of Lemma 2—with absolute values replaced by the metric on Y—up to the fourth-to-last sentence. At that step in the proof, we have that $\{f_n(x_0)\}$ is a Cauchy sequence of points in Y. Since the closure of $\{f_n(x_0) \,|\, n \text{ a natural number}\}$ is complete by Theorem 16(ii)→(i) of Section 9.5, $\{f_n(x_0)\}$ converges. Denote the limit by $f(x_0)$. Sequence $\{f_n\}$ converges pointwise on all of X to $f : X \to Y$. □

Arzelà-Ascoli theorem Let X be a compact metric space and $\{f_n\}$ an equicontinuous sequence in $C(X,Y)$ such that for each x in X, the closure of $\{f_n(x) \,|\, n \text{ a natural number}\}$ is a compact subspace of Y. Then $\{f_n\}$ has a subsequence that converges uniformly on X to a mapping in $C(X,Y)$.

Proof Modify the corresponding proof in the text. Because X is a compact metric space, X is separable. Then a subsequence of $\{f_n\}$ converges pointwise on all of X to a mapping $f : X \to Y$. For notational convenience, assume that the whole sequence $\{f_n\}$ converges pointwise on X. In particular, therefore, for each x in X, sequence $\{f_n(x)\}$ is a Cauchy sequence of points in Y. We use that and equicontinuity to show that $\{f_n\}$ is a Cauchy sequence in $C(X,Y)$.

Take a positive ϵ. Let ρ be the metric on X and σ the metric on Y. By Problem 3, $\{f_n\}$ is uniformly equicontinuous on X. Then there is a positive δ such that for all n,

$$\sigma(f_n(u), f_n(v)) < \epsilon/3 \text{ for all } u, v \text{ in } X \text{ if } \rho(u,v) < \delta.$$

Because X is a compact metric space, X is totally bounded. There are therefore a finite number of points $x_1, \ldots, x_k$ in X for which X is covered by $\{B(x_i,\delta)\}_{i=1}^{k}$. Sequence $\{f_n(x_i)\}$ is Cauchy for each i where $i = 1, \ldots, k$, so there is an index N such that if $m, n \geq N$, then

$$\sigma(f_n(x_i), f_m(x_i)) < \epsilon/3 \text{ for each } i \text{ among } 1, \ldots, k.$$

Now for each x in X, there is an i among $1, \ldots, k$ such that $\rho(x, x_i) < \delta$, and therefore if $m, n \geq N$, then

$$\begin{aligned}\sigma(f_n(x), f_m(x)) &\leq \sigma(f_n(x), f_n(x_i)) + \sigma(f_n(x_i), f_m(x_i)) + \sigma(f_m(x_i), f_m(x)) \\ &< \epsilon/3 + \epsilon/3 + \epsilon/3.\end{aligned}$$

Thus, $\{f_n\}$ is uniformly Cauchy. Because the closure of $\{f_n(x) \,|\, n \text{ a natural number}\}$ is complete, $\{f_n(x)\}$ converges for each x in X. Put f to be the mapping to which $\{f_n(x)\}$ converges for each x in X. Then $\{f_n\} \to f$ uniformly on X.

It remains to show that f is continuous. Because $\{f_n\} \to f$ uniformly on X, there is an N for which $\sigma(f_N(x), f(x)) < \epsilon/3$ at every point x in X. Because f_N is continuous, there is a δ for which if $x' \in X$ and $\rho(x', x) < \delta$, then $\sigma(f_N(x'), f_N(x)) < \epsilon/3$. In that case,

$$\begin{aligned}\sigma(f(x'), f(x)) &\leq \sigma(f(x'), f_N(x')) + \sigma(f_N(x'), f_N(x)) + \sigma(f_N(x), f(x)) && \text{triangle inequality} \\ &< \epsilon/3 + \epsilon/3 + \epsilon/3. && \rho(x', x) < \delta.\end{aligned}$$

Thus, we have a mapping f in $C(X,Y)$ such that $\{f_n\} \to f$ uniformly on X. □

9. **R** is separable by the sentence after the definition on page 152. Also, $\{f_n\}$ is pointwise bounded by our solution to Problem 2. So the Arzelà-Ascoli lemma tells us that a subsequence of $\{f_n\}$ converges pointwise on all of **R** to a real-valued function f on **R**. For notational convenience, assume the whole sequence $\{f_n\}$ converges to f pointwise on **R**.

   To see that f is continuous on **R**, take a point x in **R**. Point x is in a closed, bounded interval $[x-\delta, x+\delta]$ where δ is a positive number. That interval is compact by Theorem 20(i)→(ii) of Section 9.5. By the proof of the Arzelà-Ascoli theorem, f is continuous on $[x-\delta, x+\delta]$ and therefore at x. Because that holds for all x, function f is continuous on **R**.

   Lastly, take a bounded subset A of **R**. Subset A is contained in some closed, bounded interval I. Again by the proof of the Arzelà-Ascoli theorem, $\{f_n\} \to f$ uniformly on I. Because A is contained in I, the convergence is uniform on A too.

10. Refer to the second remark for background, notation, and definition of *equisummable*. Our solution mimics the proof of Theorem 3. First assume that a subspace $\mathcal{S}$ of ℓ^p is closed, bounded, and equisummable. From $\mathcal{S}$, take a sequence $\{\{x_n\}^m\}$ with index m (note that $\{\{x_n\}^m\}$ is a sequence of sequences). We now prove that a subsequence of $\{\{x_n\}^m\}$ converges componentwise to a sequence of real numbers, i.e., the ℓ^p counterpart of the Arzelà-Ascoli lemma (for definition of componentwise convergence, see Problem 19 of Section 8.2). Denote by x_n^m the nth real number in the mth sequence of $\{\{x_n\}^m\}$. The sequence of real numbers defined by $m \mapsto x_1^m$ is bounded because $\mathcal{S}$ is bounded. Therefore, by the Bolzano-Weierstrass theorem, this sequence has a convergent subsequence, that is, there is a strictly increasing sequence of natural numbers $\{s(1,m)\}$ and a number a_1 for which

$$\lim_{m\to\infty} x_1^{s(1,m)} = a_1.$$

   By the same argument applied to sequence $m \mapsto x_2^{s(1,m)}$, there is a subsequence $\{s(2,m)\}$ of $\{s(1,m)\}$ and a number a_2 for which $\lim_{m\to\infty} x_2^{s(2,m)} = a_2$. We inductively continue that selection process to obtain a countable collection of strictly increasing sequences of natural numbers $\{\{s(j,m)\}\}_{j=1}^{\infty}$ and a sequence of numbers $\{a_j\}$ such that for each j,

$$\{s(j+1,m)\} \text{ is a subsequence of } \{s(j,m)\} \text{ and } \lim_{m\to\infty} x_j^{s(j,m)} = a_j.$$

   Consider the diagonal subsequence $\{\{x_n\}^{m_k}\}$ obtained by setting m_k to $s(k,k)$ for each index k. For each j then, $\{m_k\}_{k=j}^{\infty}$ is a subsequence of the jth subsequence of natural numbers selected above, and therefore

$$\lim_{k\to\infty} x_j^{m_k} = a_j.$$

   Hence, $\{\{x_n\}^{m_k}\}$ converges componentwise to $\{a_n\}$.

   Next, we prove that $\{a_n\} \in \ell^p$, which is the ℓ^p counterpart of the Arzelà-Ascoli theorem. For notational convenience, assume that the whole sequence $\{\{x_n\}^m\}$ converges componentwise to $\{a_n\}$. Take a positive ϵ. By the equisummability of $\mathcal{S}$ and the fact that $\{\{x_n\}^m\} \subseteq \mathcal{S}$, there is an index N for which

$$\sum_{n=N}^{\infty} |x_n^m|^p < \epsilon \text{ for all } m.$$

   Because that inequality holds for all m, it holds also as $m \to \infty$ so that

$$\sum_{n=N}^{\infty} |a_n|^p < \epsilon.$$

Then

$$\begin{aligned}\sum_{n=1}^{\infty}|a_n|^p &= \sum_{n=1}^{N-1}|a_n|^p + \sum_{n=N}^{\infty}|a_n|^p \\ &< \sum_{n=1}^{N-1}|a_n|^p + \epsilon \\ &< \infty.\end{aligned}$$

Hence, $\{a_n\} \in \ell^p$ by the first example on page 138. Because $\mathcal{S}$ is closed, $\{a_n\}$ belongs to $\mathcal{S}$. Thus, $\mathcal{S}$ is a sequentially compact metric space and therefore is compact.

Now assume that $\mathcal{S}$ is compact. Subspace $\mathcal{S}$ is bounded and is a closed subset of ℓ^p by Problem 1. We argue by contradiction to show that $\mathcal{S}$ is equisummable. Suppose that $\mathcal{S}$ were not equisummable. Then there is a positive ϵ_0 such that for each m, there is a sequence in $\mathcal{S}$ that we label $\{x_n\}^m$ for which

$$\sum_{n=m}^{\infty}|x_n^m|^p \geq \epsilon_0.$$

Because $\mathcal{S}$ is a compact metric space, it is sequentially compact. Therefore, there is a subsequence $\{\{x_n\}^{m_k}\}$ that converges to a sequence $\{a_n\}$ in $\mathcal{S}$. Because the inequality above holds for each m, it holds with m replaced by m_k and also as $k \to \infty$. We infer that there is no N for which $\sum_{n=N}^{\infty}|a_n|^p < \epsilon_0$. That contradicts that $\{a_n\}$ is in equisummable set $\mathcal{S}$. Thus, $\mathcal{S}$ is equisummable.

11. Series $\sum_{n=1}^{\infty} c_n^2$ is summable because $\{c_n\} \in \ell^2$. Proposition 20(i) of Section 1.5 holds for each m, so it holds also as $m \to \infty$. Then for each positive ϵ, there is an index N for which

$$\sum_{n=N}^{\infty} c_n^2 < \epsilon.$$

Consequently, for all sequences $\{x_n\}$ in $\mathcal{S}$,

$$\sum_{n=N}^{\infty}|x_n|^2 \leq \sum_{n=N}^{\infty} c_n^2 < \epsilon.$$

Thus, $\mathcal{S}$ is equisummable.

12. Refer to our solutions to Problems 63(i), (ii), and (iii) of Section 9.5. We rewrite them with $\sqrt{2}$ replaced by $2^{1/p}$. For the case where p is ∞, read 1 for $2^{1/p}$. Denote by symbol $\bar{B}$ the closed unit ball $\{\{x_n\} \in \ell^p \mid \|\{x_n\}\|_p \leq 1\}$. (i) Consider sequence $\{e_n\}$. If $m \neq n$, then $\|e_n - e_m\|_p = 2^{1/p}$. Hence, no subsequence of $\{e_n\}$ is Cauchy since $2^{1/p} \not< \epsilon$ for all positive ϵ. Then by the contrapositive of Problem 38(i) of Section 9.4, no subsequence of $\{e_n\}$ converges, that is, $\bar{B}$ is not sequentially compact. Thus, $\bar{B}$ fails to be compact by the contrapositive of Theorem 16(ii)$\to$(iii) of Section 9.5.

(ii) An open cover of $\bar{B}$ is $\{B(\{x_n\}, 1/3)\}_{\{x_n\}\in\bar{B}}$, in words, the collection of open balls of radius 1/3 with centers at each sequence in $\bar{B}$. That collection covers $\bar{B}$ because each $\{x_n\}$ in $\bar{B}$ is in $B(\{x_n\}, 1/3)$. But $\bar{B}$ cannot be contained in a finite number of open balls of radius 1/3 because one of those open balls would contain two of the e_n, which are distance $2^{1/p}$ apart, and yet the open ball has diameter less than 1. Thus, $\bar{B}$ fails to be compact by definition.

(iii) Because $\bar{B}$ cannot be contained in a finite number of open balls of radius 1/3, closed unit ball $\bar{B}$ is not totally bounded. Thus, $\bar{B}$ fails to be compact by the contrapositive of Theorem 16(ii)→(i) of Section 9.5.

13. Refer to the first example of Section 8.2 for background and nomenclature. The sequence in that example is in closed unit ball $\{f \in L^p[0,1] \mid \|f\|_p \leq 1\}$, say $\bar{B}$, because $\|f_n\|_p = 1$ for every index n (see our solution to Problem 13 of Section 8.2). On the other hand, because if $m \neq n$, the value of $|f_n - f_m|$ is 2 on a set of measure 1/2, we have that $\|f_n - f_n\|_p \geq 2^{1-1/p}$. As in Problem 12, there are three ways to finish off the solution. (i) No subsequence of $\{f_n\}$ is Cauchy because $2^{1-1/p} \not< \epsilon$ for all positive ϵ. Then by the contrapositive of Problem 38(i) of Section 9.4, no subsequence of $\{f_n\}$ converges, that is, $\bar{B}$ is not sequentially compact. Thus, $\bar{B}$ fails to be compact by the contrapositive of Theorem 16(ii)→(iii) of Section 9.5.

    (ii) An open cover of $\bar{B}$ is $\{B(f, 1/3)\}_{f \in \bar{B}}$, in words, the collection of open balls of radius 1/3 with centers at each function in $\bar{B}$. That collection covers $\bar{B}$ because each f in $\bar{B}$ is in $B(f, 1/3)$. But $\bar{B}$ cannot be contained in a finite number of open balls of radius 1/3 because one of those open balls would contain two of the f_n, which are distance at least $2^{1-1/p}$ apart, and yet the open ball has diameter less than 1. Thus, $\bar{B}$ fails to be compact by definition.

    (iii) Because $\bar{B}$ cannot be contained in a finite number of open balls of radius 1/3, closed unit ball $\bar{B}$ is not totally bounded. Thus, $\bar{B}$ fails to be compact by the contrapositive of Theorem 16(ii)→(i) of Section 9.5.

14. Taking the hint in the first remark (page 209), we see that the proof is the same as the first paragraph of the proof of the Arzelà-Ascoli lemma with dense subset D replaced by S.

10.2 The Baire Category Theorem

15. The boundary of E is closed by Problem 22(i) of Section 9.2.

    Next, suppose, to get a contradiction, that $\operatorname{bd} E$ had an interior point. That point would also be an interior point of E because open balls contained in $\operatorname{bd} E$ are contained in E since $\operatorname{bd} E \subseteq E$ by Problem 22(iii) of Section 9.2. But a point cannot be both a boundary point and an interior point of E by statement (5). Thus, the interior of $\operatorname{bd} E$ is empty.

    It remains to verify statement (5). We first verify the equality. By the definitions at the beginning of the section, interior points of E are in E, exterior points of E are in $X \sim E$, and boundary points of E are in X. Hence, $\operatorname{int} E \cup \operatorname{ext} E \cup \operatorname{bd} E \subseteq X$. Conversely, let x be a point in X. If x is also in E, then every open ball centered at x contains a point, namely x itself, in E, so x is either an interior point or a boundary point. Otherwise, x is in $X \sim E$ so that x is either an exterior point or a boundary point by analogous reasoning. Hence, $X \subseteq \operatorname{int} E \cup \operatorname{ext} E \cup \operatorname{bd} E$. Thus, $X = \operatorname{int} E \cup \operatorname{ext} E \cup \operatorname{bd} E$.

    Next, we verify that the union is disjoint. Sets $\operatorname{int} E$, $\operatorname{ext} E$ are disjoint because interior points are in E and exterior points are in $X \sim E$. Set $\operatorname{bd} E$ is disjoint from each of $\operatorname{int} E$, $\operatorname{ext} E$ because if a point is in $\operatorname{bd} E$, then *every* open ball centered at that point contains points in E and $X \sim E$, so none of those balls is contained in either E or $X \sim E$. Thus, the union is disjoint.

16. Assume first that E is nowhere dense so that by definition, $\operatorname{int} \overline{E} = \emptyset$. Then a nonempty $\mathcal{O}$ cannot be contained in $\overline{E}$; otherwise, $\operatorname{int} \overline{E}$ would not be empty. It follows that $\mathcal{O} \sim \overline{E}$ is a nonempty open subset of $\mathcal{O}$ containing no points of $\overline{E}$, hence no points of E since $E \subseteq \overline{E}$. Thus, $E \cap \mathcal{O}$ is not dense in $\mathcal{O}$.

    Conversely, assume that for each $\mathcal{O}$, intersection $E \cap \mathcal{O}$ is not dense in $\mathcal{O}$. Take a nonempty $\mathcal{O}$. Then there is a nonempty open subset G of $\mathcal{O}$ containing no points of E. Neither does

G contain points of $\overline{E}$ because G is open. Hence, $\mathcal{O}$ is not contained in $\overline{E}$, that is, $\operatorname{int}\overline{E} = \emptyset$. Thus, E is nowhere dense.

17. Not necessarily. For a counterexample, put X to $\mathbf{R}$, which is complete by the sentence after the definition of completeness on page 145. By Corollary 4(ii) of Section 1.3, $\mathbf{Q}$ is the union of a countable collection of singleton sets $\{q\}$, each containing a unique rational number q. Each $\{q\}$ is nowhere dense in $\mathbf{R}$. Indeed, the closure of $\{q\}$ is $\{q\}$ itself by our solution to Problem 75 of Section 9.6. Moreover, $\{q\}$ is hollow in $\mathbf{R}$ because it has no interior points since no open interval $(q-r, q+r)$ centered at q is contained in $\{q\}$. Hence, $\mathbf{Q}$ is the union of a countable collection of nowhere dense sets. But $\mathbf{Q}$ is not nowhere dense in $\mathbf{R}$ because $\overline{\mathbf{Q}} = \mathbf{R}$ by the two sentences after the definition on page 203, and $\mathbf{R}$ is not hollow because, for example, 0 is an interior point of $\mathbf{R}$ since every open interval centered at 0 is contained in $\mathbf{R}$.

18. If $\mathcal{O} = X$ or $\emptyset$, or if $F = X$ or $\emptyset$, the corresponding claims are trivially true. So assume now that $\mathcal{O}, F$ are proper, nonempty subsets of X.

    By Problem 15, it suffices to show that $\overline{\mathcal{O}} \sim \mathcal{O} = \operatorname{bd}\overline{\mathcal{O}}$, and that $F \sim \operatorname{int} F = \operatorname{bd} F$.

    To show that $\overline{\mathcal{O}} \sim \mathcal{O} = \operatorname{bd}\overline{\mathcal{O}}$, we use statement (5) by showing that points x in $\overline{\mathcal{O}} \sim \mathcal{O}$ are not in $\operatorname{ext}\overline{\mathcal{O}}$ nor in $\operatorname{int}\overline{\mathcal{O}}$ and therefore must be in $\operatorname{bd}\overline{\mathcal{O}}$. Now $x \notin \operatorname{ext}\overline{\mathcal{O}}$ because $x \notin X \sim \overline{\mathcal{O}}$ since $x \in \overline{\mathcal{O}} \sim \mathcal{O}$. Moreover, $x \notin \operatorname{int}\overline{\mathcal{O}}$. Otherwise, there would be an open ball centered at x that is contained in $\overline{\mathcal{O}}$. Then x would be in open set $\mathcal{O}$, which contradicts that $x \in \overline{\mathcal{O}} \sim \mathcal{O}$. Consequently, x must be in $\operatorname{bd}\overline{\mathcal{O}}$ by statement (5). Thus, $\overline{\mathcal{O}} \sim \mathcal{O} = \operatorname{bd}\overline{\mathcal{O}}$.

    Likewise, points x in $F \sim \operatorname{int} F$ are not in $\operatorname{ext} F$ because $x \notin X \sim F$ since $x \in F \sim \operatorname{int} F$. Moreover, $x \notin \operatorname{int} F$ again because $x \in F \sim \operatorname{int} F$. Consequently, x must be in $\operatorname{bd} F$. Thus, $F \sim \operatorname{int} F = \operatorname{bd} F$.

19. The short answer is "yes" because the union of a countable collection of countable sets is countable (Corollary 6 of Section 1.3). Moreover, the claim is true for all metric spaces, complete or not.

    Here is a "long" answer with symbols. Let $\bigcup_{n=1}^{\infty} E_n$ be the union of a countable collection of sets E_n of the first category. Because each E_n is of the first category, each E_n is the union $\bigcup_{m=1}^{\infty} A_{nm}$ of a countable collection of nowhere dense subsets A_{nm}. As a result, $\bigcup_{n=1}^{\infty} E_n = \bigcup_{n=1}^{\infty} \bigcup_{m=1}^{\infty} A_{nm}$, which is the union of a countable collection of nowhere dense subsets and therefore of the first category.

20. In the third sentence of the problem statement, insert "absolute values of" between repeated words "the the."

    Use Proposition 6 of Section 9.2 to show that F_n is closed. Take a sequence $\{f_m\}$ of functions in F_n that converges pointwise to a function $\bar{f}$ in $C[0,1]$. Because $\{f_m\} \subseteq F_n$, for each m there is a point x_m in $[0,1]$ such that $|f_m(x) - f_m(x_m)| \leq n|x - x_m|$ for all x in $[0,1]$. Sequence $\{x_m\}$ is bounded because it is in $[0,1]$. Therefore, by the Bolzano-Weierstrass theorem, $\{x_m\}$ has a subsequence $\{x_{m_k}\}$ that converges, say to $\bar{x}$. Then for all x in $[0,1]$,

$$\begin{aligned} \left|\bar{f}(x) - \bar{f}(\bar{x})\right| &\leq \left|\bar{f}(x) - f_{m_k}(x)\right| + \left|f_{m_k}(x) - f_{m_k}(x_{m_k})\right| \\ &\quad + \left|f_{m_k}(x_{m_k}) - f_{m_k}(\bar{x})\right| + \left|f_{m_k}(\bar{x}) - \bar{f}(\bar{x})\right| && \text{tri. ineq.} \\ &\leq \left|\bar{f}(x) - f_{m_k}(x)\right| + n|x - x_{m_k}| + n|x_{m_k} - \bar{x}| + \left|f_{m_k}(\bar{x}) - \bar{f}(\bar{x})\right|. \end{aligned}$$

    Now take the limit as $k \to \infty$. The left side is independent of k, the first and fourth terms on the right vanish because $\{f_m\} \to \bar{f}$ pointwise on $[0,1]$, the third term vanishes because $\{x_{m_k}\} \to \bar{x}$, and the second term becomes $n|x - \bar{x}|$ so that

$$\left|\bar{f}(x) - \bar{f}(\bar{x})\right| \leq n|x - \bar{x}| \text{ for all } x \text{ in } [0,1],$$

which shows that $\bar{f} \in F_n$. Thus, F_n is closed.

Next, construct g as the sum $\varphi + t$ of two piecewise linear functions φ, t in $C[0,1]$. Put φ to the function of Problem 51 of Section 1.6; we can choose φ so that $\rho_{\max}(f, \varphi) < r/2$. Put t to be a triangle wave whose triangles have height $r/2$ and base length $r/(M+n+1)$ where M is a number greater than the maximum of the absolute values of the slopes of the pieces of φ. Then $\|t\|_{\max} = r/2$, and the slopes of the pieces of t are $\pm(M+n+1)$. For those who prefer a symbolic definition of t, let δ equal $r/(2(M+n+1))$ so that

$$t(x) = \begin{cases} (M+n+1)(x - i\delta) & i\delta \le x \le \min\{(i+1)\delta, 1\} \text{ where } i = 0, 2, 4, \ldots, \\ (M+n+1)((i+1)\delta - x) & i\delta < x \le \min\{(i+1)\delta, 1\} \text{ where } i = 1, 3, 5, \ldots. \end{cases}$$

Then

$$\begin{aligned} \rho_{\max}(f, g) &= \|f - g\|_{\max} && \text{definition of } \rho_{\max} \text{ (page 205)} \\ &= \|f - \varphi - t\|_{\max} && g = \varphi + t \\ &\le \|f - \varphi\|_{\max} + \|t\|_{\max} && \text{triangle inequality} \\ &< r/2 + r/2 \\ &= r. \end{aligned}$$

Moreover, the absolute value of the left or right derivative of g at point x in $[0,1]$ is equal to $|\varphi'(x) + t'(x)|$, or $|\varphi'(x) \pm (M+n+1)|$, which is greater than $n+1$ because $|\varphi'| < M$. By the prime symbol, $'$, we mean the slope of the piece on which x is located.

To see that F_n is hollow, use statement (6). Observe first that g is not in F_n because for all points x_0 in $[0,1]$, points x on the same piece of g as x_0 but not the same as x_0 are such that $|g(x) - g(x_0)| > (n+1)|x - x_0|$, as we just showed, which is greater than $n|x - x_0|$. As a result, $g \in C[0,1] \sim F_n$. It follows that $C[0,1] \sim F_n$ is dense in $C[0,1]$ because every nonempty open subset of $C[0,1]$ contains a function in $C[0,1] \sim F_n$ since for each f in $C[0,1]$ and each positive r, there is a g in $C[0,1] \sim F_n$ such that $\rho_{\max}(f, g) < r$. Thus, by statement (6), F_n is hollow in $C[0,1]$.

We conclude that $C[0,1] \neq \bigcup_{n=1}^{\infty} F_n$ because $\bigcup_{n=1}^{\infty} F_n$ is hollow by part (ii) of Baire's category theorem, but $C[0,1]$ is not hollow because $C[0,1]$ is open by Proposition 1 of Section 9.2 and nonempty, so $C[0,1]$ does have interior points.

Finally, let f be a function in $C[0,1]$ that is differentiable at a point x_0 in $(0,1)$. Then there is a real number a such that

$$\lim_{x \to x_0} \frac{f(x) - f(x_0)}{x - x_0} = a.$$

By the fifth sentence after the definition of a limit on page 188, there is a positive δ for which if $x \in [0,1]$, and $0 < |x - x_0| < \delta$, then

$$\left| \frac{f(x) - f(x_0)}{x - x_0} - a \right| < 1.$$

Therefore, if $|x - x_0| < \delta$, then

$$|f(x) - f(x_0)| \le (a+1)|x - x_0|.$$

Otherwise, if $|x - x_0| \ge \delta$, then

$$|f(x) - f(x_0)| \le \frac{2\|f\|_{\max}}{\delta}|x - x_0|.$$

Put n equal to $\lceil \max\{a+1, 2\|f\|_{\max}/\delta\}\rceil$ where $\lceil\cdot\rceil$ denotes the ceiling function. Then for all x in $[0,1]$, we have that $|f(x)-f(x_0)| \le n|x-x_0|$. Hence, $f \in F_n$. Therefore, functions in $C[0,1] \sim \bigcup_{n=1}^{\infty} F_n$ fail to be differentiable at any point in $(0,1)$. (We have shown that the set of functions in $C[0,1]$ that are nowhere differentiable on $(0,1)$ is dense in $C[0,1]$.)

21. The solution to the first part of this problem is the same as our first solution to Problem 56 of Section 1.6 with $\mathbf{R}$ replaced by X, intervals $(x-\delta, x+\delta)$ and $(x-r, x+r)$ replaced by open balls $B(x,\delta)$ and $B(x,r)$, and absolute values involving x, x', x'' replaced by the metric on X.

    We conclude that there is not a real-valued function on $\mathbf{R}$ that is continuous just at the rational numbers; otherwise, the rational numbers would be the intersection of a countable collection of open sets, which contradicts our solution to Problem 24(iii) with X chosen to be all of $\mathbf{R}$, which is closed by Proposition 5 of Section 9.2.

22. Such a set is the F of Problem 38 of Section 2.7 with α replaced by the $1/n$ of this problem. Set F is nowhere dense by Problem 24(i) with E replaced by F and X replaced by $[0,1]$, which is complete by Theorem 12 of Section 9.4. Also, the claim of Problem 38 of Section 2.7 holds if $\alpha = 1$ because in that case, F is the Cantor set, which is closed and of measure zero by Proposition 19 of Section 2.7 and nowhere dense by Problem 40 of Section 2.7.

    As for the construction, for each n, let F_n be the F described above. Set $\bigcup_{n=1}^{\infty} F_n$ is thus of the first category in $[0,1]$. The measure of $\bigcup_{n=1}^{\infty} F_n$ is no more than 1 because each F_n is in $[0,1]$. On the other hand, the measure of $\bigcup_{n=1}^{\infty} F_n$ is at least $\lim_{n\to\infty}(1-1/n)$, which is equal to 1. Thus, $\bigcup_{n=1}^{\infty} F_n$ has measure 1.

23. (i) Let X be a complete metric space without isolated points. Suppose, to get a contradiction, that X were countable. Let $\{x_n \mid n \in \mathbf{N}\}$ be an enumeration of X. Each singleton set $\{x_n\}$ is closed by our solution to Problem 75 of Section 9.6. Then because $X = \bigcup_{n=1}^{\infty}\{x_n\}$, Corollary 4 tells us that one of the $\{x_n\}$ has a nonempty interior. Then x_n is an interior point of $\{x_n\}$ because x_n is the only point in $\{x_n\}$. Then by definition, there is an open ball centered at x_n that is contained in $\{x_n\}$, which in turn means that $\{x_n\}$ is open in X. But that contradicts that x_n is not an isolated point. Thus, X has an uncountable number of points.

    (ii) Interval $[0,1]$ is complete by Theorem 12 of Section 9.4. It suffices by part (i) to prove that $[0,1]$ has no isolated points. Take a point x in $[0,1]$. Every open interval centered at x contains points in $[0,1]$ other than x. Hence, $\{x\}$ is not open in $[0,1]$. Thus, x is not an isolated point.

    The proof that $[0,1]$ is uncountable because it has positive Lebesgue measure is based on the contrapositive of the result of Section 2.2's example and Proposition 1. In both of those places, the proofs involve countable collection of subsets as does our proof of part (i).

    (iii) As explained at the beginning of the proof of Blaire's category theorem, it suffices to show that if X is a complete metric space without isolated points and $\{\mathcal{O}_n\}_{n=1}^{\infty}$ is a countable collection of open dense sets, then $\bigcap_{n=1}^{\infty}\mathcal{O}_n$ is dense and uncountable. By part (i) of Baire's category theorem, $\bigcap_{n=1}^{\infty}\mathcal{O}_n$ is dense.

    To show that $\bigcap_{n=1}^{\infty}\mathcal{O}_n$ is uncountable, suppose the contrary. Let $\{y_1, y_2, \ldots\}$ be an enumeration of $\bigcap_{n=1}^{\infty}\mathcal{O}_n$. We claim that $X \sim \{y_n\}$ is an open dense set for all n. Indeed, $X \sim \{y_n\}$ is open because it is the complement of closed singleton set $\{y_n\}$, closed since it is not an isolated point. And $X \sim \{y_n\}$ is dense because every nonempty open subset of X contains a point in $X \sim \{y_n\}$ since $\{y_n\}$ is not one of the open subsets of X. Then

$$\left(\bigcap_{n=1}^{\infty} \mathcal{O}_n\right) \cap \left(\bigcap_{n=1}^{\infty} (X \sim \{y_n\})\right) \tag{10.1}$$

is the intersection of a countable collection of open dense sets and therefore is dense by part (i) of Baire's category theorem. But that contradicts that (10.1) is empty by our supposition that $\bigcap_{n=1}^{\infty} \mathcal{O}_n = \{y_1, y_2, \dots\}$. Thus, $\bigcap_{n=1}^{\infty} \mathcal{O}_n$ is uncountable.

24. (i) By statement (6), E is hollow, that is, E has no interior points. Then neither does F because $F \subseteq E$. Thus, because F is closed and hollow, it is nowhere dense by definition. This part of the problem does not require that X be complete.

    (ii) Suppose, to get a contradiction, that each of E and $X \sim E$ were a union of a countable collection of closed sets. By part (i) and the fact that $X \sim (X \sim E)$ is dense because it is equal to E, each of those closed sets is nowhere dense. It follows that $E \cup (X \sim E)$, that is, all of X, would be the union of a countable collection of closed sets all of which are nowhere dense, i.e., have an empty interior. But that contradicts Corollary 4. Thus, at most one of E, $X \sim E$ is the union of a countable collection of closed sets.

    (iii) We verify this for all closed intervals X of $\mathbf{R}$. Interval X is complete by Theorem 12 of Section 9.4. Now the set E of rational numbers in X is dense in X by Theorem 2 of Section 1.2 as is the set $X \sim E$ of irrational numbers by Problem 12 of Section 1.2. Moreover, E is the union of a countable collection of closed sets by the same reasoning as in our solution to Problem 17. Then by part (ii), $X \sim E$ is *not* the union of a countable collection of closed sets. Because a set is open if and only if its complement is closed, we infer from De Morgan's identities that E is not the intersection of a countable collection of open sets.

25. Take the first four sentences of the proof of Theorem 6. Put $\mathcal{O}$ to $\bigcup_{n=1}^{\infty} \operatorname{int} E_n$. Set $\mathcal{O}$ is a subset of X because for each n, $\operatorname{int} E_n$ is a subset of E_n by definition, and E_n is a subset of X by construction. Set $\mathcal{O}$ is open because it is a union of open subsets (see Proposition 1 of Section 9.2 and Problem 32(i) of Section 1.4). To show that $\mathcal{O}$ is dense in X, it suffices to show that each open ball $B(x,r)$ in X contains a point in $\mathcal{O}$. Because $X = \bigcup_{n=1}^{\infty} E_n$, there is an E_n such that $B(x,r) \cap E_n$ contains a point x'. By the first paragraph of the section, either $x' \in \operatorname{int} E_n$ or $x' \in \operatorname{bd} E_n$. If $x' \in \operatorname{int} E_n$, then $x' \in \mathcal{O}$ because $\operatorname{int} E_n \subseteq \mathcal{O}$. Otherwise, $x' \in \operatorname{bd} E_n$. In that case, note that because $B(x,r)$ is open, there is another open ball $B(x',r')$ centered at x' that is contained in $B(x,r)$. Moreover, $B(x',r')$ contains a point x'' in E_n because $x' \in \operatorname{bd} E_n$. We can choose an x'' that is actually in $\operatorname{int} E_n$ because the interior of $\operatorname{bd} E_n$ is empty by Problem 15 and so does not contain $B(x',r')$. Therefore, x'' is a point in $B(x,r)$ (because $x'' \in B(x',r') \subseteq B(x,r)$) that is in $\mathcal{O}$ (because $x'' \in \operatorname{int} E_n \subseteq \mathcal{O}$). Hence, $\mathcal{O}$ is dense in X.

    Now for each x in $\mathcal{O}$, point x is some set $\operatorname{int} E_n$, that is, x is an interior point of E_n. Then we can put U to be the open ball (neighborhood) that is centered at x and contained in $\operatorname{int} E_n$. Thus, $|f| \leq n$ on U for all f in $\mathcal{F}$, that is, $\mathcal{F}$ is uniformly bounded on U.

26. Specifically, $L^2[a,b] \subseteq L^1[a,b]$ by Corollary 3 of Section 7.2.

    To show that $L^2[a,b]$ is of the first category in $L^1[a,b]$, we need to show that $L^2[a,b]$ is the union $\bigcup_{n=1}^{\infty} F_n$ of a countable collection of nowhere dense subsets F_n of $L^1[a,b]$. Define F_n by

$$F_n = \{f \in L^2[a,b] \mid \|f\|_2^2 \leq n\}.$$

    By definition of $L^2[a,b]$ (page 136), for each f in $L^2[a,b]$, there is an index n such that $\|f\|_2^2 \leq n$; that is, f belongs to F_n. Hence, $L^2[a,b] = \bigcup_{n=1}^{\infty} F_n$.

To show that F_n is nowhere dense, we need to show that closure $\overline{F_n}$ is hollow. But $\overline{F_n} = F_n$, that is, F_n is already closed. We prove that using Proposition 6 of Section 9.2: Take a sequence $\{f_m\}$ in F_n that converges to a function $\bar{f}$ in $L^1[a,b]$. By the second part of the Riesz-Fischer theorem (Section 7.3), a subsequence $\{f_{m_k}\}$ of $\{f_m\}$ converges to $\bar{f}$ pointwise a.e. on $[a,b]$. Then $\{f_{m_k}^2\}$ is a (sub)sequence of nonnegative measurable functions on $[a,b]$ that converges to $\bar{f}^2$ pointwise a.e. on $[a,b]$ by our solution to Problem 42 of Section 1.5. So

$$\begin{aligned}
\|\bar{f}\|_2^2 &= \int_a^b \bar{f}^2 && \text{definition of } \|\cdot\|_2 \\
&\le \liminf_k \left\{ \int_a^b f_{m_k}^2 \right\} && \text{Fatou's lemma (Section 4.3)} \\
&= \liminf_k \{ \|f_{m_k}\|_2^2 \} && \text{definition of } \|\cdot\|_2 \\
&\le n && \{f_{m_k}\} \subseteq F_n,
\end{aligned}$$

which shows that $\bar{f} \in F_n$. Hence, each F_n is closed.

It remains to show that each F_n is hollow, i.e., has no interior points (functions). Take a function f in F_n and an open ball $B(f,r)$ in $L^1[a,b]$ centered at f; to be clear, $B(f,r)$ is set $\{h \in L^1[a,b] \mid \|h-f\|_1 < r\}$. Take another function g in $L^1[a,b] \sim L^2[a,b]$; such a function g exists by the next-to-last example of Section 7.2. Consider function $f + \epsilon g$ where $\epsilon = r/(2\|g\|_1)$. Then $f + \epsilon g \in B(f,r)$ because

$$\|f + \epsilon g - f\|_1 = \epsilon \|g\|_1 = r/2 < r,$$

but $f + \epsilon g \notin F_n$ because

$$\begin{aligned}
\|f + \epsilon g\|_2^2 &= \int_a^b (f + \epsilon g)^2 && \text{definition of } \|\cdot\|_2 \\
&= \int_a^b f^2 + 2\epsilon \int_a^b fg + \epsilon^2 \int_a^b g^2 && \text{linearity of integration} \\
&= \infty && \textstyle\int_a^b g^2 = \infty \text{ because } g \notin L^2[a,b],
\end{aligned}$$

which shows that open ball $B(f,r)$ contains a function $f + \epsilon g$ not in F_n, that is, $B(f,r)$ is not contained in F_n. Hence, f is not an interior point which in turn shows that F_n is hollow. Thus, $L^2[a,b]$ considered as a subset of $L^1[a,b]$ is of the first category.

27. Our proof begins like that of Theorem 6. Take a positive ϵ. For each index k, define F_k by

$$F_k = \{x \in \mathbf{R} \mid |f(nx)| \le \epsilon \text{ if } n \ge k\},$$

which we can also express as intersection $\bigcap_{n=k}^{\infty} \{x \in \mathbf{R} \mid |f(nx)| \le \epsilon\}$. For each n, set $\{x \in \mathbf{R} \mid |f(nx)| \le \epsilon\}$ is closed by Problem 58 of Section 1.6 because function $x \mapsto f(nx)$ is continuous and $\{x \in \mathbf{R} \mid |f(nx)| \le \epsilon\}$ is the inverse image with respect to $x \mapsto f(nx)$ of a closed interval. Then F_k is closed since the intersection of a collection of closed sets is closed. Since $\lim_{n\to\infty} f(nx) = 0$, for each x, there is an index k such that $|f(nx)| < \epsilon$ if $n \ge k$; that is, x belongs to F_k. Hence, $\mathbf{R} = \bigcup_{k=1}^{\infty} F_k$. Since $\mathbf{R}$ is a complete metric space (see the sentence after the definition of completeness on page 192), we conclude from Corollary 4 that there is an index k for which F_k contains an open interval (a,b). Therefore,

$$\text{if } n \ge k \text{ and } x \in (a,b), \text{ then } |f(nx)| < \epsilon.$$

Upon substituting t for nx, we get the equivalent statement that

$$\text{if } n \geq k \text{ and } t \in (na, nb), \text{ then } |f(t)| < \epsilon. \tag{10.2}$$

Let K be the index that is equal to $\max\{k, \lceil a/(b-a) \rceil\} + 1$ where $\lceil \cdot \rceil$ denotes the ceiling function. Because statement (10.2) holds if $n \geq K$, we can rewrite membership $t \in (na, nb)$ as $t \in \bigcup_{n=K}^{\infty}(na, nb)$, that is, $t \in (Ka, \infty)$. Our choice of K ensures that intervals $\{(na, nb)\}_{n=K}^{\infty}$ overlap. It follows that

$$\text{if } t > Ka, \text{ then } |f(t)| < \epsilon.$$

Thus, $\lim_{t\to\infty} f(t) = 0$.

28. We can drop the hypothesis that f be continuous because a differentiable function is continuous. For the same reason, each $f^{(n)}$ is continuous, a fact we use in our solution.

    Let G be the set in $\mathbf{R}$ that is the union of all open intervals on which f is a polynomial. G is open because the union of a collection of open sets is open. Then by Proposition 9 of Section 1.4, we can express G as a union of a countable, disjoint collection of open intervals (a_k, b_k):

$$G = \bigcup_{k=1}^{K} (a_k, b_k) \text{ where } K \text{ is a natural number or } \infty. \tag{10.3}$$

    We claim that f is a polynomial on every (a_k, b_k). It suffices by our solution to Problem 37 of Section 1.4 to prove that f is a polynomial on each closed subinterval $[c, d]$ of (a_k, b_k). Subinterval $[c, d]$ is compact and therefore has a finite covering consisting of open intervals on which f is a polynomial. Choose an index N_{m} greater than the maximum of the degrees of those polynomials; then $f^{(N_{\mathrm{m}})} = 0$ on $[c, d]$. Integrating that equation N_{m} times shows that f is a polynomial (of degree less than N_{m}) on $[c, d]$.

    We want to prove that $G = \mathbf{R}$ because in that case, $K = 1$ and $(a_1, b_1) = (-\infty, \infty)$ so that by our reasoning above, f would be a polynomial on all of $\mathbf{R}$. We actually prove equivalently that $\mathbf{R} \sim G$, denoted by G^C, is empty. We first establish two lemmata:

    - G^C has no isolated points (definition of which is given in Problem 23). Otherwise, for an isolated point x_0 in G^C, there would be an open interval $(x_0 - r, x_0 + r)$ such that $(x_0 - r, x_0 + r) \cap G^C = \{x_0\}$. It follows that subintervals $(x_0 - r, x_0)$ and $(x_0, x_0 + r)$ would be in G, so each of those two subintervals would be contained in an (a_k, b_k) of equation (10.3). Then there would be natural numbers n_1, n_2 such that

$$f^{(n_1)}(x) = 0 \text{ on } (x_0 - r, x_0), \text{ and}$$
$$f^{(n_2)}(x) = 0 \text{ on } (x_0, x_0 + r).$$

      For concreteness, say that $n_1 \geq n_2$. Then

$$f^{(n_1)}(x) = 0 \text{ on } (x_0 - r, x_0 + r) \sim \{x_0\}.$$

      Because $f^{(n_1)}$ is continuous,

$$f^{(n_1)}(x) = 0 \text{ on all of } (x_0 - r, x_0 + r),$$

      and therefore $(x_0 - r, x_0 + r) \subseteq G$, which would contradict that $x_0 \in G^C$. Hence, G^C has no isolated points.

- G is dense in $\mathbf{R}$. By definition of denseness and the discussion following statement (6), it suffices to prove that every nondegenerate closed interval $[a, b]$ contains a point in G. Define set F_n by
$$F_n = \left\{x \in [a,b] \;\middle|\; f^{(n)}(x) = 0\right\}.$$
Each F_n is closed in $[a, b]$ by Problem 58 of Section 1.6 along with the facts that singleton set $\{0\}$ is closed and $f^{(n)}$ is continuous. By hypothesis, for each x in $[a, b]$, there is an index n for which $f^{(n)}(x) = 0$; that is, x belongs to F_n. Hence $[a, b] = \bigcup_{n=1}^{\infty} F_n$. Since $[a, b]$ is a complete metric space (Theorem 12(i) of Section 9.4), we conclude from Corollary 4 that there is an index n for which F_n contains an open interval (c, d). Therefore, $f^{(n)}(x) = 0$ on (c, d) so that f is a polynomial on (c, d). Hence, G is dense in $\mathbf{R}$.

Now suppose, to get a contradiction, that G^C were not empty. Define set E_n by

$$E_n = \left\{x \in G^C \;\middle|\; f^{(n)}(x) = 0\right\}. \tag{10.4}$$

As before, each E_n is closed in G^C, and $G^C = \bigcup_{n=1}^{\infty} E_n$. Moreover, G^C is closed in $\mathbf{R}$ because G^C's complement, G, is open. Set G^C is therefore a complete metric space. We conclude from Corollary 4 that there is an index N and an open interval (a, b) for which

$$(a,b) \cap G^C \subseteq E_N, \text{ and } (a,b) \cap G^C \neq \emptyset. \tag{10.5}$$

Take a point x in $(a, b) \cap G^C$ and a sequence $\{x_m\}$ in $(a, b) \cap G^C$ that converges to x where $x_m \neq x$ for all m. Such a sequence exists by our lemma that x is not an isolated point. Then

$$\begin{aligned} f^{(N+1)}(x) &= \lim_{m\to\infty} \frac{f^{(N)}(x_m) - f^{(N)}(x)}{x_m - x} && \text{definition of } (N+1)\text{st derivative} \\ &= \lim_{m\to\infty} \frac{0-0}{x_m - x} && x_m, x \in E_N \\ &= 0, \end{aligned}$$

so we can deduce inductively that

$$f^{(n)}(x) = 0 \text{ for all } x \text{ in } (a,b) \cap G^C \text{ if } n \geq N. \tag{10.6}$$

Observe that $(a, b) \cap G^C \neq (a, b)$; otherwise, f would be a polynomial of degree less than N on (a, b) so that (a, b) would be contained in G^C, which contradicts that (a, b) in that case would be in G by our definition of G. We next prove that statement (10.6) also holds if we replace G^C with G, that is,

$$f^{(n)}(x) = 0 \text{ for all } x \text{ in } (a,b) \cap G \text{ if } n \geq N. \tag{10.7}$$

Take the intersection of (a, b) with each side of equation (10.3) to get that

$$(a,b) \cap G = \bigcup_{k=1}^{K} ((a_k, b_k) \cap (a,b)) \text{ where } K \text{ is a natural number or } \infty.$$

Take an (a_k, b_k) such that $(a_k, b_k) \cap (a, b) \neq \emptyset$. Such intervals (a_k, b_k) exist by our lemma that G is dense in $\mathbf{R}$. Interval (a, b) is not contained in (a_k, b_k) because $(a_k, b_k) \subseteq G$ and $G^C \cap (a, b) \neq \emptyset$. Therefore, $(a_k, b_k) \cap (a, b)$ is of the form (c, d), and at least one of endpoints c, d is in $G^C \cap (a, b)$. For concreteness, say that $c \in G^C \cap (a, b)$. On (c, d), function f is a polynomial of some degree j so that on (c, d), the jth derivative of f is a nonzero constant. Because $f^{(j)}$ is continuous, $f^{(j)}(c) \neq 0$. Because $c \in G^C$, we deduce from statement (10.6) that $j < N$. Consequently, on (c, d), function f is a polynomial of degree less than N so that f satisfies statement (10.7).

We have shown that $f^{(N)}(x) = 0$ on both $(a, b) \cap G^C$ and $(a, b) \cap G$, which together is all of (a, b). So f is a polynomial on (a, b). In other words, (a, b) is contained in G and therefore is disjoint from G^C. That contradicts the inequation of (10.5). Hence, G^C is empty. Thus, f is a polynomial on $\mathbf{R}$.

10.3 The Banach Contraction Principle

A better definition for a contraction would be "If the Lipschitz constant less than 1, the Lipschitz mapping is called a **contraction**" where *the* (as opposed to a) Lipschitz constant is defined in Problem 4 of Section 8.1.

For clarity, replace $1/2M$ in the next-to-last sentence in the proof of Picard's local existence theorem with $1/(2M)$.

29. If the degree of p is less than 2, then $p(x) = a_0 + a_1 x$ so that for all u, v in $\mathbf{R}$,

$$\begin{aligned} |p(u) - p(v)| &= |a_0 + a_1 u - (a_0 + a_1 v)| \\ &= |a_1||u - v|, \end{aligned}$$

which shows that p is Lipschitz with Lipschitz constant $|a_1|$.

Conversely, assume that p is Lipschitz. Then there is a nonnegative c for which

$$|a_0 + a_1 u + \cdots + a_n u^n - (a_0 + a_1 v + \cdots + a_n v^n)| \leq c|u - v| \text{ for all } u, v \text{ in } \mathbf{R}. \tag{10.8}$$

In particular, that inequality holds if $u \neq 0$ and $v = 0$, in which case

$$|a_1 u + \cdots + a_n u^n| \leq c|u| \text{ for all } u \text{ in } \mathbf{R} \sim \{0\}.$$

Dividing by $|u|$,

$$\left|a_1 + \cdots + a_n u^{n-1}\right| \leq c \text{ for all } u \text{ in } \mathbf{R} \sim \{0\}. \tag{10.9}$$

If $n \geq 2$, there is a (large) u for which inequality (10.9) fails. On the other hand, inequality (10.8) holds if $n = 0$ or 1 with c equal to $|a_1|$. Thus, the degree of p is 0 or 1.

Another approach to a solution is to adapt Problem 53 of Section 6.5.

30. (i) Geometrically speaking, f is a portion of a parabola that opens down, has roots 0 and 1, is nonnegative on $[0, 1]$, and has maximum value $\alpha/4$ (when $x = 1/2$). Hence, the range $f([0, 1])$ of f is $[0, \alpha/4]$. Thus, $f([0, 1]) \subseteq [0, 1]$ precisely when $\alpha \in (0, 4]$.

    (ii) By part (i), we want to find the subset of $(0, 4]$ of values of α for which f is a contraction. By definition, f is a contraction if

$$|f(u) - f(v)| \leq c|u - v| \text{ for all } u, v \text{ in } [0, 1] \text{ where } c < 1. \tag{10.10}$$

If $v = u$, inequality (10.10) holds. Otherwise, rewrite inequality (10.10) as

$$\frac{|f(u) - f(v)|}{|u - v|} < 1 \text{ for all } u, v \text{ in } [0, 1] \text{ where } v \neq u,$$

which shows, along with the mean value theorem, that we want the maximum of the absolute slope of f to be less than 1. The absolute slope of f is $\alpha|1 - 2x|$, which has a maximum value of α on $[0, 1]$. Thus, $f([0, 1]) \subseteq [0, 1]$, and f is a contraction precisely when $\alpha \in (0, 1)$.

31. No. For a counterexample, consider Problem 30 with the domain of f changed from $[0,1]$ to $(0,1]$. Fix α at $1/2$. Function f maps $(0,1]$ into itself. Furthermore, because $(0,1] \subseteq [0,1]$ our solution to part (ii) of Problem 30 tells us that f is Lipschitz with a Lipschitz constant less than 1. But the only fixed points, i.e., those for which $x(1-x)/2 = x$, are 0 and -1, both of which are outside $(0,1]$. (This counterexample does not contradict Banach's contraction principle because $(0,1]$ is not a complete metric space.)

32. No. The authors provide a counterexample at the beginning of the section. Mapping $T : \mathbf{R} \to \mathbf{R}$ defined by $T(x) = x + 1$ has no fixed points. But T is a mapping of complete metric space $\mathbf{R}$ (see the sentence after the definition of completeness on page 192) into itself and has Lipschitz constant 1 by our solution to Problem 29.

33. Modify the proof of Banach's contraction principle. Select a point in X and label it x_0. Now define sequence $\{x_k\}$ inductively by putting x_1 to $T(x_0)$ and if k is a natural number such that x_k is defined, putting x_{k+1} to $T(x_k)$.

    By assumption, metric space X is compact. By Theorem 16(ii)$\to$(iii) of Section 9.5, $\{x_k\}$ has a subsequence that converges to a point x in X. For notational convenience, assume the whole sequence $\{x_k\}$ converges to x. Note that T is Lipschitz because $\rho(T(u), T(v)) \leq \rho(u, v)$ for all u, v in X by assumption and the case where $u = v$. Because T is Lipschitz, it is continuous. Therefore

$$T(x) = \lim_{k\to\infty} T(x_k) = \lim_{k\to\infty} x_{k+1} = x.$$

    Thus, mapping $T : X \to X$ has at least one fixed point. It remains to check that there is only one fixed point. But if u and v are different points in X such that $T(u) = u$ and $T(v) = v$, then

$$0 < \rho(u, v) = \rho(T(u), T(v)) < \rho(u, v),$$

    a contradiction. Thus, there is exactly one fixed point.

34. For simplicity, interchange u and v if necessary so that $u > v$. Then

$$\begin{aligned}
|f(u) - f(v)| &= |\pi/2 + u - \arctan u - (\pi/2 + v - \arctan v)| \\
&= |u - v - (\arctan u - \arctan v)| \\
&= (u - v)\left|1 - \frac{\arctan u - \arctan v}{u - v}\right| \\
&< u - v \qquad \text{explanation below} \\
&= |u - v|.
\end{aligned}$$

    By the mean value theorem, $(\arctan u - \arctan v)/(u - v)$ is equal to the derivative of arctan at some point w, so we can replace $(\arctan u - \arctan v)/(u - v)$ with $1/(1 + w^2)$. Then because $1/(1+w^2) \in (0, 1]$, the absolute-value factor in the third line is in $[0, 1)$ and therefore less than 1.

    Function f does not have a fixed point because equation $f(x) = x$ does not have a solution, which is apparent upon simplifying that equation to $\arctan x = \pi/2$. This does not contradict Problem 33 because $\mathbf{R}$ is not compact by the contrapositive of Theorem 20(ii)$\to$(i) of Section 9.5.

35. Modify the proof of Banach's contraction principle. Select a point in B and label it x_0. Now define sequence $\{x_k\}$ inductively by putting x_1 to $f(x_0)$ and if k is a natural number such that x_k is defined, putting x_{k+1} to $f(x_k)$.

By Theorem 20(i)→(iii) of Section 9.5, $\{x_k\}$ has a subsequence that converges to a point x in B. For notational convenience, assume the whole sequence $\{x_k\}$ converges to x. Because f is Lipschitz, it is continuous. Therefore

$$f(x) = \lim_{k\to\infty} f(x_k) = \lim_{k\to\infty} x_{k+1} = x.$$

Thus, f has a fixed point.

Note that the fixed point is not necessarily unique. For example, f defined by $f(x) = x$ maps B into B and is Lipschitz with Lipschitz constant 1, but every point in B is a fixed point of f.

36. To show that g is one-to-one, apply the definition on page 4. Take a point b in $g(\mathbf{R}^n)$. Then there is a point a in $\mathbf{R}^n$ such that $g(a) = b$, that is, $a - f(a) = b$. Let a' be another point in $\mathbf{R}^n$ such that $a' - f(a') = b$. Then $a' - f(a') = a - f(a)$, or

$$a' - a = f(a') - f(a).$$

Let c be a number in $[0, 1)$ that is a Lipschitz constant for mapping f. Then

$$0 \le \|a' - a\| = \|f(a') - f(a)\| \le c\|a' - a\|,$$

so that because $0 \le c < 1$, we have that $\|a' - a\| = 0$, that is, $a' = a$. Hence, there is exactly one a for which $g(a) = b$. Thus, g is one-to-one.

To show that g is onto, let b now be a point in $\mathbf{R}^n$. Define mapping $h : \mathbf{R}^n \to \mathbf{R}^n$ by $h(x) = f(x) + b$. Then for all u, v in $\mathbf{R}^n$,

$$\begin{aligned}\|h(u) - h(v)\| &= \|f(u) + b - (f(v) + b)\| \\ &= \|f(u) - f(v)\| \\ &\le c\|u - v\|,\end{aligned}$$

which shows that h is a contraction. By Banach's contraction principle and the fact that $\mathbf{R}^n$ is complete by the next-to-last sentence on page 192, h has a fixed point x_*. Furthermore,

$$\begin{aligned}g(x_*) &= x_* - f(x_*) && \text{definition of } g \\ &= x_* - h(x_*) + b && \text{definition of } h \\ &= b && x_* \text{ is a fixed point of } h,\end{aligned}$$

which shows that $g(\mathbf{R}^n) = \mathbf{R}^n$ because b is an arbitrary point in $\mathbf{R}^n$. Thus, g is onto.

Next, g is continuous because it is Lipschitz with a Lipschitz constant $1 + c$ as the following computation reveals; for all u, v in $\mathbf{R}^n$,

$$\begin{aligned}\|g(u) - g(v)\| &= \|u - f(u) - (v - f(v))\| && \text{definition of } g \\ &\le \|u - v\| + \|f(v) - f(u)\| && \text{triangle inequality} \\ &\le \|u - v\| + c\|v - u\| && f \text{ is a contraction} \\ &= (1 + c)\|u - v\|.\end{aligned}$$

Here is another way to show that g is continuous. Because f is Lipschitz, it is continuous. Identity function $x \mapsto x$ on $\mathbf{R}^n$ is also continuous (see our solution to Problem 26 of Section 9.3). Then our two solutions to Problem 49(i) of Section 1.6 adapted to $\mathbf{R}^n$ show that g is continuous. Adapt our first solution by replacing $|\cdot|$ around expressions involving f or

g with $\|\cdot\|$. Adapt our second solution by using the definition of continuity on page 190 instead of Proposition 21 of Section 1.6.

To show that g^{-1} is continuous, it suffices by Problem 36 of Section 9.3 to show that $g(C)$ is closed in $\mathbf{R}^n$ whenever a C is a closed subset of $\mathbf{R}^n$. Assume for the moment that C is also bounded. Then C is compact by Theorem 20(i)$\rightarrow$(ii) of Section 9.5. By Proposition 21 of Section 9.5, $g(C)$ is also compact, hence, closed. Because the boundedness that we assumed can have arbitrarily large absolute bounds, g^{-1} is continuous on all of $\mathbf{R}^n$.

Here is a second way to show that g^{-1} is continuous. It suffices by Proposition 8 of Section 9.3 to show that for each open subset G of $\mathbf{R}^n$, the inverse image under g^{-1} of G, that is, $g(G)$, is an open subset of $\mathbf{R}^n$. But that is true by the invariance of domain theorem, which is well known but has a difficult proof.

Invariance of domain theorem Let G be an open subset of $\mathbf{R}^n$. Let $g : G \to \mathbf{R}^n$ be a continuous one-to-one mapping. Then $g(G)$ is also an open subset of $\mathbf{R}^n$.

A proof is at www.math.stonybrook.edu/~cschnell/pdf/notes/mat530.pdf.

37. We first establish that $c \leq 1$. Rewrite inequality $cr + \rho(T(x_0), x_0) \leq r$ as

$$\rho(T(x_0), x_0) \leq (1 - c)r \tag{10.11}$$

to see that $c \leq 1$ since $\rho \geq 0$ by definition of a metric (page 182).

To show that $T(K) \subseteq K$, take a point x in K. Then

$$\begin{aligned}
\rho(T(x), x_0) &\leq \rho(T(x), T(x_0)) + \rho(T(x_0), x_0) && \text{triangle inequality} \\
&\leq c\rho(x, x_0) + (1 - c)r && T \text{ is Lipschitz; inequality (10.11)} \\
&\leq cr + r - cr && \rho(x, x_0) \leq r \text{ because } x \in K \\
&= r.
\end{aligned}$$

Thus, $T(K) \subseteq K$.

To show that T has a fixed point, consider two cases. If $c = 1$, inequality (10.11) simplifies to equality $\rho(T(x_0), x_0) = 0$, that is, $T(x_0) = x_0$, in which case x_0 is a fixed point. Otherwise, $c < 1$, that is, T is a contraction. Because $T(K) \subseteq K$, we can replace definition $T : K \to X$ with $T : K \to K$. Moreover, K is a complete metric space because K is a closed subset of complete metric space X (see Proposition 11 of Section 9.4 and the first part of Problem 15(i) of Section 9.2). Thus, T has a (unique) fixed point by Banach's contraction principle.

38. Let $\partial g(x, y)/\partial y$ denote the partial derivative of g with respect to y evaluated at point (x, y). By the ϵ-δ criterion for continuity (page 190), there is a positive δ for which

$$\text{if } (x, y) \in B((x_0, y_0), \delta), \text{ then } \left| \frac{\partial g(x, y)}{\partial y} - \frac{\partial g(x_0, y_0)}{\partial y} \right| < 1.$$

In other words, at all points (x, y) inside a circle centered at (x_0, y_0) and of radius δ, the partial derivative of g with respect to y is within a value of 1 of the partial derivative of g with respect to y at (x_0, y_0).

Now $B((x_0, y_0), \delta)$ will do for an open neighborhood $\mathcal{O}$ of (x_0, y_0) that we seek. For putting M equal to

$$\max\left\{ \left| \frac{\partial g(x_0, y_0)}{\partial y} - 1 \right|, \left| \frac{\partial g(x_0, y_0)}{\partial y} + 1 \right| \right\}$$

allows us to claim that

$$\left| \frac{\partial g(x, y)}{\partial y} \right| < M \text{ for all } (x, y) \text{ in } B((x_0, y_0), \delta)$$

so that by the mean value theorem, $|g(x, y_1) - g(x, y_2)|/|y_1 - y_2| < M \le M$ for all (x, y_1) and (x, y_2) in $B((x_0, y_0), \delta)$ such that $y_2 \ne y_1$, which is Lipschitz assumption (16) together with the observation that (16) is trivially true if $y_2 = y_1$.

39. Applying Leibniz's rule for differentiation under the integral sign,

$$\begin{aligned} f'(x) &= be^{b(x-x_0)}y_0 + e^{b(x-x)}h(x) + \int_{x_0}^{x} be^{b(x-t)}h(t)\,dt \\ &= h(x) + b\left(e^{b(x-x_0)}y_0 + \int_{x_0}^{x} e^{b(x-t)}h(t)\,dt \right) \\ &= h(x) + bf(x) \\ &= g(x, f(x)) \text{ for all } x \text{ in } I, \\ f(x_0) &= e^{b(x_0-x_0)}y_0 + \int_{x_0}^{x_0} e^{b(x_0-t)}h(t)\,dt \\ &= y_0. \end{aligned}$$

40. The function that is identically 0 is trivially a solution as is the second function if $x \le 0$. For the second function, if $x > 0$,

$$\begin{aligned} f'(x) &\stackrel{?}{=} 3(f(x))^{2/3}, \\ \left(x^3\right)' &\stackrel{?}{=} 3\left(x^3\right)^{2/3}, \\ 3x^2 &= 3x^2. \end{aligned}$$

This does not contradict Picard's existence theorem because Lipschitz property (16) does not hold. Indeed, as explained in the first sentence of the second paragraph on page 217, we can replace g with $3(f(x))^{2/3}$. For simplicity, denote $f(x)$ by y. Then Lipschitz property (16) becomes that

$$\left| 3y_1^{2/3} - 3y_2^{2/3} \right| \le M|y_1 - y_2| \text{ for all points } y_1 \text{ and } y_2 \text{ in } \mathbf{R}.$$

But that fails. For example, if $y_1 > 0$ and $y_2 = 0$, then there is no M such that

$$M \ge \frac{3y_1^{2/3} - 3(0)^{2/3}}{y_1 - 0} = \frac{3}{y_1^{1/3}}$$

for all y_1 since we can make the right side larger than M by choosing y_1 less than $(3/M)^3$.

41. Refer to the three sentences after differential equation (14). Function $\tan(x/\epsilon)$ is well defined and continuous on I, so if $\tan(x/\epsilon)$ satisfies the equation in the third sentence after (14), then $\tan(x/\epsilon)$ is a unique solution on I. So verify that that equation holds for all x in I with $f(x)$ equal to $\tan(x/\epsilon)$ and (x_0, y_0) equal to $(0, 0)$ and where h is given by f':

$$\begin{aligned} \tan(x/\epsilon) &\stackrel{?}{=} 0 + \int_0^x (1/\epsilon)\left(1 + \tan^2(t/\epsilon)\right) dt \\ &= \tan(x/\epsilon). \end{aligned}$$

On the other hand, function $x \mapsto \tan(x/\epsilon)$ is not defined at endpoints $-\epsilon(\pi/2)$ or $\epsilon(\pi/2)$ of I. Thus, there is no solution in an interval strictly containing I.

42. Refer to the two sentences after differential equation (14) for background and nomenclature. Suppose, to get a contradiction, that f were a solution to (14). Then f is differentiable on I and therefore continuous on I. By Problem 49(i) of Section 1.6, function $x \mapsto f(x) - cx$ is continuous on I. To find that function's minimum value on $[x_1, x_2]$, put that function's derivative to zero, $x \mapsto f'(x) - c = 0$, and solve for x. In other words, find a point x_{m} in $[x_1, x_2]$ such that $h(x_{\text{m}}) = c$ since $f' = h$. But by hypothesis, no such point x_{m} exists, which contradicts that f is continuous on I and therefore on $[x_1, x_2]$ (see the extreme value theorem on page 26). Thus, there is no solution to differential equation (14).

43. Suppose, to get a contradiction, that image $f'(I)$ of f' were not an interval. Then there are points x_1, x_2 in I where $x_1 < x_2$ and a number c such that $f'(x_1) < c < f'(x_2)$ but c does not belong to $f'(I)$. By Problem 42 with h replaced by f', there is no solution to differentiable equation (14), which contradicts the hypothesis that f is differentiable.

44. This is usually called Peano's existence theorem. It does not include a hypothesis that $\mathbf{g}$ be Lipschitz. Consequently, the solution to the system of differential equations is not necessarily unique (see Problem 40).

    To facilitate our proof of Peano's existence theorem, we first prove a different form of Picard's existence theorem.

    Picard's existence theorem For positive numbers a and b and a point $(x_0, \mathbf{y}_0)$ in $\mathbf{R} \times \mathbf{R}^n$, let R be closed region $[x_0 - a, x_0 + a] \times \overline{B}(\mathbf{y}_0, b)$ contained in $\mathbf{R} \times \mathbf{R}^n$. Assume function $\mathbf{g} : R \to \mathbf{R}^n$ is continuous, and there is a positive number M for which the following Lipschitz property in the second variable holds uniformly with respect to the first variable:

$$\|\mathbf{g}(x, \mathbf{y}_1) - \mathbf{g}(x, \mathbf{y}_2)\| \le M\|\mathbf{y}_1 - \mathbf{y}_2\| \text{ for all points } (x, \mathbf{y}_1) \text{ and } (x, \mathbf{y}_2) \text{ in } R.$$

    Then system of differential equations

$$\begin{aligned} \mathbf{f}'(x) &= \mathbf{g}(x, \mathbf{f}(x)) \text{ for all } x \text{ in } [x_0 - a, x_0 + a] \\ \mathbf{f}(x_0) &= \mathbf{y}_0 \end{aligned}$$

    has a unique solution on $[x_0 - a, x_0 + a]$.

    Proof Modify the proof in the text. For ℓ a positive number, define I_ℓ to be closed interval $[x_0 - \ell, x_0 + \ell]$. We first show that ℓ can be chosen so that there is exactly one continuous function $\mathbf{f} : I_\ell \to \mathbf{R}^n$ having the property that

$$\mathbf{f}(x) = \mathbf{y}_0 + \int_{x_0}^{x} \mathbf{g}(t, \mathbf{f}(t))\, dt \text{ for all } x \text{ in } I_\ell.$$

    Now for each positive number ℓ for which $\ell \le a$, define X_ℓ to be the subspace of metric space $C(I_\ell)$ consisting of continuous functions $\mathbf{f} : I_\ell \to \mathbf{R}^n$ such that

$$\|\mathbf{f}(x) - \mathbf{y}_0\| \le b \text{ for all } x \text{ in } I_\ell.$$

    For $\mathbf{f}$ in X_ℓ, define mapping $\mathbf{T}(\mathbf{f})$ in $C(I_\ell)$ by

$$\mathbf{T}(\mathbf{f})(x) = \mathbf{y}_0 + \int_{x_0}^{x} \mathbf{g}(t, \mathbf{f}(t))\, dt \text{ for all } x \text{ in } I_\ell.$$

    In order to choose ℓ so that $\mathbf{T}(X_\ell) \subseteq X_\ell$, first use the compactness of R together with the continuity of $\mathbf{g}$ to choose a positive number K such that

$$\|\mathbf{g}(x, \mathbf{y})\| \le K \text{ for all points } (x, \mathbf{y}) \text{ in } R.$$

Now for $\mathbf{f}$ in X_ℓ and x in I_ℓ,

$$\|\mathbf{T}(\mathbf{f})(x) - \mathbf{y}_0\| = \left\| \int_{x_0}^{x} \mathbf{g}(t, \mathbf{f}(t))\, dt \right\| \leq \ell K$$

so that

$$\mathbf{T}(X_\ell) \subseteq X_\ell \text{ provided that } \ell K \leq b.$$

Observe that for functions $\mathbf{f}_1, \mathbf{f}_2$ in X_ℓ and x in I_ℓ, we can infer from the Lipschitz-property hypothesis that

$$\|\mathbf{g}(x, \mathbf{f}_1(x)) - \mathbf{g}(x, \mathbf{f}_2(x))\| \leq M \rho_{\max}(\mathbf{f}_1, \mathbf{f}_2).$$

Consequently, by linearity and monotonicity of integration,

$$\begin{aligned}\|\mathbf{T}(\mathbf{f}_1)(x) - \mathbf{T}(\mathbf{f}_2)(x)\| &= \left\| \int_{x_0}^{x} (\mathbf{g}(t, \mathbf{f}_1(t)) - \mathbf{g}(t, \mathbf{f}_2(t)))\, dt \right\| \\ &\leq |x - x_0| M \rho_{\max}(\mathbf{f}_1, \mathbf{f}_2) \\ &\leq \ell M \rho_{\max}(\mathbf{f}_1, \mathbf{f}_2).\end{aligned}$$

That inequality and inclusion $\mathbf{T}(X_\ell) \subseteq X_\ell$ provided that $\ell K \leq b$ imply that

$$\mathbf{T} : X_\ell \to X_\ell \text{ is a contraction provided that } \ell K \leq b \text{ and } \ell M < 1.$$

Put ℓ to $\min\{b/K, 1/(2M)\}$. By Banach's contraction principle, mapping $\mathbf{T} : X_\ell \to X_\ell$ has a unique fixed point.

If $\ell < a$, we are not done because we want a solution on all of $[x_0 - a, x_0 + a]$. To extend the solution to the right all the way to $x_0 + a$, repeat the entire preceding argument with I_ℓ replaced by $[x_0, x_0 + 2\ell]$, then by $[x_0 + \ell, x_0 + 3\ell]$, and so on. The value of ℓ, that is, $\min\{b/K, 1/(2M)\}$, is the same at each iteration because constants b, M, K remain the same at each iteration. We obtain a solution on $[x_0 - \ell, x_0 + a]$ after a finite number of iterations. If necessary, reduce ℓ in the last iteration so that the right endpoint is $x_0 + a$. At each iteration, replace the center point of the interval with $x_0 + (j - 1)\ell$ where j is the iteration number, and so that solution $\mathbf{f}$ be continuous across the iteration intervals, replace $\mathbf{y}_0$ with $\mathbf{f}(x_0 + (j - 1)\ell)$. Because solution $\mathbf{f}$ is unique, $\mathbf{f}$ at each iteration coincides with $\mathbf{f}$ from the previous iteration where the iteration intervals overlap. Extension of the solution to the left all the way to $x_0 - a$ is done analogously. Thus, the system of differential equations has a unique solution on all of $[x_0 - a, x_0 + a]$. □

Peano's existence theorem Let $\mathcal{O}$ be an open subset of $\mathbf{R} \times \mathbf{R}^n$ containing point $(x_0, \mathbf{y}_0)$. Assume function $\mathbf{g} : \mathcal{O} \to \mathbf{R}^n$ is continuous. Then there is an open interval I containing x_0 on which system of differential equations

$$\begin{aligned}\mathbf{f}'(x) &= \mathbf{g}(x, \mathbf{f}(x)) \text{ for all } x \text{ in } I \\ \mathbf{f}(x_0) &= \mathbf{y}_0.\end{aligned}$$

has a (not necessarily unique) solution.

Proof Because $\mathcal{O}$ is open, we may choose positive numbers a and b such that closed region R given by $[x_0 - a, x_0 + a] \times \overline{B}(\mathbf{y}_0, b)$ is contained in $\mathcal{O}$. Denote $[x_0 - a, x_0 + a]$ by I_a.

Use the compactness of R together with the continuity of $\mathbf{g}$ to choose a positive number K such that

$$\|\mathbf{g}(x, \mathbf{y})\| \leq K/2 \text{ for all points } (x, \mathbf{y}) \text{ in } R.$$

Taking the first part of the hint, apply Weierstrass's approximation theorem (page 151). For each natural number m, there exists a polynomial mapping $\mathbf{p}_m : R \to \mathbf{R}^n$ for which

$\rho_{\max}(\mathbf{p}_m, \mathbf{g}) < (K/2)/m$ on R. Each mapping $\mathbf{p}_m$ is Lipschitz in the second variable by reasoning analogous to that of our first solution to Problem 49 of Section 9.4. To be precise, for each m, there is a positive number M_m for which the following Lipschitz property in the second variable holds uniformly with respect to the first variable:

$$\|\mathbf{p}_m(x, \mathbf{y}_1) - \mathbf{p}_m(x, \mathbf{y}_2)\| \le M_m \|\mathbf{y}_1 - \mathbf{y}_2\| \text{ for all points } (x, \mathbf{y}_1) \text{ and } (x, \mathbf{y}_2) \text{ in } R.$$

Moreover, sequence $\{\mathbf{p}_m\}$ converges to $\mathbf{g}$ uniformly on R by the second paragraph of Section 10.1. And for all m and at all points in R.

$$\begin{aligned} \|\mathbf{p}_m\| &\le \|\mathbf{p}_m - \mathbf{g}\| + \|\mathbf{g}\| && \text{triangle inequality} \\ &< (K/2)/m + K/2 \\ &\le K. \end{aligned}$$

By our Picard's existence theorem above, system of differential equations

$$\begin{aligned} \mathbf{f}_m'(x) &= \mathbf{p}_m(x, \mathbf{f}_m(x)) \text{ for all } x \text{ in } I_a \\ \mathbf{f}_m(x_0) &= \mathbf{y}_0 \end{aligned}$$

has a unique solution $\mathbf{f}_m$ on I_a for every m. Therefore,

$$\mathbf{f}_m(x) = \mathbf{y}_0 + \int_{x_0}^{x} \mathbf{p}_m(t, \mathbf{f}_m(t))\, dt \text{ for all } x \text{ in } I_a \tag{10.12}$$

for every m by reasoning analogous to that in the text (page 217) where the authors explain the equivalence between solutions of differential equation (14) and integral equation (15).

To complete our proof of Peano's existence theorem, it suffices to show that there is a continuous function $\mathbf{f} : I_a \to \mathbf{R}^n$ having the property that

$$\mathbf{f}(x) = \mathbf{y}_0 + \int_{x_0}^{x} \mathbf{g}(t, \mathbf{f}(t))\, dt \text{ for all } x \text{ in } I_a. \tag{10.13}$$

Now the hint wants us to use the Arzelà-Ascoli theorem. To do so, we must generalize it from sequences of real-valued functions to sequences of functions with values in $\mathbf{R}^n$. But it is apparent from re-reading the theorem's proof and dependent material with $|\cdot|$ replaced by $\|\cdot\|$ that the theorem holds for functions with values in $\mathbf{R}^n$. Another way to prove the $\mathbf{R}^n$ version of the Arzelà-Ascoli theorem is to apply the text's version n times to extract a subsequence that converges uniformly in the first coordinate, then a sub-subsequence that converges uniformly in the first two coordinates, and so on.

Verify that I_a and sequence $\{\mathbf{f}_m\}$ satisfy the hypotheses of the Arzelà-Ascoli theorem:

- I_a is compact by Theorem 20(i)$\to$(ii) of Section 9.5.
- $\{\mathbf{f}_m\}$ is uniformly bounded on I_a. Indeed, for all x in I_a and all m,

$$\begin{aligned} \|\mathbf{f}_m(x)\| &= \left\| \mathbf{y}_0 + \int_{x_0}^{x} \mathbf{p}_m(t, \mathbf{f}_m(t))\, dt \right\| \\ &\le \|\mathbf{y}_0\| + \left\| \int_{x_0}^{x} \mathbf{p}_m(t, \mathbf{f}_m(t))\, dt \right\| && \text{triangle inequality} \\ &< \|\mathbf{y}_0\| + |x - x_0| K && \|\mathbf{p}_m\| < K \text{ on } I_a \\ &\le \|\mathbf{y}_0\| + aK && x \in I_a. \end{aligned}$$

- $\{\mathbf{f}_m\}$ is equicontinuous on I_a. Indeed, for positive ϵ, put δ to ϵ/K. Then for every $\mathbf{f}_m$ in $\{\mathbf{f}_m\}$ and all x, x' in I_a, if $|x' - x| < \delta$, then

$$\|\mathbf{f}_m(x') - \mathbf{f}_m(x)\| = \left\| \int_x^{x'} \mathbf{p}_m(t, \mathbf{f}_m(t))\, dt \right\| < |x' - x|K < \delta K = \epsilon.$$

Hence, by the Arzelà-Ascoli theorem, $\{\mathbf{f}_m\}$ has a subsequence that converges uniformly on I_a to a continuous function on I_a. Put $\mathbf{f}$ to that continuous function. For notational convenience, assume the whole sequence $\{\mathbf{f}_m\}$ converges to $\mathbf{f}$ uniformly on I_a. As we showed above, $\{\mathbf{p}_m\} \to \mathbf{g}$ uniformly on R. Take the limit of both sides of integral equation (10.12) as $m \to \infty$, and apply Proposition 8 of Section 4.2 to get integral equation (10.13), which completes our proof. $\square$

Chapter 11

Topological Spaces: General Properties

11.1 Open Sets, Closed Sets, Bases, and Subbases

The last equation in the proof of Proposition 2 should be $\mathcal{O}_1 \cap \mathcal{O}_2 = \bigcup_{x \in \mathcal{O}_1 \cap \mathcal{O}_2} B_x$.

1. By the next-to-last sentence on page 188, for the discrete metric on X, every subset is open, so the discrete metric induces the discrete topology (see the last sentence in the description of the discrete topology (page 222)). Furthermore, because the discrete metric is indeed a metric, it induces a metric topology. Thus, the discrete topology is a metric topology.

2. Assume for this problem only that a base cannot contain the empty-set even though that is not precluded by definition of a base for a topology. Then the trivial topology for a set X has exactly one base, $\{X\}$, which follows immediately from the last sentence of the description of the trivial topology and definition of a base for a topology.

3. (i) By property (i) of the definition of a topology, X is open. By the sentence after the definition of a base for a topology, X is the union of a subcollection of $\mathcal{B}$, so certainly X is the union of all sets in $\mathcal{B}$.

   (ii) B_1 and B_2 are open because they are in $\mathcal{B}$. By property (ii) of the definition of a topology, $B_1 \cap B_2$ is open. Because $B_1 \cap B_2$ contains x, intersection $B_1 \cap B_2$ is a neighborhood of x. By definition of a base for a topology, we can put B to be a set in $\mathcal{B}_x$ for which $B \subseteq B_1 \cap B_2$. Moreover, $x \in B$ because $B \in \mathcal{B}_x$.

4. Assume first that $\mathcal{T}_1 = \mathcal{T}_2$. Then $\mathcal{T}_1$ will do for both $\mathcal{B}_1$ and $\mathcal{B}_2$. Indeed, at each point x in X, for each neighborhood $\mathcal{N}_1$ of x belonging to $\mathcal{B}_1$, there is a neighborhood $\mathcal{N}_2$ of x belonging to $\mathcal{B}_2$, namely $\mathcal{N}_1$ since $\mathcal{B}_2 = \mathcal{B}_1$, for which $\mathcal{N}_2 \subseteq \mathcal{N}_1$ since $\mathcal{N}_2 = \mathcal{N}_1$. Interchanging subscripts in the previous sentence proves the second part of the predicate.

   We now prove the converse. Refer to the definition of a base for a topology and the subsequent sentence. Take a nonempty set U in $\mathcal{T}_1$. Set U is open by definition of a topology and therefore the neighborhood of some point x in U. Then there is a neighborhood $\mathcal{N}_1$ of x belonging $\mathcal{B}_1$ for which $\mathcal{N}_1 \subseteq U$. By hypothesis, there is a neighborhood $\mathcal{N}_2$ of x belonging to $\mathcal{B}_2$ for which $\mathcal{N}_2 \subseteq \mathcal{N}_1$. Then $\mathcal{N}_2 \subseteq U$, which shows that $\mathcal{B}_2$ is a base for $\mathcal{T}_1$, which in turn implies that $\mathcal{T}_1 \subseteq \mathcal{T}_2$. Interchanging subscripts in the previous proves that $\mathcal{T}_2 \subseteq \mathcal{T}_1$. Thus, $\mathcal{T}_1 = \mathcal{T}_2$.

5. (i) If $x \in \operatorname{int} E$, then there is a neighborhood U of x that is contained in E. Because U is open, if $x' \in U$, then there is a neighborhood U' of x' that is contained in U by Proposition 1. Then $U' \subseteq E$, so $x' \in \operatorname{int} E$. So because each point x' of U is an interior point of E, we have that $U \subseteq \operatorname{int} E$. Thus, $\operatorname{int} E$ is open.

   Subset E is open if and only if for each point y in E, there is a neighborhood U of y that is contained in E if and only if $y \in \operatorname{int} E$ if and only if $E \subseteq \operatorname{int} E$. We can replace that last inclusion with equality $E = \operatorname{int} E$ if we can prove that $\operatorname{int} E \subseteq E$. But that is true for every set E because for each x in $\operatorname{int} E$, there is a neighborhood of x (and hence containing x) that is contained in E.

   (ii) Comparing definitions in this part and part (i), we see that

   $$\operatorname{ext} E = \operatorname{int}(X \sim E),$$

   which is always open. Thus, $\operatorname{ext} E$ is open.

Next,

if E is closed, then $X \sim E$ is open	Proposition 4,
then $X \sim E = \operatorname{int}(X \sim E)$	second part of part (i),
then $X \sim E = \operatorname{ext} E$,	
then $\overline{E} \sim E \subseteq \operatorname{ext} E$	$\overline{E} \subseteq X$.

Conversely, assume that $\overline{E} \sim E \subseteq \operatorname{ext} E$. Subset E is closed provided it contains all its points of closure. Let y be a point of closure of E. Then every neighborhood of y contains a point in E. If $y \in E$, then we are done. Otherwise, $y \in \overline{E} \sim E$. But that is not possible because $\overline{E} \sim E \subseteq \operatorname{ext} E$ so that there would be a neighborhood of y that is contained in $X \sim E$, which contradicts that y is a point of closure of E. Hence, all points of closure of E are in E. Thus, E is closed.

(iii) (i) Set $\operatorname{bd} E$ is closed provided it contains all its points of closure. Let x be a point of closure of $\operatorname{bd} E$. Then every neighborhood U_x of x contains a point in $\operatorname{bd} E$. Therefore every U_x contains points in E and points in $X \sim E$, and hence, $x \in \operatorname{bd} E$. Thus, $\operatorname{bd} E$ is closed.

(ii) Assume first that E is open. Then for each point y in E, there is a neighborhood U_y of y that is contained in E. Then U_y contains no points in $X \sim E$ so that $y \notin \operatorname{bd} E$. Thus, $E \cap \operatorname{bd} E = \emptyset$.

Conversely, assume that $E \cap \operatorname{bd} E = \emptyset$. Then for each y in E, there is a neighborhood U_y that does not contain points in E and points in $X \sim E$. Because $y \in E$, we can sharpen that statement by saying that there is a U_y that does not contain points in $X \sim E$. Consequently, all the points that U_y contains are in E so that $U_y \subseteq E$. Thus, E is open.

(iii) Assume first that E is closed, that is, E contains all of its points of closure. Then for each y in E, every U_y contains a point in E. Thus, $\operatorname{bd} E \subseteq E$ because every point in $\operatorname{bd} E$ contains a point in E by definition.

Conversely, assume that $\operatorname{bd} E \subseteq E$. Subset E is closed provided it contains all its points of closure. Let y be a point of closure of E. Then every U_y contains a point in E. If all of the points that U_y contains are in E, then $y \in E$. Otherwise, U_y contains not only a point in E but also points in $X \sim E$ so that $y \in \operatorname{bd} E$. Therefore $y \in E$ because $\operatorname{bd} E \subseteq E$. In either case, $y \in E$. Thus, E is closed.

6. By hypothesis and by the first sentence after the definition of *closure*, $A \subseteq B \subseteq \overline{B}$. Thus, $\overline{A} \subseteq \overline{B}$ by Proposition 3.

Next, because $A \subseteq A \cup B$, we have by the previous that $\overline{A} \subseteq \overline{A \cup B}$. Likewise, $\overline{B} \subseteq \overline{A \cup B}$. Hence, $\overline{A} \cup \overline{B} \subseteq \overline{A \cup B}$. To get the inclusion in the other direction, note that inclusions $A \subseteq \overline{A}$ and $B \subseteq \overline{B}$ together imply that $A \cup B \subseteq \overline{A} \cup \overline{B}$, which is closed by Proposition 5. Then $\overline{A \cup B} \subseteq \overline{A} \cup \overline{B}$ by Proposition 3. Thus, $\overline{A \cup B} = \overline{A} \cup \overline{B}$.

Inclusions $A \subseteq \overline{A}$ and $B \subseteq \overline{B}$ imply also that $A \cap B \subseteq \overline{A} \cap \overline{B}$, which is closed by Proposition 5. Thus, $\overline{A \cap B} \subseteq \overline{A} \cap \overline{B}$ by Proposition 3.

7. If $\mathcal{O}$ is disjoint from $\overline{E}$, then $\mathcal{O}$ is disjoint from E because $E \subseteq \overline{E}$ by the sentence after the definition of *closure*.

Conversely, assume that $\mathcal{O}$ is disjoint from E. Suppose, to get a contradiction, that there were a point x in both $\mathcal{O}$ and $\overline{E}$. Because $x \in \mathcal{O}$, there is a neighborhood of x that is contained in $\mathcal{O}$ and therefore disjoint from E by assumption. But because $x \in \overline{E}$, every neighborhood of x contains a point in E. The previous two sentences contradict each other. Thus, $\mathcal{O}$ is disjoint from $\overline{E}$.

8. Define $\mathcal{T}$ to be the collection consisting of X and $\emptyset$, intersections of finite subcollections of $\mathcal{S}$, and unions of subcollections of $\mathcal{S}$. Then $\mathcal{T}$ is a topology by definition and contains $\mathcal{S}$ by construction. Let $\mathcal{T}'$ be another topology that contains $\mathcal{S}$. Then the sets in $\mathcal{S}$ are open so that by definition, $\mathcal{T}'$ contains X and $\emptyset$, intersections of finite subcollections of $\mathcal{S}$, and unions of subcollections of $\mathcal{S}$. Thus $\mathcal{T}'$ contains $\mathcal{T}$. That is the sense in which $\mathcal{T}$ is the topology with the fewest sets that contains $\mathcal{S}$ (cf. the wording of Proposition 3).

9. Denote the collection of such intervals by $\mathcal{B}$. Check the properties of Proposition 2. The set of real numbers $\mathbf{R}$ is not empty. (i) Take a real number x. Then $x \in [x, x+1)$, which is an interval in $\mathcal{B}$. Hence, $\mathcal{B}$ covers $\mathbf{R}$. (ii) If $x \in [a_1, b_1) \cap [a_2, b_2)$, then $[\max\{a_1, a_2\}, \min\{b_1, b_2\})$ is an interval, say $[a, b)$, in $\mathcal{B}$ for which $x \in [a, b) \subseteq [a_1, b_1) \cap [a_2, b_2)$. Thus, $\mathcal{B}$ is a base for a topology for $\mathbf{R}$.

10. ("Open neighborhood" is redundant because a neighborhood is open by definition.) Denote the collection of such sets by $\mathcal{B}$. Check the properties of Proposition 2. The upper half plane $\mathbf{R}^{2,+}$ is not empty. (i) Take a point (x, y) in $\mathbf{R}^{2,+}$. Then (x, y) is in some set in $\mathcal{B}$ by construction of the sets in $\mathcal{B}$. Hence, $\mathcal{B}$ covers $\mathbf{R}^{2,+}$. (ii) Next, let B_1 and B_2 be in $\mathcal{B}$, and let (x, y) be in $B_1 \cap B_2$. If $y = 0$, then B_1 and B_2 each consist of $(x, 0)$ and all the points in an open Euclidean ball in the upper half plane that is tangent to the real line at $(x, 0)$. Put B to be whichever of B_1 or B_2 has the smaller radius. Then $(x, 0) \in B \subseteq B_1 \cap B_2$. Now consider the case for which $y > 0$. By property (ii) of a topology, $B_1 \cap B_2$ is open. Because $(x, y) \in B_1 \cap B_2$, there is a neighborhood of (x, y) that is contained in $B_1 \cap B_2$ by Proposition 1. We can make that neighborhood a usual Euclidean open ball centered at (x, y) of radius small enough to be contained in $B_1 \cap B_2$. Put B to be that ball. Then B is a set in $\mathcal{B}$ for which $(x, y) \in B \subseteq B_1 \cap B_2$. Thus, $\mathcal{B}$ is a base for a topology for $\mathbf{R}^{2,+}$.

11. (i) To simplify the notation, denote the complement of E by cE, the closure of E by kE, and the interior of E by iE. The following three identities imply that we can obtain at most fourteen different sets:

    - $ccE = E$,
    - $kkE = kE$, and
    - $kckckckcE = kckcE$, or by replacing E with cE and applying the first identity, $kckckckE = kckE$.

    For the first identity, see the equation in the sentence after the proof of Proposition 4. The second identity follows immediately from definition of *closure*.

    We break our proof of the third identity into three steps. The first step is to show that $ckcE = iE$. Indeed, kcE is a closed set that contains cE by the first sentence after the definition of *closure*. Hence, $ckcE$ is an open set by Proposition 4 that is contained in ccE, or E. Because $ckcE$ is open, for each point z in $ckcE$, there is a neighborhood of z that is contained in $ckcE$ by Proposition 1, and therefore also contained in E. Hence, $ckcE \subseteq iE$ by definition of an interior point and the interior of a set.

    To get the inclusion in the other direction, note that iE is an open subset of E by our solution to Problem 5(i). Then ciE is a closed subset that contains cE so that $kcE \subseteq ciE$ by Proposition 3. Taking complements, $cciE \subseteq ckcE$, or $iE \subseteq ckcE$ by the first identity. Hence, $ckcE = iE$.

    The second step in our proof of the third identity is to show that $kikiE = kiE$. Indeed, $ikiE \subseteq kiE$ by our solution to Problem 5(i). Then because kiE is closed, $kikiE \subseteq kiE$ by Proposition 3.

    To get the inclusion in the other direction, note that $iE \subseteq kiE$ by the sentence after the definition of *closure*. By the same reasoning as in our first step above, $iE \subseteq ikiE$, which in turn is a subset of $kikiE$. Then by the first part of Problem 6, $kiE \subseteq kkikiE$. By our second identity above, $kiE \subseteq kikiE$. Hence, $kikiE = kiE$.

The third and last step in our proof of the third identity is to apply the results of the first two steps to the left side of the third identity: $kckckckcE = kikiE = kiE = kckcE$.

The three identities allow us to show readily that we can obtain at most fourteen different sets by repeated use of complementation and closure. The first two identities tell us that we must alternate complementation and closure to obtain different sets, and the third identity tells us when alternating those operations no longer gives us a different set. The fourteen different sets therefore are E, cE, kcE, $ckcE$, $kckcE$, $ckckcE$, $kckckcE$, $ckckckcE$, kE, ckE, $kckE$, $ckckE$, $kckckE$, $ckckckE$.

(ii) A suitable E is union

$$\{(0,0)\} \cup \{(x,y) \in \mathbf{R}^2 \mid 1 < \|(x,y)\| < 3 \text{ except } \|(x,y)\| \neq 2\}$$
$$\cup \{(p,q) \in \mathbf{Q}^2 \mid 4 < \|(p,q)\| < 5\},$$

in other words, the origin, an annulus of radii 1 and 3 but without the circle of radius 2, and a rational annulus of radii 4 and 5. To simplify the listing of the fourteen different sets, we take advantage of the circular symmetry of E to consider just the fundamental domain, which is a half line, that is, consider X as interval $[0,\infty)$ instead of $\mathbf{R}^2$. Here are the redefined E and different sets (note that, for example, $(1,3)$ denotes an open interval, not a point in $\mathbf{R}^2$):

$$
\begin{aligned}
E &= \{0,\ (1,3) \sim 2,\ (4,5) \cap \mathbf{Q}\}, \\
cE &= \{(0,1],\ 2,\ [3,4],\ (4,5) \sim \mathbf{Q},\ [5,\infty)\}, \\
kcE &= \{[0,1],\ 2,\ [3,\infty)\}, \\
ckcE &= \{(1,2),\ (2,3)\}, \\
kckcE &= \{[1,3]\}, \\
ckckcE &= \{[0,1),\ (3,\infty)\}, \\
kckckcE &= \{[0,1],\ [3,\infty)\}, \\
ckckckcE &= \{(1,3)\}.
\end{aligned}
\qquad
\begin{aligned}
kE &= \{0,\ [1,3],\ [4,5]\}, \\
ckE &= \{(0,1),\ (3,4),\ (5,\infty)\}, \\
kckE &= \{[0,1],\ [3,4],\ [5,\infty)\}, \\
ckckE &= \{(1,3),\ (4,5)\}, \\
kckckE &= \{[1,3],\ [4,5]\}, \\
ckckckE &= \{[0,1),\ (3,4),\ (5,\infty)\},
\end{aligned}
$$

As a partial check on the list, note that it satisfies the third identity.

11.2 The Separation Properties

12. We first show that F is Tychonoff. Take two points x and y in F. Now because X is normal, X is Tychonoff. Then in X, there is a neighborhood U_x of x that does not contain y and a neighborhood U_y of y that does not contain x. So in F, there is a neighborhood $U_x \cap F$ of x that does not contain y and a neighborhood $U_y \cap F$ of y that does not contain x. Sets $U_x \cap F$ and $U_y \cap F$ are indeed open in F by definition of topological subspaces (page 222). Hence, F is Tychonoff.

    We can also show that F is Tychonoff using Proposition 6. In X, set $\{x\}$ is closed. Then $\{x\}$ is closed in F by Problem 18 of Section 9.2 with the X there replaced by the F here, Y replaced by the X here, and both A and F there replaced by $\{x\}$. Hence, F is Tychonoff.

    Next, let F_1 and F_2 be disjoint closed subsets of F. Because F is closed in X, subsets F_1 and F_2 are closed in X by our solution to the second part of Problem 19(ii) of Section 9.2. Then because X is normal, F_1 and F_2 can be separated by disjoint neighborhoods U_1 and U_2 in X. In turn, F_1 and F_2 can be separated by disjoint neighborhoods $U_1 \cap F$ and $U_2 \cap F$ in F. Thus, F is normal.

Is it necessary to assume that F be closed? The answer depends how one interprets the question. Here is a "no" answer: By Proposition 7 and the last paragraph on page 183, a subset—closed or not—of a metric space is a normal subspace of a normal space.

On the other hand, if F is not a closed subset of a general normal space X, then F need not be normal. We present three counterexamples, all from Lynn A. Steen and J. Arthur Seebach Jr.'s book *Counterexamples in Topology*, Holt, Rinehart and Winston, Inc., 1970; see counterexamples 82, 105, and 86 and 87.

For the first counterexample, put F to the Moore plane of Problem 10. The Moore plane is not normal. Indeed, consider subspace $\mathbf{R} \times \{0\}$ of the Moore plane. By definitions of a topological subspace and a basic neighborhood of a point $(x, 0)$ in the Moore plane, a basic neighborhood in subspace $\mathbf{R} \times \{0\}$ is a singleton set $\{(x, 0)\}$. By property (iii) of a topology, a union of a collection of such singleton sets is open. It follows from the sentence after the proof of Proposition 4 that every subset of $\mathbf{R} \times \{0\}$ is closed. In particular, the "rational" set $\mathbf{Q} \times \{0\}$ and "irrational" set $(\mathbf{R} \sim \mathbf{Q}) \times \{0\}$ are closed. They are also disjoint. We now show that they cannot be separated by disjoint neighborhoods. Take a neighborhood U of $\mathbf{Q} \times \{0\}$ and a neighborhood V of $(\mathbf{R} \sim \mathbf{Q}) \times \{0\}$. To each point (x, y) in V there corresponds a Euclidean open ball of radius $r_{(x,y)}$ contained in V and tangent to the real line at $(x, 0)$. For each natural number n, define set S_n by

$$S_n = \{(x, 0) \in (\mathbf{R} \sim \mathbf{Q}) \times \{0\} \mid r_{(x,0)} > 1/n\}.$$

Let $\{(q_n, 0)\}$ be an enumeration of $\mathbf{Q} \times \{0\}$. Then $\{(q_1, 0)\} \cup \{(q_2, 0)\} \cup \cdots \cup S_1 \cup S_2 \cup \cdots$ is a countable cover of topological space $(\mathbf{R}, \mathcal{T})$ where $\mathcal{T}$ is the usual Euclidean topology. Now refer to the paragraph after the proof of the Baire category theorem (page 211). Because $(\mathbf{R}, \mathcal{T})$ is a complete metric space that is not hollow, one of the sets in the countable cover is not nowhere dense. Each $\{(q_n, 0)\}$ is nowhere dense in $\mathbf{R} \times \{0\}$, so there is an S_m that fails to be nowhere dense. Then there is a real interval (a, b) in which set $S_m \cap (a, b)$ is dense. As a result, every neighborhood of every rational point in $(a, b) \times \{0\}$ intersects V. Hence, U and V are not disjoint. Thus, the Moore plane is not normal.

We now construct a normal space X of which the Moore plane is a subset. To do so, we merely add point $(0, -1)$ to the Moore plane so that $X = \mathbf{R}^{2,+} \cup \{(0, -1)\}$, and we define a base for the topology at $(0, -1)$ to be the collection of neighborhoods $\{(0, -1)\} \cup U$ where U is an open set in the Moore plane and contains all but finitely many points of $\mathbf{R} \times \{0\}$. Now verify that the collection $\mathcal{B}$ consisting of that base for the topology at $(0, -1)$ together with the base for the topology of the Moore plane is a base for a topology for X: Check the properties of Proposition 2. Set X is not empty. (i) $\mathcal{B}$ covers X by our solution to Problem 10 and our addition of a base for the topology at $(0, -1)$. (ii) If the point is in $\mathbf{R}^{2,+}$, we deduce from our solution to Problem 10 that property (ii) holds. Otherwise, the point is $(0, -1)$, and so the two sets in $\mathcal{B}$ are of the form $\{(0, -1)\} \cup U_1$ and $\{(0, -1)\} \cup U_2$. Then we deduce from our solution to Problem 10 that there is a set $\{(0, -1)\} \cup B$ in $\mathcal{B}$ for which $(0, -1) \in \{(0, -1)\} \cup B \subseteq (\{(0, -1)\} \cup U_1) \cap (\{(0, -1)\} \cup U_2)$. Hence, $\mathcal{B}$ is a base for a topology for X, which shows that X is indeed a topological space. We next show that X is normal. To show that X is Tychonoff, take two points in X. If both are in the Moore plane, then the Tychonoff separation property holds for those two points because the Moore plane is Hausdorff by Problem 16 and therefore Tychonoff by the rightmost inclusion at the bottom of page 226. Otherwise, one point is $(0, -1)$ and the other (x, y) in the Moore plane. In the Moore plane, there is a neighborhood U_1 of some point (x_1, y_1) that does not contain (x, y). If necessary, add all but finitely many points of $\mathbf{R} \times \{0\}$ to U_1 so that it still does not contain (x, y). Then $\{(0, -1)\} \cup U_1$ is a neighborhood of $(0, -1)$ that does not contain (x, y). There is also a neighborhood of (x, y) that lies in the Moore plane and so does not contain $(0, -1)$. Hence, X is Tychonoff. The next step in showing that X is normal is to prove that each two disjoint closed sets in X can be separated by disjoint neighborhoods. Closed sets in X are the complements of open sets. The complement of every open set in the Moore plane contains point $(0, -1)$, so no two of those complements (closed sets) are disjoint. The complement of every neighborhood of point $(0, -1)$ is the union of a closed set in the positive upper half plane and a finite set of points of $\mathbf{R} \times \{0\}$. Take two such disjoint sets F_1 and F_2. Now the positive upper half plane of the Moore plane has the usual Euclidean topology and so is normal by Proposition 7. Then the parts of F_1 and F_2 that lie in the upper half plane can be separated by disjoint neighborhoods U_1 and U_2. Add to U_1 sufficiently small basic neighborhoods of the points in F_1 that are in $\mathbf{R} \times \{0\}$, and add to U_2 sufficiently small basic neighborhoods of the points in F_2 that are in $\mathbf{R} \times \{0\}$. Then U_1 and U_2 are disjoint neighborhoods of F_1 and F_2. The remaining case to consider is the complement F_1 of an open set in the Moore plane and the complement F_2 of a neighborhood of $(0, -1)$. As before, the parts of F_1 and F_2 that lie in the positive upper half plane can be separated by disjoint neighborhoods U_1 and U_2. Add point $(0, -1)$ to U_1 as well as sufficiently small basic neighborhoods of the points in F_1 that are in $\mathbf{R} \times \{0\}$, and add to U_2 sufficiently small basic neighborhoods of the points in F_2 that are in $\mathbf{R} \times \{0\}$. Then U_1 and U_2 are disjoint neighborhoods of F_1 and F_2. Hence, X is normal. Thus, the Moore plane is a non-normal subspace of normal space X and therefore not closed in X.

The second and third counterexamples require material not yet covered in the text up to Section 11.2. For the second counterexample, denote interval $[0, 1]$ by I. Put X to be the uncountable Cartesian product I^I with the product topology, and give I the topology it inherits from $\mathbf{R}$ (see the second example of Section 11.1). Recall that I is uncountable by Theorem 7 of Section 1.3. Interval I is compact by the Heine-Borel theorem (Section 1.4). Interval I is also Hausdorff by the last sentence on page 226 since I is a metric space. Then X is compact by Tychonoff's product theorem (Section 12.2), Hausdorff by Problem 12 of Section 12.2, and so normal by Theorem 18.

Put F to be $\{1/n \mid n \in \mathbf{N}\}^I$. Set F is a subset of X that is not closed in X because F does not contain point of closure 0^I. Now the subspace topology on F is homeomorphic to $\mathbf{N}^I$ because the induced topology on $\{1/n \mid n \in \mathbf{N}\}$ is homeomorphic to the discrete topology on $\mathbf{N}$ (see page 230 for an explanation of a homeomorphism). Thus, subspace F is not normal.

For the third counterexample, put X to what is called the *Tychonoff plank*. In this case, $X = [0, \omega_1] \times [0, \omega]$ with the product topology where ω_1 is the first uncountable ordinal and ω is the first infinite ordinal and where each interval has the interval topology. (The cardinality of ω_1 is the same as that of the set of real numbers. The first infinite ordinal ω can be identified with the set of natural numbers.) Intervals $[0, \omega_1]$ and $[0, \omega]$ are compact and Hausdorff. Then X is compact by Tychonoff's product theorem (Section 12.2), Hausdorff by Problem 12 of Section 12.2, and so normal by Theorem 18.

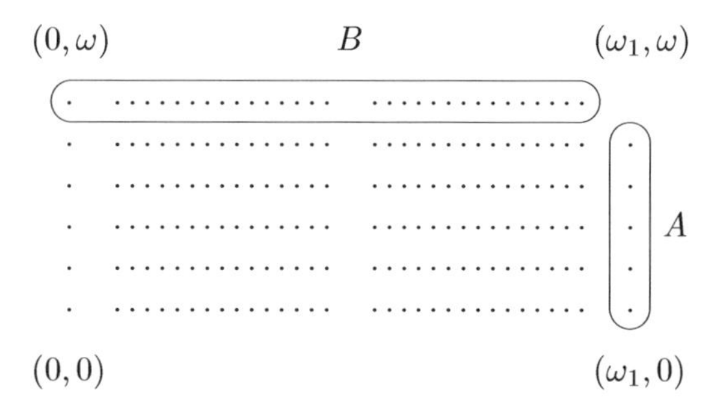

Put F to the *deleted Tychonoff plank*. In this case, $F = X \sim \{(\omega_1, \omega)\}$. Set F is a subset of X that is not closed in X because F does not contain point of closure (ω_1, ω). To show that F is not normal, consider sets A and B defined by—

$$A = \{(\omega_1, n) \in F \mid n \in [0, \omega)\},$$
$$B = \{(\alpha, \omega) \in F \mid \alpha \in [0, \omega_1)\}.$$

Then A and B are subsets of F that are closed in F, which is easier to see by observing that their complements in F are open. In F, take a neighborhood U of A. For each point (ω_1, n) in A, there is an ordinal α_n such that $\alpha_n < \omega_1$ and $\{(\alpha, n) \in F \mid \alpha \in (\alpha_n, \omega_1]\} \subseteq U$. Let $\bar{\alpha}$ be an upper bound for the α_n; we have that $\bar{\alpha} < \omega_1$ because ω_1 has uncountably many predecessors, whereas $\bar{\alpha}$ has only countably many. Hence, $(\bar{\alpha}, \omega_1] \times [0, \omega) \subseteq U$. Therefore, every neighborhood of point $(\bar{\alpha} + 1, \omega)$ in B must intersect U. Hence, every neighborhood of B intersects U. Thus, subspace F is not normal.

13. We first work out the reader exercise in the first example of Section 11.1. Check the properties of Proposition 2. Cartesian product $X \times Y$ is not empty because X and Y are not empty since $(X, \mathcal{T})$ and $(Y, \mathcal{S})$ are topological spaces. (i) The entire set X is open in X by definition. Likewise, Y is open in Y. It follows that $\mathcal{B}$ covers $X \times Y$ because $X \times Y$ itself is a set in $\mathcal{B}$. (ii) Let $B_{x1} \times B_{y1}$ and $B_{x2} \times B_{y2}$ be in $\mathcal{B}$, and let (x, y) be in $(B_{x1} \times B_{y1}) \cap (B_{x2} \times B_{y2})$, which by our solution to Problem 78 of Section 9.6 is equal to $(B_{x1} \cap B_{x2}) \times (B_{y1} \cap B_{y2})$. Now there is an open set B_x in X for which $x \in B_x \subseteq B_{x1} \cap B_{x2}$ and an open set B_y in Y for which $y \in B_y \subseteq B_{y1} \cap B_{y2}$. Hence, $B_x \times B_y$ is a set in $\mathcal{B}$ for which

$$(x, y) \in B_x \times B_y \subseteq (B_{x1} \cap B_{x2}) \times (B_{y1} \cap B_{y2}) = (B_{x1} \times B_{y1}) \cap (B_{x2} \times B_{y2}).$$

Thus, $\mathcal{B}$ is a base for a topology on $X \times Y$.

We now solve the problem. Assume first that X is Hausdorff. To show that D is a closed subset of $X \times X$, it suffices by the sentence after the proof of Proposition 4 to show that $(X \times X) \sim D$ is open. We do so using Proposition 1. Take a point (x_3, x_4) in $(X \times X) \sim D$. Because $x_4 \neq x_3$ and because X is Hausdorff, x_3 and x_4 in X can be separated by disjoint neighborhoods U_3 and U_4 in X. Then $U_3 \times U_4$ is a neighborhood of (x_3, x_4) that is contained in $(X \times X) \sim D$. Thus, $(X \times X) \sim D$ is an open subset of $X \times X$.

Conversely, assume that D is closed so that $D = \overline{D}$. Take two points x_3 and x_4 in X. Because $x_4 \neq x_3$, point (x_3, x_4) in $X \times X$ is not in $\overline{D}$, that is, (x_3, x_4) is not a point of closure of D. Then in $X \times X$, there is a neighborhood $U_3 \times U_4$ of (x_3, x_4) that does not intersect D. It follows that in X, points x_3 and x_4 are separated by disjoint neighborhoods U_3 and U_4. Thus, X is Hausdorff.

14. Take two real numbers. Name the lesser number x and the other y. A neighborhood of y has the form $(-\infty, a)$ where $a > y$. Because $x < y$, every such neighborhood contains x. Thus, the space is not Tychonoff.

Here is a second solution. It uses Proposition 6. If set $\{x\}$ were closed, then its complement $(-\infty, x) \cup (x, \infty)$ would be open by the sentence after the proof of Proposition 4. But $(-\infty, x) \cup (x, \infty)$ is not an open set in the given space. Thus, the space is not Tychonoff.

15. In the problem statement, replace X with $\mathbf{R}^n$.

We first verify that $\mathcal{B}$ is indeed a subbase for $\mathcal{T}$. Refer to the definition of a subbase (page 223). Taking p as constant polynomial 1 shows that all of $\mathbf{R}^n$ is a set in $\mathcal{B}$; therefore, $\mathcal{B}$ covers $\mathbf{R}^n$. We next need to show that the collection $\mathcal{B}_\cap$ of intersections of finite subcollections of $\mathcal{B}$ is a base for $\mathcal{T}$. Check the properties of Proposition 2. Set $\mathbf{R}^n$ is not empty. (i) As we showed above, $\mathbf{R}^n$ is a set in $\mathcal{B}$. Then $\mathbf{R}^n$ is also in $\mathcal{B}_\cap$ because $\mathbf{R}^n$ is equal to the intersection with itself. Hence, $\mathcal{B}_\cap$ covers $\mathbf{R}^n$. (ii) Let B_1 and B_2 be in $\mathcal{B}_\cap$, and let x be in $B_1 \cap B_2$. Now $B_1 \cap B_2$ has the form given by

$$B_1 \cap B_2 = \left(\bigcap_{i=1}^{k} \{x \in \mathbf{R}^n \,|\, p_i(x) \neq 0\} \right) \cap \left(\bigcap_{i=k+1}^{\ell} \{x \in \mathbf{R}^n \,|\, p_i(x) \neq 0\} \right)$$
$$= \bigcap_{i=1}^{\ell} \{x \in \mathbf{R}^n \,|\, p_i(x) \neq 0\}.$$

Consequently, $\bigcap_{i=1}^{\ell} \{x \in \mathbf{R}^n \,|\, p_i(x) \neq 0\}$ is a set in $\mathcal{B}_\cap$ that satisfies the second condition: $x \in \bigcap_{i=1}^{\ell} \{x \in \mathbf{R}^n \,|\, p_i(x) \neq 0\} = B_1 \cap B_2 \subseteq B_1 \cap B_2$. Thus, $\mathcal{B}$ is a subbase for $\mathcal{T}$.

To show that $(\mathbf{R}^n, \mathcal{T})$ is Tychonoff, take two points $(u_1, \ldots, u_n)$ and $(v_1, \ldots, v_n)$ in $\mathbf{R}^n$. Consider the polynomial defined by $p(x) = \sum_{j=1}^{n} (x_j - u_j)^2$ where $x = (x_1, \ldots, x_n)$. Then $p((u_1, \ldots, u_n)) = 0$, and $p((v_1, \ldots, v_n)) \neq 0$, which shows that $\{x \in \mathbf{R}^n \,|\, p(x) \neq 0\}$ is a neighborhood of $(v_1, \ldots, v_n)$ that does not contain $(u_1, \ldots, u_n)$. That reasoning holds upon interchanging the two points. Thus, $(\mathbf{R}^n, \mathcal{T})$ is Tychonoff.

But $(\mathbf{R}^n, \mathcal{T})$ cannot be Hausdorff because there are no disjoint nonempty open sets in $(\mathbf{R}^n, \mathcal{T})$. For consider two open sets $\{x \in \mathbf{R}^n \,|\, p_1(x) \neq 0\}$ and $\{x \in \mathbf{R}^n \,|\, p_2(x) \neq 0\}$ in $(\mathbf{R}^n, \mathcal{T})$ where p_1 and p_2 are non-constant polynomials in n variables. The number of points x for which $p_1(x) = 0$ is finite. The same holds for p_2. Hence, there is a point x in $\mathbf{R}^n$ for which $p_1(x) \neq 0$ and $p_2(x) \neq 0$. If p_1 or p_2 is a non-zero constant polynomial, then the corresponding set is all of $\mathbf{R}^n$, which again is an open set that not disjoint from another nonempty open set.

16. Take two real numbers. Name the lesser number a and the other b. Neighborhood $[a, b)$ of a is disjoint from neighborhood $[b, b+1)$ of b. Thus, the Sorgenfrey line is Hausdorff.

    Next, take two distinct points (x_1, y_1) and (x_2, y_2) in $\mathbf{R}^{2,+}$. If $y_1, y_2 = 0$, then basic neighborhoods of radius $(x_2 - x_1)/2$ for each point are disjoint. If $y_1, y_2 > 0$, then basic neighborhoods of radius $\min\{y_1, y_2, \|(x_2, y_2) - (x_1, y_1)\|/2\}$ for each point are disjoint. For the third and final case, relabel the points if necessary so that they are (x_1, y_1) and $(x_2, 0)$ where $y_1 > 0$. Then basic neighborhoods of radius $y_1/3$ for each point are disjoint. Thus, the Moore plane is Hausdorff.

11.3 Countability and Separability

The definition of *dense* should end "... provided every nonempty open set in X contains a point of E." Cf. the corresponding definition in Section 9.6.

Near the middle of page 228, "see Problem 21" can also be "see Problems 20 and 21."

17. Let $\{B_n\}_{n=1}^{\infty}$ be a countable base for X. Take an open cover $\mathcal{G}$ of X. For each B_n, choose from $\mathcal{G}$ if possible a set, which we denote G_n, that contains B_n. Collection $\mathcal{G}'$ of those sets G_n is not empty because each nonempty set in $\mathcal{G}$ contains some B_n since $\{B_n\}_{n=1}^{\infty}$ is a base for X. Collection $\mathcal{G}'$ is countable because it is indexed by a subset J of $\mathbf{N}$. Furthermore, $\mathcal{G}'$ covers X. Indeed, take a point x in X. Choose a set G from $\mathcal{G}$ that contains x. Because G is open and $\{B_n\}_{n=1}^{\infty}$ is a base for X, there is a B_m for which $x \in B_m \subseteq G$. Because B_m is in a set of $\mathcal{G}$, index m is in J, so G_m is defined. And because $B_m \subseteq G_m$, set G_m contains x. Hence, $\mathcal{G}'$ is a countable subcover of open cover $\mathcal{G}$ of X. Thus, X is Lindelöf.

18. Refer to the definition of a topology (pages 221–222). Set X is nonempty because it is uncountable. (i) The entire set X is open—i.e., in $\mathcal{T}$—because the complement of X is $\emptyset$, which is finite. The empty-set $\emptyset$ is open by hypothesis. (ii) Denote the intersection of a finite collection of open sets by $\bigcap_{i=1}^{n} G_i$. If one of the G_i is empty, then $\bigcap_{i=1}^{n} G_i = \emptyset$, which is open. Otherwise, use De Morgan's second identity on page 4 to express the complement

$X \sim \bigcap_{i=1}^{n} G_i$ of the intersection of a finite collection of open sets as $\bigcup_{i=1}^{n}(X \sim G_i)$. Each $X \sim G_i$ is finite by hypothesis, so finite union $\bigcup_{i=1}^{n}(X \sim G_i)$ is also finite. Hence, $\bigcap_{i=1}^{n} G_i$ is open. (iii) Denote the union of a collection $\mathcal{G}$ of open sets by $\bigcup_{G \in \mathcal{G}} G$. If all of the Gs are empty, then $\bigcup_{G \in \mathcal{G}} G = \emptyset$, which is open. Otherwise, use the other De Morgan identity to express the complement $X \sim \bigcup_{G \in \mathcal{G}} G$ of the union of a collection of open sets as $\bigcap_{G \in \mathcal{G}}(X \sim G)$. At least one of the $X \sim G$ is finite, so intersection $\bigcap_{G \in \mathcal{G}}(X \sim G)$ is also finite. Hence, $\bigcup_{G \in \mathcal{G}} G$ is open. Thus, $\mathcal{T}$ is a topology for X.

Next, we prove by contradiction that $(X, \mathcal{T})$ is not first countable. Suppose there were a countable base $\{B_n\}_{n=1}^{\infty}$ at a point x in X. Take a different point y in X. Set $X \sim \{y\}$ is a neighborhood of x because $X \sim \{y\}$ contains x and is open since complement $X \sim (X \sim \{y\})$, which is $\{y\}$, is finite. By definition of a base for a topology at a point (pages 222–223), there is a B_m in $\{B_n\}_{n=1}^{\infty}$ for which $B_m \subseteq X \sim \{y\}$. Then $y \notin B_m$, so neither is y in $\bigcap_{n=1}^{\infty} B_n$, which shows that no point different from x is in $\bigcap_{n=1}^{\infty} B_n$. Therefore, $\bigcap_{n=1}^{\infty} B_n = \{x\}$. Then $X \sim \{x\} = X \sim \bigcap_{n=1}^{\infty} B_n = \bigcup_{n=1}^{\infty}(X \sim B_n)$, which is countable by Corollary 6 of Section 1.3 and the fact that each $X \sim B_n$ is finite because $B_n \in \mathcal{T}$. But the implication that $X \sim \{x\}$ is countable contradicts that X is uncountable. Thus, $(X, \mathcal{T})$ is not first countable.

Here a slightly simpler way to argue. Because X is uncountable and $\bigcup_{n=1}^{\infty}(X \sim B_n)$ is only countable, there is a point y that is in X but not in $\bigcup_{n=1}^{\infty}(X \sim B_n)$. As before, there is a B_m for which $B_m \subseteq X \sim \{y\}$. By the way we chose y, it is not in $X \sim B_m$, so $y \in B_m$. But that contradicts that $B_m \subseteq X \sim \{y\}$.

19. That a second countable space is separable is the converse part of Proposition 25 of Section 9.6. Note that the proof of the converse holds for a general topological space, not just metric spaces. To tie the terminology of the proposition to the terminology of Chapter 11, refer to the definition of a second countable space and the sentence after the definition of a base for a topology (page 223).

    Next, take a subspace E of X. It is evident that $\{E \cap B_n\}_{n=1}^{\infty}$ is a countable base for E; nevertheless, we present a rigorous proof. Collection $\{E \cap B_n\}_{n=1}^{\infty}$ is countable because it is equipotent to $\mathbf{N}$. Next, we verify that $\{E \cap B_n\}_{n=1}^{\infty}$ is a base for E using the definition of a base for a topology. Each set $E \cap B_n$ is open in E by definition of a topological subspace. In E, take a point y and a neighborhood $E \cap U$ of y where U is some open set in X that contains y. Because U is a neighborhood of y in X, there is a B_m in $\{B_n\}_{n=1}^{\infty}$ for which $y \in B_m \subseteq U$. Then also $y \in (E \cap B_m) \subseteq (E \cap U)$. Hence, $\{E \cap B_n\}_{n=1}^{\infty}$ is a base for E. Thus, subspace E of X is second countable because it has a countable base.

20. The Moore plane is separable because, for example, the set $\mathbf{Q}^{2,+}$ of rational ordered pairs in the upper half plane is a countable dense subset of $\mathbf{R}^{2,+}$. Indeed, $\mathbf{Q}^{2,+}$ is countable by Section 1.3's Corollary 4(ii) and Problem 23. To see that $\mathbf{Q}^{2,+}$ is dense in $\mathbf{R}^{2,+}$, take a nonempty open set G in $\mathbf{R}^{2,+}$. By definitions of a base for a topology and the Moore plane, G contains some Euclidean open ball $B((x,y),r)$. By Theorem 2 of Section 1.2, there is a rational point (p,q) such that p is between x and $x + r/2$ and q is between y and $y + r/2$. Then (p,q) in $\mathbf{Q}^{2,+}$ is in $B((x,y),r)$ and therefore also in G. Hence, $\mathbf{Q}^{2,+}$ is dense in $\mathbf{R}^{2,+}$. Thus, the Moore plane is separable.

    Next, consider subspace $\mathbf{R} \times \{0\}$ of the Moore plane. As explained in our solution to the second part of Problem 12, open sets in $\mathbf{R} \times \{0\}$ include all singleton sets $\{(x,0)\}$ where x is a real number. Then the only set dense in $\mathbf{R} \times \{0\}$ is $\mathbf{R} \times \{0\}$ itself because by definition of denseness, every nonempty open set in $\mathbf{R} \times \{0\}$ must contain some point of the dense subset, so because open sets $\{(x,0)\}$ contain only the one point $(x,0)$, all points $(x,0)$ are in the dense subset. But $\mathbf{R} \times \{0\}$ is not countable by Theorem 7 of Section 1.3. Thus, $\mathbf{R} \times \{0\}$ is not separable.

We conclude that the Moore plane is not metrizable by the contrapositive of Proposition 26 of Section 9.6. A second way to see that the Moore plane is not metrizable is by the contrapositive of Proposition 7 and the fact that the Moore plane is not normal by our solution to the second part of Problem 12.

Next, because subspace $\mathbf{R} \times \{0\}$ of the Moore plane is not separable, $\mathbf{R} \times \{0\}$ is not second countable by the contrapositive of the first part of Problem 19. We conclude that the Moore plane is not second countable by the contrapositive of the second part of Problem 19.

21. The Sorgenfrey line is first countable because for a real number a, the countable collection $\{[a, a + 1/n)\}_{n=1}^{\infty}$ of intervals is a base at a for the topology of the Sorgenfrey line (cf. the example on page 228).

    To show that the Sorgenfrey line is not second countable, apply the definition spanning pages 222–223: A base for a topology of the Sorgenfrey line must include for each real number a a neighborhood B of a that is a subset of neighborhood $[a, a+1)$ of a. Then B must include an interval of the form $[a, a+\epsilon)$ where $0 < \epsilon \leq 1$. Such a base is uncountable because the set of real numbers is uncountable. Thus, the Sorgenfrey line is not second countable.

    The rationals are dense because every nonempty open set includes an interval of the form $[a, b)$ and because every such interval contains a rational number by Theorem 2 of Section 1.2.

    Because the rationals are dense, the Sorgenfrey line is separable. We conclude that the Sorgenfrey line is not metrizable by the contrapositive of the second sentence after the definition of a separable topological space. That sentence refers to Proposition 25 of Section 9.6.

22. For clarity, assume that $B_{s,n} = \{\omega\} \cup \{(j,k) \in X_1 \mid j \geq m_k \text{ whenever } k \geq n\}$. Furthermore, note that symbol n has two meanings: a parameter used to define $B_{s,n}$ and an index for sequence $\{x_n\}$.

    (i) Denote the collection of sets $B_{s,n}$ together with singleton sets $\{(j,k)\}$ by $\mathcal{B}$. Check the properties of Proposition 2. Set X is not empty. (i) Take a point x in X. If $x \in X_1$, then x is in some singleton set $\{(j,k)\}$. Otherwise, $x = \omega$ and therefore in every $B_{s,n}$ by its definition. Thus, $\mathcal{B}$ covers X. (ii) There are three cases:

    - If $x \in B_{s,n} \cap \{(j,k)\}$, then $x = (j,k)$. Hence, $\{(j,k)\}$ is a set in $\mathcal{B}$ for which $x \in \{(j,k)\} = B_{s,n} \cap \{(j,k)\} \subseteq B_{s,n} \cap \{(j,k)\}$.
    - If $x \in \{(j_1,k_1)\} \cap \{(j_2,k_2)\}$, then $x = (j_1,k_1)$. Hence, $\{(j_1,k_1)\}$ is a set in $\mathcal{B}$ for which $x \in \{(j_1,k_1)\} = \{(j_1,k_1)\} \cap \{(j_2,k_2)\} \subseteq \{(j_1,k_1)\} \cap \{(j_2,k_2)\}$.
    - Let x be in $B_{s_1,n_2} \cap B_{s_2,n_2}$. If $x \in X_1$, then x is a (j,k), so $\{(j,k)\}$ is a set in $\mathcal{B}$ for which $x \in \{(j,k)\} \subseteq B_{s_1,n_2} \cap B_{s_2,n_2}$. If $x = \omega$, then take set B_{s_3,n_3} where s_3 is the sequence whose kth term is the maximum of the kth terms of s_1 and s_2, and where n_3 is $\min\{n_1,n_2\}$. Then B_{s_3,n_3} is a set in $\mathcal{B}$ for which $x \in B_{s_3,n_3} = B_{s_1,n_2} \cap B_{s_2,n_2} \subseteq B_{s_1,n_2} \cap B_{s_2,n_2}$.

    Thus, $\mathcal{B}$ forms a base for a topology on X.

    (ii) Singleton set $\{\omega\}$ is not open because it is not the union of a subcollection of $\mathcal{B}$ (see the sentence after the definition of a base for a topology (page 223)). Hence, every neighborhood of ω contains some $B_{s,n}$ and therefore a point in X_1. Thus, ω is a point of closure of X_1 by definition (page 224).

    We next show that no sequence $\{x_n\}$ from X_1 converges to ω. If $\{x_n\}$ does not converge, then trivially, $\{x_n\}$ does not converge to ω. Otherwise, $\{x_n\}$ converges. Moreover, its limit is unique because X is Hausdorff (see the last sentence of the paragraph after the definition of a limit). To show that X is Hausdorff, take two points in X. If the two points are in X_1, they have the forms (j_1, k_1) and (j_2, k_2) and are separated by disjoint neighborhoods $\{(j_1, k_1)\}$ and $\{(j_2, k_2)\}$. Otherwise, one of the points is ω, and ω and (j_1, k_1) are separated by disjoint neighborhoods $B_{s_4,1}$ and $\{(j_1, k_1)\}$ where s_4 is the constant sequence $\{j_1 + 1\}$. We now have that $\{x_n\}$ converges to a unique limit. If follows that, describing $\{x_n\}$ by $((j_1, k_1), (j_2, k_2), (j_3, k_3), \ldots)$, the maximum of $j_1, j_2, \ldots$ exists—call it $j_{\max}$. Refer now to the definition of convergence. Consider neighborhood $B_{s_5,1}$ of ω where s_5 is the constant sequence $\{j_{\max} + 1\}$. Then no (j_n, k_n) belongs to $B_{s_5,1}$. Thus, $\{x_n\}$ does not converge to ω.

(iii) There are two straightforward ways to show that X is separable.

- Countable set X_1 is dense in X. Set X_1 is countable by Corollary 4(i) of Section 1.3. Set X_1 is dense in X because every nonempty open set in X contains a $B_{s,n}$ or a $\{(j, k)\}$.
- The entire set X is countable and dense in X. Set X is countable by Corollary 6 of Section 1.3. And trivially, every nonempty open set in X contains a point of X.

To see that X is not first countable, apply Proposition 9. The second sentence there is false for X by part (ii). Then X is not first countable. Nor is X second countable by the contrapositive of the sentence after the definitions of first and second countability.

It remains to prove Proposition 9, which we do by modifying the proof of Proposition 6 of Section 9.2. By definition of a closed set, it suffices to prove the first assertion. First assume that x belongs to $\overline{E}$. Because X is first countable, there is a countable base at x. By definition of a base for a topology at a point, we can choose sets from that countable base and label them so that we have another countable base $\{B_n\}_{n=1}^{\infty}$ at x that is descending, that is, $B_{n+1} \subseteq B_n$ for every n. For each n, because $B_n \cap E \neq \emptyset$, we can choose a point, which we label x_n, that belongs to $B_n \cap E$. Then $\{x_n\}$ is a sequence in E, and we claim that $\{x_n\}$ converges to x. Take a neighborhood U of x. There is a B_N in $\{B_n\}_{n=1}^{\infty}$ for which $B_N \subseteq U$. Then N is an index such that if $n \geq N$, then $x_n \in B_n \subseteq B_N \subseteq U$, that is, x_n belongs to U. Thus, $\{x_n\}$ converges to x. Conversely, if a sequence in E converges to x, then every neighborhood of x contains infinitely many terms of the sequence and therefore contains points in E. So $x \in \overline{E}$.

(iv) Yes. More generally, every countable space X, such as the X of this problem by our solution to part (iii), is Lindelöf. Take an open cover of X. For each point x in X, choose a set from that cover that contains x. The collection of those chosen sets covers X and is countable because X is countable. Thus, X is a Lindelöf space.

11.4 Continuous Mappings Between Topological Spaces

23. It suffices—by Proposition 10, the sentence after the definition of a base for a topology (page 223) where we ignore the word *nonempty* because the (open) empty-set is the union of an empty subcollection of the base, and definition of a subbase—to show that inverse images of unions of intersections of finite subcollections of $\mathcal{S}$ are unions of intersections of finite subcollections of inverse images of sets in $\mathcal{S}$. We can break the task into two parts by showing that for a finite subcollection $\{B_i\}_{i=1}^{n}$ where the B_i are sets in $\mathcal{S}$,

$$f^{-1}\left(\bigcap_{i=1}^{n} B_i\right) = \bigcap_{i=1}^{n} f^{-1}(B_i),$$

and that for a general subcollection $\{B\}_{B \in \mathcal{B}}$ where $\mathcal{B}$ is a collection of intersections B of finite subcollections of $\mathcal{S}$,

$$f^{-1}\left(\bigcup_{B \in \mathcal{B}} B\right) = \bigcup_{B \in \mathcal{B}} f^{-1}(B).$$

The authors' proof of Proposition 2 of Section 3.1 suggests that we may assume the two identities above. Nevertheless, we prove them for the sake of completeness.

$$\begin{aligned} x \in f^{-1}\left(\bigcap_{i=1}^{n} B_i\right) & \text{ if and only if } f(x) \in \bigcap_{i=1}^{n} B_i \\ & \text{ if and only if } f(x) \in B_i \text{ where } i = 1, \ldots, n \\ & \text{ if and only if } x \in f^{-1}(B_i) \text{ where } i = 1, \ldots, n \\ & \text{ if and only if } x \in \bigcap_{i=1}^{n} f^{-1}(B_i); \end{aligned}$$

$$\begin{aligned} x \in f^{-1}\left(\bigcup_{B \in \mathcal{B}} B\right) & \text{ if and only if } f(x) \in \bigcup_{B \in \mathcal{B}} B \\ & \text{ if and only if } f(x) \in B \text{ for some } B \text{ in } \mathcal{B} \\ & \text{ if and only if } x \in f^{-1}(B) \text{ for some } B \text{ in } \mathcal{B} \\ & \text{ if and only if } x \in \bigcup_{B \in \mathcal{B}} f^{-1}(B). \end{aligned}$$

The first identity holds also for general subcollections, not just finite ones.

24. (i) Mapping $f : X \to \mathbf{R}$ is continuous if and only if f is a constant map. Our proof uses Proposition 10 and the fact that the only open subsets of X are $\emptyset$ and X. Assume first that f is a constant map, that is, $f = c$ where c is some fixed real number. Take an open set G in $\mathbf{R}$. If $c \in G$, then $f^{-1}(G) = X$; otherwise, $c \notin G$ so that $f^{-1}(G) = \emptyset$. Thus, f is continuous.

    Conversely, assume that f is continuous. Suppose, to get a contradiction, that f were not a constant map. Then the image of f contains two distinct real numbers a and b. Because $\mathbf{R}$ is a metric space (sentence spanning pages 182–183), $\mathbf{R}$ is Tychonoff (bottom of page 226), so there is a neighborhood (open set) G of a that does not contain b. It follows that $f^{-1}(G)$ is neither $\emptyset$ nor X. Hence, $f^{-1}(G)$ is not an open subset of X, which contradicts that f is continuous. Thus, f is a constant map.

    (ii) Every mapping is continuous by Problem 27 of Section 9.3. In fact, our claim holds if $\mathbf{R}$ is replaced by a general topological space.

    (iii) No such mapping exists by the contrapositive of Problem 28 of Section 9.3 because not every subset of $\mathbf{R}$ is open. For example, closed interval $[0, 1]$ is not open in $\mathbf{R}$.

    (iv) Every mapping $f : \mathbf{R} \to X$, one-to-one or not, is continuous because $f^{-1}(\emptyset) = \emptyset$, and $f^{-1}(X) = \mathbf{R}$, and both $\emptyset$ and $\mathbf{R}$ are open subsets of $\mathbf{R}$ by definition (i) on page 221. Moreover, our answer holds if $\mathbf{R}$ is replaced by a general topological space.

25. Only assertions (i) and (iv) are equivalent to the continuity of f.

    (i) The proof is the same as our solution to Problem 36 of Section 9.3 with Proposition 4 now referring to that of Section 11.1, and Proposition 8 replaced by Proposition 10.

    (ii) For a counterexample, put X and Y both to $\mathbf{R}$ with the usual topology, and define f by $f(x) = x^2$. Mapping f is continuous by our solution to Problem 26 of Section 9.3. Put $\mathcal{O}$ to open interval $(-1, 1)$ so that $f(\mathcal{O})$ is interval $[0, 1)$, which is not open in $\mathbf{R}$ because there is no positive r for which open interval $(-r, r)$ centered at point 0 in $[0, 1)$ is contained in $[0, 1)$.

(iii) For a counterexample, put X and Y both to $\mathbf{R}$ with the usual topology, and define f by continuous function $f(x) = \arctan x$. Put F to all of $\mathbf{R}$, which is closed by Proposition 12 of Section 1.4. Then $f(F)$ is open interval $(-\pi/2, \pi/2)$, which is not closed in $\mathbf{R}$.

Another counterexample function is given in our solution to Problem 26 of Section 1.3. The image of that function is real interval $(0, 1)$.

(iv) Assume first that f is continuous. By the sentence after the definition of *closure* on page 224, $f(A) \subseteq \overline{f(A)}$, that is, $A \subseteq f^{-1}(\overline{f(A)})$. By assertion (i), $f^{-1}(\overline{f(A)})$ is closed in X. Then by Proposition 3, $\overline{A} \subseteq f^{-1}(\overline{f(A)})$. Thus, $f(\overline{A}) \subseteq \overline{f(A)}$.

A second way to verify that $f(\overline{A}) \subseteq \overline{f(A)}$ is to show that if $x \in \overline{A}$, then $f(x) \in \overline{f(A)}$. Take a neighborhood $\mathcal{O}$ of $f(x)$. Then $f^{-1}(\mathcal{O})$ is open in X and contains x and so must intersect A in some point y (see the definition on page 224). Then $\mathcal{O}$ intersects $f(A)$ in point $f(y)$ so that $f(x) \in \overline{f(A)}$.

We now verify the converse. Assume that $f(\overline{A}) \subseteq \overline{f(A)}$. To prove that f is continuous, it suffices to prove assertion (i). Take a closed subset C of Y. Put A to $f^{-1}(C)$. We now show that A is closed in X by showing that $A = \overline{A}$. By elementary set theory, $f(A) = f(f^{-1}(C)) \subseteq C$. Therefore, if $x \in \overline{A}$, then $f(x) \in f(\overline{A}) \subseteq \overline{f(A)} \subseteq \overline{C} = C$ so that $x \in f^{-1}(C) = A$. Thus, $\overline{A} \subseteq A$ so that $A = \overline{A}$. Note that $\overline{f(A)} \subseteq \overline{C}$ by Problem 6.

26. The proof is the same as that of Proposition 9 of Section 9.3 with "metric" replaced by "topological" and "the preceding proposition" replaced by "Proposition 10."

27. The paragraph spanning pages 223–224 justifies the second sentence of Proposition 12. Our solution to Problem 8 and definition of a weak topology justify the third sentence of Proposition 12. Note that in our solution to Problem 8, we specified that X be included in $\mathcal{T}$, but that is not required here because $\mathcal{S}$ covers X.

28. Take two such functions f_1 and f_2. The first step of our proof is to show that mapping $g : X \to \mathbf{R} \times \mathbf{R}$ defined by $g(x) = (f_1(x), f_2(x))$ is continuous. Take a point x in X and a neighborhood $\mathcal{O}_1 \times \mathcal{O}_2$ of $g(x)$. Because f_1 and f_2 are continuous, there are neighborhoods $\mathcal{U}_1$ and $\mathcal{U}_2$ of x such that $f_1(\mathcal{U}_1) \subseteq \mathcal{O}_1$, and $f_1(\mathcal{U}_2) \subseteq \mathcal{O}_2$. Let $\mathcal{U}$ be the intersection $\mathcal{U}_1 \cap \mathcal{U}_2$. Intersection $\mathcal{U}$ is a neighborhood of x because $\mathcal{U}_1$ and $\mathcal{U}_2$ each contain x and because the intersection of two opens sets is open by part (ii) of the definition of a topology (page 221). Furthermore, $f_1(\mathcal{U}) \subseteq \mathcal{O}_1$ and $f_2(\mathcal{U}) \subseteq \mathcal{O}_2$ because $\mathcal{U} \subseteq \mathcal{U}_1$ and $\mathcal{U} \subseteq \mathcal{U}_2$ so that $g(\mathcal{U}) \subseteq \mathcal{O}_1 \times \mathcal{O}_2$. Hence, mapping g is continuous.

Now $f_1 + f_2$ is the composition of g and the sum operation $+ : \mathbf{R} \times \mathbf{R} \to \mathbf{R}$. Thus, $f_1 + f_2$ is continuous by Proposition 11 and by Problem 49(i) of Section 1.6. Likewise, $f_1 \cdot f_2$ is the composition of g and the product operation $\cdot : \mathbf{R} \times \mathbf{R} \to \mathbf{R}$ and therefore also continuous.

29. By Proposition 6, it suffices to find necessary and sufficient conditions on $\mathcal{F}$ in order that every set in X consisting of a single point is closed. So let us characterize our weak topology in terms of closed sets. By Proposition 13 with each X_λ replaced by $\mathbf{R}$, the weak topology induced by $\mathcal{F}$ is the topology on X that has the fewest number of sets among the topologies on X for which each mapping $f_\lambda : X \to \mathbf{R}$ is continuous. Now f_λ is continuous if and only if $f_\lambda^{-1}(C)$ is closed in X whenever C is closed in $\mathbf{R}$ by Problem 25(i). As a result, we can define our weak topology as the weakest topology for X that contains the collection of sets $\{f_\lambda^{-1}(C) \mid f_\lambda \in \mathcal{F},\ C \text{ closed in } \mathbf{R}\}$. Thus, necessary and sufficient conditions on $\mathcal{F}$ in order that every set in X consisting of a single point is closed is that for every point x in X, there is a f_λ in $\mathcal{F}$ such that $f_\lambda^{-1}(C) = \{x\}$ for some closed set C in $\mathbf{R}$.

30. We first prove that assertions (i) and (ii) are equivalent, which is the fourth sentence after the definition of a homeomorphism. By Proposition 10, f^{-1} is continuous if and only if

for every open subset E in X, inverse image $f(E)$ is an open subset of Y. Likewise, f is continuous if and only if for every open subset G in Y, inverse image $f^{-1}(G)$ is an open subset of X; in particular, if $f(E)$ is open in Y, then $f^{-1}(f(E))$, or E, is open in X. Now the first parts of those two statements, i.e., that f and f^{-1} are continuous, along with the hypothesis that f is one-to-one and onto are the same as assertion (i) by definition of a homeomorphism. The second parts of those two statements are the same as assertion (ii). Furthermore, our two statements are *if-and-only-if* statements. Thus, assertions (i) and (ii) are equivalent.

Next, we prove that assertion (ii) implies assertion (iii). Applying Proposition 4 and the sentence following its proof, a subset E of X is closed in X if and only if $X \sim E$ is open in X if and only if $f(X \sim E)$ is open in Y if and only if $Y \sim f(X \sim E)$ is closed in Y. Now because f is one-to-one and onto, $Y \sim f(X \sim E) = f(E)$. Thus, assertion (ii) implies assertion (iii).

Interchanging words *open* and *closed* in the previous shows that assertion (iii) implies assertion (ii).

We have now shown that assertions (i), (ii), and (iii) are equivalent.

Next, we show that assertion (iii) implies assertion (iv). So assume assertion (iii). Then because $\overline{A}$ is closed, $f(\overline{A})$ is closed, that is, $f(\overline{A}) = \overline{f(\overline{A})}$. Furthermore, because assertion (iii) is equivalent to assertion (i), f is continuous, so by our solution to Problem 25(iv), $f(\overline{A}) \subseteq \overline{f(A)}$. Altogether, $f(\overline{A}) \subseteq \overline{f(A)} \subseteq \overline{f(\overline{A})} = f(\overline{A})$, which shows that $f(\overline{A}) = \overline{f(A)}$.

Finally, we show that assertion (iv) implies assertion (iii). Assume first that a subset E of X is closed in X, that is, $E = \overline{E}$. Then $f(E) = f(\overline{E}) = \overline{f(E)}$, which shows that $f(E)$ is closed in Y. Conversely, assume that $f(E)$ is closed in Y, that is, $f(E) = \overline{f(E)}$. Then $f(E) = f(\overline{E})$ by assertion (iv). Hence, $E = \overline{E}$ because f is one-to-one. Thus, E is closed in X.

31. If X is Hausdorff, then Y is not necessarily Hausdorff. For example, let X have at least two points. Put Y to be the same set as X but with the trivial topology, which guarantees that f is continuous by our solution to Problem 24(iv). Topological space Y is not Hausdorff because the only neighborhood of a point in Y is the whole set Y (see the description of the trivial topology on page 222), so there are no disjoint neighborhoods.

    If X is normal, our answer above holds also if we replace "Hausdorff" with "normal."

32. No, the metrics are not necessarily equivalent. See our solution to Problem 25 of Section 9.2.

33. Although this is "clear" by the sentence after the definition of a homeomorphism, we provide a proof. Take a homeomorphism $f : X \to Y$. By definition, f is continuous, one-to-one, maps X onto Y, and has a continuous inverse f^{-1}. Now check that f^{-1} satisfies the definition of a homeomorphism from Y onto X. Function f^{-1} is continuous. Because f is one-to-one and onto, there is a one-to-one correspondence between X and Y; therefore, f^{-1} is one-to-one and maps Y onto X. Finally, f^{-1} has a continuous inverse because $(f^{-1})^{-1} = f$. Thus, the inverse of a homeomorphism is a homeomorphism.

    Take another homeomorphism $g : Y \to Z$. Composition $g \circ f$ and its inverse $(g \circ f)^{-1}$, which is $f^{-1} \circ g^{-1}$, are continuous by Proposition 11. Moreover, $g \circ f$ is one-to-one because for each member z of $(g \circ f)(X)$, there is exactly one member y of Y for which $z = g(y)$, and in turn, there is exactly one member x of X for which $y = f(x)$; hence, for each member z of $(g \circ f)(X)$, there is exactly one member x of X for which $z = (g \circ f)(x)$. Finally, $g \circ f$ is onto because $(g \circ f)(X) = g(f(X)) = g(Y) = Z$. Thus, the composition of homeomorphisms, when defined, is again a homeomorphism.

34. Take a topological space Y that is homeomorphic to X, and take a continuous real-valued function f on Y. There is a homeomorphism h from X onto Y. Then $f \circ h$ is a real-valued function on X and also continuous by Proposition 11, so $f \circ h$ takes a minimum value. Now the image of f is equal to the image of $f \circ h$ because $(f \circ h)(X) = f(h(X)) = f(Y)$. Thus, f also takes a minimum value (the same one as $f \circ h$).

35. The proof is the same as our solution to Problem 34 with "takes a minimum value" replaced by "has an interval as its image."

36. $\mathbf{R}$ is homeomorphic to $(0, 1)$ using homeomorphism $x \mapsto 1/(1+e^{-x})$ from $\mathbf{R}$ onto $(0,1)$; see our solution to Problem 26 of Section 1.3. Three other homeomorphisms, now from $(0, 1)$ onto $\mathbf{R}$, are $x \mapsto 1/x + 1/(x-1)$; $x \mapsto \ln x + x/(1-x)$; and $x \mapsto \tan(\pi x - \pi/2)$.

    Next, every continuous real-valued function on $[0, 1]$ takes a minimum value by the extreme value theorem (Section 1.6). If $\mathbf{R}$ were homeomorphic to $[0, 1]$, then $\mathbf{R}$ would also possess that property by Problem 34. But, for example, function $x \mapsto x$ is a continuous real-valued function on $\mathbf{R}$ that does not take a minimum value. Thus, $\mathbf{R}$ is not homeomorphic to $[0, 1]$.

37. Put X and Y both to $\mathbf{R}$ with the usual topology; put f to the Dirichlet function $\chi_{\mathbf{Q}}$, that is, $\chi_{\mathbf{Q}}(x) = 1$ if x is rational, and $\chi_{\mathbf{Q}}(x) = 0$ if x is irrational (see the first example of Section 4.2); put X_1 to $\mathbf{Q}$; and put X_2 to the subset $\mathbf{R} \sim \mathbf{Q}$ of irrational numbers in $\mathbf{R}$. Then $\mathbf{R} = \mathbf{Q} \cup (\mathbf{R} \sim \mathbf{Q})$, and the restrictions of $\chi_{\mathbf{Q}}$ to topological subspaces $\mathbf{Q}$ and $\mathbf{R} \sim \mathbf{Q}$ are continuous because the restrictions are constant functions: $\chi_{\mathbf{Q}}|_{\mathbf{Q}} = 1$ and $\chi_{\mathbf{Q}}|_{\mathbf{R}\sim\mathbf{Q}} = 0$. Dirichlet function $\chi_{\mathbf{Q}}$ is not continuous at any point x in $\mathbf{R}$ because there is no positive δ for which if x' is a real number and $|x' - x| < \delta$, then $|\chi_{\mathbf{Q}}(x') - \chi_{\mathbf{Q}}(x)| < 1/2$. For if x is rational, there is an irrational number x' in open interval $(x - \delta, x + \delta)$ by Problem 12 of Section 1.2; else if x is irrational, there is a rational number x' in $(x-\delta, x+\delta)$ by Theorem 2 of Section 1.2. In either case, $|x' - x| < \delta$, but $|\chi_{\mathbf{Q}}(x') - \chi_{\mathbf{Q}}(x)| = 1 \not< 1/2$.

    Return now to the general case. Assume that X_1 and X_2 are open. Take a point x in X and a neighborhood $\mathcal{O}$ of $f(x)$. Assume without loss of generality that $x \in X_1$. Because the restriction of f to X_1 is continuous, there is a neighborhood $\mathcal{U}$ of x for which $\mathcal{U} \subseteq X_1$, and $f|_{X_1}(\mathcal{U}) \subseteq \mathcal{O}$. We want to show that $\mathcal{U}$ is open in X. Indeed, by definition of a topological subspace (page 222), $\mathcal{U}$ has the form $X_1 \cap G$ where G is open in X. Hence, $\mathcal{U}$ is open in X because $\mathcal{U}$ is the intersection of two open sets in X. We can now state that there is a neighborhood $\mathcal{U}$ of x for which $f(\mathcal{U}) \subseteq \mathcal{O}$. Thus, f is continuous on X.

    Lastly, refer to Proposition 5(ii) of Section 3.1. Inheritance of measurability from measurability of restrictions is more robust compared with inheritance of continuity from continuity of restrictions because no additional conditions need be placed on the subsets in the measurable-function case whereas in the continuous-function case, some condition, such as the subsets' being open, is required. See also Problem 2 of Section 3.1.

38. This is Problem 47 of Section 7.4 with p equal to 2.

    Because the inequality is used in the Mazur example, here is an appropriate place to verify other details of the example. Regarding the first inequality in the example, our solution to Problem 46 of Section 7.4 shows that we can in fact tighten the inequality there by replacing 2^p with 2^{p-1}. Then putting p equal to 2 gives us first inequality in the Mazur example. Most of the other details are given in our solution to Problem 48 of Section 7.4. Note that the different definitions of a continuous mapping used in Problem 48 of Section 7.4 and the Mazur example are equivalent. The last sentence of the example follows directly from definition of a homeomorphism.

11.5 Compact Topological Spaces

In the proof of Proposition 15, replace F with K.

Proposition 19 is evidently an inadvertent copy of Proposition 21.

39. Assume first that X is compact. Then every open cover of X has a finite subcover. In particular, every countable open cover has a finite subcover. Thus, X is countably compact. Note that this direction of the *if-and-only-if* statement does not require that X be second countable.

    Conversely, assume that X is countably compact. Take an open cover of X. Because X is second countable, there is a countable base $\mathcal{B}$ for the topology. By the sentence after the definition of a base for a topology (page 223), each set in the open cover is the union of a subcollection of $\mathcal{B}$. It follows that the collection of sets in the open cover is contained in $\mathcal{B}$ and therefore countable. Hence, because X is countably compact, the open cover has a finite subcover. Thus, X is compact.

40. We first prove the countable analogue of Proposition 14: A topological space X is countably compact if and only if every countable collection of closed subsets of X that possesses the finite intersection property has a nonempty intersection. Our proof proceeds along the lines of the proof of Proposition 14 of Section 9.5 and then adds details to the last step: If $\mathcal{T}$ is a countable collection of open subsets of X, then the collection $\mathcal{F}$ of complements of sets in $\mathcal{T}$ is a countable collection of closed sets. Moreover, $\mathcal{T}$ is a cover if and only if $\mathcal{F}$ has an empty intersection. So by De Morgan's first identity on page 4, X is countably compact if and only if every countable collection of closed sets with an empty intersection has a finite subcollection whose intersection is also empty. Now the condition for countable compactness is equivalent to the condition's contrapositive, which is "If no finite subcollection of $\mathcal{T}$ covers X, then $\mathcal{T}$ does not cover X," which in turn is equivalent to "If every countable collection $\mathcal{F}$ of closed subsets has the finite intersection property, then $\mathcal{F}$ has a nonempty intersection."

    We are now ready to prove the intersection theorem of Fréchet. Assume first that X is countably compact. Sequence $\{F_n\}$ is a countable collection of closed subsets. Moreover, $\{F_n\}$ possesses the finite intersection property because $\{F_n\}$ is descending and each F_n is nonempty. Thus, by the countable analogue of Proposition 14, $\bigcap_{n=1}^{\infty} F_n$ is nonempty.

    Conversely, assume that whenever $\{F_n\}$ is a descending sequence of nonempty closed subsets of X, intersection $\bigcap_{n=1}^{\infty} F_n$ is nonempty. Take a countable open cover $\{\mathcal{O}_n\}_{n=1}^{\infty}$ of X. Suppose, to get a contradiction, that there were no finite subcover. Then $X \sim \bigcup_{i=1}^{n} \mathcal{O}_i \neq \emptyset$ for each n. By De Morgan's first identity, $\bigcap_{i=1}^{n}(X \sim \mathcal{O}_i) \neq \emptyset$. Put F_n to $\bigcap_{i=1}^{n}(X \sim \mathcal{O}_i)$. Then $\{F_n\}$ is a descending sequence of nonempty closed subsets of X, so intersection $\bigcap_{n=1}^{\infty} F_n$ is nonempty. But that contradicts the fifth sentence of our solution (the sentence that begins "Moreover, ..."). Thus, X is countably compact.

41. Suppose, to get a contradiction, that there were no such N. Then $F_n \sim \mathcal{O}$ is nonempty for all n so that descending sequence $\{F_n \sim \mathcal{O}\}$ has the finite intersection property. Moreover, each $F_n \sim \mathcal{O}$ is a closed subset of X, which we prove by showing that complement $X \sim (F_n \sim \mathcal{O})$ is open: Using a set identity, $X \sim (F_n \sim \mathcal{O}) = (X \sim F_n) \cup \mathcal{O}$, which is open by definition (page 221) since $X \sim F_n$, being the complement of a closed set, is open. Then by Proposition 14, $\bigcap_{n=1}^{\infty}(F_n \sim \mathcal{O}) \neq \emptyset$. Now $\bigcap_{n=1}^{\infty}(F_n \sim \mathcal{O}) = (\bigcap_{n=1}^{\infty} F_n) \sim \mathcal{O} \subseteq \bigcap_{n=1}^{\infty} F_n \subseteq \mathcal{O}$ where the equality comes from another set identity, and the last inclusion is by hypothesis. But the result that nonempty set $(\bigcap_{n=1}^{\infty} F_n) \sim \mathcal{O}$ is a subset of $\mathcal{O}$ is absurd. Thus, there is an N such that $F_n \subseteq \mathcal{O}$ if $n \geq N$.

    Note that we do not need the hypothesis that X is Hausdorff (I think that the F_n in a previous edition of the text were taken to be compact instead of closed, in which case we would use Proposition 16 to get that the F_n are also closed).

42. Take a closed, bounded interval $[a, b]$ of real numbers. If $b = a$, then the claim is true trivially. Otherwise, $b > a$. It suffices to prove the claim for open interval (a, b); for if $[a, b]$ is a pairwise disjoint union of a collection of closed, bounded intervals, then we can remove the end intervals to obtain a cover of open interval (c, d) where $a < c < d < b$.

    Suppose, to get a contradiction, that (a, b) were a pairwise disjoint union $\bigcup_{n=1}^{\infty}[a_n, b_n]$ of a countable collection of closed, bounded intervals. The union is necessarily infinite because the union of a finite number of closed intervals is closed by Proposition 12 of Section 1.4, but (a, b) is open. Consider the set $\bigcup_{n=1}^{\infty}\{a_n, b_n\}$ of endpoints of those intervals. Denote that set by F. Set F is countably infinite. Furthermore, because $\bigcup_{n=1}^{\infty}[a_n, b_n]$ is a disjoint cover of (a, b), set F is the complement in (a, b) of open set $\bigcup_{n=1}^{\infty}(a_n, b_n)$ and is therefore closed (see the sentence after the proof of Proposition 4 of Section 9.2 where we take X to be (a, b), which is a metric space by the last sentence on page 183). Lastly, we claim that every point in F is an accumulation point of F (recall from Problem 30 of Section 1.4 that d is an accumulation point of F provided that d is a point of closure of $F \sim \{d\}$). Indeed, take a point d in F. Then d is either a left or a right endpoint of an interval $[a_n, b_n]$ but not both. Take an open interval that contains d. If d is a right endpoint, the open interval contains the left endpoint of some $[a_n, b_n]$ to the right of d. Hence, d is a point of closure of $F \sim \{d\}$ and therefore an accumulation point of F. If d is a left endpoint, the claim holds analogously.

    Now consider a closed interval $[c_1 - r_1, c_1 + r_1]$ in (a, b) such that $c_1 \in F$. Because c_1 is an accumulation point, there is a closed interval $[c_2 - r_2, c_2 + r_2]$ that is inside $[c_1 - r_1, c_1 + r_1]$, does not contain c_1, has center c_2 in F, and has radius r_2 less than $r_1/4$. Because c_2 is an accumulation point, there is a closed interval $[c_3 - r_3, c_3 + r_3]$ that is inside $[c_2 - r_2, c_2 + r_2]$, does not contain c_2, has center c_3 in F, and has radius r_3 less than $r_2/4$. Continuing in that fashion, we obtain a descending sequence $[c_1 - r_1, c_1 + r_1]$, $[c_2 - r_2, c_2 + r_2]$, $[c_3 - r_3, c_3 + r_3]$, $[c_4 - r_4, c_4 + r_4]$, $\ldots$ of closed intervals such that each c_n is in F, and $\{r_n\} \to 0$. Then $\{c_n\}$ is a Cauchy sequence. By Theorem 17 of Section 1.5, $\{c_n\}$ converges to some real number c. Because F is closed, c belongs to F by the second part of Proposition 6 of Section 9.2. Now $c \in \bigcap_{n=1}^{\infty}[c_n - r_n, c_n + r_n]$ (in fact, $\bigcap_{n=1}^{\infty}[c_n - r_n, c_n + r_n] = \{c\}$ by Cantor's intersection theorem). But that contradicts that $\bigcap_{n=1}^{\infty}[c_n - r_n, c_n + r_n]$ contains no point of F by construction. Thus, it is not possible to express a closed, bounded interval of real numbers as a pairwise disjoint union of a countable collection—having more than one member—of closed, bounded intervals.

43. We use the fact that g is continuous if and only if $g^{-1}(F)$ is closed in Y whenever F is closed in Z. We proved that for metric spaces in our solution to Problem 36 of Section 9.3—the proof is the same for topological spaces.

    Take a closed subset F in Z. Because $g \circ f$ is continuous, $(g \circ f)^{-1}(F)$ is closed in X. Because X is compact, $(g \circ f)^{-1}(F)$ is compact by Proposition 15. Because f is continuous, $f((g \circ f)^{-1})(F)$, which is $g^{-1}(F)$, is compact in Y by Proposition 20. Because Y is Hausdorff, $g^{-1}(F)$ is closed in Y by Proposition 16. Thus, g is continuous.

44. (i) Take an open cover of $(X, \mathcal{T}_1)$. Because $\mathcal{T}_1 \subseteq \mathcal{T}$, the open cover is contained in $\mathcal{T}$. So because $(X, \mathcal{T})$ is compact, the open cover has a finite subcover. Thus, $(X, \mathcal{T}_1)$ is compact.

    (ii) Take two points in $(X, \mathcal{T}_2)$. Because $(X, \mathcal{T})$ is Hausdorff, the two points can be separated by disjoint neighborhoods taken from $\mathcal{T}$. So because $\mathcal{T} \subseteq \mathcal{T}_2$, the two points can be separated by disjoint neighborhoods taken from $\mathcal{T}_2$. Thus, $(X, \mathcal{T}_2)$ is Hausdorff.

(iii) Take a topology $\mathcal{T}_1$ that is strictly weaker than topology $\mathcal{T}$. Consider identity mapping $id_X : (X, \mathcal{T}) \to (X, \mathcal{T}_1)$. Mapping id_X is one-to-one and onto. Moreover, id_X is continuous by Propsition 10 because the inverse image of an open subset in $(X, \mathcal{T}_1)$ is an open subset of $(X, \mathcal{T})$ since $\mathcal{T}_1 \subseteq \mathcal{T}$. Now if $(X, \mathcal{T}_1)$ were Hausdorff, then id_X would be a homeomorphism by Proposition 21. But that contradicts Problem 30(i)$\to$(ii) because there is an open subset E in $(X, \mathcal{T})$ such that $id_X(E)$, or E, is not open in $(X, \mathcal{T}_1)$ since $\mathcal{T}_1$ is strictly weaker than $\mathcal{T}$. Thus, $(X, \mathcal{T}_1)$ is not Hausdorff.

Next, take a topology $\mathcal{T}_2$ that is strictly stronger than $\mathcal{T}$. If $(X, \mathcal{T}_2)$ were compact, then by reasoning analogous to the previous, $id_X : (X, \mathcal{T}_2) \to (X, \mathcal{T})$ would be a homeomorphism, which contradicts that $\mathcal{T}_2$ is strictly stronger than $\mathcal{T}$. Thus, $(X, \mathcal{T}_2)$ is not compact.

45. (i) Take a point x in X. Put K to $\{x\}$. Set K is compact because every open cover of $\{x\}$ has a finite subcover, namely, one set in the cover that contains x. Take a neighborhood $\mathcal{O}$ of $f(x)$. Because $\{f_n\} \to f$ in Y^X with respect to the compact-open topology, there is an index N such that for neighborhood $\mathcal{U}_{K,\mathcal{O}}$ of f, if $n \geq N$, then f_n belongs to $\mathcal{U}_{K,\mathcal{O}}$, that is, $f_n(K) \subseteq \mathcal{O}$. In other words, if $n \geq N$, then $f_n(K)$, which is $f_n(x)$, belongs to $\mathcal{O}$. Thus, $\{f_n\} \to f$ pointwise on X.

(ii) Assume first that $\{f_n\} \to f$ with respect to the compact-open topology. Take a compact subset K of X. Take a positive ϵ. Because f is continuous, for each x in K, there is a neighborhood G_x of x for which $f(G_x) \subseteq B(f(x), \epsilon/3)$. Then

$$\begin{aligned} f(\overline{G_x}) &\subseteq \overline{f(G_x)} && \text{Problem 25(iv)} \\ &\subseteq \overline{B(f(x), \epsilon/3)} && \text{Problem 6} \\ &\subseteq B(f(x), \epsilon/2) && \text{continuity of metric (bottom of page 210).} \end{aligned}$$

Moreover, $\{G_x\}_{x \in K}$ is an open cover of K. But K is compact. So there is a finite subcover $\{G_{x_1}, G_{x_2}, \ldots, G_{x_m}\}$. For i equal to $1, \ldots, m$, define K_{x_i} by $K_{x_i} = \overline{G_{x_i}} \cap K$. Each K_{x_i} is compact by Proposition 15 because $\overline{G_{x_i}} \cap K$ is closed by Proposition 5 since K is closed by Proposition 16. Define set $\mathcal{U}$ by $\mathcal{U} = \bigcap_{i=1}^m \mathcal{U}_{K_{x_i}, B(f(x_i), \epsilon/2)}$. Set $\mathcal{U}$ is open in Y^X by definition of a subbase (page 223) and contains f because as we have shown above, $f(K_{x_i}) \subseteq f(\overline{G_{x_i}}) \subseteq B(f(x_i), \epsilon/2)$ for each i.

Now because $\{f_n\} \to f$ in Y^X with respect to the compact-open topology, there is an index N such that for neighborhood $\mathcal{U}$ of f, if $n \geq N$, then f_n belongs to $\mathcal{U}$, that is, $f_n(K_{x_i}) \subseteq B(f(x_i), \epsilon/2)$ for each i.

Lastly, take a point x in K. Point x is in some G_{x_j} and therefore also in K_{x_j}. Then if $n \geq N$,

$$\begin{aligned} \|f(x) - f_n(x)\| &\leq \|f(x) - f(x_j)\| + \|f(x_j) - f_n(x)\| && \text{triangle inequality} \\ &< \epsilon/2 + && f(K_{x_j}) \subseteq B(f(x_j), \epsilon/2) \\ &\quad \epsilon/2 && f_n(K_{x_j}) \subseteq B(f(x_j), \epsilon/2), n \geq N \\ &= \epsilon. \end{aligned}$$

Because that holds for all points in K, sequence $\{f_n\}$ converges to f uniformly on K. Conversely, assume that $\{f_n\} \to f$ uniformly on each compact subset of X. Take a neighborhood $\mathcal{U}$ of f. That neighborhood contains some neighborhood $\mathcal{U}_{K,\mathcal{O}}$ of f because (1) $\mathcal{U}$ contains the intersection of a finite collection of subbase sets by definitions of a subbase and a base; (2) the intersection of compact subsets of X is compact by Propositions 16, 5, and 15; and (3) the intersection of open subsets of Y is open by definition (page 221). Now because f is continuous, $f(K)$ is compact by Proposition 20.

By Problem 74 of Section 9.5, there is an open set H of Y for which $f(K) \subseteq H \subseteq \overline{H} \subseteq \mathcal{O}$. There is a positive ϵ small enough such that we can put H to set $\{y \in Y \mid \text{dist}(y, K) < \epsilon\}$ where we are using the definition and notation in the proof of Proposition 7 as well as the fact stated there that H so defined is indeed open.

Now because $\{f_n\} \to f$ uniformly on K, there is an N for which $\|f_n - f\| < \epsilon$ on K if $n \geq N$. Hence, if $n \geq N$, then $f_n(K) \subseteq H \subseteq \mathcal{O}$. In terms of the topology on Y^X, if $n \geq N$, then f_n belongs to $\mathcal{U}_{K,\mathcal{O}}$, which is a subset of the arbitrary neighborhood $\mathcal{U}$ of f. Thus, $\{f_n\} \to f$ in Y^X with respect to the compact-open topology.

46. (Cf. Problem 22 of Section 3.2.) Take a positive ϵ. For each n, let G_n be the set of points x in X such that $f_n(x) < \epsilon$. Set G_n is open by Proposition 10 since G_n is the inverse image under continuous function f_n of open set $(-1, \epsilon)$. Because $\{f_n(x)\}$ decreases monotonically to zero for each x, sequence $\{G_n\}$ is ascending, that is, $G_n \subseteq G_{n+1}$. We claim moreover that this countable collection $\{G_n\}$ of open sets covers X. Indeed, take a point x in X. Sequence $f_n(x)$ converges to zero by the monotone convergence criterion for real sequences (Section 1.5), so there is an index N for which if $n \geq N$, then $f_n(x) < \epsilon$. Hence, $x \in G_n$ when $n \geq N$, which proves our claim. Now because X is countably compact, $\{G_n\}$ has a finite subcover. Because $\{G_n\}$ is ascending, the largest set G_M in the subcover covers X, that is, $X \subseteq G_M$. Hence, $f_n < \epsilon$ on X if $n \geq M$. Thus, $\{f_n\}$ converges to zero uniformly.

11.6 Connected Topological Spaces

In the last sentence of the proof of Proposition 24, replace f with X.

47. Suppose, to get a contradiction, that the union of $\{C_\lambda\}_{\lambda \in \Lambda}$ were not connected. Let $\mathcal{O}_1$ and $\mathcal{O}_2$ be a separation of $\{C_\lambda\}_{\lambda \in \Lambda}$. Take a pair of sets $C_{\lambda'}$ and $C_{\lambda''}$ from $\{C_\lambda\}_{\lambda \in \Lambda}$. By hypothesis, they have a point x in common. Point x is in either $\mathcal{O}_1$ or $\mathcal{O}_2$ by definition of a separation. If $x \in \mathcal{O}_1$, then both $C_{\lambda'}$ and $C_{\lambda''}$ are contained in $\mathcal{O}_1$ by the second paragraph of the section because $C_{\lambda'}$ and $C_{\lambda''}$ are connected subsets of X. In fact, because every C_λ has a point in common with $C_{\lambda'}$, every C_λ is contained in $\mathcal{O}_1$. Hence, $\mathcal{O}_2 = \emptyset$, which contradicts definition of a separation. Thus, the union of $\{C_\lambda\}_{\lambda \in \Lambda}$ is connected. We reach the same conclusion if $x \in \mathcal{O}_2$.

48. Suppose, to get a contradiction, that B were not connected. Let $\mathcal{O}_1$ and $\mathcal{O}_2$ be a separation of B. Then A is contained in either $\mathcal{O}_1$ or $\mathcal{O}_2$ by the second paragraph of the section because A is a connected subset of X, and $A \subseteq B = \mathcal{O}_1 \cup \mathcal{O}_2$. If $A \subseteq \mathcal{O}_1$, then $B = B \cap \overline{A} \subseteq B \cap \overline{\mathcal{O}_1} = \mathcal{O}_1$ where the inclusion is true by Problem 6. Hence, $\mathcal{O}_2 = \emptyset$, which contradicts definition of a separation. Thus, B is connected. We reach the same conclusion if $A \subseteq \mathcal{O}_2$.

49. Denote the first set in the union by I and the second by S. Set S is connected by Proposition 23 because S is the image of continuous function $x \mapsto \sin 1/x$ of connected space $(0, 1]$. Our strategy is to show that $I \subseteq \overline{S}$ so that $S \subseteq X = I \cup S \subseteq \overline{S}$ and then apply Problem 48. So take a point $(0, y_0)$ in I and a neighborhood of $(0, y_0)$. By definition (page 186), that neighborhood contains an open ball $B((0, y_0), r)$. Choose a natural number m such that $1/(2m\pi) < r$. On interval $(0, 1/(2m\pi)]$ function $x \mapsto \sin 1/x$ takes every value between -1 and 1 because $\sin((4m+3)\pi/2) = -1$ and $\sin((4m+1)\pi/2) = 1$ and because interval $(0, 1/(2m\pi)]$ contains interval $[2/((4m+3)\pi), 2/((4m+1)\pi)]$. In particular, there is an x_0 in interval $(0, 1/(2m\pi)]$ for which $\sin 1/x_0 = y_0$, that is, $(x_0, y_0) \in S$. Furthermore,

$$\|(x_0, y_0) - (0, y_0)\| = x_0 \leq 1/(2m\pi) < r,$$

so (x_0, y_0) is also in $B((0, y_0), r)$. Consequently, every neighborhood of point $(0, y_0)$ contains a point in S. Hence, $(0, y_0)$ is a point of closure of S. Moreover, because our choice of $(0, y_0)$ is arbitrary, every point in I is a point of closure of S, that is, $I \subseteq \overline{S}$. Thus, X is connected.

Next, suppose, to get a contradiction, that X were arcwise connected. Then there is a continuous map $f : [0,1] \to X$ for which $f(0) = (0,0)$, and $f(1) = (1, \sin 1)$. The set of real numbers t for which $f(t) \in I$ is closed by Problem 25(i) and so has a maximum value t_{m}. There is another continuous map $g : [0,1] \to X$ for which $g(0) = (0, f(t_{\mathrm{m}}))$, and $g(1) = (1, \sin 1)$. If we express $g(s)$ as $(x(s), y(s))$, then $x(0) = 0$, and when $0 < s \leq 1$, we have that $0 < x(s) \leq 1$ and $y(s) = \sin 1/x(s)$. (Note that x and y with or without subscript are coordinates whereas $x(\cdot)$ and $y(\cdot)$ are continuous, real-valued functions.)

We now define a sequence $\{s_n\}$ of real numbers. It follows from our solution to the first part of this problem that for each n, there is a real number x_n between 0 and $x(1/n)$ for which $\sin 1/x_n = (-1)^n$. The intermediate value theorem of Section 1.6 allows us to choose s_n such that $x(s_n) = x_n$ where $0 < s_n < 1/n$. Then $\{s_n\} \to 0$, and $y(s_n) = (-1)^n$. But sequence $\{(-1)^n\}$ does not converge. That contradicts the continuity of function $s \mapsto y(s)$ (see Proposition 21 of Section 1.6). Thus, X is not arcwise connected.

50. We are proving the fourth-to-last sentence of the section. Suppose, to get a contradiction, that X were not connected. Let $\mathcal{O}_1$ and $\mathcal{O}_2$ be a separation of X. Take a point u in $\mathcal{O}_1$ and another point v in $\mathcal{O}_2$. There is a continuous map $f : [0,1] \to X$ for which $f(0) = u$ and $f(1) = v$. Because $[0,1]$ is connected, $f([0,1])$ is connected by Proposition 23. Then $f([0,1])$ is contained in either $\mathcal{O}_1$ or $\mathcal{O}_2$ by the second paragraph of the section. If $f([0,1]) \subseteq \mathcal{O}_1$, then v is also in $\mathcal{O}_1$, which contradicts that $\mathcal{O}_1$ and $\mathcal{O}_2$ are disjoint. Thus, X is connected. We reach the same conclusion if $f([0,1]) \subseteq \mathcal{O}_2$.

    Next, if $\mathcal{O}$ is empty, then it is true vacuously that $\mathcal{O}$ is arcwise connected. So assume that $\mathcal{O}$ is not empty. Proceed as in the hint. Set C is not empty because x itself can be connected to x by a (trivial) linear arc.

    Set C is open in $\mathcal{O}$ provided for every point c in C, there is an open ball centered at c that is contained in C (refer to the definition on page 186). Because $C \subseteq \mathcal{O}$ and $\mathcal{O}$ is open, there is an open ball $B(c,r)$ that is contained in $\mathcal{O}$. For a point c' in $B(c,r)$, define map $g : [0,1] \to \mathbf{R}^n$ by $g(t) = (1-t)c' + tc$. Then g is a continuous map for which $g(0) = c'$ and $g(1) = c$, that is, g is a linear arc that connects c' to c. Moreover, image $g([0,1])$ is contained in $B(c,r)$ because if $0 \leq t \leq 1$, then $\|((1-t)c' + tc) - c\| = (1-t)\|c' - c\| < (1-t)r \leq r$. Then $g([0,1])$ is also contained in $\mathcal{O}$ because $B(c,r) \subseteq \mathcal{O}$. Now because $c \in C$, there is a piecewise linear arc $h : [0,1] \to \mathcal{O}$ for which $h(0) = c$ and $h(1) = x$. Consequently, map $i : [0,1] \to \mathcal{O}$ defined by

$$i(t) = \begin{cases} g(2t) & 0 \leq t \leq 1/2, \\ h(2t-1) & 1/2 < t \leq 1 \end{cases}$$

    is a piecewise linear arc in $\mathcal{O}$ that connects c' to x, which shows that $c' \in C$. Then $B(c,r) \subseteq C$ because c' is an arbitrary point in $B(c,r)$. Hence, C is open in $\mathcal{O}$.

    Set C is closed in $\mathcal{O}$ if C contains all of its points of closure (refer to the definition on page 187 and the subsequent two sentences). So take a point c in $\overline{C}$. Because c is also in $\mathcal{O}$ by definition, there is an open ball $B(c,r)$ that is contained in $\mathcal{O}$. And because c is a point of closure of C, neighborhood $B(c,r)$ of c contains a point c'' in C. As before, there is a linear arc in $\mathcal{O}$ that connects c to c'' and a piecewise linear arc in $\mathcal{O}$ that connects c'' to x, so there is a piecewise linear arc in $\mathcal{O}$ that connects c to x, which shows that $c \in C$. Hence, C is closed in $\mathcal{O}$.

    To conclude, the fourth sentence of the section tells us that $C = \mathcal{O}$ because $\mathcal{O}$ is connected and C is both open and closed in $\mathcal{O}$ and not empty. Now the choice of x in $\mathcal{O}$ is arbitrary, and every point in C can be connected in $\mathcal{O}$ to x by a piecewise linear arc. In other words, for each pair of points in $\mathcal{O}$, there is a continuous map that connects the two points. Thus, $\mathcal{O}$ is arcwise connected.

51. Apply Proposition 23. Mapping $f : [0, 2\pi) \to \mathbf{R}^2$ defined by $f(x) = (\cos x, \sin x)$ is a continuous mapping of connected space $[0, 2\pi)$ to topological space $\mathbf{R}^2$, and f's image $f([0, 2\pi))$ is C. Thus, C is connected.

52. Take a pair of points u, v in $\mathbf{R}^n$. Define map $f : [0, 1] \to \mathbf{R}^n$ by $f(t) = (1 - t)u + tv$; in words, $f([0, 1])$ is the line segment that connects u and v. Then f is a continuous map for which $f(0) = u$ and $f(1) = v$. Hence, $\mathbf{R}^n$ is arcwise connected. By Problem 50, $\mathbf{R}^n$ is connected.

53. Assume first that (X, ρ) fails to be connected. By the fourth sentence of the section, we can put A to a nontrivial subset of X that is both open and closed. By Problem 59 of Section 9.5, A is compact. Put B to $X \sim A$. Then B is closed by the sentence after the proof of Proposition 4 of Section 9.2, and A and B are disjoint, nonempty subsets whose union is X. Put ϵ to $\inf\{\rho(u, v) \mid u \in A,\ v \in B\}$. Then by Problem 73 of Section 9.5, $\epsilon > 0$, and $\rho(u, v) \geq \epsilon$ for all u in A, v in B.

    Conversely, assume the *if* part of the problem statement. By definition of a separation, it remains to show that A and B are open in X. We do so by appealing to the definition on page 186. Take a point x in A. Open ball $B(x, \epsilon/2)$ contains no points of subset B by assumption. Then $B(x, \epsilon/2)$ is contained in A because $B(x, \epsilon/2) \subseteq X$ by definition of an open ball and because $A \cup B = X$ by assumption. Hence, A is open in X. Likewise, B is open in X. Thus, (X, ρ) fails to be connected. Note that for this direction of the *if-and-only-if* statement, it is not necessary that (X, ρ) be compact.

    A counterexample for the noncompact case is the union $(0, 1) \cup (1, 2)$, say Y, of two open intervals in $\mathbf{R}$. Topological space Y is not compact by the contrapositive of Theorem 20(ii)$\to$(i) of Section 9.5 because Y is not closed in $\mathbf{R}$. Moreover, Y fails to be connected because $(0, 1)$ and $(1, 2)$ separate Y. Put A to $(0, 1)$ and B to $(1, 2)$ so that A and B are disjoint, nonempty subsets whose union is Y. Take an ϵ. Put u to $1 - \epsilon/4$ and v to $1 + \epsilon/4$. Then $|v - u| = |1 + \epsilon/4 - (1 - \epsilon/4)| = \epsilon/2 \not\geq \epsilon$. Thus, the claim is not necessarily true for noncompact spaces.

54. (i) Our solution is like our solution to the second part of Problem 50. Let u belong to X. Take an ϵ. Define C to be the set of points v in X for which there is a finite number of points $x_0, x_1, \ldots, x_{n-1}, x_n$ in X such that $x_0 = u$, $x_n = v$, and $\rho(x_{i-1}, x_i) < \epsilon$ where $i = 1, \ldots, n$. Set C is not empty because it contains u: Just put v to u and n to 1 so that $\rho(x_0, x_1) = \rho(u, v) = \rho(u, u) = 0 < \epsilon$. Hence, $u \in C$.

    Set C is open provided for every point v in C, open ball $B(v, \epsilon)$ is contained in C. Because $v \in C$, there is a finite number of points $x_0, x_1, \ldots, x_{n-1}, x_n$ in X such that $x_0 = u$, $x_n = v$, and $\rho(x_{i-1}, x_i) < \epsilon$ where $i = 1, \ldots, n$. For a point v' in $B(v, \epsilon)$, let x_{n+1} equal v'. Then there is a finite number of points $x_0, x_1, \ldots, x_n, x_{n+1}$ in X such that $x_0 = u$, $x_n = v$, $x_{n+1} = v'$, and $\rho(x_{i-1}, x_i) < \epsilon$ where $i = 1, \ldots, n+1$, which shows that $v' \in C$. Then $B(v, \epsilon) \subseteq C$ because v' is an arbitrary point in $B(v, \epsilon)$. Hence, C is open.

    Set C is closed if it contains all of its points of closure. So take a point v in $\overline{C}$. Because v is a point of closure of C, neighborhood $B(v, \epsilon)$ of v contains a point v'' in C. As before, there is a finite number of points $x_0, x_1, \ldots, x_n, x_{n+1}$ in X such that $x_0 = u$, $x_n = v''$, $x_{n+1} = v$, and $\rho(x_{i-1}, x_i) < \epsilon$ where $i = 1, \ldots, n+1$, which shows that $v \in C$. Hence, C is closed.

    To conclude, the fourth sentence of the section tells us that $C = X$ because X is connected and C is both open and closed and not empty. Now the choices of u and ϵ are arbitrary. Hence, for each pair of points u, v in X and each ϵ, there is a finite number of points $x_0, x_1, \ldots, x_{n-1}, x_n$ in X such that $x_0 = u$, $x_n = v$, and $\rho(x_{i-1}, x_i) < \epsilon$ where $i = 1, \ldots, n$. Thus, X is well chained.

For a counterexample to the converse, put X to the union $(0,1) \cup (1,2)$ of two open intervals in $\mathbf{R}$. To show that X is well chained, take a pair of real numbers u, v in X. If $v < u$, rename them so that $u < v$. Take an ϵ. Iteratively put x_i to $x_{i-1} + \epsilon/2$ where $i = 1, 2, \ldots$ until some x_i, say x_j, reaches or surpasses v or until x_j equals 1. If x_j reaches or surpasses v, then redefine, if necessary, x_j as v so that $n = j$. If x_j equals 1, then redefine x_j as $x_{j-1} + \epsilon/3$, and continue as before, iteratively putting x_i to $x_{i-1} + \epsilon/2$ where $i = j+1, j+2, \ldots$ until some x_i, say x_k, reaches or surpasses v. Then redefine, if necessary, x_k as v so that $n = k$. We therefore have a finite number of real numbers $x_1, \ldots, x_{n-1}, x_n$ in X such that $|x_i - x_{i-1}| < \epsilon$ where $i = 1, \ldots, n$. Hence, X is well chained. But X is not connected because $(0,1)$ and $(1,2)$ separate X.

(ii) Suppose, to get a contradiction, that X were not connected. By the fourth sentence of the section, there is a nontrivial subset A of X that is both open and closed. By Problem 59 of Section 9.5, A is compact. Define subset B by $B = X \sim A$. Then B is closed by the sentence after the proof of Proposition 4 of Section 9.2, and $A \cap B = \emptyset$. We claim that there is a point u in A and a point v in B for which

$$\rho(u, v) = \min\{\rho(a, b) \mid a \in A,\ b \in B\} > 0.$$

Indeed, by Problem 73 of Section 9.5,

$$\operatorname{dist}(A, B) = \inf\{\rho(a, b) \mid a \in A,\ b \in B\} > 0.$$

It remains to show that $\operatorname{dist}(A, B)$ takes a minimum value on $A \cup B$. By Problem 33(i) of Section 9.3, dist is continuous at each point in A, so dist is continuous. Union $A \cup B$ is compact because $A \cup B$ is X, which is compact by hypothesis. Then dist takes a minimum value by the extreme value theorem of Section 9.5. That proves our claim. (Another way to prove our claim uses sequences in A and B.)

Because X is well chained, there is a finite number of points $x_0, x_1, \ldots, x_{n-1}, x_n$ in X such that $x_0 = u$, $x_n = v$, and $\rho(x_{i-1}, x_i) < \rho(u, v)$ where $i = 1, \ldots, n$. Denote by i_{m} the greatest index i such that $x_i \in A$. Then $x_{i_{\mathrm{m}}} \in A$, and $x_{i_{\mathrm{m}}+1} \in B$. Our contradiction is that $\rho(u, v) \le \rho(x_{i_{\mathrm{m}}}, x_{i_{\mathrm{m}}+1}) < \rho(u, v)$. Thus, X is connected.

55. Take a pair of points (u_1, u_2) and (v_1, v_2) in $\mathbf{R}^2 \sim \{(x, y)\}$. If point (x, y) does not lie on the line segment that connects (u_1, u_2) and (v_1, v_2), then as in our solution to Problem 52, define map $f : [0,1] \to \mathbf{R}^2 \sim \{(x, y)\}$ by $f(t) = ((1-t)u_1 + tv_1, (1-t)u_2 + tv_2)$. Otherwise, choose a point (w_1, w_2) that is not on the line through points (u_1, u_2) and (v_1, v_2). Define f instead by

$$f(t) = \begin{cases} ((1-2t)u_1 + 2tw_1, (1-2t)u_2 + 2tw_2) & 0 \le t \le 1/2, \\ ((2-2t)w_1 + (2t-1)v_1, (2-2t)w_2 + (2t-1)v_2) & 1/2 < t \le 1; \end{cases}$$

in words, $f([0, 1/2])$ is the line segment that connects points (u_1, u_2) and (w_1, w_2), and $f([1/2, 1])$ is the line segment that connects points (w_1, w_2) and (v_1, v_2). In either situation, f is a continuous map for which $f(0) = (u_1, u_2)$ and $f(1) = (v_1, v_2)$. Hence, $\mathbf{R}^2 \sim \{(x, y)\}$ is arcwise connected. By Problem 50, $\mathbf{R}^2 \sim \{(x, y)\}$ is connected.

Now suppose, to get a contradiction, that mapping $g : \mathbf{R}^2 \to \mathbf{R}$ were a homeomorphism. The restriction g_{r} of g to $\mathbf{R}^2 \sim \{(0,0)\}$ would then also be a homeomorphism from $\mathbf{R}^2 \sim \{(0,0)\}$ onto $\mathbf{R} \sim \{g((0,0))\}$. Because space $\mathbf{R}^2 \sim \{(0,0)\}$ is connected by the first part of this problem, image $g_{\mathrm{r}}(\mathbf{R}^2 \sim \{(0,0)\})$, which is $\mathbf{R} \sim \{g((0,0))\}$, would be connected by Proposition 23. But that contradicts that $\mathbf{R} \sim \{g((0,0))\}$ is *not* connected by the contrapositive of Problem 56(iii)→(i). Thus, $\mathbf{R}$ is not homeomorphic to $\mathbf{R}^2$.

56. We first verify that assertions (i) and (ii) are equivalent. Take two real numbers x_1, x_2 in C. Recall (page 9) that C is an interval provided all real numbers that lie between x_1 and x_2 also belong to C. On the other hand, C is convex provided for each λ where $0 \leq \lambda \leq 1$, real number $\lambda x_1 + (1-\lambda)x_2$ is in C. But $\lambda x_1 + (1-\lambda)x_2$ where λ runs from 0 to 1 represents all real numbers between x_1 and x_2 (see the first sentence of the third paragraph of Section 6.6). Definitions of an interval and convexity are therefore equivalent. Hence, assertions (i) and (ii) are equivalent.

Next, we show that an open interval (a, b) of real numbers is connected. Suppose, to get a contradiction, that (a, b) were not connected. Let $\mathcal{O}_1$ and $\mathcal{O}_2$ be a separation of (a, b). Take a real number x_1 in $\mathcal{O}_1$ and another one x_2 in $\mathcal{O}_2$. Interchange subscripts 1 and 2 if necessary so that $x_1 < x_2$. Denote $\sup(\mathcal{O}_1 \cap (a, x_2))$ by c. Then $a < c < b$ so that $c \in (a, b) = \mathcal{O}_1 \cup \mathcal{O}_2$. If $c \in \mathcal{O}_1$, then $c \neq x_2$ because $\mathcal{O}_1$ and $\mathcal{O}_2$ are disjoint, and $c \not> x_2$ by definition of c; therefore, $a < c < x_2$ so that $c \in \mathcal{O}_1 \cap (a, x_2)$. Because $\mathcal{O}_1 \cap (a, x_2)$ is open, there is a positive r_1 for which $(c - r_1, c + r_1) \subseteq \mathcal{O}_1 \cap (a, x_2)$. But that is absurd because although c is the supremum of $\mathcal{O}_1 \cap (a, x_2)$, a real number greater than c (a number between c and $c + r_1$) is in $\mathcal{O}_1 \cap (a, x_2)$. Consequently, c is not in $\mathcal{O}_1$ but rather in $\mathcal{O}_2$. It follows that $c = x_2$. Because $\mathcal{O}_2$ is open, there is a positive r_2 for which $(c - r_2, c + r_2) \subseteq \mathcal{O}_2$. That results in absurdity $c = \sup(\mathcal{O}_1 \cap (a, x_2)) \leq c - r_2$ because $c \notin \mathcal{O}_1$. Our contradiction, then, is that c is in (a, b) but not in $\mathcal{O}_1$ or $\mathcal{O}_2$. Interval (a, b) is therefore connected.

By Problem 48, $(a, b]$, $[a, b)$, and $[a, b]$ are also connected because $\overline{(a, b)} = [a, b]$, and

$$
\begin{aligned}
(a, b) &\subseteq (a, b] \subseteq [a, b], \\
(a, b) &\subseteq [a, b) \subseteq [a, b], \\
(a, b) &\subseteq [a, b] \subseteq [a, b].
\end{aligned}
$$

Hence, assertion (i) implies assertion (iii).

Here is a simpler proof that an interval C of real numbers is connected. By Problem 54 of Section 1.6, C has the intermediate value property. Interval C is therefore connected by Proposition 24.

Next, we show that a connected set C of real numbers is an interval. If C has only a single real number c, then C is (degenerate) interval $[c, c]$. Assume now that C has more than one real number. Take two of them. If there is a real number d between them that is not in C, then consider sets $C \cap (-\infty, d)$ and $C \cap (d, \infty)$. They are open in the topology that C inherits from $\mathbf{R}$ (page 222) and also satisfy the other conditions to be a separation of C. But that contradicts that C is connected. So for each pair of real numbers in C, all real numbers that lie between that pair also belong to C. Connected set C is therefore an interval by definition. Hence, assertion (iii) implies assertion (i).

Thus, the three assertions in (1) are equivalent.

Chapter 12

Topological Spaces: Three Fundamental Theorems

12.1 Urysohn's Lemma and the Tietze Extension Theorem

In the fourth line on page 240, replace the second occurrence of $f^{-1}(I_1)$ with $f^{-1}(I_2)$.

1. This is Problem 33 of Section 9.3. Note that d_C is continuous even if C is not closed.

2. Function $f : (0,1) \to (1,\infty)$ defined by $f(x) = 1/x$ is continuous by our solution to Problem 65(ii) of Section 9.5. If f had a continuous extension to $\mathbf{R}$, then, in particular, $f(0)$ would be some positive real number. Refer now to the definition of continuity at the beginning of Section 1.6. No δ responds to the ϵ challenge when x is 0 because if $x' = \min\{1/(f(0)+\epsilon), \delta/2\}$, then $|x' - x| = \min\{1/(f(0)+\epsilon), \delta/2\} < \delta$, but $|f(x') - f(x)| \geq f(0) + \epsilon - f(0) = \epsilon$. Thus, f does not have a continuous extension to $\mathbf{R}$. That does not contradict Tietze's extension theorem because $(0,1)$ is not a closed subset of $\mathbf{R}$.

   Another counterexample is function $x \mapsto \sin(1/x)$. See the last part of our solution to Problem 49 of Section 11.6.

3. Union $A \cup B$ is a closed subset of X by Proposition 5 of Section 11.1. Define real-valued function $f : A \cup B \to [a,b]$ by $f = a$ on A and $f = b$ on B. Function f is continuous according to the third sentence after Urysohn's lemma. We justify that claim. Take a closed subset C of $[a,b]$. Then

$$f^{-1}(C) = \begin{cases} \emptyset & a \notin C,\ b \notin C, \\ A & a \in C,\ b \notin C, \\ B & a \notin C,\ b \in C, \\ A \cup B & a \in C,\ b \in C. \end{cases}$$

   In all cases, the inverse image under f of C is closed in X. Hence, f is continuous by Problem 25(i) of Section 11.4. By Tietze's extension theorem, f has a continuous extension to all of X that takes values in $[a,b]$.

4. A function f on a topological space F with values in $\mathbf{R}^n$ is called a *real-vector-valued function* and has the form $(f_1, \ldots, f_n)$ where the $f_i : F \to \mathbf{R}$ are called *coordinate functions*.

   We first prove a lemma.

   Lemma Let F be a topological space. Let f be a real-vector-valued function on F. Then f is continuous if and only if each coordinate function f_i is continuous.

   Proof Take a point x in F. Assume first that f is continuous. Take a neighborhood $\mathcal{O}_i$ of $f_i(x)$ for each i. Because f is continuous, there is neighborhood $\mathcal{U}$ of x for which $f(\mathcal{U}) \subseteq \mathcal{O}_1 \times \cdots \times \mathcal{O}_n$. Then $\mathcal{U}$ is a neighborhood of x for which $f_i(\mathcal{U}) \subseteq \mathcal{O}_i$ for each i. Hence, each f_i is continuous.

   Conversely, assume that each f_i is continuous. Then f is continuous by n applications of the first step in our solution to Problem 28 of Section 11.4. □

   We now state and prove two versions of Tietze's extension theorem.

   Tietze's extension theorem, bounded version Let X be a normal topological space, F a closed subset of X, and f a continuous real-vector-valued function on F that takes values in closed, bounded region $[a_1, b_1] \times \cdots \times [a_n, b_n]$. Then f has a continuous extension to all of X that also takes values in $[a_1, b_1] \times \cdots \times [a_n, b_n]$.

Proof Each coordinate function f_i is a continuous real-valued function on F (by the lemma) that takes values in closed, bounded interval $[a_i, b_i]$. By Tietze's extension theorem, each f_i has a continuous extension to all of X that also takes values in $[a_i, b_i]$. Thus, by the lemma's converse direction, f has a continuous extension to all of X that takes values in $[a_1, b_1] \times \cdots \times [a_n, b_n]$. □

Tietze's extension theorem, not-necessarily-bounded version (generalization of Problem 8) Let X be a normal topological space, F a closed subset of X, and f a continuous real-vector-valued function on F. Then f has a continuous extension to a real-vector-valued function $\overline{f}$ on all of X.

Proof Each coordinate function f_i is a continuous real-valued function on F by the lemma. By Problem 8, each f_i has a continuous extension to a real-vector-valued function $\overline{f_i}$ on all of X. Thus, by the lemma's converse direction, f has a continuous extension to a real-vector-valued function $\overline{f}$ on all of X. □

5. Justifications of some of the statements in this solution are in the paragraph after Urysohn's lemma and in our solution to Problem 3. Take two disjoint closed subsets A and B of X. Define real-valued function $f : A \cup B \to [0, 4]$ by $f = 1$ on A and $f = 3$ on B. Because f is a continuous, bounded real-valued function on closed subset $A \cup B$, function f has a continuous extension to all of X by hypothesis. Consider disjoint open intervals $(0, 2)$ and $(2, 4)$. Inverse images $f^{-1}((0, 2))$ and $f^{-1}((2, 4))$ are disjoint neighborhoods of A and B—disjoint by the contrapositive of Proposition 23 of Section 11.6 and neighborhoods, i.e., open, by Proposition 10 of Section 11.4. Thus, X is normal.

6. By Proposition 13 of Section 11.4, the weak topology induced by $\mathcal{F}$ is a subset of $\mathcal{T}$.

   To get the inclusion in the other direction, we show that an open set G in $\mathcal{T}$ is also open in the weak topology induced by $\mathcal{F}$. If X consists of just one point, then it is true trivially that G is open in the weak topology induced by $\mathcal{F}$. Assume now that X has more than one point. Apply Proposition 1 of Section 11.1. Take a point x in G. Singleton set $\{x\}$ is closed by Proposition 6 of Section 11.2 because X is Tychonoff by definition of normality. Subset $X \sim G$ is closed by the sentence after the proof of Proposition 4 of Section 11.1. Then $\{x\}$ and $X \sim G$ are two nonempty, disjoint (because $x \in G$), closed subsets of X. By Urysohn's lemma, there is a function f in $\mathcal{F}$ that takes values in $[0, 2]$ while $f = 0$ on $X \sim G$ and $f(x) = 2$. Consider inverse image $f^{-1}((1, 2])$. It is an open subset of X by Proposition 10 of Section 11.4 because $(1, 2]$ is an open subset in $[0, 2]$. Then, $f^{-1}((1, 2])$ is a neighborhood of x. Moreover, $f^{-1}((1, 2])$ is contained in G. Hence, G is open in the weak topology induced by $\mathcal{F}$, that is, $\mathcal{T}$ is a subset of the weak topology induced by $\mathcal{F}$. Thus, $\mathcal{T}$ *is* the weak topology induced by $\mathcal{F}$.

7. First verify that ρ satisfies the properties of a metric—refer to its definition on page 182. Set X is nonempty because a topological space is nonempty by definition (top of page 222). Function ρ is from $X \times X$ to $\mathbf{R}$ by definition of ρ.

   (i) $\rho(x, y) \geq 0$ because ρ is a sum of nonnegative values.

   (ii) If $x = y$, then $f_{n,m}(x) - f_{n,m}(y) = 0$ for each (n, m) in A. Hence, $\rho(x, y) = 0$. Conversely, assume that $\rho(x, y) = 0$. Suppose, to get a contradiction, that $x \neq y$. Then singleton sets $\{x\}$ and $\{y\}$ are disjoint. Moreover, they are closed by Proposition 6 of Section 11.2. Then because X is normal, there is an open set that contains x but not y. By definition of a base for a topology (pages 222–223), there is a basic neighborhood $\mathcal{U}_m$ of x that is contained in that open set. By Proposition 8 of Section 11.2, there is an open set $\mathcal{O}$ for which $\{x\} \subseteq \mathcal{O} \subseteq \overline{\mathcal{O}} \subseteq \mathcal{U}_m$. There is another basic neighborhood $\mathcal{U}_n$ of x for which $\mathcal{U}_n \subseteq \mathcal{O}$. By Problem 6 of Section 11.1, $\overline{\mathcal{U}_n} \subseteq \overline{\mathcal{O}}$. Altogether, $x \in \mathcal{U}_n \subseteq \overline{\mathcal{U}_n} \subseteq \overline{\mathcal{O}} \subseteq \mathcal{U}_m$, so $(n,m) \in A$, and $f_{n,m}(x) = 0$. Furthermore, $f_{n,m}(y) = 1$ because $y \notin \mathcal{U}_m$, that is, $y \in X \sim \mathcal{U}_m$. Hence, $|f_{n,m}(x) - f_{n,m}(y)| = 1$, so $\rho(x,y) \neq 0$, which contradicts the assumption that $\rho(x,y) = 0$. Thus, $x = y$.

(iii) $\rho(x,y) = \rho(y,x)$ because $|f_{n,m}(x) - f_{n,m}(y)| = |f_{n,m}(y) - f_{n,m}(x)|$ for each (n,m) in A.

(iv) The triangle inequality for ρ follows from the triangle inequality for absolute value.

The previous shows that ρ is a metric on X and therefore that (X,ρ) is a metric space. It remains to show that ρ defines the same topology as the given topology. We do so by verifying the two assertions at the end of the proof of Urysohn's metrization theorem.

(i) By the same reasoning as in our verification of property (ii) of a metric, there is a $\mathcal{U}_m$ for which $x \in \mathcal{U}_m \subseteq \overline{\mathcal{U}_m} \subseteq \mathcal{U}_n$ (but note the reversal of the roles of m and n). Then $(m,n) \in A$. Our approach is to show that for each point x' in $B_\rho(x,\epsilon)$, we have that $f_{m,n}(x') < 1$, in which case $x' \notin X \sim \mathcal{U}_n$, that is, $x' \in \mathcal{U}_n$. Put ϵ to $1/2^{m+n}$. Then

$$\begin{aligned}
f_{m,n}(x') &= f_{m,n}(x') - f_{m,n}(x) && f_{m,n}(x) = 0 \text{ because } x \in \overline{\mathcal{U}_m} \\
&= \frac{1}{\epsilon}\cdot\frac{1}{2^{m+n}}(f_{m,n}(x') - f_{m,n}(x)) && \frac{1}{\epsilon}\cdot\frac{1}{2^{m+n}} = 1 \\
&\leq \frac{1}{\epsilon}\sum_{(i,j)\in A}\frac{1}{2^{i+j}}|f_{i,j}(x') - f_{i,j}(x)| && (m,n) \in A \\
&= (1/\epsilon)\rho(x',x) && \text{definition (6)} \\
&< (1/\epsilon)\epsilon && \rho(x',x) < \epsilon \text{ since } x' \in B_\rho(x,\epsilon) \text{ (page 186)} \\
&= 1.
\end{aligned}$$

Thus, because $x' \in \mathcal{U}_n$ for each x' in $B_\rho(x,\epsilon)$, we have that $B_\rho(x,\epsilon) \subseteq \mathcal{U}_n$.

(ii) Distance to $\{x\}$ function $d_{\{x\}} : X \to [0,\infty)$ is continuous by Problem 1. By definition of a continuous function (page 229), for each neighborhood $[0,\epsilon)$ of $d_{\{x\}}(x)$, there is a neighborhood U of x for which $d_{\{x\}}(U) \subseteq [0,\epsilon)$. Then $U \subseteq d_{\{x\}}^{-1}([0,\epsilon))$. Now there is a basic neighborhood $\mathcal{U}_n$ of x for which $\mathcal{U}_n \subseteq U$. Observe also that $d_{\{x\}}^{-1}([0,\epsilon)) = B_\rho(x,\epsilon)$. Hence, $\mathcal{U}_n \subseteq B_\rho(x,\epsilon)$. Thus, for each positive ϵ, there is a $\mathcal{U}_n$ such that $x \in \mathcal{U}_n$, and $\mathcal{U}_n \subseteq B_\rho(x,\epsilon)$.

8. (i) Function $f\cdot(1+|f|)^{-1}$ is a real-valued function on F that takes values in closed, bounded interval $[-1,1]$. To apply Tietze's extension theorem to obtain h, it remains to show that $f\cdot(1+|f|)^{-1}$ is continuous. Indeed, $|f|$ is continuous by Proposition 11 of Section 11.4 because $|f|$ is the composition of real-valued, continuous function f and the absolute value function, which is continuous. Then $1+|f|$ is continuous by Problem 28 of Section 11.4. In turn, $(1+|f|)^{-1}$ is continuous because it is the composition of positive, continuous function $1+|f|$ and function $x \mapsto x^{-1}$, which is continuous on $(0,\infty)$ by our solution to Problem 65(ii) of Section 9.5. Finally, $f\cdot(1+|f|)^{-1}$ is continuous by another application of Problem 28 of Section 11.4.

(ii) Singleton sets $\{1\}$ and $\{-1\}$ are closed in $[-1,1]$ by Proposition 6 of Section 11.2, so $h^{-1}(1)$ and $h^{-1}(-1)$ are closed in X by Problem 25(i) of Section 11.4. Therefore $F \cup h^{-1}(1) \cup h^{-1}(-1)$ is a closed subset of X by Proposition 5 of Section 11.1. Moreover, F is disjoint from $h^{-1}(1) \cup h^{-1}(-1)$ because $f\cdot(1+|f|)^{-1} \neq 1$ or -1. Define function $\psi : F \cup h^{-1}(1) \cup h^{-1}(-1) \to [0,1]$ by $\psi = 1$ on F and $\psi = 0$ on $h^{-1}(1)$ and $h^{-1}(-1)$.

Function ψ is continuous because it is constant on each of disjoint, closed subsets F and $h^{-1}(1) \cup h^{-1}(-1)$. By Tietze's extension theorem (or Urysohn's lemma), ψ has a continuous extension ϕ to all of X that takes values in $[0,1]$.

(iii) Function $\overline{f}$ is continuous because the functions that comprise its definition are continuous and because sums, products, and compositions, when defined, of continuous functions are continuous (see our solution to part (i) above). On F,

$$\overline{f} = \frac{1 \cdot h}{1 - 1 \cdot |h|} = \frac{f \cdot (1 + |f|)^{-1}}{1 - |f \cdot (1 + |f|)^{-1}|} = f.$$

Denominator $1 - |h|$ is not 0 because $|h| \neq 1$ on F as we noted in our solution to part (ii) above. It remains to show that $\overline{f}$ is real valued on $X \sim F$, which is the case if denominator $1 - \phi \cdot |h|$ is not 0 on $X \sim F$. Now $1 - \phi \cdot |h| = 0$ if and only if both ϕ and $|h|$ equal 1. But if $|h| = 1$, then $\phi = 0$. So $\overline{f}$ is real-valued on $X \sim F$, hence all of X.

Thus, f has a continuous extension to real-valued function $\overline{f}$ on all of X.

9. The first part of this problem is equivalent to Problem 23 of Section 11.4.

For the second part, assume first that for each real number c in (a,b), sets $\{x \in X \mid f(x) < c\}$ and $\{x \in X \mid f(x) > c\}$ are open. Observe that—

$$\{x \in X \mid f(x) < c\} = f^{-1}([a,c)),$$
$$\{x \in X \mid f(x) > c\} = f^{-1}((c,b]).$$

Then by the second example on page 223, there is a subbase for the topology on $[a,b]$ such that the preimage under f of each set in the subbase is open in X. Thus, f is continuous.

Conversely, assume that f is continuous. Then by Proposition 10 of Section 11.4, the preimage under f of every open subset in $[a,b]$—in particular, subsets of the form $[a,c)$ and $(c,b]$—is an open subset of X. Thus, for each real number c in (a,b), sets $\{x \in X \mid f(x) < c\}$ and $\{x \in X \mid f(x) > c\}$ are open.

10. We first establish two elementary facts:

$$\text{If } x \notin \mathcal{O}_\lambda, \text{ then } f(x) \geq \lambda. \tag{12.1}$$
$$\text{If } x \in \overline{\mathcal{O}_\lambda}, \text{ then } f(x) \leq \lambda. \tag{12.2}$$

To prove the first, observe that if $x \notin \mathcal{O}_\lambda$, then $x \notin \mathcal{O}_{\lambda_1}$ for every λ_1 less than λ. Set $\{\lambda \in \Lambda \mid x \in \mathcal{O}_\lambda\}$ therefore contains no real numbers in Λ that are less than λ. Hence, $\inf\{\lambda \in \Lambda \mid x \in \mathcal{O}_\lambda\} \geq \lambda$.

To prove the second, observe that if $x \in \overline{\mathcal{O}_\lambda}$, then $x \in \mathcal{O}_{\lambda_2}$ for every λ_2 greater than λ. Set $\{\lambda \in \Lambda \mid x \in \mathcal{O}_\lambda\}$ therefore contains all real numbers in Λ that are greater than λ. Hence, by definition of the infimum function, $\inf\{\lambda \in \Lambda \mid x \in \mathcal{O}_\lambda\} \leq \lambda$.

Now use the second part of Problem 9. Because Λ is dense in (a,b), it suffices to consider each real number λ in Λ instead of each real number in (a,b). By the contrapositive of fact (12.1), $\{x \in X \mid f(x) < \lambda\} = \mathcal{O}_\lambda$, which is open. And by the contrapositive of fact (12.2), $\{x \in X \mid f(x) > \lambda\} = X \sim \overline{\mathcal{O}_\lambda}$, which is open by Proposition 4 of Section 11.1. Thus, f is continuous.

12.2 The Tychonoff Product Theorem

The authors call x_λ a *component* in the first paragraph but use the term *coordinate* in the proof of the theorem and in Problems 22 and 23. I will stick with *component*.

11. Take the product $\prod_{\lambda\in\Lambda} X_\lambda$ of an arbitrary collection of Tychonoff spaces. Take two points u and v in $\prod_{\lambda\in\Lambda} X_\lambda$. Because $v \neq u$, there is an index λ' in Λ for which $v_{\lambda'} \neq u_{\lambda'}$. Because $X_{\lambda'}$ is Tychonoff, there is a neighborhood $U_{\lambda'}$ of $u_{\lambda'}$ in $X_{\lambda'}$ that does not contain $v_{\lambda'}$ and a neighborhood $V_{\lambda'}$ of $v_{\lambda'}$ in $X_{\lambda'}$ that does not contain $u_{\lambda'}$. For all other indices, i.e., for each λ in $\Lambda \sim \{\lambda'\}$, define sets U_λ and V_λ to be equal to X_λ. Then $\prod_{\lambda\in\Lambda} U_\lambda$ is a neighborhood of u that does not contain v, and $\prod_{\lambda\in\Lambda} V_\lambda$ is a neighborhood of v that does not contain u. Thus, $\prod_{\lambda\in\Lambda} X_\lambda$ is Tychonoff.

12. Take the product $\prod_{\lambda\in\Lambda} X_\lambda$ of an arbitrary collection of Hausdorff spaces. Take two points u and v in $\prod_{\lambda\in\Lambda} X_\lambda$. Because $v \neq u$, there is an index λ' in Λ for which $v_{\lambda'} \neq u_{\lambda'}$. Because $X_{\lambda'}$ is Hausdorff, $u_{\lambda'}$ and $v_{\lambda'}$ can be separated in $X_{\lambda'}$ by disjoint neighborhoods $U_{\lambda'}$ of $u_{\lambda'}$ and $V_{\lambda'}$ of $v_{\lambda'}$. For all other indices, i.e., for each λ in $\Lambda \sim \{\lambda'\}$, define sets U_λ and V_λ to be equal to X_λ. Then $\prod_{\lambda\in\Lambda} U_\lambda$ and $\prod_{\lambda\in\Lambda} V_\lambda$ are disjoint neighborhoods that separate u and v. Thus, $\prod_{\lambda\in\Lambda} X_\lambda$ is Hausdorff.

13. Refer to the definitions of open sets in $\mathbf{R}$ (page 16); Euclidean norm (page 183); open balls, open subsets, and neighborhoods in a metric space (page 186); Euclidean topology (page 222); a base for a topology (pages 222–223); and product topology (page 244).

    We compare bases as at the end of the proof of Urysohn's metrization theorem. A base for $\mathbf{R}^n$ in the product topology is the collection of Cartesian products $\prod_{k=1}^n (x_k - r_k, x_k + r_k)$ of open intervals where x_k is the kth component of a point x in $\mathbf{R}^n$, and $r_k > 0$ for k equal to $1, \ldots, n$.

    A base for $\mathbf{R}^n$ in Euclidean topology is the collection of open balls $B(x, r)$ where $x \in \mathbf{R}^n$, and $r > 0$.

    We first show that for each $\prod_{k=1}^n (x_k - r_k, x_k + r_k)$, there is a $B(x, r)$ for which inclusion $B(x, r) \subseteq \prod_{k=1}^n (x_k - r_k, x_k + r_k)$ holds. That happens when we put r to $\min\{r_1, \ldots, r_n\}$. For, take a point x' in $B(x, r)$. Then for the first components of x' and x,

$$\begin{aligned} |x_1' - x_1| &= \left((x_1' - x_1)^2 + (x_2 - x_2)^2 + \cdots + (x_n - x_n)^2\right)^{1/2} && x_2 - x_2, \ldots, x_n - x_n = 0 \\ &= \|(x_1', x_2, \ldots, x_n) - x\| && \text{definition of Euclidean norm} \\ &< r && (x_1', x_2, \ldots, x_n) \in B(x, r) \\ &\leq r_1 && r = \min\{r_1, \ldots, r_n\}, \end{aligned}$$

    which shows that $x_1' \in (x_1 - r_1, x_1 + r_1)$. By analogous computations, $x_k' \in (x_k - r_k, x_k + r_k)$ where $k = 2, \ldots, n$. Then $x' \in \prod_{k=1}^n (x_k - r_k, x_k + r_k)$. Because x' is an arbitrary point in $B(x, r)$, we have that $B(x, r) \subseteq \prod_{k=1}^n (x_k - r_k, x_k + r_k)$.

    We now show that the inclusion holds in the other direction, i.e., for each $B(x, r)$, there is a $\prod_{k=1}^n (x_k - r_k, x_k + r_k)$ for which $\prod_{k=1}^n (x_k - r_k, x_k + r_k) \subseteq B(x, r)$. That happens when we put each r_k to $r/\sqrt{n}$ because for every point x' in $\prod_{k=1}^n (x_k - r_k, x_k + r_k)$,

$$\begin{aligned} \|x' - x\| &= \left((x_1' - x_1)^2 + \cdots + (x_n' - x_n)^2\right)^{1/2} && \text{definition of Euclidean norm} \\ &< \left(r_1^2 + \cdots + r_n^2\right)^{1/2} && x_k' \in (x_k - r_k, x_k + r_k) \text{ where } k = 1, \ldots, n \\ &= \left((r/\sqrt{n})^2 + \cdots + (r/\sqrt{n})^2\right)^{1/2} && r_k = r/\sqrt{n} \text{ where } k = 1, \ldots, n \\ &= r, \end{aligned}$$

    which shows that $x' \in B(x, r)$. Hence, $\prod_{k=1}^n (x_k - r_k, x_k + r_k) \subseteq B(x, r)$.

    Thus, the product topology is the same as Euclidean topology on $\mathbf{R}^n$.

14. In the problem statement, replace "X and X" with "X and Y".

    Proceed as in our solution to Problem 13. A base for (X, ρ_1) is the collection of open balls $B_1(x, r_1)$ where $x \in X$ and $r_1 > 0$. Likewise, a base for (Y, ρ_2) is the collection of open balls $B_2(y, r_2)$ where $y \in Y$ and $r_2 > 0$. Then a base for $X \times Y$ in the product topology is the collection of Cartesian products $B_1(x, r_1) \times B_2(y, r_2)$ of those open balls.

    A base for $X \times Y$ in the topology induced by ρ is the collection of open balls $B((x, y), r)$ where $(x, y) \in X \times Y$ and $r > 0$.

    We first show that for each $B_1(x, r_1) \times B_2(y, r_2)$, there is a $B((x, y), r)$ for which inclusion $B((x, y), r) \subseteq B_1(x, r_1) \times B_2(y, r_2)$ holds. That happens when we put r to $\min\{r_1, r_2\}$. For, take a point (x', y') in $B((x, y), r)$. Then

$$\begin{aligned} \rho_1(x', x) &= \sqrt{\rho_1(x', x)^2 + \rho_2(y, y)^2} && \rho_2(y, y) = 0 \\ &= \rho((x', y), (x, y)) && \text{definition of } \rho \\ &< r && (x', y) \in B((x, y), r) \\ &\le r_1 && r = \min\{r_1, r_2\}, \end{aligned}$$

    which shows that $x' \in B_1(x, r_1)$. By an analogous computation, $y' \in B_2(y, r_2)$. Then $(x', y') \in B_1(x, r_1) \times B_2(y, r_2)$. Because (x', y') is an arbitrary point in $B((x, y), r)$, we have that $B((x, y), r) \subseteq B_1(x, r_1) \times B_2(y, r_2)$.

    We now show that the inclusion holds in the other direction, i.e., for each $B((x, y), r)$, there is a $B_1(x, r_1) \times B_2(y, r_2)$ for which $B_1(x, r_1) \times B_2(y, r_2) \subseteq B((x, y), r)$. That happens when we put r_1 and r_2 each to $r/\sqrt{2}$ because for every point (x', y') in $B_1(x, r_1) \times B_2(y, r_2)$,

$$\begin{aligned} \rho((x', y'), (x, y)) &= \sqrt{\rho_1(x', x)^2 + \rho_2(y', y)^2} && \text{definition of } \rho \\ &< \sqrt{r_1^2 + r_2^2} && x' \in B_1(x, r_1),\ y' \in B_2(y, r_2) \\ &= \sqrt{\left(r/\sqrt{2}\right)^2 + \left(r/\sqrt{2}\right)^2} && r_1, r_2 = r/\sqrt{2} \\ &= r, \end{aligned}$$

    which shows that $(x', y') \in B((x, y), r)$. Hence, $B_1(x, r_1) \times B_2(y, r_2) \subseteq B((x, y), r)$.

    Thus the product topology on $X \times Y$ is the same as the topology induced by the product metric.

15. This is true by Problem 25 of Section 9.2. Note that because convergence of sequences with respect to ρ and ρ^* is the same, we conclude also that sets that are open with respect to ρ^* are open with respect to ρ.

16. Function ρ is a metric on X by Problem 10 of Section 9.1.

    Use Problem 15 to simplify notation by replacing $\rho_n/(1 + \rho_n)$ with ρ_n^*. Note that $\rho_n^* < 1$ for all n.

    Proceed as in our solutions to Problems 13 and 14. A base for X in the product topology is the collection of Cartesian products $\prod_{n=1}^{\infty} \mathcal{O}_n$ where $\mathcal{O}_n = X_n$ except for finitely many n. For those finitely many n, subset $\mathcal{O}_n$ is an open ball $B_n^*(x_n, r_n^*)$ where x_n is the nth component of a point x in X, and $0 < r_n^* < 1$ (we indicate for clarity that $r_n^* < 1$ because if $r_n^* \ge 1$, then $B_n^*(x_n, r_n^*) = X_n$).

    A base for X in the topology induced by ρ is the collection of open balls $B(x, r)$ where $x \in X$, and $r > 0$.

We first show that for each $\prod_{n=1}^{\infty} \mathcal{O}_n$, there is a $B(x,r)$ for which $B(x,r) \subseteq \prod_{n=1}^{\infty} \mathcal{O}_n$. If $\mathcal{O}_n \neq X_n$, then $\mathcal{O}_n$ is some $B_n^*(x_n, r_n^*)$. Put r to $\min\limits_{\mathcal{O}_n \neq X_n} \{r_n^*/2^n\}$. Now take a point x' in $B(x,r)$. For those n such that $\mathcal{O}_n \neq X_n$, compute as follows.

$$\begin{aligned}
\rho_n^*(x_n', x_n) &= 2^n \left(\frac{\rho_1^*(x_1,x_1)}{2^1} + \cdots + \frac{\rho_{n-1}^*(x_{n-1},x_{n-1})}{2^{n-1}} \right. \\
&\qquad \left. + \frac{\rho_n^*(x_n', x_n)}{2^n} + \frac{\rho_{n+1}^*(x_{n+1},x_{n+1})}{2^{n+1}} + \cdots \right) && \rho_m^*(x_m, x_m) = 0 \\
&= 2^n \rho((x_1, \ldots, x_{n-1}, x_n', x_{n+1}, \ldots), x) && \text{definition of } \rho \\
&< 2^n r && \text{first point is in } B(x,r) \\
&\leq r_n^* && r = \min_{\mathcal{O}_m \neq X_m} \{r_m^*/2^m\},
\end{aligned}$$

which shows that $x_n' \in B_n^*(x_n, r_n^*)$. For those n such that $\mathcal{O}_n = X_n$, we have that $x_n' \in \mathcal{O}_n$. In summary, for an arbitrary point x' in $B(x,r)$ and for all n, we have that $x_n' \in \mathcal{O}_n$. Hence, $B(x,r) \subseteq \prod_{n=1}^{\infty} \mathcal{O}_n$.

We now show that the inclusion holds in the other direction, i.e., for each $B(x,r)$, there is a $\prod_{n=1}^{\infty} \mathcal{O}_n$ for which $\prod_{n=1}^{\infty} \mathcal{O}_n \subseteq B(x,r)$. Choose an index N such that $1/2^N \leq r/2$. Put $\prod_{n=1}^{\infty} \mathcal{O}_n$ to

$$\prod_{n=1}^{N} B_n^*(x_n, r/2) \times \prod_{n=N+1}^{\infty} X_n.$$

For every point x' in that Cartesian product,

$$\begin{aligned}
\rho(x', x) &= \sum_{n=1}^{N} \frac{\rho_n^*(x_n', x_n)}{2^n} + \sum_{n=N+1}^{\infty} \frac{\rho_n^*(x_n', x_n)}{2^n} && \text{definition of } \rho \text{ split into two sums} \\
&< \sum_{n=1}^{N} \frac{r/2}{2^n} && x_n' \in B_n^*(x_n, r/2) \\
&\quad + \sum_{n=N+1}^{\infty} \frac{1}{2^n} && \rho_n^* < 1 \\
&= (1 - 1/2^N)(r/2) + 1/2^N && \text{sums of geometric series} \\
&< r/2 + r/2 && 1 - 1/2^N < 1;\ 1/2^N \leq r/2 \\
&= r,
\end{aligned}$$

which shows that $x' \in B(x,r)$. Hence, $\prod_{n=1}^{\infty} \mathcal{O}_n \subseteq B(x,r)$.

Thus, ρ is a metric on X that induces the product topology on X.

17. Here is why f_0 is a point of closure of E. Take a neighborhood $\prod_{x \in \mathbf{R}} U_x$ of f_0. By definition of a product topology, $U_x = \mathbf{R}$ except for finitely many x. By definition of E, there is a function f in E that takes the value 0 on those finitely many x and elsewhere takes the value 1; consequently, $f(x) \in U_x$ for those finitely many x, and $f(x) \in \mathbf{R}$ for the other x. Therefore $f \in \prod_{x \in \mathbf{R}} U_x$, which shows that every neighborhood of f_0 contains a function in E. Hence, f_0 is a point of closure of E.

    Here is more explanation of why there is no sequence $\{f_n\}$ in E that converges to f_0. Sequence $\{f_n\}$ is a countable collection of functions f_n, and each f_n takes the value 0 on only a countable set, so by Corollary 6 of Section 1.3, the set on which $\{f_n\}$ takes the value 0 is countable. Since $\mathbf{R}$ is uncountable (Theorem 7 of Section 1.3), there is some x_0 in $\mathbf{R}$

such that $f_n(x_0) = 1$ for all n, and so sequence $\{f_n(x_0)\}$ does not converge to $f_0(x_0)$ because $\{f_n(x_0)\} \to 1$, but $f_0(x_0) = 0$. Hence, by the contrapositive of Proposition 3, there is no sequence $\{f_n\}$ in E that converges to f_0.

Space $\mathbf{R}^{\mathbf{R}}$ is not first countable by the contrapositive of Proposition 9 of Section 11.3 and therefore not metrizable by the contrapositive of the example preceding that proposition.

It remains to verify Proposition 3. We first prove a lemma.

Lemma, continuity of projection mappings Projection mapping π_{λ_0} described in the first paragraph of the section is continuous.

Proof Apply Proposition 10 of Section 11.4. Take an open subset $\mathcal{O}_{\lambda_0}$ in X_{λ_0}. Inverse image $\pi_{\lambda_0}^{-1}(\mathcal{O}_{\lambda_0})$ is subbasic set $\mathcal{O}_{\lambda_0} \times \prod_{\lambda \in \Lambda \sim \{\lambda_0\}} X_\lambda$, which is an open subset of $\prod_{\lambda \in \Lambda} X_\lambda$. Thus, π_{λ_0} is continuous. □

Proof of Proposition 3 Refer to the definitions of convergence (page 227) and continuity (page 229). Assume first that $\{f_n\}$ converges to f. Take a λ_0 in Λ and a neighborhood $\mathcal{O}_{\lambda_0}$ of $f(\lambda_0)$. Express $f(\lambda_0)$ as $\pi_{\lambda_0}(f)$. Because π_{λ_0} is continuous, there is a neighborhood $\prod_{\lambda \in \Lambda} \mathcal{U}_\lambda$ of f for which $\pi_{\lambda_0}(\prod_{\lambda \in \Lambda} \mathcal{U}_\lambda) \subseteq \mathcal{O}_{\lambda_0}$, that is, $\mathcal{U}_{\lambda_0} \subseteq \mathcal{O}_{\lambda_0}$. Because $\{f_n\}$ converges to f, there is an index N such that if $n \geq N$, then f_n belongs to $\prod_{\lambda \in \Lambda} \mathcal{U}_\lambda$ so that in particular, $f_n(\lambda_0)$ belongs to $\mathcal{U}_{\lambda_0}$. The two previous sentences together say that for each neighborhood $\mathcal{O}_{\lambda_0}$ of $f(\lambda_0)$, if $n \geq N$, then $f_n(\lambda_0)$ belongs to $\mathcal{O}_{\lambda_0}$. Thus, $\{f_n(\lambda)\}$ converges to $f(\lambda)$ for each λ in Λ because λ_0 is an arbitrary parameter in Λ. (Cf. the first half of the proof of Proposition 21 in our solution to Problem 49(i) of Section 1.6.)

Conversely, assume that $\{f_n(\lambda)\}$ converges to $f(\lambda)$ for each λ in Λ. Take a neighborhood $\prod_{\lambda \in \Lambda} \mathcal{U}_\lambda$ of f. If $\mathcal{U}_\lambda = X$, then $f_n(\lambda)$ belongs to $\mathcal{U}_\lambda$ for all n. If $\mathcal{U}_\lambda \neq X$, then proceed as follows. Because $\{f_n(\lambda)\}$ converges to $f(\lambda)$, there is an index N_λ such that if $n \geq N_\lambda$, then $f_n(\lambda)$ belongs to $\mathcal{U}_\lambda$. Denote the maximum of those (finitely many) indices N_λ by N. So if $n \geq N$, then $f_n(\lambda)$ belongs to $\mathcal{U}_\lambda$. Returning now to all the $\mathcal{U}_\lambda$ in $\prod_{\lambda \in \Lambda} \mathcal{U}_\lambda$, if $n \geq N$, then f_n belongs to $\prod_{\lambda \in \Lambda} \mathcal{U}_\lambda$. Thus, $\{f_n\}$ converges to f. □

Although this problem does not need Proposition 4, here is a good place to verify it (as requested by the authors) because it also involves projection mappings.

Proof of Proposition 4 (Reviewing relevant material on page 230 might be helpful.) Take a topology $\mathcal{T}$ on $\prod_{\lambda \in \Lambda} X_\lambda$ for which all the π_λ are continuous. Each basic set $\prod_{\lambda \in \Lambda} \mathcal{O}_\lambda$ in the product topology is a finite intersection of inverse images $\mathcal{O}_{\lambda_0} \times \prod_{\lambda \in \Lambda \sim \{\lambda_0\}} X_\lambda$ under π_{λ_0} of sets $\mathcal{O}_{\lambda_0}$, each of which is open in a corresponding X_{λ_0}. Hence, each basic set in the product topology is in $\mathcal{T}$, and so the product topology is a subset of $\mathcal{T}$. Thus, the product topology on $\prod_{\lambda \in \Lambda} X_\lambda$ has the fewest number of sets among the topologies for which all the π_λ are continuous. □

18. Replace superscript N in the problem statement with $\mathbf{N}$, the set of natural numbers.

Let Y denote the discrete topological space $\{0, 2\}$. Then $Y^{\mathbf{N}}$ is homeomorphic to $X^{\mathbf{N}}$. Because homeomorphism is an equivalence relation, it suffices to prove that $Y^{\mathbf{N}}$ is homeomorphic to the Cantor set $\mathbf{C}$. We do so using Proposition 21 of Section 11.5.

Topological space Y is compact because an open cover of Y has a subcover consisting of at most two sets. Then $Y^{\mathbf{N}}$ also is compact by Tychonoff's product theorem. On the other hand, $\mathbf{C}$ is Hausdorff by the last sentence on page 226 and the fact that $\mathbf{C}$ is a subset of $\mathbf{R}$ and therefore a metric space by the last sentence on page 183.

Now $Y^{\mathbf{N}}$ is the collection of sequences

$$\{a_n\} \text{ where } a_n \text{ is 0 or 2,}$$

and from the construction of **C** described in Section 2.7, we deduce that **C** is the set of real numbers of the form

$$\sum_{n=1}^{\infty} \frac{a_n}{3^n} \text{ where } a_n \text{ is 0 or 2,}$$

so there is a natural mapping f of $Y^{\mathbf{N}}$ onto **C** defined by $f(\{a_n\}) = \sum_{n=1}^{\infty} a_n/3^n$. Mapping f merely relabels the members of $Y^{\mathbf{N}}$, so f is one-to-one and onto.

It remains to show that f is continuous. Apply the definition of continuity (page 229). Take a sequence $\{a_n\}$ in $Y^{\mathbf{N}}$ and a neighborhood of $f(\{a_n\})$, i.e., a neighborhood of $\sum_{n=1}^{\infty} a_n/3^n$. That neighborhood contains a basic set

$$\left(\sum_{n=1}^{\infty} \frac{a_n}{3^n} - r, \sum_{n=1}^{\infty} \frac{a_n}{3^n} + r\right) \cap \mathbf{C}$$

by definitions of an open set in **R** (page 16), a base for a topology (pages 222–223), and topological subspaces (page 222) with **C** a subspace of **R**. Choose an index N large enough so that $\sum_{n=N}^{\infty} 2/3^n < r$. Then Cartesian product

$$\prod_{n=1}^{N-1} \{a_n\} \times \prod_{n=N}^{\infty} Y$$

is a neighborhood of $\{a_n\}$ for which

$$f\left(\prod_{n=1}^{N-1} \{a_n\} \times \prod_{n=N}^{\infty} Y\right) \subseteq \left(\sum_{n=1}^{\infty} \frac{a_n}{3^n} - r, \sum_{n=1}^{\infty} \frac{a_n}{3^n} + r\right) \cap \mathbf{C}$$

because for all sequences $\{a'_n\}$ in that neighborhood of $\{a_n\}$,

$$\begin{aligned}
\left| f(\{a'_n\}) - \sum_{n=1}^{\infty} \frac{a_n}{3^n} \right| &= \left| \sum_{n=1}^{\infty} \frac{a'_n}{3^n} - \sum_{n=1}^{\infty} \frac{a_n}{3^n} \right| && \text{definition of } f \\
&= \left| \sum_{n=N}^{\infty} \frac{a'_n - a_n}{3^n} \right| && a'_n = a_n \text{ for } n \text{ equal to } 1, \ldots, N-1 \\
&\leq \sum_{n=N}^{\infty} \frac{|a'_n - a_n|}{3^n} && \text{triangle inequality} \\
&\leq \sum_{n=N}^{\infty} \frac{2}{3^n} && |a'_n - a_n| \text{ is 0 or 2} \\
&< r.
\end{aligned}$$

Hence, f is continuous. Note that the triangle inequality holds for the infinite series in our case because the sum converges.

19. Let Cartesian product $F_1 \times \cdots \times F_n$ be a closed, bounded subset of $\mathbf{R}^n$ where $F_1, \ldots, F_n$ are subsets of **R**. We claim that F_1 is closed. Otherwise, it would not contain one of its points $\overline{x_1}$ of closure. For take a point $(x_1, \ldots, x_n)$ in $F_1 \times \cdots \times F_n$, and replace the first component x_1 with $\overline{x_1}$. Then $(\overline{x_1}, x_2, \ldots, x_n)$ would be a point of closure of $F_1 \times \cdots \times F_n$ but would not be contained in $F_1 \times \cdots \times F_n$, which contradicts that $F_1 \times \cdots \times F_n$ is closed. Hence, F_1 is closed.

Furthermore, we claim that F_1 is bounded. Otherwise, there would not be any real number a such that $|x_1| \le a$ for all x_1 in F_1. But then there would not be any real number b such that $\left(x_1^2 + \cdots + x_n^2\right)^{1/2} \le b$ for all $(x_1, \ldots, x_n)$ in $F_1 \times \cdots \times F_n$, which contradicts that $F_1 \times \cdots \times F_n$ is bounded. Hence, F_1 is bounded.

By analogous reasoning, $F_2, \ldots, F_n$ are also closed and bounded set of real numbers.

By the Heine-Borel theorem (Section 1.4), $F_1, \ldots, F_n$ are compact. Thus, $F_1 \times \cdots \times F_n$ also is compact by Tychonoff's product theorem.

Note that compactness of each closed, bounded *interval* of real numbers is used in the proof of the Heine-Borel theorem.

20. The hint assumes (without loss of generality) that $I = [0, 1]$.

    We use the first paragraph of the proof of the Heine-Borel theorem (Section 1.4) to flesh out the hint. Start with the first sentence of the hint. Such a t exists because $X \times [0, 0]$ can be covered by a finite number of sets in $\mathcal{U}$ since X is compact and since there is a set in $\mathcal{U}$ that covers $A \times [0, 0]$ where A is some set in X. Because t belongs to $[0, 1]$, there is an $\mathcal{O}$ in $\mathcal{U}$ that contains $B \times \{t\}$ where B is some set in X. Because $\mathcal{O}$ is open, there is a positive ϵ such that $B \times (t - \epsilon, t + \epsilon)$ is contained in $\mathcal{O}$. Because $t - \epsilon < t$, set $X \times [0, t - \epsilon]$ can be covered by a finite collection $\{\mathcal{O}_1, \ldots, \mathcal{O}_k\}$ of sets in $\mathcal{U}$. Consequently, finite collection $\{\mathcal{O}_1, \ldots, \mathcal{O}_k, \mathcal{O}\}$ covers $X \times [0, t + \epsilon)$. Hence, $t = 1$ for otherwise, $t < 1$ and there would be some t'' greater than t (and less than $t + \epsilon$) such that $X \times [0, t'']$ could be covered by a finite number of sets in $\mathcal{U}$. Thus, $X \times [0, 1]$ can be covered by a finite number of sets from $\mathcal{U}$, proving that $X \times I$ is compact.

21. Take a sequence $\{x_n\}$ of points x_n in a product $\prod_{j=1}^{\infty} X_j$ of a countable number of sequentially compact topological spaces X_j. We use a diagonal argument (first remark on page 209) to construct a subsequence $\{x_{n_k}\}$ of $\{x_n\}$ that converges to a point x of $\prod_{j=1}^{\infty} X_j$.

    To avoid double subscripts in our notation, we use $x_n(j)$ to denote the jth component of x_n. Likewise, $x(j)$ denotes the jth component of x.

    Now because X_1 is sequentially compact, sequence $n \mapsto x_n(1)$ of points in X_1 has a convergent subsequence, that is, there is a strictly increasing sequence $\{s(1, n)\}$ of natural numbers and a point $x(1)$ in X_1 for which

    $$\lim_{n \to \infty} x_{s(1,n)}(1) = x(1).$$

    By the same argument applied to sequence $n \mapsto x_{s(1,n)}(2)$, there is a subsequence $\{s(2, n)\}$ of $\{s(1, n)\}$ and a point $x(2)$ in X_2 for which $\lim_{n \to \infty} x_{s(2,n)}(2) = x(2)$. We inductively continue that selection process to obtain a countable collection of strictly increasing sequences $\{\{s(j, n)\}\}_{j=1}^{\infty}$ of natural numbers and a sequence $\{x(j)\}$ such that for each j,

    $$\{s(j+1, n)\} \text{ is a subsequence of } \{s(j, n)\}, \text{ and } \lim_{n \to \infty} x_{s(j,n)}(j) = x(j).$$

    Note that sequence $\{x(j)\}$ defines a point x of $\prod_{j=1}^{\infty} X_j$. Consider the "diagonal" subsequence $\{x_{n_k}\}$ obtained by putting n_k to $s(k, k)$ for each index k. For each j, sequence $\{n_k\}_{k=j}^{\infty}$ is a subsequence of the jth subsequence of natural numbers selected above and therefore

    $$\lim_{k \to \infty} x_{n_k}(j) = x(j).$$

    By Proposition 3, $\lim_{k \to \infty} x_{n_k} = x$, that is, $\{x_{n_k}\}$ is a subsequence of $\{x_n\}$ that converges to x. Thus, $\prod_{j=1}^{\infty} X_j$ is sequentially compact.

22. If X has only one point, then the claim is true trivially, so assume X has at least two points.

    Let $\mathcal{F}$ be the family of continuous real-valued functions on X with values in $[0,1]$. Family $\mathcal{F}$ is not empty because constant functions are continuous. Denote cube $\prod_{f\in\mathcal{F}} I_f$ by Q. The mapping g of X into Q that takes x to the point whose fth component is $f(x)$ is one-to-one because if x and x' are distinct points in X, then $g(x') \neq g(x)$ since there is an f in $\mathcal{F}$ such that $f(x') \neq f(x)$. For example, Urysohn's lemma (Section 12.1) assures us that there is an f in $\mathcal{F}$ such that $f(x) = 0$ and $f(x') = 1$. We can apply Urysohn's lemma here because X is normal by Theorem 18 of Section 11.5 and because singleton sets $\{x\}$ and $\{x'\}$ are closed subsets of X by Proposition 6 of Section 11.2.

    To show that g is continuous, take a point x in X and a neighborhood of $g(x)$. That neighborhood contains a basic set $\prod_{f\in\mathcal{F}} \mathcal{O}_f$ where $\mathcal{O}_f = I$ except for finitely many f. Because each f is continuous, there is a neighborhood $\mathcal{U}_f$ of x for which $f(\mathcal{U}_f) \subseteq \mathcal{O}_f$ for each f. Denote finite intersection $\bigcap_{\substack{f\in\mathcal{F}\\ \mathcal{O}_f\neq I}} \mathcal{U}_f$ by $\mathcal{U}$. Subset $\mathcal{U}$ is open because it is an intersection of a finite collection of open sets. Subset $\mathcal{U}$ is therefore a neighborhood of x. Moreover, $g(\mathcal{U}) \subseteq \prod_{f\in\mathcal{F}} \mathcal{O}_f$. Hence, g is continuous. Note that in our construction of $\mathcal{U}$, we ignored indices f for which $\mathcal{O}_f = I$ because the image under such f of a subset of X is contained in $\mathcal{O}_f$ since $\mathcal{O}_f = I$.

    A second way to show that g is continuous uses Problem 23 of Section 11.4. Take a basic subset $\prod_{f\in\mathcal{F}} \mathcal{O}_f$ of Q. Then

$$
\begin{aligned}
g^{-1}\left(\prod_{f\in\mathcal{F}} \mathcal{O}_f\right) &= \left\{x\in X \,\middle|\, g(x)\in \prod_{f\in\mathcal{F}} \mathcal{O}_f\right\} && \text{definition of inverse}\\
&= \{x\in X \mid f(x)\in \mathcal{O}_f \text{ for each } f \text{ in } \mathcal{F}\}\\
&= \bigcap_{\substack{f\in\mathcal{F}\\ \mathcal{O}_f\neq I}} f^{-1}(\mathcal{O}_f),
\end{aligned}
$$

    which is open because it is the intersection of a finite collection of open sets. Hence, g is continuous. Note that in the intersection, we can ignore indices f for which $\mathcal{O}_f = I$ because $f^{-1}(\mathcal{O}_f) = X$ for such f.

    Next, image $g(X)$ is compact by Proposition 20 of Section 11.5. Moreover, Q is Hausdorff by Problem 12 because I is Hausdorff by the last sentence on page 226 since I is a metric space. Hence, by Proposition 16 of Section 11.5, $g(X)$ is closed.

    Now unit interval I, which is $[0,1]$, is compact by the Heine-Borel theorem (Section 1.4). By Tychonoff's product theorem, Q also is compact. By Theorem 18 of Section 11.5, Q is normal. By Problem 12 of Section 11.2, $g(X)$ is normal. By the last sentence on page 226, $g(X)$ is Hausdorff. By Proposition 21 of Section 11.5, g as a mapping from X onto $g(X)$ is a homeomorphism. Thus, X is homeomorphic to closed subset $g(X)$ of cube Q.

23. Cube Q is compact by our solution to Problem 22. Range $f(Q)$ of f is compact by Proposition 20 of Section 11.5. By Section 9.5's Theorem 16(ii)→(i) and definition of a totally bounded metric space, we can cover $f(Q)$ by a finite number of open intervals $J_1, \ldots, J_n$ of length ϵ. Look at inverse images $f^{-1}(J_k)$ of those intervals where $k = 1, \ldots, n$. Those sets $f^{-1}(J_k)$ cover Q. Moreover, each $f^{-1}(J_k)$ is open by Proposition 10 of Section 11.4. Then $f^{-1}(J_k)$ has the form $\prod_{\alpha\in A} \mathcal{O}_{k\alpha}$ for each k where $\mathcal{O}_{k\alpha}$ is open in I, and $\mathcal{O}_{k\alpha} = I$ except for finitely many α. Define a subset S of indices in A by $S = \bigcup_{k=1}^{n} \{\alpha\in A \mid \mathcal{O}_{k\alpha} \neq I\}$. Subset S is finite. Define mapping $h: Q\to Q$ by

$$
\pi_\alpha(h(x)) = \begin{cases} x_\alpha & \alpha\in S,\\ 1/2 & \alpha\notin S. \end{cases}
$$

Mapping h is continuous because with respect to a component, h's projection is either the identity mapping or a constant mapping. Put g to $f \circ h$. Then g is a real-valued function on Q and is a function of only a finite number of components—those represented by the indices in S. Function g is continuous by Proposition 11 of Section 11.4. Finally, $|f - g| < \epsilon$, because for each x in Q, values $f(x)$ and $g(x)$ are in the same J_k since for indices α that are not in S, projection $\pi_\alpha(f^{-1}(J_k))$ is all of I.

12.3 The Stone-Weierstrass Theorem

The maximum norm is used throughout the text in this section and our solutions, so it might be helpful to recall definition and properties of $\|\cdot\|_{\max}$ in the second paragraph of Section 10.1, but read "topological space X" for "metric space X."

In footnote 2, the first author is B. Brosowski. We use his article to solve Problem 32.

24. Take two distinct points u and v in X. There is a continuous real-valued function f on X for which $f(u) \neq f(v)$. Because $\mathbf{R}$ is Hausdorff—by the last sentence on page 226 and the fact that $\mathbf{R}$ is a metric space—numbers $f(u)$ and $f(v)$ can be separated in $\mathbf{R}$ by disjoint neighborhoods U and V. By Proposition 10 of Section 11.4, $f^{-1}(U)$ and $f^{-1}(V)$ are open subsets of X. Furthermore, $f^{-1}(U)$ is a neighborhood of u, and $f^{-1}(V)$ is a neighborhood of v. Lastly, $f^{-1}(U)$ and $f^{-1}(V)$ are disjoint; otherwise, f would take a point in $f^{-1}(U) \cap f^{-1}(V)$ to two distinct numbers in $\mathbf{R}$ because U and V are disjoint, which contradicts definition of a function. Thus, X is Hausdorff because u and v are separated by disjoint neighborhoods $f^{-1}(U)$ and $f^{-1}(V)$.

25. If $\mathcal{A}$ separates points in X, then $\mathcal{A}$ is dense in $C(X)$ by the Stone-Weierstrass approximation theorem.

    Conversely, assume that $\mathcal{A}$ is dense in $C(X)$. Take two distinct points u and v in X. By our solution to Problem 22, there is a function f in $C(X)$ such that $f(u) = 0$ and $f(v) = 1$. By the denseness of $\mathcal{A}$ in $C(X)$, there is a function g in $\mathcal{A}$ for which $|g - f| < 1/2$ on X. Then $g(u) \neq g(v)$ since $g(u) < 1/2$ and $g(v) > 1/2$. Thus, $\mathcal{A}$ separates points in X.

26. Assume for the sake of argument that α is nonzero; otherwise, the claim is false.

    By Proposition 6 of Section 9.2, $\alpha(f + c)$ is the limit of a sequence $\{g_n\}$ in $\mathcal{A}$. Then by definition of convergence, for each positive ϵ, there is an index N for which if $n \geq N$, then

    $$\|g_n - \alpha(f + c)\|_{\max} < \epsilon|\alpha|,$$

    or

    $$\|(g_n/\alpha - c) - f\|_{\max} < \epsilon \qquad \text{positive homogeneity (page 137).}$$

    Sequence $\{g_n/\alpha - c\}$ is in $\mathcal{A}$ because $\mathcal{A}$ is an algebra that contains the constant functions. Thus, by another appeal to Proposition 6 of Section 9.2, f belongs to $\overline{\mathcal{A}}$.

27. If $f = g$, then the integrals are the same.

    Conversely, assume that $\int_a^b x^n f(x)\,dx = \int_a^b x^n g(x)\,dx$ for all n in $\{0, 1, 2, \ldots\}$. Then for all polynomials p, we have by linearity of integration (Theorem 17 of Section 4.4) that

    $$\int_a^b p(x)(f(x) - g(x))\,dx = 0.$$

By the fourth and first sentences on page 151, there is a sequence $\{p_m\}$ of polynomials for which $\lim_{m\to\infty} p_m = f - g$ in $[a, b]$. Then

$$\begin{aligned} 0 &= \lim_{m\to\infty} \int_a^b p_m(x)(f(x) - g(x))\, dx && \text{integral is 0 for each } m \\ &= \int_a^b (f(x) - g(x))^2\, dx \end{aligned}$$

where we can pass the limit under the integral sign because (i) each $p_m(x)(f(x) - g(x))$ is Riemann integrable by Problem 6 of Section 4.1 and (ii) a basic theorem for Reimann integrals allows such a passage since convergence is uniform on compact interval $[a, b]$. By the Reimann-integral analogue of Proposition 9 of Section 4.3, $(f - g)^2 = 0$. Thus, $f = g$.

28. Verify that the conditions of the Stone-Weierstrass approximation theorem hold for this problem. Closed, bounded interval $[a, b]$ is compact by the Heine-Borel theorem (Section 1.4) and Hausdorff by the last sentence on page 226.

Next, let $\mathcal{A}$ be the collection of functions

$$x \mapsto c_0 + \sum_{k=1}^{n} c_k e^{kx} \text{ for all } x \text{ in } [a, b] \text{ and for all } n.$$

Then $\mathcal{A}$ is a collection of real-valued functions on $[a, b]$.

Functions in $\mathcal{A}$ are continuous because constant and exponential functions are continuous and because compositions, products, and sums of continuous functions are continuous by Section 11.4's Proposition 11 and Problem 28.

Collection $\mathcal{A}$ contains the constant functions because n can be 0 so that functions $x \mapsto c_0$ are in $\mathcal{A}$.

We show that $\mathcal{A}$ is an algebra by first showing that $\mathcal{A}$ is a linear subspace of $C[a, b]$, which we do by checking the properties of a linear space described in the first paragraph of Section 13.1. Set $\mathcal{A}$ is an abelian group. Indeed, the sum of two functions in $\mathcal{A}$ is again in $\mathcal{A}$. Furthermore, because $\mathcal{A}$ contains the constant functions, $\mathcal{A}$ inherits from $C[a, b]$ the other properties required to be an abelian group: commutative, associative, additive-identity, and additive-inverse properties of addition. Likewise, $\mathcal{A}$ inherits from $C[a, b]$ the properties having to do with multiplication by a real number. Hence, $\mathcal{A}$ is a linear space. To show that $\mathcal{A}$ is an algebra, we need to show that the product of two functions in $\mathcal{A}$ also belongs to $\mathcal{A}$. But such a product

$$\left(c_0 + \sum_{k=1}^{n} c_k e^{kx}\right)\left(b_0 + \sum_{j=1}^{m} b_j e^{jx}\right)$$

is the sum of a real number and a finite sum of products of a real number and an exponential function because $e^{kx}e^{jx} = e^{(k+j)x}$ and therefore also belongs to $\mathcal{A}$. Hence, $\mathcal{A}$ is an algebra.

Lastly, $\mathcal{A}$ separates points in $[a, b]$. For if u and v are two distinct points in $[a, b]$, then $e^u \neq e^v$.

According to the Stone-Weierstrass approximation theorem, $\mathcal{A}$ is dense in $C[a, b]$. Then the problem's claim follows directly from definition of denseness (page 150) in linear space $C[a, b]$ with the maximum norm.

29. The solution is the same as that of our solution to Problem 28 with $[a, b]$ replaced by $[0, \pi]$, the exponential function replaced by cosine, and identity $e^{kx}e^{jx} = e^{(k+j)x}$ replaced by $\cos kx \cdot \cos jx = (\cos|k - j|x + \cos(k + j)x)/2$.

30. Taking the hint, we can assume that f is a continuous real-valued function on the unit circle X in the plane, that is, $f \in C(X)$. Unit circle X is compact by Theorem 20(i)→(ii) of Section 9.5 because X is bounded by 1 and closed. To see why X is closed, refer to pages 186–187, and note that $X = \overline{B}(0,1) \sim B(0,1)$, that is, X is the complement in metric space $\overline{B}(0,1)$ of open set $B(0,1)$. Unit circle X also is Hausdorff by the last sentence on page 226 and the fact that X is a metric space by the last sentence on page 183.

    The remainder of our solution is the same as that of our solution to Problem 29 with $[0,\pi]$ replaced by X, adding terms $b_k \sin kx$ in the summation, further identities

$$\sin kx \cdot \sin jx = (\cos|k-j|x - \cos(k+j)x)/2,$$
$$\sin kx \cdot \cos jx = (\sin(k+j)x \pm \sin|k-j|x)/2,$$

    and the fact that for distinct u, v in X, either $\cos u \neq \cos v$, or $\sin u \neq \sin v$.

31. Verify that the conditions of the Stone-Weierstrass approximation theorem hold for this problem. Space $X \times Y$ is compact by Tychonoff's product theorem and Hausdorff by Problem 12.

    Next, let $\mathcal{A}$ be the collection of functions

$$(x,y) \mapsto \sum_{k=1}^{n} f_k(x) \cdot g_k(y) \text{ for all } (x,y) \text{ in } X \times Y \text{ and for all } n.$$

    Then $\mathcal{A}$ is a collection of real-valued functions on $X \times Y$.

    We show that the functions in $\mathcal{A}$ are continuous by modifying our solution to Problem 28 of Section 11.4. The first step is to show that mapping $h : X \times Y \to \mathbf{R} \times \mathbf{R}$ defined by $h((x,y)) = (f_k(x), g_k(y))$ is continuous. Take a point (x,y) in $X \times Y$ and a neighborhood $\mathcal{O}_1 \times \mathcal{O}_2$ of $h((x,y))$. Because f_k and g_k are continuous, there are neighborhoods $\mathcal{U}_x$ of x in X such that $f_k(\mathcal{U}_x) \subseteq \mathcal{O}_1$ and $\mathcal{U}_y$ of y in Y such that $g_k(\mathcal{U}_y) \subseteq \mathcal{O}_2$. Then $h(\mathcal{U}_x \times \mathcal{U}_y) \subseteq \mathcal{O}_1 \times \mathcal{O}_2$. Hence, h is continuous. Now $f_k \cdot g_k$ is the composition of h and the product operation $\cdot : \mathbf{R} \times \mathbf{R} \to \mathbf{R}$. Products $f_k \cdot g_k$ are therefore continuous by Proposition 11 of Section 11.4 and Problem 49(i) of Section 1.6. The latter problem tells us also that finite sums of the $f_k \cdot g_k$ are continuous. Hence, the functions in $\mathcal{A}$ are continuous.

    Collection $\mathcal{A}$ contains the constant functions. Indeed, take a constant real number c. The product of constant function 1 in $C(X)$ and constant function $y \mapsto c$ in $C(Y)$ is a function in $\mathcal{A}$ that takes constant value c. Hence, $\mathcal{A}$ contains the constant functions.

    Collection $\mathcal{A}$ is a linear subspace of $C(X \times Y)$ by the same reasoning as in our solution to Problem 28. To show that $\mathcal{A}$ is an algebra, we need to show that the product of two functions in $\mathcal{A}$ also belongs to $\mathcal{A}$. But such a product

$$\left(\sum_{k=1}^{n} f_k g_k\right)\left(\sum_{j=1}^{m} f_j g_j\right)$$

    is a finite sum of products $f_k g_k f_j g_j$, or $(f_k f_j)(g_k g_j)$, and because $C(X)$ and $C(Y)$ are algebras, $f_k f_j$ is in $C(X)$ and $g_k g_j$ is in $C(Y)$. Consequently, the product of two functions in $\mathcal{A}$ is a finite sum of products of a function in $C(X)$ and a function in $C(Y)$ and therefore also belongs to $\mathcal{A}$. Hence, $\mathcal{A}$ is an algebra.

    Lastly, $\mathcal{A}$ separates points in $X \times Y$. Indeed, let (x_1, y_1) and (x_2, y_2) be two distinct points in $X \times Y$. Take the constant function 1 from $C(X)$. From $C(Y)$, take a function g for which $g(y_1) \neq g(y_2)$—such a g exists by the last two sentences of the second paragraph on page 248. Then $1g$ is a function in $\mathcal{A}$ for which $(1g)((x_1, y_1)) \neq (1g)((x_2, y_2))$. Hence, $\mathcal{A}$ separates points in $X \times Y$.

According to the Stone-Weierstrass approximation theorem, $\mathcal{A}$ is dense in $C(X \times Y)$. Then the problem's claim follows directly from definition of denseness (page 150) in linear space $C(X \times Y)$ with the maximum norm.

32. By inspection, $p(0) = 0$ and $p(1) = 1$. Moreover, $dp(x)/dx = mnx^{n-1}(1 - x^n)^{m-1}$, which shows that $p' > 0$ on $(0, 1)$. Consequently, $0 \leq p \leq 1$ on $[0, 1]$, and p increases on $(0, 1)$. Hence, for p to satisfy (14), it suffices to choose m and n such that $p(c/2) < \epsilon$ and $p(c) > 1-\epsilon$.

    Let k be the smallest integer that is greater than $1/c$. Then $k - 1 \leq 1/c$, which implies that $k \leq (1 + c)/c < 2/c$. Together, $1 < kc < 2$. Now put m to k^n. Then

$$\begin{aligned} p(c/2) &= 1 - (1 - (c/2)^n)^{k^n} \\ &\leq 1 - (1 - k^n (c/2)^n) && \text{our solution to Problem 14 of Section 1.2, } -(c/2)^n > -1 \\ &= (kc/2)^n \\ &< \epsilon && \text{for large enough } n \text{ because } kc < 2, \end{aligned}$$

and

$$\begin{aligned} p(c) &= 1 - (1 - c^n)^{k^n} \\ &= 1 - \frac{(1 - c^n)^{k^n}}{k^n c^n} (k^n c^n) && kc \neq 0 \\ &> 1 - \frac{(1 - c^n)^{k^n}}{k^n c^n} (1 + k^n c^n) \\ &\geq 1 - \frac{(1 - c^n)^{k^n}}{k^n c^n} (1 + c^n)^{k^n} && \text{Problem 14 of Section 1.2} \\ &= 1 - \frac{\left(1 - c^{2n}\right)^{k^n}}{k^n c^n} \\ &> 1 - \frac{1}{(kc)^n} && 0 < c < 1 \\ &\geq 1 - \epsilon && \text{for large enough } n \text{ because } kc > 1. \end{aligned}$$

Thus, there is an n large enough so that together, $p(c/2) < \epsilon$ and $p(c) > 1 - \epsilon$ where $m = k^n$.

33. Our solution provides details to the first part of the proof of Borsuk's theorem. But I do not know why function f_0 is needed in that proof because the collection of polynomials already contains the constant functions.

    Let $\mathcal{P}$ be the collection of polynomials in a finite number of functions of $\mathcal{A}$. Verify that $\mathcal{P}$ satisfies the conditions of the Stone-Weierstrass approximation theorem. The functions in $\mathcal{P}$ are indeed real-valued functions on X.

    Because polynomial functions are continuous, the functions in $\mathcal{P}$ are continuous by Section 11.4's Proposition 11 and Problem 28.

    Collection $\mathcal{P}$ contains the constant functions because it contains the constant polynomials.

    Collection $\mathcal{P}$ is a linear subspace of $C(X)$ because (i) the sum of two functions in $\mathcal{P}$ is also a polynomial in a finite number of functions of $\mathcal{A}$ and so is again in $\mathcal{P}$ and (ii) since $\mathcal{P}$ contains the constant functions, it inherits from $C(X)$ the other properties required to be a linear space as described in the first paragraph of Section 13.1. Then $\mathcal{P}$ is an algebra because the product of two functions in $\mathcal{P}$ is also a polynomial in a finite number of functions of $\mathcal{A}$ and so also belongs to $\mathcal{P}$.

Lastly, $\mathcal{P}$ separates points in X. Indeed, take two distinct points u and v in X. By hypothesis, there is an f in $\mathcal{A}$ for which $f(u) \neq f(v)$. But f is also in $\mathcal{P}$ because f is polynomial x in function f itself. Hence, $\mathcal{P}$ separates points in X.

According to the Stone-Weierstrass approximation theorem, $\mathcal{P}$ is dense in $C(X)$. Then the problem's claim follows directly from definition of denseness (page 150) in linear space $C(X)$ with the maximum norm.

34. It suffices to show that the sum $f+g$ and product fg of two functions f and g in $\overline{\mathcal{A}}$ also belong to $\overline{\mathcal{A}}$ because it inherits from $\mathcal{A}$ and $C(X)$ the other properties—namely, those described in the second paragraph of Section 13.1—required for $\overline{\mathcal{A}}$ to be an algebra.

    By Proposition 6 of Section 9.2, f is the limit of a sequence $\{f_n\}$ in $\mathcal{A}$; likewise, g is the limit of a sequence $\{g_n\}$ in $\mathcal{A}$. By Problem 24 of Section 7.3, $\{f_n + g_n\} \to f + g$ in $C(X)$. Now each $f_n + g_n$ is in $\mathcal{A}$ because $\mathcal{A}$ is an algebra. Sum $f+g$ is therefore the limit of a sequence in $\mathcal{A}$. Hence, $f+g$ is in $\overline{\mathcal{A}}$ by another appeal to Proposition 6 of Section 9.2.

    We next show that fg is in $\overline{\mathcal{A}}$. Take a positive ϵ. There is an index N for which if $n \geq N$, then all of the following four inequalities hold:

$$\begin{aligned} &\|f - f_n\| < \epsilon/(3\|g\|), && \|g - g_n\| < \epsilon/(3\|f\|), \\ &\|f - f_n\| < \sqrt{\epsilon/3}, && \|g - g_n\| < \sqrt{\epsilon/3} \end{aligned}$$

    where $\|\cdot\|$ is an abbreviation of the maximum norm symbol $\|\cdot\|_{\max}$. Then if $n \geq N$,

$$\begin{aligned} \|fg - f_n g_n\| &\leq \|fg - f_n g\| + \|f_n g - f_n g_n\| && \text{triangle inequality} \\ &\leq \|f - f_n\|\|g\| + \|g - g_n\|\|f_n\| \\ &\leq \|f - f_n\|\|g\| + \|g - g_n\|(\|f_n - f\| + \|f\|) && \text{triangle inequality} \\ &= \|f - f_n\|\|g\| + \|g - g_n\|\|f - f_n\| + \|g - g_n\|\|f\| \\ &< \epsilon/(3\|g\|)\|g\| + \sqrt{\epsilon/3} \cdot \sqrt{\epsilon/3} + \epsilon/(3\|f\|)\|f\| \\ &= \epsilon. \end{aligned}$$

    Now each $f_n g_n$ is in $\mathcal{A}$ because $\mathcal{A}$ is an algebra. Product fg is therefore the limit of sequence $\{f_n g_n\}$ in $\mathcal{A}$. Hence, fg is in $\overline{\mathcal{A}}$. Thus, $\overline{\mathcal{A}}$ is an algebra. Note that it is not necessary that X be Hausdorff.

    For our solution to Problem 35, we also want to know that the functions in $\overline{\mathcal{A}}$ are continuous. The proof is like our solution to Problem 59 of Section 1.6, but here we need to show that for a point x in X, there is a neighborhood U of x for which if $x' \in U$, then $|f(x') - f(x)| < \epsilon$. Indeed, because $\{f_n\} \to f$ uniformly, there is an N such that

$$\begin{aligned} |f_N(x) - f(x)| &< \epsilon/3, \\ |f(x') - f_N(x')| &< \epsilon/3 \end{aligned}$$

    for all x, x' in X. Also, because f_N is continuous, there a neighborhood U of x for which if $x' \in U$, then

$$|f_N(x') - f_N(x)| < \epsilon/3.$$

    Altogether, if $x' \in U$, then

$$\begin{aligned} |f(x') - f(x)| &\leq |f(x') - f_N(x')| + |f_N(x') - f_N(x)| + |f_N(x) - f(x)| && \text{triangle inequality} \\ &< \epsilon/3 + \epsilon/3 + \epsilon/3 \\ &= \epsilon. \end{aligned}$$

    Hence, the functions in $\overline{\mathcal{A}}$ are continuous.

35. If $1 \in \overline{\mathcal{A}}$, then $\overline{\mathcal{A}}$ contains the constant functions because $\overline{\mathcal{A}}$ is an algebra (Problem 34). That problem's solution also shows that the functions in $\overline{\mathcal{A}}$ are continuous. Furthermore, $\overline{\mathcal{A}}$ separates points because $\mathcal{A}$ separates points and $\mathcal{A} \subseteq \overline{\mathcal{A}}$ by the sentence after the definition of *closure* on page 187. Then according to the Stone-Weierstrass approximation theorem, $\overline{\mathcal{A}}$ is dense in $C(X)$. Thus, we are done because $\overline{\mathcal{A}} = C(X)$ by the sentence after the definition on page 203.

Now assume that for each x in X, there is an f_x in $\mathcal{A}$ for which $f_x(x) \neq 0$. Function f_x^2 belongs to $\mathcal{A}$ and satisfies inequality $f_x^2 > 0$ (cf. proof of Lemma 7). By the continuity of f_x^2, there is a neighborhood $\mathcal{N}_x$ of x on which f_x^2 takes only positive values. Because X is compact, we may choose a finite collection $\{\mathcal{N}_{x_1}, \ldots, \mathcal{N}_{x_n}\}$ of those neighborhoods that covers X. Put g to $f_{x_1}^2 + \cdots + f_{x_n}^2$. Function g is in $\mathcal{A}$ and is positive on X, that is, $g > 0$ on X. But a continuous function on a compact set takes a minimum value, so we may choose a positive c for which $g \geq c$ on X. To show that such a g implies that $1 \in \overline{\mathcal{A}}$, we show that $1/g$ is in $\overline{\mathcal{A}}$ so that because $\overline{\mathcal{A}}$ is an algebra, product $g \cdot 1/g$, which is 1, is in $\overline{\mathcal{A}}$.

Define real-valued function h on $\overline{g(X)} \cup \{0\}$ by

$$h(y) = \begin{cases} 1/y & y \in \overline{g(X)}, \\ 0 & y = 0. \end{cases}$$

The closure $\overline{g(X)}$ of the range of g does not contain 0 because $g \geq c > 0$ on X. Moreover, $\overline{g(X)}$ is compact by Proposition 20 of Section 11.5 and Theorem 20 of Section 9.5. Then $\overline{g(X)} \cup \{0\}$ also is compact. Furthermore, $\overline{g(X)} \cup \{0\}$ is Hausdorff by the last sentence on page 226 and the fact that $\overline{g(X)} \cup \{0\}$ is a subset of $\mathbf{R}$ and therefore a metric space by the last sentence on page 183. Then because h is continuous (see our solution to Problem 65(ii) of Section 9.5), we can find a sequence $\{p_m\}$ of polynomials on $\overline{g(X)} \cup \{0\}$ such that $\{p_m\} \to h$. We can assume that each polynomial p_m does not have a constant term, i.e., $p_m(0) = 0$. Here is why: For every positive ϵ, there is an index N for which if $m \geq N$, then $\|p_m - h\|_{\max} < \epsilon/2$. Then $|p_m(0) - h(0)| < \epsilon/2$, that is, $|p_m(0)| < \epsilon/2$. Consequently, if we remove the constant term from the p_m, then $\|p_m - h\|_{\max} < \epsilon$ if $m \geq N$, so we still have that $\{p_m\} \to h$ on $\overline{g(X)} \cup \{0\}$. Then because the p_m are polynomials without constant terms and g belongs to the algebra $\mathcal{A}$, each composition $p_m \circ g$ also belongs to $\mathcal{A}$. Furthermore, $\{p_m \circ g\} \to 1/g$. Then $1/g$ is in $\overline{\mathcal{A}}$ by Proposition 6 of Section 9.2. Finally, because $\overline{\mathcal{A}}$ is an algebra, product $g \cdot 1/g$, which is 1, is in $\overline{\mathcal{A}}$, and we are done.

36. (i) Because $\mathcal{A}$ separates points, there is a function g in $\mathcal{A}$ for which $g(u) \neq g(v)$. Define f by

$$f(x) = a\,\frac{g(x) - g(v)}{g(u) - g(v)} + b\,\frac{g(x) - g(u)}{g(v) - g(u)}.$$

Then f is in $\mathcal{A}$ because $\mathcal{A}$ is an algebra, and $f(u) = a$ and $f(v) = b$.

(ii) No, it is not necessarily the case. For example, put—

- X to closed, bounded interval $[0, 3]$, and $\mathcal{A}$ to the collection of polynomial functions on $[0, 3]$ (see the sentence after the Stone-Weierstrass approximation theorem),
- a to 0, and b to 1, and
- A to $[0, 1]$, and B to $[2, 3]$.

A function f in $\mathcal{A}$ for which $f = 0$ on $[0, 1]$ has an infinite number of zeros. But by the fundamental theorem of algebra, a nonzero polynomial has only a finite number of zeros, so f is constant polynomial 0. Then $f \neq 1$ on $[2, 3]$. Thus, there is no function f in $\mathcal{A}$ for which $f = 0$ on $[0, 1]$ and $f = 1$ on $[2, 3]$.

Chapter 13

Continuous Linear Operators Between Banach Spaces

13.1 Normed Linear Spaces

1. Assume first that S is a subspace. Then by definition, every linear combination of vectors in S also belongs to S. In particular, if x and y are vectors in S, then $x+y$ and λx belong to S, so $S+S \subseteq S$ and $\lambda S \subseteq S$. To get the first inclusion in the other direction, note that the zero vector 0 is in S because the zero vector is linear combination $0x$, and S is nonempty. Then $S \subseteq S+S$ because $x = x+0$. Next, note that linear combination $(1/\lambda)x$ belongs to S, so $x \in \lambda S$, that is, $S \subseteq \lambda S$. Thus, $S+S=S$, and $\lambda S = S$.

   Conversely, assume that $S+S=S$ and $\lambda S = S$. Take vectors $x_1, \ldots, x_n$ in S and real numbers $\alpha_1, \ldots, \alpha_n$, and consider linear combination $\sum_{k=1}^{n} \alpha_k x_k$. Each $\alpha_k x_k$ for which $\alpha_k \neq 0$ is in S because $\lambda S = S$. If an α_k is 0, then $\alpha_k x_k = 0$, so we need to verify that the zero vector is in S. Indeed, $-1x_k$, or $-x_k$, is in S because $\lambda S = S$; so $-x_k + x_k$, or 0, is in S because $S+S=S$. Hence, each $\alpha_k x_k$ is in S. Then $\sum_{k=1}^{n} \alpha_k x_k$ also belongs to S by a finite number of applications of the hypothesis that $S+S=S$. Thus, S is a subspace.

2. Take vectors $x_1, \ldots, x_n$ in $Y+Z$ and real numbers $\lambda_1, \ldots, \lambda_n$, and consider linear combination $\sum_{k=1}^{n} \lambda_k x_k$. Now

$$\begin{aligned}
\sum_{k=1}^{n} \lambda_k x_k &= \sum_{k=1}^{n} \lambda_k (y_k + z_k) && \text{for some } y_k \text{ in } Y \text{ and } z_k \text{ in } Z \\
&= \sum_{k=1}^{n} \lambda_k y_k + \sum_{k=1}^{n} \lambda_k z_k \\
&\in Y+Z && Y \text{ and } Z \text{ are subspaces.}
\end{aligned}$$

   Thus, $Y+Z$ also is a subspace.

   To show that $Y+Z = \text{span}[Y \cup Z]$, we first satisfy the authors request (sentence after the definition of *span* on page 254) to show that for a nonempty subset S of X, span$[S]$ is the smallest linear subspace of X that contains S in the sense that span$[S]$ is contained in each linear subspace that contains S. That span$[S]$ is a linear subspace follows directly from definitions of *span* and *linear subspace*. And every vector in span$[S]$ is a linear combination of vectors in S and therefore belongs to each linear subspace that contains S. Hence, span$[S]$ is the smallest linear subspace of X that contains S.

   Returning the problem, note that $Y \subseteq Y+Z$ and $Z \subseteq Y+Z$ by our solution to Problem 1. Then together, $Y \cup Z \subseteq Y+Z$. Because $Y+Z$ is a linear subspace that contains $Y \cup Z$, span$[Y \cup Z] \subseteq Y+Z$. To get the inclusion in the other direction, take a vector x in $Y+Z$. Vector x is $y+z$ for some y in Y and z in Z. Because x is a linear combination of vectors in $Y \cup Z$, we have that $x \in \text{span}[Y \cup Z]$, which shows that $Y+Z \subseteq \text{span}[Y \cup Z]$. Thus, $Y+Z = \text{span}[Y \cup Z]$.

3. (i) Take a collection $\{Y_\mu\}$ of linear subspaces of X, and let Y be their intersection $\bigcap_\mu Y_\mu$. Intersection Y is not empty because it contains the zero vector since every linear subspace Y_μ contains the zero vector by our solution to Problem 1. Take vectors $y_1, \ldots, y_n$ in Y and real numbers $\lambda_1, \ldots, \lambda_n$, and consider linear combination $\sum_{k=1}^{n} \lambda_k y_k$. Because $Y = \bigcap_\mu Y_\mu$, each y_k is in every Y_μ. Then because every Y_μ is a linear subspace, $\sum_{k=1}^{n} \lambda_k y_k$ belongs to every Y_μ. Then $\sum_{k=1}^{n} \lambda_k y_k$ also belongs to Y. Thus, Y is a linear subspace of X.

(ii) The intersection Y of all linear subspaces of X that contain S is a linear subspace by part (i). Moreover, Y is the smallest linear subspace of X that contains S. Hence, span$[S]$ is Y and is therefore a linear subspace.

(iii) We first justify the fifth sentence after the definition of *span*. Closure $\overline{\text{span}}[S]$ is a linear subspace of X by Problem 6(ii). And Proposition 3 of Section 9.2 tells us that $\overline{\text{span}}[S]$ is the smallest closed subset of X containing span$[S]$.

Thus, the claim of this part (iii) is true by the same reasoning as in our solution to part (ii) along with the fact that the intersection of a collection of closed subsets is closed by Proposition 5 of Section 9.2.

4. Function $\|\cdot\|$ is the same as function $\rho(\,\cdot\,,0)$ where ρ is the metric induced by $\|\cdot\|$. Then $\|\cdot\|$ is (uniformly) continuous by Problem 30 of Section 9.3.

   We reword our solution to that problem using the norm notation: Take vectors x, y in X. By the reverse triangle inequality (see our solution to Problem 49(iv) of Section 1.6), $|\|x\| - \|y\|| \leq \|x - y\|$. Function $\|\cdot\|$ is uniformly continuous because δ equal to ϵ responds to all ϵ challenges for if $\|x - y\| < \delta$, then $|\|x\| - \|y\|| \leq \|x - y\| < \delta = \epsilon$.

5. We first show that $\|\cdot\|$ satisfies the properties of a norm listed at the beginning of the section. Cartesian product $X \times Y$ is a linear space because the properties of a linear space carry over componentwise from linear spaces X and Y. Function $\|\cdot\|$ is nonnegative and real-valued because it is the sum of two norms. Moreover,

   - $\|(x, y)\| = 0$ if and only if $\|x\|_1 + \|y\|_2 = 0$ if and only if x, y both equal 0 if and only if $(x, y) = (0, 0)$,
   - $\|(x_1, y_1) + (x_2, y_2)\| = \|x_1 + x_2\|_1 + \|y_1 + y_2\|_2 \leq \|x_1\|_1 + \|x_2\|_1 + \|y_1\|_2 + \|y_2\|_2 = \|(x_1, y_1)\| + \|(x_2, y_2)\|$, and
   - $\|\lambda(x, y)\| = \|\lambda x\|_1 + \|\lambda y\|_2 = |\lambda|\|x\|_1 + |\lambda|\|y\|_2 = |\lambda|\|(x, y)\|$.

   Hence, $\|\cdot\|$ defines a norm on $X \times Y$.

   Next, assume that $\{(x_n, y_n)\} \to (x, y)$ with respect to $\|\cdot\|$. Then for every positive ϵ, there is an index N for which if $n \geq N$, then $\|(x_n, y_n) - (x, y)\| < \epsilon$, that is, $\|x_n - x\|_1 + \|y_n - y\|_2 < \epsilon$. Consequently, if $n \geq N$, then $\|x_n - x\|_1 < \epsilon$ and $\|y_n - y\|_2 < \epsilon$. Hence, $\{x_n\} \to x$, and $\{y_n\} \to y$.

   Conversely, assume that $\{x_n\} \to x$ and $\{y_n\} \to y$. Then there is an index J for which if $n \geq J$, then $\|x_n - x\|_1 < \epsilon/2$ and $\|y_n - y\|_2 < \epsilon/2$. Add these two inequalities to get that $\|x_n - x\|_1 + \|y_n - y\|_2 < \epsilon$, that is, $\|(x_n, y_n) - (x, y)\| < \epsilon$, if $n \geq J$. Thus, $\{(x_n, y_n)\} \to (x, y)$.

   Normed linear space $X \times Y$ is a Banach space provided every Cauchy sequence in $X \times Y$ converges to a vector pair in $X \times Y$. Take a Cauchy sequence $\{(x_n, y_n)\}$ in $X \times Y$. There is an index K such that $\|(x_n, y_n) - (x_m, y_m)\| < \epsilon$, that is, $\|x_n - x_m\|_1 + \|y_n - y_m\|_2 < \epsilon$, if $m, n \geq K$. Consequently, $\|x_n - x_m\|_1 < \epsilon$ and $\|y_n - y_m\|_2 < \epsilon$ if $m, n \geq K$, which show that $\{x_n\}$ and $\{y_n\}$ are Cauchy sequences. Because X and Y are Banach spaces, $\{x_n\}$ converges to a vector x in X, and $\{y_n\}$ converges to a vector y in Y. Then $\{(x_n, y_n)\}$ converges to vector pair (x, y) in $X \times Y$. Thus, $X \times Y$ is a Banach space. Cf. Problem 43 of Section 9.4.

6. (i) This is Problem 24 of Section 7.3.

   (ii) Take vectors $\bar{y}_1, \ldots, \bar{y}_n$ in $\overline{Y}$ and real numbers $\lambda_1, \ldots, \lambda_n$, and consider linear combination $\sum_{k=1}^{n} \lambda_k \bar{y}_k$. By Proposition 6 of Section 9.2, for each k, there is a sequence $\{y_k(m)\}$ in Y such that $\{y_k(m)\} \to \bar{y}_k$ where m is the index for the sequences. By a finite number of applications of part (i), $\{\sum_{k=1}^{n} \lambda_k y_k(m)\} \to \sum_{k=1}^{n} \lambda_k \bar{y}_k$ in X. Now for all m, linear combination $\sum_{k=1}^{n} \lambda_k y_k(m)$ belongs to Y because Y is a linear subspace. Linear combination $\sum_{k=1}^{n} \lambda_k \bar{y}_k$ is therefore the limit of a sequence in Y. Hence, $\sum_{k=1}^{n} \lambda_k \bar{y}_k$ belongs to $\overline{Y}$ by another appeal to Proposition 6 of Section 9.2. Thus, $\overline{Y}$ is a linear subspace of X. Cf. Problem 34 of Section 12.3.

(iii) For convenience, assume that $X \times X$ has the norm of Problem 5. Take a vector pair (x, y) in $X \times X$ and a sequence $\{(x_n, y_n)\}$ in $X \times X$.

$$\begin{aligned} &\text{If } \{(x_n, y_n)\} \to (x, y), \text{ then } \{x_n\} \to x \text{ and } \{y_n\} \to y && \text{Problem 5,} \\ &\text{then } \{x_n + y_n\} \to x + y && \text{part (i).} \end{aligned}$$

Thus, vector summation is continuous by definition (page 190).
Next, take a pair (α, x) in $\mathbf{R} \times X$ and a sequence $\{(\alpha, x_n)\}$ in $\mathbf{R} \times X$.

$$\begin{aligned} &\text{If } \{(\alpha, x_n)\} \to (\alpha, x), \text{ then } \{x_n\} \to x && \text{Problem 5,} \\ &\text{then } \{\alpha x_n\} \to \alpha x && \text{part (i) with } \beta \text{ equal to } 0. \end{aligned}$$

Thus, scalar multiplication is continuous.

7. Set $\mathcal{P}$ is a linear space because $\mathcal{P}$ is a nonempty subset of linear space $C[a, b]$ (last example on page 138) and because every linear combination of polynomials on $[a, b]$ is also a polynomial on $[a, b]$.

   Next, the Weierstrass approximation theorem tells us that $\mathcal{P}$ is dense in $C[a, b]$. Then $\overline{\mathcal{P}} = C[a, b]$ by the sentence after the definition on page 203. But there is a function in $C[a, b]$ that is not a polynomial. For example, function $x \mapsto 1/(x - a + 1)$ is continuous on $[a, b]$ but is not a polynomial because none of its nth derivatives vanishes. Thus, $\mathcal{P}$ as a subset of $C[a, b]$ fails to be closed.

   As we showed above, $\mathcal{P}$ does not contain all of its points of closure in $C[a, b]$, so because $C[a, b] \subseteq L^1[a, b]$, space $\mathcal{P}$ does not contain all of its points of closure in $L^1[a, b]$. Thus, $\mathcal{P}$ as a subset of $L^1[a, b]$ fails to be closed. (We proved that $C[a, b] \subseteq L^1[a, b]$ in our solution to Problem 1 of Section 7.1.)

8. See the last paragraph of Section 9.1. Relation $\cong$ is—

   - reflexive because $\|x - x\| = 0$,
   - symmetric because if $\|x - y\| = 0$, then $\|y - x\| = 0$, and
   - transitive because if $\|x - y\| = 0$ and $\|y - z\| = 0$, then $\|x - z\| \leq \|x - y\| + \|y - z\| = 0$.

   Thus, $\cong$ defines an equivalence relation.

   Set $X/\cong$ is a vector space because the properties of a vector space carry over from X. In particular, linear combinations $\alpha[x] + \beta[y]$ are in $X/\cong$ because $\alpha[x] + \beta[y] = [\alpha x + \beta y]$, and $\alpha x + \beta y$ is in vector space X. Function $\|\cdot\|$ on $X/\cong$ is nonnegative, real-valued, and satisfies the properties of a norm:

$$\begin{aligned} \big\|[x]\big\| = 0 &\text{ if and only if } \|x\| = 0 \\ &\text{ if and only if } \|x - 0\| = 0 \\ &\text{ if and only if } x \cong 0 \\ &\text{ if and only if } [x] = [0]. \end{aligned}$$

   The remaining two properties of a norm (top of page 254) follow directly from definition $\big\|[x]\big\| = \|x\|$. Thus, $X/\cong$ is a normed vector space.

The authors have already illustrated this procedure with $L^p[a,b]$. See the first two paragraphs and first example of Section 7.1 with E replaced by $[a,b]$ and the first three paragraphs of Section 7.2. The pseudonorm is defined by

$$\|f\|_p = \left(\int_a^b |f|^p\right)^{1/p}.$$

And $f \cong g$ provided $f(x) = g(x)$ for almost all x in $[a,b]$.

13.2 Linear Operators

In the proof of Theorem 1, the authors state, "Since T is linear, $T(0) = 0$." That is because $T(0) = T(0+0) = T(0) + T(0)$.

9. (i) We apply the definition on page 190. If T is continuous, then it is continuous at every point in X, in particular, at u_0. Conversely, assume that T is continuous at u_0. Take a point u in X and a sequence $\{u_n\}$ in X. If $\{u_n\} \to u$, then $\{u_n - u + u_0\} \to u_0$. Because $\{u_n - u + u_0\}$ is a sequence that converges to u_0, and T is continuous at u_0,

$$\lim_{n\to\infty} T(u_n - u + u_0) = T(u_0),$$

but because T is linear,

$$\begin{aligned}\lim_{n\to\infty} T(u_n - u + u_0) &= \lim_{n\to\infty} (T(u_n) - T(u) + T(u_0)) \\ &= \lim_{n\to\infty} T(u_n) - T(u) + T(u_0) \qquad \text{Problem 24 of Section 7.3.}\end{aligned}$$

Equating the two expressions for $\lim_{n\to\infty} T(u_n - u + u_0)$, we get that $\lim_{n\to\infty} T(u_n) = T(u)$, in other words, $\{T(u_n)\} \to T(u)$. Thus, T is continuous.

(ii) If T is Lipschitz, then it is (uniformly) continuous by the example at the end of Section 9.3. Conversely, assume that T is continuous. Then T is bounded by Theorem 1. Take two vectors u and v in X.

$$\begin{aligned}\|T(u) - T(v)\| &= \|T(u - v)\| && T \text{ is linear} \\ &\le M\|u - v\| && \text{inequality (1)}, u - v \in X,\end{aligned}$$

which shows that T is Lipschitz. Cf. Problem 4 of Section 8.1.

(iii) By our solution to Problem 2 of Section 12.1, function $x \to 1/x$ from $\mathbf{R}$ to $\mathbf{R} \cup \pm\infty$ is continuous at $1/2$ but not at 0.

We exhibit a (uniformly) continuous mapping that is not Lipschitz in our solution to Problem 50 of Section 1.6. Another counterexample is Problem 29 of Section 10.3.

10. Mapping T is bounded by hypothesis, so we can use the second part of our solution to Problem 9(ii). Then $\|T\|$ is the smallest Lipschitz constant for T by definitions of a Lipschitz constant and $\|T\|$.

11. Cf. Problem 1 of Section 8.1. Denote $\sup\{\|T(u)\| \mid u \in X, \ \|u\| \le 1\}$ by $\|T\|'$. Take a nonzero v in X. We have that $\|T(v/\|v\|)\| \le \|T\|'$ because $v/\|v\|$ has norm 1. Then $\|T(v)\| \le \|T\|'\|v\|$ by linearity of T and positive homogeneity of a norm. That inequality holds also if $v = 0$ because $T(0) = 0$. Hence, $\|T\| \le \|T\|'$ by definition of an operator norm.

We now get the opposite inequality. As stated in the text, it is easy to see that inequality (1) holds if $M = \|T\|$, that is, $\|T(v)\| \le \|T\|\|v\|$ for all v in X. In particular, if $\|u\| \le 1$, then $\|T(u)\| \le \|T\|\|u\| \le \|T\|$. Hence, $\|T\|' \le \|T\|$. Thus, $\|T\| = \|T\|'$.

12. Starting with the triangle inequality,

$$\begin{aligned}\|T_n(u_n) - T(u)\| &\le \|T_n(u_n) - T(u_n)\| + \|T(u_n) - T(u)\| \\ &= \|(T_n - T)(u_n)\| + \|T(u_n - u)\| && \text{definition (4); linearity of } T \\ &\le \|T_n - T\|\|u_n\| + \|T\|\|u_n - u\| && \text{definition (1).}\end{aligned}$$

Take the limit of the left and right sides to get that $\lim_{n\to\infty} \|T_n(u_n) - T(u)\| = 0$ because—

- $\lim_{n\to\infty} \|T_n - T\| = 0$, and $\lim_{n\to\infty} \|u_n - u\| = 0$ by hypothesis,
- $\{u_n\}$ is bounded by the normed-linear-space analogue of Proposition 14 of Section 1.5 (the proof that we offer at the beginning of our Section 1.5 solutions holds if we replace $|\cdot|$ with $\|\cdot\|$),
- T and $\{T_n - T\}$ are bounded by hypothesis and Proposition 2, and
- Problem 24 of Section 7.3 allows us to apply the limit linearly to the right side.

Thus, $\{T_n(u_n)\} \to T(u)$ in Y.

13. (i) Operator $I - T$ is in $\mathcal{L}(X, X)$ because $I \in \mathcal{L}(X, X)$ and $\mathcal{L}(X, X)$ is a linear space by Proposition 2.

Operator T is a contraction by definition (page 215) and Problem 10, so by Banach's contraction principle, T has exactly one fixed point. Here, it is the zero vector because $T(0) = 0$ since T is linear. To show that $I - T$ is one-to-one, apply the next-to-last sentence of the section: $(I - T)(x) = 0$ if and only if $x - T(x) = 0$ if and only if $x = 0$. Hence, $\ker(I - T) = \{0\}$.

To show that $I - T$ is onto, take a vector v in X. Define operator T_v in $\mathcal{L}(X, X)$ by $T_v(u) = T(v+u)$. Operator T_v is a contraction because T is a contraction and therefore has exactly one fixed point w. Then

$$\begin{aligned}(I - T)(v + w) &= v + w - T_v(w) \\ &= v + w - w \\ &= v.\end{aligned}$$

Thus, $I - T$ is onto.

(ii) By the next-to-last paragraph of the section, it suffices to show that $(I - T)^{-1}$ is bounded. Now $(I - T)^{-1}(v) = u$ for some u. Then $v = u - T(u)$, or $u = v + T(u)$. By the triangle inequality and definition of an operator norm, $\|u\| \le \|v\| + \|T\|\|u\|$. Solve for $\|u\|$ to get that

$$\|u\| \le (1 - \|T\|)^{-1}\|v\|,$$

that is,

$$\left\|(I - T)^{-1}(v)\right\| \le (1 - \|T\|)^{-1}\|v\|,$$

which shows that $(I - T)^{-1}$ is bounded. Thus, $I - T$ is an isomorphism.

Another proof is to note that by the next-to-last paragraph of the section, it suffices to show that $(I - T)^{-1}$ is continuous. But that is true by Corollary 9.

A second solution follows directly from Problem 14: Because $(I - T)^{-1}$ exists and is in $\mathcal{L}(X, X)$, operator $I - T$ is one-to-one, onto, and an isomorphism.

14. Note that Id is the identity map I on X—see the proof of Corollary 10.

(i) We show that the sequence $\{S_k\}$ of partial sums $\sum_{n=0}^{k} T^n$ (see page 24) is a Cauchy sequence. Take a positive ϵ. Choose index K for which $\|T\|^{K+1}/(1-\|T\|) < \epsilon$; such a K exists because $\|T\| < 1$. Then if $j \geq k \geq K$,

$$
\begin{aligned}
\|S_j - S_k\| &= \left\| \sum_{n=k+1}^{j} T^n \right\| \\
&\leq \sum_{n=k+1}^{j} \|T^n\| && \text{triangle inequality} \\
&\leq \sum_{n=k+1}^{j} \|T\|^n && \text{Problem 19} \\
&\leq \sum_{n=k+1}^{\infty} \|T\|^n \\
&= \|T\|^{k+1}/(1-\|T\|) && \text{sum of geometric series} \\
&< \epsilon && \text{choice of } K,
\end{aligned}
$$

which shows that $\{S_k\}$ is Cauchy and hence converges in $\mathcal{L}(X,X)$ because $\mathcal{L}(X,X)$ is complete by Theorem 3. Thus, $\sum_{n=0}^{\infty} T^n$ converges in $\mathcal{L}(X,X)$.

(ii) We need to show that $(I-T)\sum_{n=0}^{\infty} T^n = I$, and $\sum_{n=0}^{\infty} T^n(I-T) = I$. Because powers of T commute, it suffices to prove that the second equality holds. Observe that

$$\sum_{n=0}^{\infty} T^n(I-T) = \lim_{k\to\infty}\left(\sum_{n=0}^{k} T^n - \sum_{n=0}^{k} T^{n+1}\right) = \lim_{k\to\infty}\left(I - T^{k+1}\right).$$

Now refer to the first paragraph of the proof of Banach's contraction principle. Express sequence $\{x_k\}$ there equivalently as $\{T^k(x_0)\}$. That sequence converges to the fixed point of T. Here, the fixed point is the zero vector because $T(0) = 0$ since T is linear. Hence, we can continue our chain of equations above as follows:

$$\lim_{k\to\infty}\left(I - T^{k+1}\right) = I - 0 = I.$$

Thus, the inverse of $I - T$ is $\sum_{n=0}^{\infty} T^n$.

15. Refer to the "Mappings between sets" subsection (page 4) and the next-to-last paragraph of this section. Assume first that T is an isomorphism. Then T has an inverse in $\mathcal{L}(Y,X)$. Put S to that inverse. Then $S(T(u)) = u$ and $T(S(v)) = v$.

    Conversely, assume that there is an operator S in $\mathcal{L}(Y,X)$ such that $S(T(u)) = u$ and $T(S(v)) = v$. Then T is invertible so that it is one-to-one and onto. Thus, T is an isomorphism.

16. To show that $\ker T$ is a subspace of X, take vectors $u_1, \ldots, u_n$ in $\ker T$ and real numbers $\lambda_1, \ldots, \lambda_n$, and consider linear combination $\sum_{k=1}^{n} \lambda_k u_k$. By linearity of T, we have that $T(\sum_{k=1}^{n} \lambda_k u_k) = \sum_{k=1}^{n} \lambda_k T(u_k)$, which is 0 because each $T(u_k)$ is 0 since each u_k is in $\ker T$. Hence, linear combination $\sum_{k=1}^{n} \lambda_k u_k$ belongs to $\ker T$. Thus, $\ker T$ is a subspace.

To show that $\ker T$ is closed, apply Proposition 6 of Section 9.2. Take a sequence $\{u_m\}$ in $\ker T$ that converges to a limit u in X. By property (3), $\{T(u_m)\} \to T(u)$ in Y. But each $T(u_m)$ is 0 because each u_m is in $\ker T$. Hence, $T(u) = 0$ so that $u \in \ker T$. We have shown that whenever a sequence in $\ker T$ converges to a limit in X, the limit belongs to $\ker T$. Thus, $\ker T$ is closed in X.

Next,

$$\begin{aligned} T \text{ is one-to-one} &\text{ if and only if equality } T(v) = T(w) \text{ implies that } v = w && \text{definition} \\ &\text{ if and only if equality } T(v - w) = 0 \text{ implies that } v = w && \text{linearity of } T \\ &\text{ if and only if } \ker T = \{0\}. \end{aligned}$$

17. Verify the conditions of a normed linear space in the first paragraph of Section 13.1. For f, g in $\mathrm{Lip}_0(X)$, function $f + g$ is Lipschitz, and

$$(f+g)(x_0) = f(x_0) + g(x_0) = 0 + 0 = 0,$$

so $f + g \in \mathrm{Lip}_0(X)$. Furthermore, for a real number α, function αf is Lipschitz, and $\alpha f(x_0) = \alpha 0 = 0$, so αf also is in $\mathrm{Lip}_0(X)$. Set $\mathrm{Lip}_0(X)$ satisfies the remaining conditions of a linear space because $\mathrm{Lip}_0(X)$ inherits them from linear space $C(X)$ since $\mathrm{Lip}_0(X) \subseteq C(X)$ by Problem 9(ii). Next, $\|\cdot\|$ is a nonnegative real-valued function defined on $\mathrm{Lip}_0(X)$, and—

- $\|f\| = 0$ if and only if $f(x) = f(y)$ for all x, y where $x \neq y$, in particular when $x = x_0$, if and only if $f = 0$,
- $\|f + g\| \le \|f\| + \|g\|$ because

$$|(f+g)(x) - (f+g)(y)| \le |f(x) - f(y)| + |g(x) - g(y)|$$

by the triangle inequality and because a general property of the supremum function is that $\sup(\alpha + \beta) \le \sup \alpha + \sup \beta$ where β is another real number, and

- $\|\alpha f\| = \sup\limits_{x \neq y} \dfrac{|\alpha f(x) - \alpha f(y)|}{\rho(x,y)} = |\alpha| \sup\limits_{x \neq y} \dfrac{|f(x) - f(y)|}{\rho(x,y)} = |\alpha| \|f\|.$

Thus, $\mathrm{Lip}_0(X)$ is a linear space that is normed by $\|\cdot\|$.

Parts of our proof that $\mathrm{Lip}_0(X)$ is a Banach space are like the proof of Theorem 3. Let $\{f_n\}$ be a Cauchy sequence in $\mathrm{Lip}_0(X)$. Then for each positive ϵ, there is an index N for which

$$\text{if } m, n \ge N, \text{ then } \|f_n - f_m\| = \sup_{x \neq y} \frac{|(f_n - f_m)(x) - (f_n - f_m)(y)|}{\rho(x,y)} < \epsilon. \tag{13.1}$$

Now put y to x_0 so that if $x \neq x_0$ and $m, n \ge N$, then $|f_n(x) - f_m(x)| < \epsilon \rho(x, x_0)$. Therefore, $\{f_n(x)\}$ is a Cauchy sequence of real numbers for each x (including x_0) in X. By Theorem 17 of Section 1.5, $\{f_n(x)\}$ converges to a real number, which we denote by $h(x)$. That defines a mapping $h : X \to \mathbf{R}$. We must show that h belongs to $\mathrm{Lip}_0(X)$, and $\{f_n\} \to h$ in $\mathrm{Lip}_0(X)$. Indeed, $h(x_0) = 0$ because sequence $\{f_n(x_0)\}$ is the constant zero sequence $\{0\}$. Moreover, h is Lipschitz because for all x, y in X,

$$\begin{aligned} |h(x) - h(y)| &= \left| \lim_{n\to\infty} (f_n(x) - f_n(y)) \right| && \text{Theorem 18 (linearity) of Section 1.5} \\ &= \lim_{n\to\infty} |f_n(x) - f_n(y)| \\ &\le \lim_{n\to\infty} \|f_n\| \rho(x,y) && \text{definition of } \|\cdot\|\text{; Th. 18 (monotonicity) of Sec. 1.5} \\ &\le c\rho(x,y) && \text{Problem 38(ii) of Section 9.4} \end{aligned}$$

where positive number c is a bound for Cauchy sequence $\{f_n\}$. Hence, $h \in \mathrm{Lip}_0(X)$. Next, we establish the convergence of $\{f_n\}$ to h in $\mathrm{Lip}_0(X)$. If $n \geq N$, then

$$\begin{aligned}
\|f_n - h\| &= \sup_{x\neq y} \frac{|(f_n-h)(x)-(f_n-h)(y)|}{\rho(x,y)} && \text{definition of } \|\cdot\| \\
&= \sup_{x\neq y} \frac{\lim\limits_{m\to\infty} |(f_n-f_m)(x)-(f_n-f_m)(y)|}{\rho(x,y)} && \{f_m\}\to h \text{ pointwise} \\
&\leq \sup_{x\neq y} \frac{\lim\limits_{m\to\infty} \|f_n-f_m\|\rho(x,y)}{\rho(x,y)} && \text{definition of } \|\cdot\|\text{; monotonicity} \\
&< \epsilon && \text{inequality (13.1).}
\end{aligned}$$

Thus, $\{f_n\} \to h$ in $\mathrm{Lip}_0(X)$.

To show that F_x belongs to $\mathcal{L}(\mathrm{Lip}_0(X), \mathbf{R})$, we first show that F_x is indeed linear by applying definition of linearity (page 256):

$$F_x(\alpha f + \beta g) = (\alpha f + \beta g)(x) = \alpha f(x) + \beta g(x) = \alpha F_x(f) + \beta F_x(g).$$

Secondly, F_x is bounded because $\|F_x(f)\| = |f(x)| = |f(x) - f(x_0)| \leq \rho(x, x_0)\|f\|$ by definition of $\|\cdot\|$. Thus, $F_x \in \mathcal{L}(\mathrm{Lip}_0(X), \mathbf{R})$.

We next prove equality $\|F_x - F_y\| = \rho(x,y)$ by proving the corresponding inequalities. For all f, we have that $\|(F_x - F_y)(f)\| = \|F_x(f) - F_y(f)\| = |f(x) - f(y)| \leq \rho(x,y)\|f\|$. Then $\|F_x - F_y\| \leq \rho(x,y)$ by definition of an operator norm. We now get the inequality in the opposite direction. If $\|f\| \leq 1$, then for each positive ϵ, there are x, y in X such that $|f(x) - f(y)| > (1-\epsilon)\rho(x,y)$ by definition of $\|\cdot\|$. Then

$$\begin{aligned}
\|F_x - F_y\| &= \sup\{\|(F_x - F_y)(f)\| \mid f \in \mathrm{Lip}_0(X),\ \|f\| \leq 1\} && \text{Problem 11} \\
&= \sup\{|f(x) - f(y)| \mid f \in \mathrm{Lip}_0(X),\ \|f\| \leq 1\} \\
&> (1-\epsilon)\rho(x,y).
\end{aligned}$$

Because that holds for ϵ arbitrarily small, $\|F_x - F_y\| \geq \rho(x,y)$. Here is a second way to get that inequality: For each x in X, define function d_x on X by $d_x(y) = \rho(y,x) - \rho(x_0,x)$. We have that $d_x(x_0) = 0$ and

$$\begin{aligned}
\|d_x\| &= \sup_{y\neq z} \frac{|d_x(y) - d_x(z)|}{\rho(y,z)} && \text{definition of } \|\cdot\| \\
&= \sup_{y\neq z} \frac{|\rho(y,x) - \rho(x_0,x) - \rho(z,x) + \rho(x_0,x)|}{\rho(y,z)} && \text{definition of } d_x \\
&\leq \sup_{y\neq z} \frac{\rho(y,z)}{\rho(y,z)} && \text{reverse triangle inequality} \\
&= 1,
\end{aligned}$$

which shows not only that $\|d_x\| \leq 1$ but also that d_x is Lipschitz (with Lipschitz constant no greater than 1) so that $d_x \in \mathrm{Lip}_0(X)$. (Regarding the reverse triangle inequality, see our solution to Problem 49(iv) of Section 1.6.) Then

$$\begin{aligned}
\|F_x - F_y\| &= \sup\{|f(x) - f(y)| \mid f \in \mathrm{Lip}_0(X),\ \|f\| \leq 1\} \\
&\geq |d_x(x) - d_x(y)| \\
&= |\rho(x,x) - \rho(x_0,x) - \rho(y,x) + \rho(x_0,x)| \\
&= \rho(x,y).
\end{aligned}$$

Thus, $\|F_x - F_y\| = \rho(x, y)$.

Next, refer to the page-184 definition of an isometry. Metric space X is isometric to subset $\{F_x \mid x \in X\}$ of $\mathcal{L}(\mathrm{Lip}_0(X), \mathbf{R})$ because mapping $x \mapsto F_x$ from X to $\{F_x \mid x \in X\}$ is onto and $\|F_x - F_y\| = \rho(x, y)$. Note that $\mathcal{L}(\mathrm{Lip}_0(X), \mathbf{R})$ is indeed a Banach (complete metric) space by Theorem 3 and the sentence after the definition on page 145. Hence, the closure of subset $\{F_x \mid x \in X\}$ yields a completion for metric space X.

18. By the paragraph after the page-184 definition of an isometry, two isometric spaces are exactly the same. Then by our solution to Problem 17, $\overline{X} = \overline{\{F_x \mid x \in X\}}$, and $\overline{\{F_x \mid x \in X\}}$ is a Banach space. Thus, normed linear space X is a dense subspace of a Banach space by the sentence after the definition on page 203 (note that the X in that sentence is not the X of this problem but rather $\overline{\{F_x \mid x \in X\}}$).

19. We state and solve a more general version of this problem:

For X, Y, Z normed linear spaces and $T \in \mathcal{L}(X, Y)$ and $S \in \mathcal{L}(Y, Z)$, show that composition $S \circ T$ belongs to $\mathcal{L}(X, Z)$ and $\|S \circ T\| \leq \|S\| \cdot \|T\|$.

Composition $S \circ T$ is linear because it satisfies the definition at the beginning of the section:

$$\begin{aligned}(S \circ T)(\alpha u + \beta v) &= S(\alpha T(u) + \beta T(v)) && \text{linearity of } T\\ &= \alpha S(T(u)) + \beta S(T(v)) && \text{linearity of } S\\ &= \alpha(S \circ T)(u) + \beta(S \circ T)(v).\end{aligned}$$

To show that $S \circ T$ is bounded, apply twice the second definition of the section:

$$\begin{aligned}\|(S \circ T)(u)\| &\leq \|S\|\|T(u)\| && \text{boundedness of } S\\ &\leq \|S\|\|T\|\|u\| && \text{boundedness of } T.\end{aligned}$$

Thus, $S \circ T \in \mathcal{L}(X, Z)$. The result above also shows that $\|S \circ T\| \leq \|S\|\|T\|$.

20. Function $\|\cdot\|_1$ is nonnegative and real-valued on X because $\|\cdot\|$ is. For all x, x' in X and each positive ϵ, there are y, y' in Y such that

$$\begin{aligned}\|x - y\| &< \|x\|_1 + \epsilon,\\ \|x' - y'\| &< \|x'\|_1 + \epsilon\end{aligned}$$

by definition of inf. Then

$$\begin{aligned}\|x + x'\|_1 &\leq \|x + x' - (y + y')\| && \text{definition of } \|\cdot\|_1;\ y + y' \in Y \text{ since } Y \text{ is a linear space}\\ &\leq \|x - y\| + \|x' - y'\| && \text{triangle inequality}\\ &< \|x\|_1 + \|x'\|_1 + 2\epsilon.\end{aligned}$$

Because that holds for ϵ arbitrarily small, $\|x + x'\|_1 \leq \|x\|_1 + \|x'\|_1$. Also,

$$\|\alpha x\|_1 = \inf_{y \in Y} \|\alpha x - y\| = |\alpha| \inf_{y \in Y} \|x - y\| = |\alpha|\|x\|_1$$

where we once more used that Y is a linear space so that real multiples of vectors in Y are again in Y. Thus, $\|\cdot\|_1$ is a pseudonorm on X.

We remark that Y is specified to be closed so that $\inf_{y \in Y}\|x - y\| = 0$ if and only if $x \in Y$ (see Problem 33(ii) of Section 9.3). Now the natural map φ of X onto X/Y is the map that takes a vector in X to its equivalence class, that is, $\varphi(x) = [x]$. Then because Y is closed, $\|\varphi(x)\| = 0$ if and only if $x \in Y$.

Take an open set $\mathcal{O}$ in X and a vector x in $\mathcal{O}$. By the page-186 definition of an open set and definition of $\|\cdot\|_1$, there is a positive r such that if $\|x' - x\|_1 < r$, then $x' \in \mathcal{O}$. Now if $\|z - \varphi(x)\| < r$ where $z = \varphi(x')$ for some x' in X, then $\|x' - x\|_1 < r$, so $x' \in \mathcal{O}$ and $z \in \varphi(\mathcal{O})$. Hence, $\varphi(\mathcal{O})$ is open. Thus, φ maps open sets into open sets.

21. Apply our solution to Problem 32 and the notation of Problem 20. Take a sequence $\{\varphi(x_n)\}$ in X/Y such that $\sum\|\varphi(x_n)\|$ is summable where index n of summation runs from 1 to ∞. Then $\sum\|x_n\|_1$ is summable. Take a positive ϵ. For each n, there is a y_n in Y such that $\|x_n - y_n\| < \|x_n\|_1 + (1/2)^n$ by definition of inf. Then $\sum\|x_n - y_n\| < \sum\|x_n\|_1 + 1$, which shows that $\sum\|x_n - y_n\|$ is summable. By Problem 32, $\sum(x_n - y_n) = s$ where s is some vector in X. Now φ is continuous by Theorem 1 because φ is linear by the next-to-last sentence of Problem 8 and bounded by the second part of this problem. Also, the kernel of φ is Y because $\{x \in X \mid \inf_{y\in Y}\|x - y\| = 0\} = Y$. Then

$$\begin{aligned}
\sum \varphi(x_n) &= \sum(\varphi(x_n) - \varphi(y_n)) && \varphi(y_n) = 0 \text{ because } y_n \in \ker\varphi \\
&= \varphi\Big(\sum(x_n - y_n)\Big) && \text{linearity of } \varphi \\
&= \varphi(s),
\end{aligned}$$

which shows that $\sum\varphi(x_n)$ is summable. Thus, by the converse part of our solution to Problem 32, X/Y is a Banach space.

To show that $\|\varphi\| = 1$, we prove the corresponding inequalities. Since $\|\varphi(x)\| = \|x\|_1 \le \|x\|$ for all x in X, we have that $\|\varphi\| \le 1$ by definition of an operator norm. Before getting the inequality in the opposite direction, we remark that Y is specified to be a proper subspace of X because if $Y = X$, then $\ker\varphi = X$ so that $\varphi = 0$ and $\|\varphi\| = 0$. Now for an x in $X \sim Y$ and each positive ϵ, there is a y in Y such that $\|x\|_1 > (1-\epsilon)\|x - y\|$. Then $\|\varphi(x - y)\| = \|\varphi(x)\| = \|x\|_1 > (1-\epsilon)\|x - y\|$. Hence, $\|\varphi\| > 1 - \epsilon$. Because that holds for ϵ arbitrarily small, $\|\varphi\| \ge 1$. Thus, φ has norm 1. Another way to get that inequality is to apply Riesz's lemma (Section 13.3, but proof of the lemma does not rely on concepts introduced after Section 13.2). There is a unit vector u in X for which $\|u - y\| > 1 - \epsilon$ for all y in Y. Then $\|\varphi(u)\| = \|\varphi(u - y)\| = \|u - y\|_1 > 1 - \epsilon$. By Problem 11, $\|\varphi\| > 1 - \epsilon$.

22. We verify that S defined by $(S \circ \varphi)(x) = T(x)$ is well defined, linear, unique, and bounded.

We show that S is well defined by showing that if $\varphi(x_1) = \varphi(x_2)$, then $T(x_1) = T(x_2)$. Indeed, $\varphi(x_1) = \varphi(x_2)$ only if $\varphi(x_1 - x_2) = 0$ only if $x_1 - x_2 \in \ker T$ only if $T(x_1 - x_2) = 0$ only if $T(x_1) = T(x_2)$ where we used linearity of both φ (next-to-last sentence of Problem 8) and T (hypothesis). Note that S is one-to-one because our reasoning holds if we replace *only if* with *if and only if.*

Applying the first definition of the section, S is linear because φ and T are linear:

$$\begin{aligned}
S(\alpha\varphi(x_1) + \beta\varphi(x_2)) = S(\varphi(\alpha x_1 + \beta x_2)) = T(\alpha x_1 + \beta x_2) &= \alpha T(x_1) + \beta T(x_2) \\
&= \alpha S(\varphi(x_1)) + \beta S(\varphi(x_2)).
\end{aligned}$$

To show uniqueness, let $U : X/Z \to Y$ be another operator such that $U \circ \varphi = T$. For all y in image $T(X)$ of T, there is an x in X such that $T(x) = y$. Then $(U\circ\varphi)(x) = T(x) = (S\circ\varphi)(x)$, and as $\varphi(x)$ is uniquely defined, we must have that $U([x]) = S([x])$. As that is true for all $U([x])$ in $T(X)$, it follows that $U = S$. (Recall that $[x]$ is the equivalence class of x.)

Finally, S is bounded because $\|S\| = \|T\|$, which we show by proving the corresponding inequalities. Starting with our solution to Problem 19, $\|T\| = \|S \circ \varphi\| \le \|S\|\|\varphi\| \le \|S\|$

where $\|\varphi\|$ is 0 or 1 by our solution to Problem 21. To get the inequality in the opposite direction, note that

$$\begin{aligned}
\|(S\circ\varphi)(x)\| = \|S(\varphi(x))\| &= \|T(x)\| \\
&= \inf_{z\in Z}\|T(x)-T(z)\| && T(z)=0 \text{ because } z\in\ker T \\
&\le \inf_{z\in Z}\|T\|\|x-z\| && \text{definition of operator norm} \\
&= \|T\|\|\varphi(x)\|_1 && \text{notation of Problem 20,}
\end{aligned}$$

so $\|S\| \le \|T\|$, again by definition of an operator norm. Thus, $\|S\| = \|T\|$.

13.3 Compactness Lost: Infinite Dimensional Normed Linear Spaces

In the proof of Theorem 4, *subadditivity* refers to the triangle inequality. See page 277.

23. The claim follows from the proof of Corollary 5, Section 9.5's Theorem 20(i)$\leftrightarrow$(ii) and Proposition 21, and Section 9.3's Problem 36. (Those references justify some statements in the proof of Corollary 7.)

24. If $0 < \epsilon < 1$, then the proof is the same as the author's proof if we replace numeral 2 in inequality (8) with $1/(1-\epsilon)$ and both instances of $1/2$ in the last line with $1-\epsilon$.

    Now assume that $\epsilon \ge 1$. By the author's proof of the lemma, there is a unit vector x_0 for which $\|x_0 - y\| > 1/2$ for all y in Y. But $1/2 > 1-\epsilon$.

25. The ℓ^2 and $L^2[0,1]$ cases are Problems 63(ii) and 64(ii) of Section 9.5.

    An open cover of closed unit ball B of $C[0,1]$ is $\{B(f,1/4)\}_{f\in B}$, that is, the collection of open balls of radius $1/4$ with centers at each function in B. That collection covers B because each f in B is in $B(f,1/4)$. We have two ways to show why that open cover of B has no finite subcover—one using tent functions (see the last full sentence on page 167) and another using power functions.

    (i) For each natural number n, define tent function f_n on $[0,1]$ to vanish outside interval $(1/(n+1),1/n)$, be linear on intervals $[1/(n+1),x_n]$ and $[x_n,1/n]$ where x_n is the midpoint of $(1/(n+1),1/n)$, and take the value 1 at x_n. Then each f_n is in B because $\|f_n\|_{\max} = 1$ and because f_n is continuous by Problem 51 of Section 1.6. Now observe that if $m \ne n$, then $\|f_n - f_m\|_{\max} = 1$. Then B cannot be contained in a finite number of balls of radius $1/4$ because one of those balls would contain two of the f_n, which are distance 1 apart, and yet the ball has diameter less than 1. Thus, B has no finite subcover.

    (ii) Consider sequence $\{g_n\}$ of power functions on $[0,1]$ defined by $g_n(x) = x^n$. We claim that for each natural number m, there is an n greater than m for which $\|g_m - g_n\|_{\max} > 1/2$. To prove our claim, we first use calculus to find the point $x_{\max}$ in $[0,1]$ at which $x^m - x^n$ is a maximum on $[0,1]$. Take the derivative of $x^m - x^n$ with respect to x, put that derivative equal to 0, and solve for x to get that $x_{\max} = (m/n)^{1/(n-m)}$. Then

    $$\|g_m - g_n\|_{\max} = x_{\max}^m - x_{\max}^n = \left(\frac{m}{n}\right)^{m/(n-m)} - \left(\frac{m}{n}\right)^{n/(n-m)}.$$

    Because m is fixed, the limit as $n\to\infty$ of the right side is $1-0$, or 1. Hence, there is an n greater than m for which $\|g_m - g_n\|_{\max} > 1/2$. Now that we have proved our claim, we can use it to construct a subsequence $\{g_{n_k}\}$ for which $\left\|g_{n_j} - g_{n_k}\right\|_{\max} > 1/2$ if $k \ne j$.

26. Let $\{e_1, \ldots, e_n\}$ be a basis for X. Express a vector x in X as $x_1e_1 + \cdots + x_ne_n$. Then we can define T by $T(x) = x_1T(e_1) + \cdots + x_nT(e_n)$ because T is linear. In the compound inequality after inequality (5), replace $\|x\|$ with $\|T(x)\|$ and $\|e_i\|$ with $\|T(e_i)\|$ to get that

$$\|T(x)\| \leq \sum_{i=1}^{n} |x_i| \|T(e_i)\| \leq c_1\|x\|_* \text{ where } c_1 = \sqrt{\sum_{i=1}^{n} \|T(e_i)\|^2}.$$

Then by inequality (6), $\|T(x)\| \leq c_1c_2\|x\|$ for all x in X, which shows that T is bounded by definition (page 256). Thus, T is continuous by Theorem 1.

For the second part of the problem, assume first that T is continuous. Then T is also bounded by Theorem 1, so $T \in \mathcal{L}(X, Y)$. Thus, $\ker T$ is closed by Problem 16. Note that this direction of the *if-and-only-if* statement is true also if Y is infinite dimensional.

Conversely, assume that $\ker T$ is closed. Then $X/\ker T$ is a normed linear space by Problem 20. By our solution to Problem 22, there is a linear operator $S : X/\ker T \to Y$ such that $T = S \circ \varphi$. Note that we cannot use that solution to say that S is bounded because our proof there that S is bounded depends on T's being bounded, which is equivalent to what we are trying to prove here. But we do show in that solution that S is one-to-one, so because Y is finite dimensional, we deduce that $X/\ker T$ is finite dimensional. Then by the first part of this Problem 26, S is continuous. Natural map φ also is continuous by our solution to Problem 21. Thus, T is continuous by Proposition 9 of Section 9.3.

27. Observe that $\text{span}[\{x_1, \ldots, x_n\}]$ does indeed satisfy the conditions of Y in Riesz's lemma because $\text{span}[\{x_1, \ldots, x_n\}]$ is—

- a linear subspace of X by Problem 3(ii),
- closed by Corollary 6 since $\text{span}[\{x_1, \ldots, x_n\}]$ is a finite dimensional subspace of X, and
- a proper subspace of X since $\dim X = \infty$.

By Riesz's lemma, we can choose x_0 in B such that $\|x_0 - x_i\| > 1/2$ for all i in $\{1, \ldots, n\}$. But that contradicts that x_0 is in some $x_i + B_0/3$ and so is less than a distance $1/3$ from x_i. Thus, B is not compact.

Here is a longer proof; it does not use Riesz's lemma. We first show that $\text{span}[\{x_1, \ldots, x_n\}]$ contains B. Take a vector x in B. Then $x = x_{i_1} + u_1/3$ for some i_1 in $\{1, \ldots, n\}$ and u_1 in B_0. But $B_0/3$ is covered by $\{(x_i + B_0/3)/3\}_{1 \leq i \leq n}$, so $x = x_{i_1} + x_{i_2}/3 + u_2/9$ where $u_2 \in B_0$. Continuing in this way, we have that $x \in x_{i_1} + x_{i_2}/3 + \cdots + x_{i_m}/3^{m-1} + B_0/3^m$ for each natural number m. Define vector y_m by $y_m = x_{i_1} + x_{i_2}/3 + \cdots + x_{i_m}/3^{m-1}$. Then $y_m \in \text{span}[\{x_1, \ldots, x_n\}]$. Moreover, $x - y_m \in B_0/3^m$, so as $m \to \infty$, we have that sequence $\{x - y_m\} \to 0$, that is, $\{y_m\} \to x$. Now refer to Proposition 6 of Section 9.2. Because $\text{span}[\{x_1, \ldots, x_n\}]$ is closed and sequence $\{y_m\}$ in $\text{span}[\{x_1, \ldots, x_n\}]$ converges to x, vector x belongs to $\text{span}[\{x_1, \ldots, x_n\}]$. Hence, $B \subseteq \text{span}[\{x_1, \ldots, x_n\}]$.

Now take a vector x' in X. Vector x' is in $\text{span}[\{x_1, \ldots, x_n\}]$ because x' is linear combination $\|x'\|(x'/\|x'\|)$ of unit vector $x'/\|x'\|$, which is in $\text{span}[\{x_1, \ldots, x_n\}]$. Hence, X is finite dimensional because it is spanned by n vectors. But that contradicts that X is infinite dimensional. Thus, B is not compact.

28. Refer to Section 9.6, "Separable Metric Spaces." Assume first that X is separable. Let $\{x_n\}$ be an enumeration of a countable dense subset of X. Then $\overline{\{x_n\}} = X$ because $\{x_n\}$ is dense in X. Define sequence $\{y_n\}$ in X by $y_n = x_n/(n\|x_n\|)$. Then $\{y_n\} \to 0$ because $\|y_n\| = 1/n$. Put K to $\{y_n\} \cup \{0\}$. To show that K is compact, take an open cover of K.

In that cover, there is an open set G that contains 0. Then contained in G is an open ball $B(0,r)$ centered at 0 of radius r. Choose an index N such that $1/N < r$. Then $B(0,r)$ contains vectors $0, y_N, y_{N+1}, \ldots$. A finite subcover consists of G and the at most $N-1$ open-cover sets that contain $y_1, y_2, \ldots, y_{N-1}$. Hence, K is compact (another way to show that K is compact is to show that K is sequentially compact). Now $\{x_n\} \subseteq \text{span}[K]$ because x_n is linear combination $n\|x_n\|y_n$ of y_n for each n. Then $\overline{\{x_n\}} \subseteq \overline{\text{span}}[K]$ (Problem 23 of Section 9.2), that is, $X \subseteq \overline{\text{span}}[K]$. On the other hand, $\overline{\text{span}}[K] \subseteq X$ because $K \subseteq X$, and X is a closed linear space. Thus, $\overline{\text{span}}[K] = X$.

Conversely, assume that $\overline{\text{span}}[K] = X$. Because K is compact, there is a countable subset D of K such that $\overline{D} = K$. Consider set $\text{span}_{\mathbf{Q}}[D]$ where subscript $\mathbf{Q}$ indicates that the linear combinations are over $\mathbf{Q}$ instead of $\mathbf{R}$. Because $\mathbf{Q}$ and D are countable, $\text{span}_{\mathbf{Q}}[D]$ is a countable subset of X (Corollaries 4(ii) and 6 of Section 1.3). It remains to show that $\overline{\text{span}_{\mathbf{Q}}}[D] = X$:

$$\begin{aligned}
\overline{\text{span}_{\mathbf{Q}}}[D] &= \overline{\text{span}}[D] && \text{see explanation (i) below} \\
&= \overline{\text{span}}[\overline{D}] && \text{see explanation (ii) below} \\
&= \overline{\text{span}}[K] \\
&= X && \text{assumption.}
\end{aligned}$$

(i) Because $\mathbf{Q} \subseteq \mathbf{R}$, we have that $\overline{\text{span}_{\mathbf{Q}}}[D] \subseteq \overline{\text{span}}[D]$. On the other hand, because $\mathbf{Q}$ is dense in $\mathbf{R}$ (Theorem 2 of Section 1.2), it follows that $\text{span}[D] \subseteq \overline{\text{span}_{\mathbf{Q}}}[D]$. Then Proposition 3 of Section 9.2 tells us that $\overline{\text{span}}[D] \subseteq \overline{\text{span}_{\mathbf{Q}}}[D]$.

(ii) Because $D \subseteq \overline{D}$, we have that $\overline{\text{span}}[D] \subseteq \overline{\text{span}}[\overline{D}]$. On the other hand, the third-to-last sentence on page 254 tells us that $\overline{\text{span}}[\overline{D}]$ is contained in every closed linear subspace that contains $\overline{D}$. But $\overline{\text{span}}[D]$ contains $\overline{D}$. Hence, $\overline{\text{span}}[\overline{D}] \subseteq \overline{\text{span}}[D]$.

Thus, X is separable because countable subset $\text{span}_{\mathbf{Q}}[D]$ of X is dense in X.

13.4 The Open Mapping and Closed Graph Theorems

29. Operator T is continuous by Problem 26.

We now show that T is open. We can redefine T as $T : X \to T(X)$. Let $\{e_1, \ldots, e_n\}$ be a basis for X. Then by our solution to Problem 26, $T(X) = \text{span}[\{T(e_1), \ldots, T(e_n)\}]$. Hence, $T(X)$ is a linear space by Problem 3(ii) and finite dimensional by definition (page 259). By Corollary 6, X and $T(X)$ are Banach spaces, and $T(X)$, in particular, is closed. Thus, T is open by the open mapping theorem because we defined T as $T : X \to T(X)$.

Here is another solution. We want to show that the image of an open set in X is open in $T(X)$. Because the set of open balls is a base for the topology for X, it suffices by scaling and translation to show that the image $T(B_X)$ of the open unit ball B_X in X is open in $T(X)$. Now $\ker T$ is a linear subspace of Y by Problem 16. Let $\mathcal{W}$ be the collection of all subspaces W of X for which $W \oplus \ker T = X$. Collection $\mathcal{W}$ is not empty by Problem 35. Then

$$B_X = (B_X \cap \ker T) \cup \bigcup_{W \in \mathcal{W}} (B_X \cap W).$$

Now $T(X)$ and each W in $\mathcal{W}$ are Banach spaces by the same reasoning as in our first solution. Furthermore, restriction operator $T|_W : W \to T(X)$ is in $\mathcal{L}(W, T(X))$ by Theorem 1, onto, and one-to-one by Problem 16 because $\ker T|_W = \{0\}$. Hence, $T|_W$ is open by Corollary 9 and its proof. So $T|_W(B_X \cap W)$ is open in $T(X)$ because $B_X \cap W$ is open in W by Proposition 2 of Section 9.2. Then $T(B_X)$ is open in $T(X)$ because

$$\begin{aligned} T(B_X) &= T(B_X \cap \ker T) \cup \bigcup_{W \in \mathcal{W}} T(B_X \cap W) \\ &= \{0\} \cup \bigcup_{W \in \mathcal{W}} T|_W(B_X \cap W) \end{aligned}$$

along with the observation that $\bigcup_{W \in \mathcal{W}} T|_W(B_X \cap W)$ is open (Proposition 1 of Section 9.2) and contains $\{0\}$. Thus, T is open.

30. Projection P is bounded and linear because it is in $\mathcal{L}(X, X)$. Then P is continuous by Theorem 1. By the first three sentences of the proof of Theorem 11, $P(X)$ is closed. Thus, P is open by the open mapping theorem.

31. Let B_X and B_Y denote the open unit balls in X and Y. Let U denote the neighborhood in which $T(B_X)$ is dense. Then $\overline{T(B_X)} = \overline{U}$ by the sentence after the definition on page 203. Because U is open, there is a positive radius r_1 such that $r_1 B_Y$ is contained in U and therefore contained in $\overline{U}$. Then $r_1 B_Y \subseteq \overline{T(B_X)}$. Hence, if we denote $1/r_1$ by r, then $B_Y \subseteq r\overline{T(B_X)}$ since $\overline{T(B_X)}$ is closed. That gives us the third sentence on page 264 (in the proof of Theorem 8). The remainder of that proof tells us that there is an M for which statement (9) holds. Thus, T is open by the first sentence of the proof of the open mapping theorem.

32. This is true by the normed-linear-space analogue of Proposition 20(ii) of Section 1.5. Note that in adapting our proof of that proposition (Problem 45 of Section 1.5), we need to replace our appeal to Theorem 17 with the hypothesis that X is a Banach space.

    We also prove this converse: If for each $\{u_n\}$ in normed linear space X, summability of $\sum_{k=1}^{\infty} \|u_k\|$ implies summability of $\sum_{k=1}^{\infty} u_k$, then X is a Banach space. Indeed, take a Cauchy sequence $\{x_n\}$ in X. As in the proof of Proposition 5 of Section 7.3, choose a strictly increasing sequence $\{n_k\}$ of natural numbers for which

$$\|x_{n_{k+1}} - x_{n_k}\| \leq (1/2)^k \text{ for all } k.$$

    Then applying the formula for the sum of a geometric series, $\sum_{k=1}^{\infty} \|x_{n_{k+1}} - x_{n_k}\|$ is no more than 1 and therefore summable. By hypothesis, $\sum_{k=1}^{\infty}(x_{n_{k+1}} - x_{n_k})$ also is summable, say to vector s in X. But because $\sum_{k=1}^{\infty}(x_{n_{k+1}} - x_{n_k})$ is a telescoping series,

$$s = \lim_{m \to \infty} \sum_{k=1}^{m} (x_{n_{k+1}} - x_{n_k}) = \lim_{m \to \infty} x_{n_{m+1}} - x_{n_1},$$

    which shows that $\{x_{n_k}\}$ converges to $s + x_{n_1}$. Hence, $\{x_n\}$ has a convergent subsequence. By Proposition 4 of Section 7.3, $\{x_n\}$ converges. Thus, X is a Banach space.

33. This is the second part of Problem 26.

34. No, T_0 is not necessarily open by the remark at the end of the section. Put X to ℓ^∞, which is a Banach space by Problem 34 of Section 7.3. Define operator $T : \ell^\infty \to \ell^\infty$ by $T(\{x_n\}) = \{x_n/n\}$ for $\{x_n\}$ in ℓ^∞. Then $T \in \mathcal{L}(\ell^\infty, \ell^\infty)$ because T is linear and also bounded since $\|T(\{x_n\})\|_\infty = \|\{x_n/n\}\|_\infty \leq \|\{x_n\}\|_\infty$. So T is continuous by Theorem 1. Furthermore, observe that $T(\ell^\infty) = \ell^\infty$, which is closed by Proposition 5 of Section 9.2. Hence, T is open by the open mapping theorem.

Put X_0 to ℓ^2, which is a closed subspace of ℓ^∞ by Proposition 11 of Section 9.4. To show that image $T_0(\ell^2)$ is not closed, apply the contrapositive of Proposition 6 of Section 9.2. Sequence

$$((1,0,0,0,\dots),\ (1,1/2,0,0,\dots),\ (1,1/2,1/3,0,\dots),\ \dots)$$

is in $T_0(\ell^2)$ because $T^{-1}((1,1/2,\dots,1/n,0,\dots)) = (1,1,\dots,1,0,\dots)$ and $\sum_{k=1}^{n} 1^2 < \infty$. The limit $\{1/n\}$ of the sequence, however, is not in $T_0(\ell^2)$ because $T^{-1}(\{1/n\}) = \{1\}$, and $\{1\}$ is not in ℓ^2 since $\sum_{k=1}^{\infty} 1^2 = \infty$. Hence, $T_0(\ell^2)$ is not closed. Thus, T_0 is not open by the contrapositive of the open mapping theorem.

35. If $V = X$, then $X = V \oplus \{0\}$, and we are done. So assume now that $V \neq X$.

(i) **Lemma** Let $\{e_i\}_{i=1}^n$ be a linearly-independent subset of a linear space X. If e_{n+1} is a vector in X that is not in the subspace spanned by $\{e_i\}_{i=1}^n$, then $\{e_i\}_{i=1}^{n+1}$ is linearly independent.

Proof If

$$x_1 e_1 + \cdots + x_n e_n + x_{n+1} e_{n+1} = 0,$$

then $x_{n+1} = 0$ for otherwise,

$$e_{n+1} = -\frac{x_1}{x_{n+1}} e_1 - \cdots - \frac{x_n}{x_{n+1}} e_n,$$

so e_{n+1} would be in the subspace spanned by $\{e_i\}_{i=1}^n$, a contradiction. Then because $x_{n+1} = 0$, we have that $x_1 e_1 + \cdots + x_n e_n = 0$. But because $\{e_i\}_{i=1}^n$ is linearly independent, $x_1, \dots, x_n$ also equal 0. Thus, $\{e_i\}_{i=1}^{n+1}$ is linearly independent. □

Now use the lemma to extend basis $\{e_i\}_{i=1}^n$ for V to a basis for X. Choose a vector e_{n+1} in $X \sim V$, i.e., in $X \sim \text{span}[\{e_i\}_{i=1}^n]$. By the lemma, $\{e_i\}_{i=1}^{n+1}$ is linearly independent. If $\{e_i\}_{i=1}^{n+1}$ spans X, then $\{e_i\}_{i=1}^{n+1}$ is a basis for X. Otherwise, choose a vector e_{n+2} in $X \sim \text{span}\big[\{e_i\}_{i=1}^{n+1}\big]$ so that $\{e_i\}_{i=1}^{n+2}$ is linearly independent. Continuing in that way, in $\dim X - n$ steps, we reach a basis $\{e_i\}_{i=1}^{n+k}$ for X.

Now W is a linear subspace of X by Problem 3(ii). Express a vector x in X as $(x_1 e_1 + \cdots + x_n e_n) + (x_{n+1} e_{n+1} + \cdots + x_{n+k} e_{n+k})$, which shows that $X = V + W$. It remains to show that $V \cap W = \{0\}$. So assume that $x \in V \cap W$, in other words, x is in both V and W. Then

$$x_1 e_1 + \cdots + x_n e_n = x_{n+1} e_{n+1} + \cdots + x_{n+k} e_{n+k},$$

or

$$x_1 e_1 + \cdots + x_n e_n - x_{n+1} e_{n+1} - \cdots - x_{n+k} e_{n+k} = 0.$$

Because $\{e_i\}_{i=1}^{n+k}$ is linearly independent, $x_1, \dots, x_{n+k} = 0$. Hence, $x = 0$ so that $V \cap W = \{0\}$. Thus, W is a linear complement of V in X.

(ii) Collection $\mathcal{F}$ is not empty because it contains $\{0\}$. Now refer to page 6. Collection $\mathcal{F}$ is a partially ordered set. Take a totally ordered subcollection $\{Z_\lambda\}_{\lambda \in \Lambda}$ of $\mathcal{F}$. Denote $\text{span}[\{Z_\lambda\}_{\lambda \in \Lambda}]$ by Z_s. We want to show that $Z_\text{s} \in \mathcal{F}$, i.e., that $V \cap Z_\text{s} = \{0\}$. So take a vector x in $V \cap Z_\text{s}$. Because $x \in Z_\text{s}$, we can express x as a linear combination of vectors in the Z_λ. Because $\{Z_\lambda\}_{\lambda \in \Lambda}$ is totally ordered, there is a $Z_{\lambda'}$ that contains all of those vectors, so $x \in Z_{\lambda'}$. But because $x \in V$ too, $x \in V \cap Z_{\lambda'}$, which is $\{0\}$ since $Z_{\lambda'}$ is a Z. Therefore, $V \cap Z_\text{s} = \{0\}$, and so $Z_\text{s} \in \mathcal{F}$. Hence, every totally ordered subcollection $\{Z_\lambda\}_{\lambda \in \Lambda}$ of $\mathcal{F}$ has an upper bound Z_s. By Zorn's lemma, $\mathcal{F}$ has a maximal subspace W. And because $W \in \mathcal{F}$, we have that $V \cap W = \{0\}$.

It remains to show that $V + W = X$. Now $V + W \subseteq X$ because $V, W \subseteq X$. To get the inclusion in the opposite direction, suppose the contrary, i.e., that $X \not\subseteq V + W$. Then there is a nonzero vector y in $X \sim (V+W)$. We claim that $(W+\text{span}[\{y\}]) \cap V = \{0\}$. If $w + \alpha y \in V$ for some vector w in W and real number α, then $\alpha y \in V+W$. But then $\alpha = 0$ because $y \neq 0$. So $w \in V$. But then $w \in V \cap W$, which is $\{0\}$, so $w = 0$, which completes the proof of our claim. Then $W + \text{span}[\{y\}] \in \mathcal{F}$. But that contradicts the maximality of W. Therefore, $X \subseteq V+W$. Hence, $X = V+W$. Thus, W is a linear complement of V in X.

36. Referring to the definition of a linear operator (page 256), P is linear because

$$P(\alpha x_1 + \beta x_2) = P(\alpha(v_1 + w_1) + \beta(v_2 + w_2)) = \alpha v_1 + \beta v_2 = \alpha P(x_1) + \beta P(x_2).$$

Secondly, $P^2 = P$ because $P^2(x) = P(P(x)) = P(v) = P(v+0) = v = P(x)$.

Thirdly, $P(X) = V$ because for all v in V, vector $v + 0$ in X is such that $P(v+0) = v$.

Fourthly, $(\text{Id} - P)(X) = W$ because for all w in W, vector $0 + w$ in X gives us that

$$(\text{Id} - P)(0+w) = \text{Id}(0+w) - P(0+w) = 0 + w - 0 = w.$$

To verify decomposition (16), observe that $X = \text{Id}(X) = P(X) + (\text{Id} - P)(X)$. It therefore remains to show that $P(X) \cap (\text{Id} - P)(X) = \{0\}$. Take a v in $P(X)$. Then $v = P(x)$ for some x in X, and because $P^2 = P$, it follows that $P(v) = P^2(x) = P(x) = v$. But $\text{Id} - P$ is a projection too (top of page 267), so by the same reasoning, if v is also in $(\text{Id} - P)(X)$, then $(\text{Id} - P)(v) = v$, that is, $P(v) = 0$. Hence, $v = 0$ so that $P(X) \cap (\text{Id} - P)(X) = \{0\}$. Thus, $X = P(X) \oplus (\text{Id} - P)(X)$.

37. Assume first that Y is a Banach space. Put X to Y. Then Id is a continuous, linear, open mapping of Y onto Y.

Our proof of the converse mimics the first part of the proof of Theorem 8. Assume that there is a Banach space X and a continuous, linear, open mapping T of X onto Y. Let $\{y_n\}$ be a Cauchy sequence in Y. We must show that $\{y_n\}$ converges to a vector in Y. By selecting a subsequence if necessary, we can assume that

$$\|y_n - y_{n-1}\| \leq 1/2^n \text{ for } n \text{ equal to } 2, 3, \ldots. \tag{13.2}$$

We claim that for n equal to $2, 3, \ldots$, there is a positive M and a vector u_n in X for which

$$T(u_n) = y_n - y_{n-1}, \text{ and } \|u_n\| \leq M/2^n.$$

We justify our claim using the reasoning and nomenclature in the proof of the open mapping theorem. (Note that because T is onto, $T(X) = Y$, so we do not need to intersect sets in Y with $T(X)$ as is done in that proof.) Because T is open, $T(B_X)$ is an open subset of $T(X)$ and contains 0. Hence, there is a positive r for which $rB_Y \subseteq T(B_X)$. By homogeneity, that inclusion is equivalent to the existence of a constant M for which $\overline{B_Y} \subseteq MT(\overline{B_X})$. We infer from homogeneity of T and of the norms that that inclusion is equivalent to statement (9). Applying inequality (13.2) to statement (9) completes the proof of our claim.

Therefore, for n equal to $2, 3, \ldots$, if we define x_n by $x_n = \sum_{j=2}^{n} u_j$, then by linearity of T,

$$T(x_n) = y_n - y_1, \tag{13.3}$$

and

$$\|x_{n+k} - x_n\| \leq M \sum_{j=n}^{\infty} 1/2^j \text{ for } k \text{ equal to } 1, 2, \ldots.$$

But X is a Banach space, and therefore Cauchy sequence $\{x_n\}$ converges to a vector x_* in X. Take the limit as $n \to \infty$ in formula (13.3) and use the continuity of T to infer that $\{y_n\}$ converges to vector $T(x_*) + y_1$ in Y. If necessary, apply the second part of Proposition 4 of Section 7.3 to conclude that the Cauchy sequence with which we started converges. Thus, Y is a Banach space.

13.5 The Uniform Boundedness Principle

38. **Theorem** (Cf. Theorem 7 of Section 10.2.) Let X be a complete metric space and (Y, ρ) a metric space. Let $\{f_n : X \to Y\}$ be a sequence of continuous mappings that converges pointwise to mapping $f : X \to Y$. Then there is a dense subset D of X for which f is continuous at each point in D.

    Proof Let m and n be natural numbers. Define set $E(m, n)$ by

$$E(m, n) = \{x \in X \mid \rho(f_j(x), f_k(x)) \leq 1/m \text{ where } j, k = n, n+1, \ldots\}.$$

Since each function $x \mapsto \rho(f_j(x), f_k(x))$ is continuous, $E(m, n)$, being the intersection of a collection of closed sets, is closed. Put D to

$$X \sim \left(\bigcup_{m,n \in \mathbf{N}} \operatorname{bd} E(m, n) \right),$$

which is dense in X according to Corollary 5 of Section 10.2. Observe that if point x in D belongs to $E(m, n)$, then x belongs to the interior of $E(m, n)$. We claim that f is continuous at each point in D. Indeed, take a point x_0 in D. Take a positive ϵ. Choose an m for which $1/m \leq \epsilon/3$. Since $\{f_n(x_0)\}$ converges to a point in Y, sequence $\{f_n(x_0)\}$ is Cauchy. Choose an index N for which

$$\rho(f_j(x_0), f_k(x_0)) < 1/m \text{ when } j, k \geq N. \tag{13.4}$$

Therefore, x_0 belongs to $E(m, N)$. As we observed above, x_0 belongs to the interior of $E(m, N)$. Choose a radius r such that $B(x_0, r) \subseteq E(m, N)$, that is,

$$\rho(f_j(x), f_k(x)) \leq 1/m \text{ for all } x \text{ in } B(x_0, r) \text{ when } j, k \geq N. \tag{13.5}$$

Function f_N is continuous at x_0. Therefore, there is a radius δ less than r for which

$$\rho(f_N(x), f_N(x_0)) < 1/m \text{ for all } x \text{ in } B(x_0, \delta). \tag{13.6}$$

Observe that for every j and point x in X, we have by the triangle inequality that

$$\rho(f_j(x), f_j(x_0)) \leq \rho(f_j(x), f_N(x)) + \rho(f_N(x), f_N(x_0)) + \rho(f_N(x_0), f_j(x_0)).$$

We infer from inequalities (13.4), (13.5), and (13.6) that

$$\rho(f_j(x), f_j(x_0)) < 3/m \leq \epsilon \text{ for all } x \text{ in } B(x_0, \delta) \text{ when } j \geq N.$$

Take the limit as $j \to \infty$ to obtain that

$$\rho(f(x), f(x_0)) < \epsilon \text{ for all } x \text{ in } B(x_0, \delta).$$

Thus, f is continuous at x_0. □

Now under the conditions of the theorem, assume further that the f_n are linear. Then f is also linear by the Banach-Saks-Steinhaus theorem. Thus, f is continuous on all of X by Problem 9(i).

39. Verify that the conditions of the Banach-Saks-Steinhaus theorem hold: Space $L^1[a, b]$ is a Banach space by the Riesz-Fischer theorem (Section 7.3). The set $\mathbf{R}$ of real numbers is a normed linear space (page 183). Define the sequence $\{T_n : L^1[a, b] \to \mathbf{R}\}$ of operators by $T_n(g) = \int_a^b g f_n$. By Proposition 2 of Section 8.1 (interchange f and g for clarity), each T_n

is a bounded linear functional on $L^1[a, b]$. By Theorem 1, each T_n is continuous. Then by the Banach-Saks-Steinhaus theorem, the operator $T : L^1[a, b] \to \mathbf{R}$ defined by

$$T(g) = \lim_{n\to\infty} \int_a^b g f_n \text{ for all } g \text{ in } L^1[a, b]$$

is linear and continuous. By Theorem 1, T is bounded. By Theorem 5 of Section 8.1 (interchange f and g for clarity), there is a function f in $L^\infty[a, b]$ for which

$$T(g) = \int_a^b g f \text{ for all } g \text{ in } L^1[a, b].$$

Equating our two expressions for $T(g)$ concludes our solution.

40. Express p as $a_0 + a_1 x + a_2 x^2 + \cdots + a_r x^r$ so that $\|p\| = \sum_{i=0}^{r} |a_i|$. Verify the conditions of a norm at the beginning of the chapter. Function $\|\cdot\|$ is nonnegative and real-valued.

- We have that $\|p\| = 0$ if and only if each a_i is 0 if and only if $p = 0$.
- Express another polynomial q as $b_0 + b_1 x + b_2 x^2 + \cdots + b_s x^s$. If $r \le s$, then by the triangle inequality,

$$\|p + q\| = \sum_{i=0}^{r} |a_i + b_i| + \sum_{i=r+1}^{s} |b_i| \le \sum_{i=0}^{r} |a_i| + \sum_{i=0}^{s} |b_i| = \|p\| + \|q\|.$$

  If $r > s$, then an analogous calculation gives us again that $\|p + q\| \le \|p\| + \|q\|$.
- Lastly,

$$\|\alpha p\| = \sum_{i=0}^{r} |\alpha a_i| = |\alpha| \sum_{i=0}^{r} |a_i| = |\alpha| \|p\|.$$

Thus, $\|\cdot\|$ is a norm on X.

Next, we show that X is not a Banach space by showing that $\{\psi_n\}$ satisfies the hypotheses of the Banach-Saks-Steinhaus theorem but is not uniformly bounded. Indeed, $\mathbf{R}$ is a normed linear space (page 183). Each ψ_n is continuous because for every p in X, if a sequence $\{p_m\}$ in X converges to p, then

$$\{\psi_n(p_m)\} = \left\{p_m^{(n)}(0)\right\} \to p^{(n)}(0) = \psi_n(p)$$

where m is the index for the sequences. Furthermore, each ψ_n is linear because for each p, q in X and real numbers α and β,

$$\psi_n(\alpha p + \beta q) = (\alpha p + \beta q)^{(n)}(0) = \alpha p^{(n)}(0) + \beta q^{(n)}(0) = \alpha \psi_n(p) + \beta \psi_n(q).$$

Lastly, for each p in X, we have that $\lim_{n\to\infty} \psi_n(p) = 0$ because $p^{(n)}(0) = 0$ when n is greater than the degree of p. Hence, $\{\psi_n\}$ satisfies the hypotheses of the Banach-Saks-Steinhaus theorem. But $\{\psi_n\}$ is not uniformly bounded because

$$\frac{|\psi_n(x^n)|}{\|x^n\|} = \frac{|(x^n)^{(n)}(0)|}{|1|} = n!,$$

which shows that $\|\psi_n\| \ge n!$ and therefore that there is no constant M for which $\|\psi_n\| \le M$ for all n (see the second definition on page 256). Thus, X is not a Banach space by the contrapositive of the Banach-Saks-Steinhaus theorem. (Cf. Problem 1(iii) of Section 14.1.)

Chapter 14

Duality for Normed Linear Spaces

14.1 Linear Functionals, Bounded Linear Functionals, and Weak Topologies

1. (i) This is Problem 17.

   (ii) Take an infinite dimensional Banach space X. Suppose, to get a contradiction, that X has a countable Hamel basis $\{e_1, e_2, \dots\}$. Then x is a vector in X if and only if there is a natural number n and real numbers $\lambda_1, \dots, \lambda_n$ such that $x = \lambda_1 e_1 + \cdots + \lambda_n e_n$. It follows that

$$X = \bigcup_{n=1}^{\infty} F_n \text{ where } F_n = \operatorname{span}[\{e_1, \dots, e_n\}].$$

Furthermore, each F_n is—

- a linear subspace of X by Problem 3(ii) of Section 13.1,
- finite dimensional ($\dim F_n = n$), and therefore
- closed by Corollary 6 of Section 13.3.

By Corollary 4 of Section 10.2, one of the F_n, say F_j, has nonempty interior. So there is an open ball $B(y, r)$ that is contained in F_j. By definition of a linear subspace (page 254), $B(y,r) - y$ also is contained in F_j so that $B(0, r) \subseteq F_j$. Then for all vectors x in X,

$$\frac{r}{2\|x\|} x \in F_j.$$

Appealing again to the fact that F_j is a linear subspace, $x \in F_j$ so that $X = F_j$. But that contradicts that X is infinite dimensional and so is not spanned by a finite collection of vectors (see the first paragraph of Section 13.3). Thus, a Hamel basis for X is uncountable.

This second solution comes from Nam-Kiu Tsing's article "Infinite-dimensional Banach spaces must have uncountable basis—an elementary proof," *The American Mathematical Monthly*, 96 (1984) 505–506 (jstor.org/stable/2322577). We begin as our first solution begins except that we assume without loss of generality that the e_n are unit vectors. Define real numbers r_n by

$$r_n = \inf\{\|x + e_n\| \mid x \in F_{n-1}\} \text{ when } n \geq 2.$$

By our first solution, F_{n-1} contains an open ball centered at 0, so $0 \in F_{n-1}$. Then $r_n \leq \|0 + e_n\| = 1$. Moreover, $r_n > 0$ because $-e_n \notin F_{n-1}$. Define another set of positive numbers s_n by

$$s_1 = 1,\ s_2 = 1/3,\ s_3 = r_2 s_2/3,\ \dots,\ s_{n+1} = r_n s_n/3,\ \dots.$$

Then for each natural number k,

$$0 < s_{n+k} \leq r_n s_n/3^k \leq 1/3^{n+k-1} \text{ when } n \geq 2. \tag{14.1}$$

Define a sequence $\{x_n\}$ of vectors in X by $x_n = \sum_{i=1}^{n} s_i e_i$. Note that $\{x_n\}$ is a Cauchy sequence in X. We claim that $\{x_n\}$ does not converge in X. Indeed, because

$\{e_1, e_2, \dots\}$ is a Hamel basis for X, we can express a vector x in X as $\sum_{i=1}^{m-1} \lambda_i e_i$. Then when $n > m$,

$$\begin{aligned}
\|x_n - x\| &= \left\| \sum_{i=1}^{m-1} (s_i - \lambda_i) e_i + s_m e_m + \sum_{i=m+1}^{n} s_i e_i \right\| \\
&\geq \left\| \sum_{i=1}^{m-1} (s_i - \lambda_i) e_i + s_m e_m \right\| - \left\| \sum_{i=m+1}^{n} s_i e_i \right\| \\
&\geq s_m \left\| \sum_{i=1}^{m-1} \frac{s_i - \lambda_i}{s_m} e_i + e_m \right\| - \sum_{i=m+1}^{n} s_i && \text{triangle inequality, } \|s_i e_i\| = s_i \\
&\geq s_m r_m - \sum_{i=1}^{n-m} \frac{r_m s_m}{3^i} && \text{definition of } r_m\text{; inequal. (14.1)} \\
&> \frac{s_m r_m}{2} && \sum_{i=1}^{n-m} \frac{1}{3^i} < \sum_{i=1}^{\infty} \frac{1}{3^i} = \frac{1}{2},
\end{aligned}$$

from which it follows that $\lim_{n\to\infty} \|x_n - x\| > 0$. Then $\{x_n\}$ does not converge in X, which contradicts that X is a Banach space. Thus, a Hamel basis for X is uncountable.

(iii) Express a polynomial in X as $a_0 + a_1 x + a_2 x^2 + \cdots + a_n x^n$. Linear space X is infinite dimensional because linear combinations of a finite set of polynomials cannot have degree greater than the greatest degree of those polynomials. Now a Hamel basis for X is $\{1, x, x^2, \dots\}$, which is countable. Thus, by the contrapositive of part (ii), there is no norm on X with respect to which X is a Banach space.

2. We need to verify that $\psi(x) = \lambda_1 \psi_1(x)$ for all x in X. Because $X = \ker \psi_1 \oplus \text{span}[x_0]$, we can express x as $y + \alpha x_0$ where $y \in \ker \psi_1$, and α is a real number. Then

$$\begin{aligned}
\psi(x) &\stackrel{?}{=} \lambda_1 \psi_1(x), \\
\psi(y + \alpha x_0) &\stackrel{?}{=} \lambda_1 \psi_1(y + \alpha x_0), \\
\psi(y) + \alpha \psi(x_0) &\stackrel{?}{=} \lambda_1 \psi_1(y) + \lambda_1 \alpha \psi_1(x_0) && \text{linearity of } \psi \text{ and } \psi_1, \\
\alpha &= \alpha && y \in \ker \psi_1 \subseteq \ker \psi;\ \psi(x_0) = 1;\ \lambda_1 = 1/\psi_1(x_0).
\end{aligned}$$

Thus, $\psi = \lambda_1 \psi_1$. Using the same procedure but with y in $\ker \psi_k$, that is, $y \in Y$,

$$\begin{aligned}
\psi(x) &\stackrel{?}{=} \sum_{i=1}^{k} \lambda_i \psi_i(x) \\
&= \sum_{i=1}^{k-1} \lambda_i \psi_i(x) + \lambda_k \psi_k(x), \\
\psi(y + \alpha x_0) &\stackrel{?}{=} \sum_{i=1}^{k-1} \lambda_i \psi_i(y + \alpha x_0) + \left(\psi(x_0) - \sum_{i=1}^{k-1} \lambda_i \psi_i(x_0) \right) \psi_k(y + \alpha x_0), \\
\psi(y) + \alpha \psi(x_0) &\stackrel{?}{=} \sum_{i=1}^{k-1} \lambda_i \psi_i(y) + \alpha \sum_{i=1}^{k-1} \lambda_i \psi_i(x_0) + \left(\psi(x_0) - \sum_{i=1}^{k-1} \lambda_i \psi_i(x_0) \right) \alpha, \\
\psi(y) + \alpha \psi(x_0) &= \psi(y) + \alpha \psi(x_0)
\end{aligned}$$

where in the fourth step, $\psi_k(y + \alpha x_0) = \alpha$ because $y \in \ker \psi_k$ and $\psi_k(x_0) = 1$; and in the fifth step, $\sum_{i=1}^{k-1} \lambda_i \psi_i(y) = \psi(y)$ because $\psi = \sum_{i=1}^{k-1} \lambda_i \psi_i$ on Y.

3. By Proposition 1, $X_0 = \ker \psi$ for some ψ in $X^\sharp$. Assume first that $\psi \in X^*$. By Theorem 1 of Section 13.2, ψ is bounded. Then $\psi \in \mathcal{L}(X, \mathbf{R})$. Thus, X_0 is closed by Problem 16 of Section 13.2.

   Conversely, assume that X_0 is closed. Now $\mathbf{R}$ is finite dimensional ($\dim \mathbf{R} = 1$), so the second part of Problem 26 of Section 13.3 tells us that ψ is continuous. Thus, $\psi \in X^*$.

4. This is the special case of the first part of Problem 26 of Section 13.3 where $Y = \mathbf{R}$.

5. There is a typo in the problem statement. The expression for x should be $x_1e_1 + \cdots + x_ne_n$.

   Because X is finite dimensional, we can use the first paragraph of Section 13.3. Subset $\{\psi_1, \ldots, \psi_n\}$ spans X^* because for each ψ in X^*,

$$\begin{aligned}
\psi(x) &= \psi(x_1e_1 + \cdots + x_ne_n) \\
&= x_1\psi(e_1) + \cdots + x_n\psi(e_n) && \text{linearity of } \psi \\
&= \psi_1(x)\psi(e_1) + \cdots + \psi_n(x)\psi(e_n) && x_i = \psi_i(x),
\end{aligned}$$

   which shows that ψ is linear combination $\psi(e_1)\psi_1 + \cdots + \psi(e_n)\psi_n$.

   We next show that $\{\psi_1, \ldots, \psi_n\}$ is linearly independent. Assume that

$$\lambda_1\psi_1 + \lambda_2\psi_2 + \cdots + \lambda_n\psi_n = 0$$

   where the λ_i are real numbers. Apply both sides of that equation to e_1, and simplify:

$$\begin{aligned}
0 &= \lambda_1\psi_1(e_1) + \lambda_2\psi_2(e_1) + \cdots + \lambda_n\psi_n(e_1) \\
&= \lambda_1 1 + \lambda_2 0 + \cdots + \lambda_n 0 && \psi_i(e_j) = \delta_{ij} \text{ (Kronecker delta)} \\
&= \lambda_1.
\end{aligned}$$

   Repeat for $e_2, \ldots, e_n$ to see that $\lambda_2, \ldots, \lambda_n$ also equal 0. Hence, $\{\psi_1, \ldots, \psi_n\}$ is linearly independent. Thus, $\{\psi_1, \ldots, \psi_n\}$ is a basis for X^* so that $\dim X^* = n$.

6. By the second displayed line on page 275, the weak topology on $X \subseteq$ the strong topology on X. We prove the inclusion in the opposite direction by showing that a base for the strong topology is a base for the weak topology (refer to the definition on pages 222–223 and subsequent paragraph). A base at a vector x in X for the strong topology on X comprises open balls $B(x, \epsilon)$. By Section 13.3's Theorem 4 and Corollary 5 and Section 9.1's Problem 2, we can define, without loss of generality, the norm on X by $\|x\| = \max\{|x_1|, \ldots, |x_n|\}$ in the notation of Problem 5. Then,

$$\begin{aligned}
B(x, \epsilon) &= \{x' \in X \mid \|x' - x\| < \epsilon\} && \text{definition of } B \\
&= \{x' \in X \mid \max\{|x_1' - x_1|, \ldots, |x_n' - x_n|\} < \epsilon\} && \text{definition of } \|\cdot\| \\
&= \{x' \in X \mid \max\{|\psi_1(x') - \psi_1(x)|, \ldots, |\psi_n(x') - \psi_n(x)|\} < \epsilon\} && \text{Problem 5} \\
&= \{x' \in X \mid |\psi_k(x' - x)| < \epsilon \text{ where } 1 \le k \le n\} && \text{linearity of } \psi_k,
\end{aligned}$$

   which has the form of set (5). Moreover, $\{\psi_1, \ldots, \psi_n\}$ is a basis for X^* by Problem 5. Hence, $B(x, \epsilon)$ is a base set also in the weak topology. Thus, the weak and strong topologies on X are the same.

7. Let X be an infinite dimensional normed linear space. By definition of a weak topology (page 274) and subsequent paragraph, every weakly open subset of X that contains a vector x includes a base set $\mathcal{N}$ of the form $\{x' \in X \mid |\psi_k(x' - x)| < \epsilon \text{ where } 1 \le k \le n\}$.

   We claim that $\bigcap_{k=1}^n \ker \psi_k \neq \{0\}$. Indeed, consider linear operator $T : X \to \mathbf{R}^n$ defined by $T(x) = (\psi_1(x), \ldots, \psi_n(x))$. Linear operator T is continuous by our lemma in our solution to Problem 4 of Section 12.1, so $T \in \mathcal{L}(X, \mathbf{R}^n)$ by Theorem 1 of Section 13.2. Now observe that $\bigcap_{k=1}^n \ker \psi_k = \ker T$. But $\ker T \neq \{0\}$; otherwise, $T : X \to \mathbf{R}^n$ would be one-to-one by Problem 16 of Section 13.2, which contradicts that X is infinite dimensional whereas $\mathbf{R}^n$ has finite dimension (n). Hence, there is a nonzero vector y in $\bigcap_{k=1}^n \ker \psi_k$.

It follows that $\lambda y + x \in \mathcal{N}$ for all real numbers λ because $\psi_k(\lambda y + x - x) = \lambda \psi_k(y) = 0 < \epsilon$ where $1 \leq k \leq n$. Moreover, because λ is arbitrary, $\|\lambda y + x\|$ is arbitrarily large. Hence, $\mathcal{N}$ is unbounded. Thus, every nonempty weakly open subset of X is unbounded.

8. Linear operator J is continuous by Problem 26 of Section 13.3, so $J \in \mathcal{L}(X, X^{**})$ by Theorem 1 of Section 13.2. Then J is one-to-one if $\ker J = \{0\}$ (Problem 16 of Section 13.2). So assume that $J(x) = 0$, which means that $\psi(x) = 0$ if $\psi \in X^*$. Suppose, to get a contradiction, that $x \neq 0$. Then expressing x as in Problem 5, one of the x_i, say x_j, is nonzero so that $\psi_j(x) \neq 0$. But $\psi_j \in X^*$, so $\psi_j(x) = 0$, a contradiction. Hence, $x = 0$, that is, $\ker J = \{0\}$. Thus, J is one-to-one.

   Apply Problem 5 to both X and X^* to get that $\dim X^* = \dim X$, and $\dim X^{**} = \dim X^*$. Then $\dim X^{**} = \dim X$. Thus, J is onto because $J : X \to X^{**}$ is one-to-one and X and X^{**} have the same dimension. By definition of *onto*, $J(X) = X^{**}$. So X is reflexive.

9. Refer to the proofs of Propositions 2 and 1. There is a linear subspace X_0 of $\mathbf{R}^n$ for which $\mathbf{R}^n = \text{span}[v] \oplus X_0$. Express vectors in $\mathbf{R}^n$ as $\lambda v + x_0$ where λ is a real number and $x_0 \in X_0$. Define ψ by $\psi(\lambda v + x_0) = \lambda$. Then $\psi(v) = \psi(1v + 0) = 1$, and ψ is linear.

17. If $X = \{0\}$, then the empty-set $\emptyset$ is a (the only) Hamel basis for X because the only linear combination of points in $\emptyset$ is the empty sum, and it is 0 by convention.

    If $X \neq \{0\}$, then $\mathcal{F}$ is not empty because a nonzero point y in X constitutes linearly-independent set $\{y\}$. Now refer to page 6. Take a totally ordered subcollection $\{E_\lambda\}_{\lambda \in \Lambda}$ of $\mathcal{F}$. We claim that $\bigcup_{\lambda \in \Lambda} E_\lambda$ is an upper bound E for $\{E_\lambda\}_{\lambda \in \Lambda}$. Indeed, every E_λ is a subset of E. We also need to show that $E \in \mathcal{F}$, which we do by showing that E is linearly independent. Take a point x in E and a finite number of points $e_1, \ldots, e_n$ in $E \sim \{x\}$. Each e_i is in some E_λ, say $e_1 \in E_{\lambda_1}, \ldots, e_n \in E_{\lambda_n}$. Because the E_λ are totally ordered, one of the E_{λ_i} contains the others. That means that the e_i are all in a common E_{λ_j}, so if $x \in E_{\lambda_j}$, then x fails to be a finite linear combination of points in $E_{\lambda_j} \sim \{x\}$. Because that reasoning holds for every choice of a finite number of points in $E \sim \{x\}$, point x fails to be a finite linear combination of points in $E \sim \{x\}$. Hence, E is linearly independent so that $E \in \mathcal{F}$. That proves our claim that E is an upper bound for $\{E_\lambda\}_{\lambda \in \Lambda}$. Then by Zorn's lemma, $\mathcal{F}$ has a maximal (linearly independent) set $\mathcal{B}$.

    Subset $\mathcal{B}$ of X is a Hamel basis for X provided that we can express each point in X as a unique finite linear combination of points in $\mathcal{B}$, i.e., that $X = \text{span}[\mathcal{B}]$. Suppose, to get a contradiction, that $X \neq \text{span}[\mathcal{B}]$. Then there would be a nonzero y in $X \sim \text{span}[\mathcal{B}]$. By our lemma in our solution to Problem 35(i) of Section 13.4, $\mathcal{B} \cup \{y\}$ would be linearly independent. But that contradicts the maximality of $\mathcal{B}$. Hence, $X = \text{span}[\mathcal{B}]$. Thus, $\mathcal{B}$ is a Hamel basis for X.

Index

Following are some page corrections and additions to the index.